Vladislav G. Bagrov, Dmitry Gitman
The Dirac Equation and its Solutions

De Gruyter Studies in Mathematical Physics

Volume 4

Vladislav G. Bagrov, Dmitry Gitman

The Dirac Equation and its Solutions

DE GRUYTER

Physics and Astronomy Classification Scheme 2010
02; 03.65.Pm; 04.20.Jb

Authors
Prof. Vladislav G. Bagrov
Tomsk State University
Faculty of Physics
Department of Quantum Field Theory
Lenin Prospekt 36
634050 Tomsk
Russian Federation
E-mail: bagrov@phys.tsu.ru

Prof. Dmitry Gitman
Universidade de São Paulo
Instituto de Física
Departamento de Física Nuclear
Rua do Matão Travessa R 187
05508-090 São Paulo
Brazil
E-mail: gitman@dfn.if.usp.br

and

P. N. Lebedev Physical Institute
53 Leninskiy prospect
119991 Moscow
Russian Federation

and

Tomsk State University, Faculty of Physics
Lenin Prospekt 36
634050 Tomsk
Russian Federation

ISBN 978-3-11-026292-6
e-ISBN 978-3-11-026329-9
Set-ISBN 978-3-11-916378-1
ISSN 2194-3532

Library of Congress Cataloging-in-Publication Data
A CIP catalog record for this book has been applied for at the Library of Congress.

Bibliographic information published by the Deutsche Nationalbibliothek
The Deutsche Nationalbibliothek lists this publication in the Deutsche Nationalbibliografie; detailed bibliographic data are available in the Internet at http://dnb.dnb.de.

Typesetting: le-tex publishing services GmbH, Leipzig
Printing and binding: CPI books GmbH, Leck
♾ Printed on acid-free paper
Printed in Germany

www.degruyter.com

Preface

Ever since the wave mechanics has been invented in the early 20-th century, the nonrelativistic and relativistic wave equations responsible for motion of quantum particles are making an essential part in the mathematical description of the physical reality. Their status remains unshaken no matter how far away we move from the physics of these days or how far we deviate from what is now called the Standard Model, or how large and how small the dimensionality is prescribed to the space-time manifold. The relativistic wave equations have quite righteously shared the triumph of Quantum Electrodynamics in the fifties. The methods elaborated for handling them in the context of quantum electrodynamics at the time when it was the only reliable relativistic quantum theory, were later readily extended to nonAbelian gauge field theories that, together with quantum electrodynamics are now valued as a basis for dynamic description of electromagnetic, strong and weak interactions.

The most updated review of the Dirac equation and its solutions that includes also the related topics of the Schrödinger, Klein–Gordon and other associated equations, make the subject of this book. We list all known cases of the external fields and potentials when exact solutions can be found. We present different complete sets of such solutions. Among them there are nonrelativistic and relativistic coherent states that pave the bridge to classical motion.

The knowledge of exact solutions is especially important as it supplies a researcher with the tool of a model study of various situations, with a clear idea about the happenings in front of his/her eyes. On the other hand, many of exact solutions serve perfect mathematical idealizations to important physical problems, like, say, behavior of a physical system in strong background fields. They are laid into the mathematical apparatus adequate for describing these problems.

It goes without saying that exact solutions also present an excellent training material for the physical education that enables a student to gain the necessary skill in handling the basic equations of mathematical physics and supplies him or her with an example for understanding general facts that underlie them.

The present book is intended both to serve as a fundamental text book on the wave equations and as a handbook, a sort of a dictionary, of their exact solutions.

Tomsk, Russia, São Paulo, Brazil *Vladislav Bagrov, Dmitry Gitman*

Acknowledgements

We are grateful to our friends and coauthors V. Kuchin, A. Shabad, I. Tyutin, B. Voronov, V. Man'ko, J. P. Gazeau, V. Shapovalov, S. Gavrilov, Sh. Shvartsman, P. Lavrov, I. Buchbinder, A. Shelepin, and S. Zlatev, for fruitful and stimulating discussions. We thank the Brazilian foundation FAPESP whose financial support allowed the authors to work together in Brazil for a long time to finish this book. Gitman is grateful to the Brazilian foundation CNPq for permanent support, to his family and true friends who supported him during the writing of this book, in particular, to George Keros and J. Geraldo Beggiato.

Contents

1 Introduction

The Dirac and Klein–Gordon (K–G) equations describing the motion of charged spinning and spinless relativistic particles in external electromagnetic backgrounds (fields) provide a basis for relativistic quantum mechanics and quantum electrodynamics (QED), as well as for their possible extensions covering other fields and interactions, or other space-time dimensions both within and beyond Lorentz-invariant and local theories. Exact solutions of these equations are of special physical interest. In relativistic quantum mechanics, the Dirac and K–G equations are referred to as one-particle wave equations for spinning and spinless particles respectively [8, 82, 192, 277, 292]. In QED, exact solutions made it possible to develop the perturbation expansion known as the Furry picture which incorporates the interaction of particles with the external backgrounds exactly, while treating the interaction with the quantized electromagnetic field perturbatively [145, 149, 171, 292]. Knowing exact solutions is crucial within the Green function technique in the vacuum and also in a medium. In particular, all propagators of particles, various Green's functions, can be constructed by using exact solutions of the Dirac and K–G equations.

The physically most important exact solutions of these equations were obtained and analyzed in the early years of relativistic quantum theory. These are the solution for an electron in a Coulomb field [116, 120, 188, 189, 247, 277, 291], in a uniform magnetic field [208, 260, 266, 268, 288], in the field of a plane wave [336, 338] and in some simple one-dimensional electric fields [219, 255, 288].

During the subsequent three decades, only one solution of the Dirac and K–G equations was found, namely that for a charge placed in the field of a magnetic monopole [196]. Beginning in the mid-sixties however, new works appeared. The following solutions were found: for an electron in the field of a plane wave combined with a uniform collinear magnetic field (the Redmond field) [273]; for an electron in some nonuniform fields [10, 12, 197–199, 316, 317]; stationary solutions in constant and uniform electric and magnetic fields that are equal in magnitude and mutually orthogonal (crossed fields) [11, 107, 233, 234]; and in the field of a wave with an isotropic four-potential [296].

In the early seventies there appeared important works in which the problem of listing all external fields that allow a complete separation of variables in the Dirac [26, 299] and K–G [27, 318] equations was solved. This progress became possible following an essential advance in the theory of separation of variables in second-order differential equations and in systems of first-order differential equations [26, 27, 299–303, 318].

Solving the Dirac or K–G equation means either solving the Cauchy problem, i.e. finding the wave function at any instant of time using the data given at an initial time, or determining the complete system of solutions that are, simultaneously, eigenfunctions of a complete set of operator-valued integrals of motion. Most of the solutions known belong to the latter type.

It is known that for the K–G equation in Minkowski space the complete set contains three operator-valued integrals of motion. These three integrals of motion usually have direct classical analogues which makes the parallel solution of the corresponding classical problem especially interesting. For the Dirac equation, the complete set contains four integrals of motion, with at least one of them not admitting classical interpretation (the so-called spin-operator-valued integral of motion). All operators in the complete set should commute among themselves and with operators of the equations (K–G or Dirac operators) and be functionally independent. The method of separation of variables contains in itself the finding of complete sets of integrals of motion (to be more precise, of complete sets of symmetry operators[1] that are also integrals of motion). It should be mentioned that all known exact solutions of the Dirac and K–G equations were obtained for fields belonging to the class found in [27], when separation of variables is possible.

The explicit description of all external fields allowing separation of variables avalanched new exact solutions of the Dirac and K–G equations. At present, hundreds of such solutions are known. In our previous book [55] we had reviewed, at least briefly, the solutions known at that time, to highlight their common features and to distinguish their peculiarities. The present book is a natural continuation and significant extension of the previous one. It includes new solutions and new topics related to the Dirac and K–G equations.

In particular, the material on the motion in a magnetic field and in its combination with other fields is essentially expanded (Sections 3.3 and 3.4), completely new stuff being added on the motion in the so-called magneto-solenoid field, which is a superposition of magnetic and Aharonov–Bohm field (Section 3.6). Path-integral representations for propagators of relativistic particles are given for the first time, and their applicability for calculating these propagators is demonstrated (Chapter 6). A new Chapter 8 contains the so far most complete presentation of known and original results on the spin equation that usually appears in the course of solving the Dirac equation. The material on the coherent states is also essentially expanded and occupies now the complete Chapter 10. We have significantly broadened appendices (Chapter A and B) to help the reader to achieve a deeper apprehension of the basic matter.

1.1 Book content

Chapter 2 begins with recalling the Maxwell equations for electromagnetic fields. Next, it presents the relativistic equations of motion for a charged massive particle, starting with the relativistic-mechanical equations in the Lorentz and Hamilton–

1 A symmetry operator of a given equation is an operator that maps every solution of this equation into a solution of the same equation. Details can be found in [26, 27, 300–303, 318].

Jacobi form. The relativistic wave equations K–G and Dirac are written in arbitrary curvilinear and light-cone coordinates. Complete systems of solutions and the Cauchy problem are discussed. The useful reduction schemes of the Dirac equation to lesser-number but higher-order differential equations are given. Special attention is paid to the choice of the spin operator.

Chapters 3 and 4 of the book collect those exact solutions of the Dirac and K–G equations 3 known at the present time, which can be expressed as combinations of a finite number of elementary and transcendental functions. We have chosen to classify the solutions according to the type of external fields involved. This means the following: For many external fields (e.g. the plane wave or homogeneous magnetic field) different complete systems of solutions are known, depending on different choices of the sets of integrals of motion. For the homogeneous magnetic field, for example, there are various complete systems of solutions, widely used in the literature and presented also in this book. They result from separation of variables performed in Cartesian and cylindrical reference frames. There are also coherent states and some other more special complete systems. However, it is not our intention to list all the complete systems of solutions ever described in the literature for every given field. Our purpose has been to indicate all the external electromagnetic fields allowing exact solutions of the Dirac or K–G equations and to present at least one complete system of solutions for each such field. The physically most important and most often used solutions are presented in Chapter 3.

It is evident that once an exact solution of a relativistic-covariant equation of motion is known for a certain external field, its solution is thereby known for the whole class of fields obtained from the given field by Lorentz boosts. In every case, we pick up the simplest representative from a given class of fields. We have done our best to cite original papers wherever possible, although we have not always mentioned complementary results which have appeared later.

It should be noted that, in general, solving the Dirac equation does not always reduce to solving the K–G equation for the same electromagnetic field. (The class of fields, for which a connection between solutions of these equations is established is described in Section 2.3).

Nevertheless, in most cases, whenever an explicit solution of the K–G equation is known, one succeeds in finding the explicit solution of the Dirac equation too. This observation is illustrated by the three known cases studied in Ref. [53] and described in Section 4.4.3 in items Types VIII, IX, and X. The following statement is also true: If the K–G equation allows complete separation of variables for a certain electromagnetic field, then the solution of the classical Lorentz equations can be reduced to performing quadratures and the classical action can be found as a quadrature as well. In view of these facts, for every class of electromagnetic fields considered, we present, in succession, solutions of the classical Lorentz and Hamilton–Jacobi equations, and then solutions of the K–G and Dirac equations. As far as the latter two are concerned, we present their complete systems of solutions. Exact solutions of classical problems are of independent value; they are also useful in quantum theory, for example for interpretation of integrals of motion.

The reader who may only be interested in the solution for a special electromagnetic field is advised to acquaint himself not only with the text that relates directly to this solution, but also to consult the beginning of the corresponding chapter and section, since, as a rule, in those places notations to be used specifically for a given chapter and section are introduced, and other information that belongs to the case under consideration is given.

It should be said that, as a rule, solution of the K–G and Dirac equations is usually reduced to solving a one-dimensional stationary Schrödinger equation with a certain potential. Therefore, in order to avoid repetition, all the known solutions of the one-dimensional stationary Schrödinger equation and all the potentials that allow solutions of this equation are listed in Chapter 9.

The contents of Chapters 5–6 are closely associated with the previous text on the exact solutions of relativistic wave equations in external electromagnetic fields.

In Chapter 5, we present exact solutions of the extended Dirac equation, the so-called Dirac–Pauli equation, that takes into account the fact that spinor particles may have anomalous magnetic and electric moments. A consistent treatment of the anomalous magnetic moment of an electron is possible within the scope of QED, whereas the anomalous electric moment requires for its description a quantum field model with parity nonconservation. Nonetheless, the Dirac–Pauli equation can be an approximate theoretical foundation, as well as a phenomenological justification within QED, and is used describing the influence of the anomalous moments on the physical parameters of various processes. We also indicate here all the external electromagnetic fields, which enable one to obtain solutions to the Dirac–Pauli equation, and we present one (out of many possible) complete system(s) of such solutions for each field.

Chapter 7 is concerned with exact solutions of the Dirac and K–G equations with an operator-valued electromagnetic potential, which represents the plane-wave quantized electromagnetic field. From the point of view of quantum field theory this equation describes the model problem of the interaction of a charge with the quantized electromagnetic field of photons having collinear momenta.

1.2 Notation

Coordinates of the Minkowski space in 3 + 1 dimensions are denoted by

$$x = (x^\mu) = \left(x^0, x^i\right), \quad \mu = 0, 1, 2, 3, \quad i = 1, 2, 3 \, .$$

We write sometimes (setting $c = 1$)

$$x^0 = t, \quad x^1 = x, \quad x^2 = y, \quad x^3 = z, \quad \mathbf{r} = \left(x^i\right) = \left(x, y, z\right), \quad x = (t, \mathbf{r}) \, .$$

In addition,

$$dx = dx^0 dx^1 dx^2 dx^3 = dx^0 d\mathbf{r}, \quad d\mathbf{r} = dx^1 dx^2 dx^3 .$$

Greek vector and tensor indices take on values 0, 1, 2, 3 and Latin indices take on values 1, 2, 3. The summation convention over repeated sub- and upper-scripts is assumed throughout, unless otherwise explicitly stated. The metric in a flat the space-time is determined by the *Minkowski tensors*

$$\eta_{\mu\nu} = \mathrm{diag}(1, -1, -1, -1), \quad \eta_{\mu\alpha}\eta^{\alpha\nu} = \delta^{\nu}_{\mu} ,$$
$$\eta^{\mu\nu} = \mathrm{diag}(1, -1, -1, -1) ,$$

where, for example,

$$\mathrm{diag}\,(a, b) = \begin{pmatrix} a & 0 \\ 0 & b \end{pmatrix} .$$

Contravariant and covariant four-vectors are often represented in the form

$$a^{\mu} = \left(a^0, a^i\right) = \left(a^0, \mathbf{a}\right), \ \mathbf{a} = \left(a^i\right), \quad a^1 = a_x, \quad a^2 = a_y, \quad a^3 = a_z ,$$
$$a_{\mu} = \eta_{\mu\nu} a^{\nu}, \quad a^{\mu} = \eta^{\mu\nu} a_{\nu} .$$

Three-vectors are indicated by letters in boldface.

Via **i**, **j**, and **k** we denote three mutually orthogonal unit vectors along the axes x, y, and z respectively. Sometimes, $\mathbf{i} = \mathbf{e}_1, \mathbf{j} = \mathbf{e}_2, \mathbf{k} = \mathbf{e}_3$. Then

$$\mathbf{r} = x\mathbf{i} + y\mathbf{j} + z\mathbf{k} = x^i \mathbf{e}_i , \quad \mathbf{a} = a_x\mathbf{i} + a_y\mathbf{j} + a_z\mathbf{k} = a^i \mathbf{e}_i .$$

Scalar products of three- and four-vectors are

$$(\mathbf{ab}) = a^i b^i = -a_i b^i = -a^i b_i, \qquad \mathbf{a}^2 = (\mathbf{aa}) = \left(a^i\right)^2 , \tag{1.1}$$
$$ab = a^{\mu} b_{\mu} = \eta_{\mu\nu} a^{\mu} b^{\nu} = a^0 b^0 - (\mathbf{ab}) , \quad a^2 = a^{\mu} a_{\mu} = \left(a^0\right)^2 - \mathbf{a}^2 . \tag{1.2}$$

The vector product of three-vectors **a** and **b** is defined as

$$\mathbf{c} = [\mathbf{a} \times \mathbf{b}] \Longrightarrow c^i = \epsilon^{ijk} a^j b^k ,$$

where $\epsilon^{ijk} = -\epsilon_{ijk}$ is the fully antisymmetric Levi-Civita tensor in three dimensions with the normalization $\epsilon^{123} = 1$.

The line of a matrix M is denoted by the index which is situated on the left, and the column is denoted by the index which is situated on the right. This agreement does not depend on the nature of the indices, are they co- or contravariant ones. For example, μ is a line and ν is a column of a matrix M in all the cases below

$$M^{\mu}{}_{\nu}, \quad M_{\mu}{}^{\nu}, \quad M_{\mu\nu}, \quad M^{\mu\nu} .$$

By I we usually denote the unite 2×2 matrix, and by $\mathbb{I}$ the unite 4×4 matrix, and δ_{ij}, $i, j = 1, 2, 3$, is the Kronecker symbol in three dimensions, while δ_μ^ν is the Kronecker symbol in $3 + 1$ dimensions.

If A is a matrix or a vector, then A^*, A^T; and $A^+ = A^{*T}$ are used for the complex conjugate, transpose, and Hermitian adjoint objects respectively.

Via $\epsilon^{\mu\nu\alpha\beta}$, we denote the fully antisymmetric Levi-Civita tensor in $3+1$ dimensions with the normalization $\epsilon^{0123} = 1$. Its covariant partner $\epsilon_{\mu\nu\alpha\beta}$ has then the normalization $\epsilon_{0123} = -1$, such that $\epsilon_{\mu\nu\alpha\beta} = -\epsilon^{\mu\nu\alpha\beta}$. In three dimensions, we define the fully antisymmetric Levi-Civita tensor $\epsilon^{ijk} = \epsilon^{0ijk}$, thus, $\epsilon^{123} = 1$. The tensor ϵ_{ijk} is defined as $\epsilon_{ijk} = \epsilon_{0ijk}$, then $\epsilon_{123} = -1$, and $\epsilon_{ijk} = -\epsilon^{ijk}$.

The (pseudo) tensor dual to an antisymmetric tensor $T^{\mu\nu}$ is denoted by $\tilde{T}^{\mu\nu}$,

$$\tilde{T}^{\mu\nu} = \frac{1}{2}\epsilon^{\mu\nu\alpha\beta}T_{\alpha\beta} .$$

The notations of the type $\gamma p = \gamma^\mu p_\mu$ are used throughout.

An overline denotes the complex conjugation unless otherwise specified. Sometimes by $*$ the complex conjugation is denoted as well.

The ordinary derivative in x of the order k of a function $\psi(x)$ is commonly denoted by $\psi^{(k)}(x)$. In addition, we also use the following notation:

$$d_x = d/dx, d_x f(x) = f'(x), \ldots, d_x^n f(x) = f^{(n)}(x) .$$

Partial derivatives:

$$\frac{\partial A}{\partial x^\mu} = \partial_\mu A = A_{,\mu}, \quad \frac{\partial}{\partial t} = \partial_t = \partial_0, \quad \partial_1 = \partial_x, \quad \partial_2 = \partial_y, \quad \partial_3 = \partial_z ,$$
$$\varphi \overleftrightarrow{\partial}_\mu \psi = \varphi\partial_\mu\psi - \left(\partial_\mu\varphi\right)\psi .$$

The operator ∇ the Laplacian $\triangle$, and the d'Alembertian $\square$ are defined as follows:

$$\nabla = \mathbf{i}\frac{\partial}{\partial x} + \mathbf{j}\frac{\partial}{\partial y} + \mathbf{k}\frac{\partial}{\partial z} = \mathbf{e}_i\partial_i = (\partial_i) = \left(\partial_x, \partial_y, \partial_z\right) ,$$
$$\triangle = \nabla^2 = \partial_i^2, \quad \square = \partial_\mu\partial^\mu = \partial_0^2 - \triangle .$$

In addition:

$$\operatorname{div}\mathbf{a} = (\nabla\mathbf{a}) = \partial_i a^i = \partial_x a_x + \partial_y a_y + \partial_z a_z ,$$
$$\operatorname{grad}\varphi = \nabla\varphi = \mathbf{i}\frac{\partial\varphi}{\partial x} + \mathbf{j}\frac{\partial\varphi}{\partial y} + \mathbf{k}\frac{\partial\varphi}{\partial z} = \mathbf{e}_i\partial_i\varphi ,$$
$$\operatorname{rot}\mathbf{a} = [\nabla \times \mathbf{a}] = \mathbf{e}_i(\operatorname{rot}\mathbf{a})^i, \quad (\operatorname{rot}\mathbf{a})^i = -\frac{1}{2}\epsilon^{ijk}\partial_j a_k .$$

The commutator and anticommutator of two operators (or matrices) $\hat{A}$ and $\hat{B}$ are denoted as

$$\left[\hat{A}, \hat{B}\right]_- = \left[\hat{A}, \hat{B}\right] = \hat{A}\hat{B} - \hat{B}\hat{A}, \quad \left[\hat{A}, \hat{B}\right]_+ = \hat{A}\hat{B} + \hat{B}\hat{A} .$$

The following abbreviations and notation are used:
- QED: quantum electrodynamics
- QM: quantum mechanics, or quantum mechanical, and so on
- MSF: magnetic-solenoid field
- SE: spin equation
- ESP: exactly solvable potential
- $L^2(a, b)$: a space of functions square-integrable on (a, b)
- I is the unit 2×2 matrix and $\mathbb{I}$ is the unit 4×4 matrix
- $\mathbb{N} = \{1, 2, \dots\}$: the set of natural numbers
- $\mathbb{Z} = \{0, \pm 1, \dots\}$: the set of integers
- $\mathbb{Z}_+ = \{0, 1, 2, \dots\}$: the set of nonnegative integers
- $\mathbb{Z}_- = \{0, -1, -2, \dots\}$: the set of nonpositive integers
- $\mathbb{R} = (-\infty, \infty)$: the set of all real numbers, the real axis
- $\mathbb{R}_+ = [0, \infty)$: the set of nonnegative real numbers
- $\mathbb{R}^n$: n-dimensional real linear space, the set of all real n-tuples $(x^1, \dots, x^n)$
- $\mathbb{C}$: the set of all complex numbers, the complex plane.

The following designations are adopted for the special functions (these designations are in agreement with those used in the reference book [191]):

$\Phi(x)$ is the Airy function
$J_\mu(x)$ is the Bessel function of the first kind
$I_\mu(x)$ is the Bessel function of imaginary argument
$K_\mu(x)$ is the McDonald function (the first Hankel function of imaginary argument)
$\Phi(\alpha, \gamma; x)$ is the confluent hypergeometric function
$W_{\lambda,\mu}(x)$ is the Whittaker function
$D_\mu(x)$ is the Weber parabolic cylinder function
$F(\alpha, \beta; \gamma; x)$ is the Gauss hypergeometric function
$H_n(x)$ are Hermite polynomials
$L_n^\alpha(x)$ are Laguerre polynomials
$P_n^{(\alpha,\beta)}(x)$ are Jacobi polynomials.

We shall be using throughout the special system of units, where $\hbar = c = 1$ and the Coulomb law takes the form $F = e_1 e_2 / 4\pi r^2$ (it is supposed that e is an algebraic charge of a particle, for an electron $e < 0$). We call this the Heaviside system, following convention. For convenience, we list below some equations that relate various quantities in the Gauss and Heaviside system of units. The corresponding quantities are distinguished by the subscripts G and H, respectively.

$$\mathbf{r}_{\mathrm{G}} = \mathbf{r}_{\mathrm{H}}, m_{\mathrm{G}} = \frac{\hbar}{c} m_{\mathrm{H}}, t_{\mathrm{G}} = \frac{1}{c} t_{\mathrm{H}} \ \left(t_{\mathrm{H}} = x^0\right) ,$$

$$j^{\mu}_{\mathrm{G}} = \sqrt{\frac{\hbar c^3}{4\pi}} j^{\mu}_{\mathrm{H}} \ p^{\mu}_{\mathrm{G}} = \hbar p^{\mu}_{\mathrm{H}} \ S_{\mathrm{G}} = \hbar S_{\mathrm{H}}$$

$$A^{\mu}_{\mathrm{G}} = \sqrt{4\pi\hbar c} A^{\mu}_{\mathrm{H}} \ F^{\mu\nu}_{\mathrm{G}} = \sqrt{4\pi\hbar c} F^{\mu\nu}_{\mathrm{H}}$$

$$e_{\mathrm{G}} = \sqrt{\frac{\hbar c}{4\pi}} e_{\mathrm{H}} \ \alpha = \frac{e^2_{\mathrm{G}}}{\hbar c} = \frac{e^2_{\mathrm{H}}}{4\pi} = \frac{1}{137} .$$

The dimensions of all physical quantities in the Heaviside system are expressed in terms of the dimension of length l alone:

$$|e| = |S| = 0, \quad |\mathbf{r}| = \left|x^0\right| = l ,$$

$$|m| = \left|A_{\mu}\right| = \left|p_{\mu}\right| = l^{-1} ,$$

$$|\mathbf{E}| = |\mathbf{H}| = \left|F_{\mu\nu}\right| = l^{-2}, \quad \left|j^{\mu}\right| = l^{-3} .$$

2 Relativistic equations of motion

2.1 Classical equations

2.1.1 Maxwell equations

Classical electromagnetic (electric $\mathbf{E}(x)$ and magnetic $\mathbf{H}(x)$) fields (in the presence of external charges and currents) obey the *Maxwell equations*, see for example [231]. These equations can be written both in three-dimensional and four-dimensional (relativistic covariant) forms:

$$\left.\begin{aligned} \operatorname{rot}\mathbf{E} + \partial_t \mathbf{H} &= 0 \\ \operatorname{div}\mathbf{H} &= 0 \end{aligned}\right\} \Longleftrightarrow \partial_\nu \tilde{F}^{\nu\mu} = 0\,;$$

$$\left.\begin{aligned} \operatorname{rot}\mathbf{H} - \partial_t \mathbf{E} &= \mathbf{j} \\ \operatorname{div}\mathbf{E} &= \rho \end{aligned}\right\} \Longleftrightarrow \partial_\nu F^{\nu\mu} = j^\mu\,. \tag{2.1}$$

Here $\rho(x)$ is the charge density and $\mathbf{j}(x)$ is the three-curent density, while $j^\mu(x) = (\rho, \mathbf{j})$ is the four-current density. The *electromagnetic stress tensor* $F_{\mu\nu}(x)$ and its dual (pseudo) tensor $\tilde{F}^{\nu\mu}(x)$ are defined as

$$F_{\mu\nu} = \partial_\mu A_\nu - \partial_\nu A_\mu, \quad \tilde{F}^{\mu\nu} = \frac{1}{2}\epsilon^{\mu\nu\alpha\beta}F_{\alpha\beta} \quad \left(F^{\mu\nu} = -\frac{1}{2}\epsilon^{\mu\nu\alpha\beta}\tilde{F}_{\alpha\beta}\right), \tag{2.2}$$

where $A^\mu(x) = (A^0, \mathbf{A})$ are potentials of the electromagnetic field and $\epsilon^{\mu\nu\alpha\beta}$ is the completely antisymmetric tensor of Levi-Civita in four dimensions with the normalization $\epsilon^{0123} = 1$. The electric and the magnetic fields, the electromagnetic potentials, and the stress tensors, are related as follows:

$$\begin{aligned} &\mathbf{E} = -\partial_t \mathbf{A} - \nabla A_0, \quad \mathbf{H} = \operatorname{rot}\mathbf{A} = [\nabla \times \mathbf{A}]\,, \\ &E^i = F^{i0}, \quad H^i = -\frac{1}{2}\epsilon^{ijk}F_{jk} = \tilde{F}^{i0}, \quad F^{ij} = -\epsilon^{ijk}H^k\,, \end{aligned} \tag{2.3}$$

where $\epsilon^{ijk} = -\epsilon_{ijk}$ is the completely antisymmetric tensor of Levi-Civita in three dimensions with the normalization $\epsilon^{123} = 1$.

The current density $j^\mu(x)$ obeys the continuity equation

$$\partial_t \rho + \operatorname{div}\mathbf{j} = 0 \Longleftrightarrow \partial_\mu j^\mu = 0\,, \tag{2.4}$$

which, in particular, follows from the Maxwell equations.

The stress tensors can also be represented in the following form:

$$\begin{aligned} F_{\mu\nu} &= F(\mathbf{E}, \mathbf{H}), \quad & F^{\mu\nu} &= F(-\mathbf{E}, \mathbf{H})\,, \\ \tilde{F}_{\mu\nu} &= F(\mathbf{H}, -\mathbf{E}), \quad & \tilde{F}^{\mu\nu} &= -F(\mathbf{H}, \mathbf{E})\,, \end{aligned} \tag{2.5}$$

where the matrix $F(\mathbf{a},\mathbf{b})$ is defined as

$$F(\mathbf{a},\mathbf{b}) = \begin{pmatrix} 0 & a_x & a_y & a_z \\ -a_x & 0 & -b_z & b_y \\ -a_y & b_z & 0 & -b_x \\ -a_z & -b_y & b_x & 0 \end{pmatrix} \tag{2.6}$$

for any three-dimensional vectors **a** and **b**. The stress tensors also obey the relations

$$\tilde{F}_{\mu\alpha}\tilde{F}^{\alpha\nu} = F_{\mu\alpha}F^{\alpha\nu} + \frac{1}{2}\left(F_{\alpha\beta}F^{\alpha\beta}\right)\delta^{\nu}_{\mu}, \quad F^{\nu\alpha}\tilde{F}_{\alpha\mu} = (\mathbf{EH})\,\delta^{\nu}_{\mu}\,. \tag{2.7}$$

The electromagnetic relativistic invariants I_1 and I_2 have the form

$$\begin{aligned} I_1 &= \frac{1}{2}F_{\mu\nu}F^{\mu\nu} = -\frac{1}{2}\tilde{F}_{\mu\nu}\tilde{F}^{\mu\nu} = \mathbf{H}^2 - \mathbf{E}^2\,, \\ I_2 &= -\frac{1}{4}\tilde{F}_{\mu\nu}F^{\mu\nu} = -\frac{\partial}{\partial x^{\mu}}\left(\epsilon^{\mu\nu\alpha\beta}A_{\nu}\frac{\partial}{\partial x^{\alpha}}A_{\beta}\right) = (\mathbf{EH})\,. \end{aligned} \tag{2.8}$$

The antisymmetric matrix $F_{\mu\nu}$ formed by the components of the stress tensor has in the general case of nonvanishing invariants four isotropic eigenvectors, namely,

$$\begin{aligned} F_{\mu\nu}n^{\nu} &= -\mathcal{E}n_{\mu}, & F_{\mu\nu}\bar{n}^{\nu} &= \mathcal{E}\bar{n}_{\mu}\,, \\ F_{\mu\nu}m^{\nu} &= i\mathcal{H}m_{\mu}, & F_{\mu\nu}\bar{m}^{\nu} &= -i\mathcal{H}\bar{m}_{\mu}\,, \\ \tilde{F}_{\mu\nu}m^{\nu} &= -i\mathcal{E}m_{\mu}, & \tilde{F}_{\mu\nu}\bar{m}^{\nu} &= i\mathcal{E}\bar{m}_{\mu}\,, \\ \tilde{F}_{\mu\nu}n^{\nu} &= -\mathcal{H}n_{\mu}, & \tilde{F}_{\mu\nu}\bar{n}^{\nu} &= \mathcal{H}\bar{n}_{\mu}\,, \end{aligned} \tag{2.9}$$

so that

$$\mathcal{E} = \left[\left(I_1^2 + I_2^2\right)^{1/2} - I_1^2\right]^{1/2}, \quad \mathcal{H} = \left[\left(I_1^2 + I_2^2\right)^{1/2} + I_1^2\right]^{1/2}. \tag{2.10}$$

The eigenvectors are supposed to be normalized, $\bar{n}^{\mu}n_{\mu} = -\bar{m}^{\mu}m_{\mu} = 2$, while all other scalar products vanish,

$$n^2 = \bar{n}^2 = m^2 = \bar{m}^2 = nm = n\bar{m} = \bar{n}m = \bar{n}\bar{m} = 0\,.$$

Then the matrix $F_{\mu\nu}$ can be presented in the form

$$F_{\mu\nu} = \frac{\mathcal{E}}{2}\left(\bar{n}_{\mu}n_{\nu} - n_{\mu}\bar{n}_{\nu}\right) + \frac{i\mathcal{H}}{2}\left(\bar{m}_{\mu}m_{\nu} - m_{\mu}\bar{m}_{\nu}\right), \tag{2.11}$$

and its square F^2 has the spectral decomposition

$$F^2 = \mathcal{E}^2 P_{\mathcal{E}} - \mathcal{H}^2 P_{\mathcal{H}}\,, \tag{2.12}$$

where

$$\left(P_{\mathcal{E}}\right)_{\mu\nu} = \frac{1}{2}\left(\bar{n}_{\mu}n_{\nu} + n_{\mu}\bar{n}_{\nu}\right), \quad \left(P_{\mathcal{H}}\right)_{\mu\nu} = \frac{1}{2}\left(\bar{m}_{\mu}m_{\nu} + m_{\mu}\bar{m}_{\nu}\right) \tag{2.13}$$

are orthogonal projection operators onto some two-dimensional subspaces,

$$\begin{aligned} &P_{\mathcal{E}}^2 = P_{\mathcal{E}}, \quad P_{\mathcal{H}}^2 = P_{\mathcal{H}}, \quad P_{\mathcal{E}}P_{\mathcal{H}} = P_{\mathcal{H}}P_{\mathcal{E}} = 0, \\ &P_{\mathcal{E}} + P_{\mathcal{H}} = \eta, \quad \mathrm{tr}P_{\mathcal{E}} = \mathrm{tr}P_{\mathcal{H}} = 2\,. \end{aligned} \tag{2.14}$$

2.1.2 Equations of motion for a charge in an electromagnetic field

Classical motion of a relativistic particle is represented by a trajectory in the Minkowski space. In general, the trajectory is given in the parametrized form $x^\mu = x^\mu(\theta)$, where θ is a parameter along the trajectory.

Classical relativistic equations of motion for a massive (of a mass m) and charged (of a charge e) particle in an external electromagnetic field with potentials $A_\mu(x)$ can be derived from the action principle. The action of such particle has the form (see e.g. [91, 230, 231])

$$S = \int \left(-m\sqrt{dx_\mu dx^\mu} - eA_\mu dx^\mu\right) = \int L_\theta d\theta \,,$$
$$L_\theta = -m\sqrt{\dot{x}_\mu \dot{x}^\mu} - e\dot{x}^\mu A_\mu \,, \quad \dot{x}^\mu = \frac{dx^\mu}{d\theta} \,, \tag{2.15}$$

where L_θ is the Lagrange function in the θ-parametrization. The corresponding Euler–Lagrange equations are

$$m\frac{d}{d\theta}\left(\frac{\dot{x}^\mu}{\sqrt{\dot{x}_\mu \dot{x}^\mu}}\right) = eF^{\mu\nu}\dot{x}_\nu \,. \tag{2.16}$$

Choosing the proper-time τ as the parameter ($\theta = \tau$),

$$x^\mu = x^\mu(\tau) \,, \quad d\tau = \sqrt{dx_\mu dx^\mu} = dt\sqrt{1-v^2}, \quad v = |v| \,, \quad v = \frac{d\mathbf{r}}{dt} \,, \tag{2.17}$$

we obtain relativistic equations of motion in the following form:

$$m\frac{d^2x^\mu}{d\tau^2} = eF^{\mu\nu}\frac{dx_\nu}{d\tau}, \quad \frac{dx^\mu}{d\tau}\frac{dx_\mu}{d\tau} = 1 \,. \tag{2.18}$$

It should be noted that, in fact, in such a form the relativistic classical equations of motion were first obtained by Plank [265]. We, nevertheless, shall call these equations *the Lorentz equations* following the majority of works.

The quantity $F^{\mu\nu}(dx_\nu/d\tau)$ may be treated as the Lorentz force four-vector.

Taking the laboratory time t as the parametrization parameter, $x^\mu = x^\mu(t)$, we obtain from (2.15)

$$S = \int L dt, \quad L = -m\sqrt{1-v^2} + e(\mathbf{A}v) - eA_0 \,, \tag{2.19}$$

where L is the Lagrange function in the t-parametrization. The three-dimensional generalized momentum $\mathbf{p}$ is defined as

$$\mathbf{p} = \frac{\partial L}{\partial v} = \frac{mv}{\sqrt{1-v^2}} + e\mathbf{A} \,. \tag{2.20}$$

Respectively, the three-dimensional kinetic momentum $\mathbf{P}$ is

$$\mathbf{P} = \mathbf{p} - e\mathbf{A} = \frac{mv}{\sqrt{1-v^2}} \,. \tag{2.21}$$

The energy of the relativistic particle reads

$$\mathcal{E} = \frac{\partial L}{\partial v} v - L = \frac{m}{\sqrt{1-v^2}} + eA_0 = \sqrt{m^2 + (\mathbf{p} - e\mathbf{A})^2} + eA_0 \,. \tag{2.22}$$

In the proper-time parametrization, we define the kinetic four-momentum P_μ as

$$\begin{aligned} P^\mu &= m\dot{x}^\mu = (P_0, \mathbf{P}) \,, \qquad P^2 = m^2 \,, \\ P_0 &= \frac{m}{\sqrt{1-v^2}} = \mathcal{E} - eA_0, \quad \mathbf{P} = \frac{m\mathbf{v}}{\sqrt{1-v^2}} \,. \end{aligned} \tag{2.23}$$

In this parametrization, the generalized four-momentum p_μ is

$$p_\mu = P_\mu + eA_\mu = (\mathcal{E}, -\mathbf{p}) \,, \quad \mathbf{p} = \mathbf{P} + e\mathbf{A} \,. \tag{2.24}$$

Then it follows from (2.23) that

$$\left(p_\mu - eA_\mu\right)\left(p^\mu - eA^\mu\right) = m^2 \,. \tag{2.25}$$

2.1.3 Hamilton–Jacobi equation

Let us consider the action (2.15) of a relativistic charged particle on genuine trajectories (trajectories that obey the Lorentz equations) that have the final point $x = x_{(1)}$. Such an action appears to be a function of the final point x. We denote it by $S(x)$. The generalized momentum p_μ defined by (2.24) is related to the function $S(x)$ as follows:

$$p_\mu = -\partial_\mu S(x) \Longrightarrow \begin{cases} \mathbf{p} = \frac{\partial S(x)}{\partial \mathbf{r}} = \nabla S(x) \\ p_0 = \mathcal{E} = -\frac{\partial S(x)}{\partial t} \end{cases} . \tag{2.26}$$

The function $S(x)$ obeys the relativistic *Hamilton–Jacobi equation*

$$(\partial_\mu S + eA_\mu)(\partial^\mu S + eA^\mu) - m^2 = 0 \,, \tag{2.27}$$

which follows from (2.25) with the use of (2.26), see for example [231].

2.2 K–G equation

2.2.1 General

The quantum motion of spinless relativistic particles is described by a scalar wave function $\varphi(x)$, the latter obeying the *Klein–Gordon equation*[1] *(K–G)*

$$\hat{\mathcal{K}}\varphi(x) = 0, \quad \hat{\mathcal{K}} = \hat{P}^2 - m^2 \,, \tag{2.28}$$

1 In fact, equation (2.28) was already known to Schrödinger [291], in its present form it appeared in the works [126, 140, 187, 220, 226], see also [292] for historical references.

where the kinetic momentum operators $\hat{P}_\mu$ have the form

$$\hat{P}_\mu = \hat{p}_\mu - eA_\mu(x), \quad \hat{p}_\mu = i\partial_\mu = (i\partial_0, -\hat{\mathbf{p}}), \quad \hat{\mathbf{p}} = -i\nabla ,$$
$$\left[\hat{P}_\mu, \hat{P}_\nu\right] = -ieF_{\mu\nu} , \tag{2.29}$$

and the operator $\hat{\mathcal{K}}$ is henceforth referred to as the *K–G operator*.

The K–G equation is invariant under the charge conjugation operation $e \to -e$, $\varphi \to \varphi^*$. This means that if $\varphi(x)$ is a wave function of a spinless particle with the charge e, then $\varphi^*(x)$ is a wave function of a spinless particle with the charge $-e$.

The Lorentz-invariant scalar product of two scalar wave functions φ_1 and φ_2 is defined on an arbitrary space-like hypersurface σ as follows [292]:

$$(\varphi_1, \varphi_2) = \int_\sigma \left[\varphi_1^* \left(\hat{P}^\mu \varphi_2\right) + \left(\hat{P}^\mu \varphi_1\right)^* \varphi_2\right] d\sigma_\mu , \tag{2.30}$$

where

$$d\sigma_\mu = dx/dx^\mu = \left(dx^1dx^2dx^3, dx^0dx^2dx^3, dx^0dx^1dx^3, dx^0dx^1dx^2\right) . \tag{2.31}$$

On the hyperplane $x^0 = \text{const}$, equation (2.30) is reduced to the commonly used scalar product

$$\begin{aligned}(\varphi_1, \varphi_2) &= \int \left[\varphi_1^* \left(\hat{P}_0 \varphi_2\right) + \left(\hat{P}_0 \varphi_1\right)^* \varphi_2\right] d\mathbf{r} \\ &= \int \varphi_1^*(x) \left(i\overleftrightarrow{\partial_0} - 2eA_0\right) \varphi_2(x) d\mathbf{r} ,\end{aligned} \tag{2.32}$$

where $\overleftrightarrow{\partial_0} = \overrightarrow{\partial_0} - \overleftarrow{\partial_0}$.

If φ_1 and φ_2 are solutions of the K–G equation that decrease fast enough at infinity, the scalar product (2.30) does not depend on the choice of the hypersurface σ, and (2.32) does not, correspondingly, depend on time x^0. This is due to the fact that the continuity equation

$$\partial_\mu \left[\left(\hat{P}_\mu \varphi\right)^* \varphi + \varphi^* \left(\hat{P}_\mu \varphi\right)\right] = 0 \tag{2.33}$$

holds provided that φ_1 and φ_2 are solutions of the K–G equation.

If φ is a solutions of the K–G equation, then the quantity

$$j_\mu = e\left[\left(\hat{P}_\mu \varphi\right)^* \varphi + \varphi^* \left(\hat{P}_\mu \varphi\right)\right] \tag{2.34}$$

obeys the continuity equation $\partial_\mu j^\mu = 0$ (which follows from (2.33)) and transforms under Lorentz transformations as a four-vector. One can interpret j_μ as the four-current density induced by the wave function φ. Then the continuity equation for j_μ is, in fact, the charge conservation law.

Consider curvilinear coordinates $u^\mu = u^\mu(x)$ such that the equation of the hypersurface σ has the form $u^0 = \text{const}$ and let $\mathbf{u} = (u^1, u^2, u^3)$ be coordinates on the hypersurface. With the use of these coordinates the scalar product (2.30) takes the form

$$(\varphi_1, \varphi_2) = \int_\sigma \left[\varphi_1^* \left(\hat{P}^\mu \varphi_2\right) + \left(\hat{P}^\mu \varphi_1\right)^* \varphi_2\right] \frac{\partial u^0}{\partial x^\mu} \sqrt{-g} d\mathbf{u} \,,$$

$$g = \det \left\|g_{\mu\nu}\right\|, \quad g_{\mu\nu} = \frac{\partial x^\alpha}{\partial u^\mu} \frac{\partial x_\alpha}{\partial u^\nu} \,. \tag{2.35}$$

The so-called *light-cone variables (coordinates)* are often useful in QED and QFT, see for example [77, 99, 170, 229, 276]. These variables u^μ are related to the Cartesian coordinates x^μ as follows:

$$u^0 = nx, \quad u^1 = x^1, \quad u^2 = x^2, \quad u^3 = \bar{n}x,$$
$$n^\mu = (1, \mathbf{n}), \quad \bar{n}^\mu = (1, -\mathbf{n}), \quad \mathbf{n}^2 = 1, \quad n^2 = \bar{n}^2 = 0, \quad n_3 \neq 0 \,. \tag{2.36}$$

If the axis x^3 is chosen to be directed along the vector $\mathbf{n}$, then

$$u^0 = x^0 - x^3, \quad u^1 = x^1, \quad u^2 = x^2, \quad u^3 = x^0 + x^3 \,. \tag{2.37}$$

For the light-cone coordinates (2.37), the scalar product (2.35) is reduced to the following form

$$(\varphi_1, \varphi_2)_{u^0} = 2 \int \varphi_1^* \hat{\tilde{P}}_3 \varphi_2 d\mathbf{u},$$

$$\hat{\tilde{P}}_3 = \frac{1}{2} n\hat{P} = i\frac{\partial}{\partial u^3} - e\tilde{A}_3, \quad \tilde{A}_3 = \frac{1}{2} nA \,, \tag{2.38}$$

where the tilde denotes the vector components in the light-cone coordinate system (2.37).

For an operator $Q = (Q^{\alpha_1 \cdots \alpha_r}_{\beta_1 \cdots \beta_s})$, let us consider the form

$$Q^\mu = \varphi_1^* \hat{P}^\mu Q \varphi_2 + \left(\hat{P}^\mu \varphi_1\right)^* Q \varphi_2 \,, \tag{2.39}$$

where φ_1 and φ_2 are solutions of the K–G equation. If Q^μ is an $(r+1)$-times contravariant and an s-times covariant tensor, the operator Q is referred to as a rank $r+s$ tensor of the same covariance. In this case, the matrix elements of the operator Q relative to the scalar product (2.30) have the same tensor dimensionality as the operator Q itself. The same tensor dimensionality is inherent in matrix elements of Q relative to the scalar product (2.32).

Recall that an operator Q is called an *integral of motion* if its average value calculated using any wave function satisfying the equation of motion does not depend on time (if the scalar product defined on an arbitrary space-time hypersurface σ is used for calculating the matrix element, the latter should not depend on the choice of σ).

For an operator (2.39) to be an integral of motion, it is sufficient that the condition $\partial_\mu Q^\mu = 0$ be fulfilled, or the operator Q has to commute with the K–G operator $\hat{\mathcal{K}}$ given by (2.28). If the operator Q is an integral of motion, one can impose the condition that, apart from satisfying the K–G equation, the wave function should be an eigenfunction of Q.

2.2.2 Evolution function and completeness relations

By convention, we shall refer to an orthogonal set of solutions $\varphi_n(x)$ as complete, provided that an arbitrary solution $\varphi(x)$ of the K–G equation can be represented as a linear combination (the Fourier series) of $\varphi_n(x)$,

$$\varphi(x) = \sum_n c_n \varphi_n(x), \quad c_n = \frac{(\varphi_n, \varphi)}{(\varphi_n, \varphi_n)}, \quad (\varphi_n, \varphi_m) = 0, \quad n \neq m\,. \tag{2.40}$$

Let the scalar product be defined by (2.32). Then coefficients c_n do not depend on the time x^0 and they can be calculated via $\varphi(x)$ and $\varphi_n(x)$ at any time instant, say, x'^0. Thus, we can write

$$c_n = \frac{1}{(\varphi_n, \varphi_n)} \int \varphi_n^*\left(x'\right)\left(i\frac{\overleftrightarrow{\partial}}{\partial x'^0} - 2eA_0\left(x'\right)\right)\varphi\left(x'\right) d\mathbf{r}'\,, \tag{2.41}$$

and therefore,

$$\varphi(x) = \int G\left(x, x'\right)\left(i\frac{\overleftrightarrow{\partial}}{\partial x'^0} - 2eA_0\left(x'\right)\right)\varphi\left(x'\right) d\mathbf{r}'\,, \tag{2.42}$$

where

$$G\left(x, x'\right) = \sum_n \frac{\varphi_n(x)\varphi_n^*\left(x'\right)}{(\varphi_n, \varphi_n)}\,. \tag{2.43}$$

We call the function $G(x, y)$ *the evolution function of the K–G equation* in what follows.

The evolution function (2.43) satisfies the K–G equation, $\hat{\mathcal{K}}G(x, x') = 0$, and the conditions

$$\begin{gathered}
G\left(x, x'\right)\Big|_{x^0 = x'^0} = 0\,, \\
i\frac{\partial}{\partial x^0} G\left(x, x'\right)\Big|_{x^0 = x'^0} = \delta\left(\mathbf{r} - \mathbf{r}'\right)\,, \\
\int G\left(x, x'\right)\left(i\frac{\overleftrightarrow{\partial}}{\partial x'^0} - 2eA_0\left(x'\right)\right) G\left(x', x''\right) d\mathbf{r}' = G\left(x, x''\right)\,,
\end{gathered} \tag{2.44}$$

which can be verified using equations (2.43) and (2.42). The first two conditions in (2.44) can be rewritten explicitly in terms of the solutions $\varphi_n(x)$ to give

$$\sum_n \frac{\varphi_n(x^0, \mathbf{r})\,\varphi_n^*(x^0, \mathbf{r}')}{(\varphi_n, \varphi_n)} = 0\,,$$
$$\sum_n \frac{\varphi_n(x^0, \mathbf{r})\,\partial_0\varphi_n^*(x^0, \mathbf{r}')}{(\varphi_n, \varphi_n)} = i\delta\left(\mathbf{r} - \mathbf{r}'\right)\,. \tag{2.45}$$

They are equivalent to the completeness relations for the set $\varphi_n(x)$ in the space of solutions of the K–G equation.

As follows from (2.42), the solution of the *Cauchy problem* for the K–G equation with the initial data, say at $x^0 = 0$,

$$\varphi(x)\big|_{x^0=0} = u(\mathbf{r})\,, \quad \partial_0\varphi(x)\big|_{x^0=0} = v(\mathbf{r})\,, \tag{2.46}$$

can be written using the evolution function as follows:

$$\varphi\left(x'\right) = \int \left\{ iG\left(x', x\right)\big|_{x^0=0} v(\mathbf{r}) - \left[i\frac{\partial G\left(x', x\right)}{\partial x^0} + 2eG\left(x', x\right) A_0(x) \right]_{x^0=0} u(\mathbf{r}) \right\} d\mathbf{r}\,. \tag{2.47}$$

2.2.3 Hamiltonian forms of the K–G equation

The K–G equation (2.28) can be rewritten in the form of a first order equation in time, which can be interpreted as a Schrödinger equation. This can be done in different ways. For example, it is easy to see that in terms of two-component wave functions $\Phi(x)$,

$$\Phi(x) = \begin{pmatrix} \chi(x) \\ \varphi(x) \end{pmatrix}, \quad \chi(x) = \hat{P}_0\varphi(x)\,, \tag{2.48}$$

the K–G equation takes the form of the Schrödinger equation (*Hamiltonian form of the K–G equation*)

$$i\frac{\partial\Phi(x)}{\partial x^0} = \hat{H}\Phi(x), \quad \hat{H} = \begin{pmatrix} eA_0 & \hat{P}_k^2 + m^2 \\ 1 & eA_0 \end{pmatrix}. \tag{2.49}$$

The scalar product (2.32) induces the following scalar product for two-component wave functions:

$$(\Phi_1, \Phi_2) = \int \bar{\Phi}_1\Phi_2 d\mathbf{r}, \quad \bar{\Phi} = \Phi^+\sigma^1\,. \tag{2.50}$$

The Hamiltonian $\hat{H}$ is formally self-adjoint with respect to this scalar product.

Sometimes, the Hamiltonian form of the K–G equation is written for a two-component wave function with components that are certain linear combinations of $\varphi(x)$ and $\chi(x)$, for example [113, 192].

Another possibility to represent the K–G equation as the Schrödinger one is related to the use of the light-cone variables (2.37). In such variables the K–G operator $\hat{\mathfrak{K}}$ (2.28) takes the form

$$\hat{\mathfrak{K}} = \left(n\hat{P}\right)\left(\bar{n}\hat{P}\right) + ieF_{30} - \hat{P}_1 - \hat{P}_2 - m^2 , \qquad (2.51)$$

where $n^\mu = (1,0,0,1)$, $\bar{n}^\mu = (1,0,0,-1)$, and

$$n\hat{P} = 2i\frac{\partial}{\partial u^3} - enA, \quad \bar{n}\hat{P} = 2i\frac{\partial}{\partial u^0} - e\bar{n}A .$$

Then, the K–G equation can be formally written in the form of the Schrödinger equation, where the role of the time is played by the light-cone variable u^0,

$$i\frac{\partial\varphi(u)}{\partial u^0} = \hat{H}_{u^0}\varphi(u) , \qquad (2.52)$$

where

$$\hat{H}_{u^0} = \frac{1}{2n\hat{P}}\left(\hat{P}_1 + \hat{P}_2 + m^2 + ieF_{03}\right) + \frac{e}{2}\bar{n}A . \qquad (2.53)$$

Since equation (2.52) is first-order with respect to the time u^0, one should take the value of a function at $u^0 = \text{const}$ for initial data in the Cauchy problem. This leads, in the end, to the fact that the completeness of a system of solutions $\varphi_n(u)$ of the K–G equation with the scalar product (2.38) should imply the completeness of the system of solutions, now at each time moment u^0, in the space of functions that depend solely on $\mathbf{u}$,

$$2\sum_n \hat{\bar{P}}_3(u)\varphi_n(u)\varphi_n^*\left(u'\right)\Big|_{u'^0=u^0} = \delta\left(\mathbf{u}-\mathbf{u}'\right) . \qquad (2.54)$$

2.3 Dirac equation

2.3.1 General

The equation which most exactly describes the quantum motion of relativistic spin one-half particles (in particular the electron motion) was proposed by Dirac [120, 121] and bears his name[2]. In 3 + 1 dimensions the corresponding wave function (*the Dirac wave function*) is a four-component bispinor $\Psi(x)$, which satisfies *the Dirac equation*

$$\hat{\mathfrak{D}}\Psi(x) = 0, \quad \hat{\mathfrak{D}} = \gamma^\mu\hat{P}_\mu - m , \qquad (2.55)$$

2 See also [116, 188] and [292, 330] for historical review.

where the matrix operator $\hat{\mathfrak{D}}$ is henceforth referred to as *the Dirac operator*. The latter is written in terms of Dirac *γ-matrices (gamma-matrices)* and in terms of the *kinetic momentum operator* $\hat{P}_\mu$ which has the form (2.29). The γ-matrices in 3 + 1 dimensions, γ^μ, $\mu = 0, 1, 2, 3$, are defined by the following relations:

$$\{\gamma^\mu, \gamma^\nu\} = 2\eta^{\mu\nu}\mathbb{I}\,, \quad \eta^{\mu\nu} = \mathrm{diag}(1, -1, -1, -1)\,, \tag{2.56}$$

$$(\gamma^\mu)^+ = \gamma^0\gamma^\mu\gamma^0\,, \tag{2.57}$$

where $\mathbb{I}$ is the unit 4×4 matrix. Their properties and different representations are described in Section A.2.

If the γ-matrices are taken in the standard representation, then the Dirac operator has the following box form:

$$\hat{\mathfrak{D}} = \begin{pmatrix} \hat{P}_0 - m & -(\boldsymbol{\sigma}\hat{\mathbf{P}}) \\ (\boldsymbol{\sigma}\hat{\mathbf{P}}) & -\hat{P}_0 - m \end{pmatrix}, \tag{2.58}$$

where $\boldsymbol{\sigma}$ are *Pauli matrices*. The latter are taken in the Pauli representation (see Section A.1.1)

$$\sigma_1 = \begin{pmatrix} 0 & 1 \\ 1 & 0 \end{pmatrix}, \quad \sigma_2 = \begin{pmatrix} 0 & -i \\ i & 0 \end{pmatrix}, \quad \sigma_3 = \begin{pmatrix} 1 & 0 \\ 0 & -1 \end{pmatrix}. \tag{2.59}$$

The operator $\hat{\mathfrak{D}}_{\mathrm{H}} = \gamma^0\hat{\mathfrak{D}}$ is often used. In the standard representation it has the form

$$\hat{\mathfrak{D}}_{\mathrm{H}} = \gamma^0\hat{\mathfrak{D}} = \hat{P}_0 - \rho_1\left(\boldsymbol{\sigma}\hat{\mathbf{P}}\right) - m\rho_3 = \begin{pmatrix} \hat{P}_0 - m & -(\boldsymbol{\sigma}\hat{\mathbf{P}}) \\ -(\boldsymbol{\sigma}\hat{\mathbf{P}}) & \hat{P}_0 + m \end{pmatrix}. \tag{2.60}$$

Although γ^μ do not form a Lorentz vector (the γ-matrices are not transformed under the Lorentz transformations), we often use relativistic notations like $\gamma p = \gamma^\mu p_\mu$, and $\gamma^\mu = (\gamma^0, \boldsymbol{\gamma})$, $\boldsymbol{\gamma} = (\gamma^k)$, as well as γ-matrices with lower indices defined by $\gamma_\mu = \eta_{\mu\nu}\gamma^\nu$.

The Dirac equation (2.55) is invariant under the charge conjugation operation

$$e \to -e, \quad \Psi \to \Psi^c = C\overline{\Psi}^T\,, \quad \overline{\Psi} = \Psi^+\gamma^0\,, \tag{2.61}$$

where the matrix C satisfies the relation $C\gamma_\mu^T C^{-1} = -\gamma_\mu$ (such a matrix exists for any representation of γ-matrices) and $\overline{\Psi}$ is the Dirac conjugate bispinor. The bispinor Ψ^c is called *the charge conjugate bispinor* to Ψ. Thus, if Ψ is a Dirac wave function of a particle with a charge e then Ψ^c is a Dirac wave function of a particle with the charge $-e$.

The Dirac equation can be written in the form of the Schrödinger equation as follows:

$$i\frac{\partial\Psi(x)}{\partial t} = \mathcal{H}\Psi(x), \quad \mathcal{H} = \left(\boldsymbol{\alpha}\hat{\mathbf{P}}\right) + eA_0 + m\gamma^0, \quad \boldsymbol{\alpha} = \gamma^0\boldsymbol{\gamma}\,, \tag{2.62}$$

where the operator $\mathcal{H}$ is called the *Dirac Hamiltonian.* In the standard representation of γ-matrices, the Dirac Hamiltonian $\mathcal{H}$ takes the form

$$\mathcal{H} = \begin{pmatrix} eA_0 + m & (\sigma\hat{\mathbf{P}}) \\ (\sigma\hat{\mathbf{P}}) & eA_0 - m \end{pmatrix} . \tag{2.63}$$

The scalar product of any two bispinors Ψ_1 and Ψ_2 is defined in a Lorentz-invariant form on an arbitrary space-time hypersurface σ as follows [292]:

$$(\Psi_1, \Psi_2) = \int_\sigma \overline{\Psi}_1 \gamma^\mu \Psi_2 d\sigma_\mu , \tag{2.64}$$

where $d\sigma_\mu = (dx^1dx^2dx^3, dx^0dx^2dx^3, dx^0dx^1dx^3, dx^0dx^1dx^2)$. If, for the surface σ in (2.64), one chooses the plane $x^0 = \text{const}$, which implies $d\sigma_\mu = (dx^1dx^2dx^3 = d\mathbf{r}, 0, 0, 0)$, one obtains a more standard form of the scalar product for Dirac wave functions,

$$(\Psi_1, \Psi_2) = \int \overline{\Psi}_1 \gamma^0 \Psi_2 d\mathbf{r} = \int \Psi_1^+ \Psi_2 d\mathbf{r} . \tag{2.65}$$

It is known that if Dirac wave functions decrease fast enough at infinity, the scalar product (2.64) does not depend on the choice of the hypersurface σ, and (2.65) does not, correspondingly, depend on time x^0. In particular, this is due to the fact that the continuity equation

$$\partial_\mu \left(\overline{\Psi}_1 \gamma^\mu \Psi_2\right) = 0$$

holds provided that Ψ_1 and Ψ_2 obey the Dirac equation.

If Ψ is a solution of the Dirac equation, then the quantity

$$j^\mu = \overline{\Psi} \gamma^\mu \Psi , \tag{2.66}$$

can be interpreted as the four-current density induced by the wave function Ψ. This density j^μ obeys the continuity equation $\partial_\mu j^\mu = 0$ and transforms under the Lorentz transformations as a contravariant four-vector [100].

Let us now write the scalar product (2.64) in arbitrary curvilinear coordinates $u^\mu = u^\mu(x)$,

$$(\Psi_1, \Psi_2) = \int_\sigma \overline{\Psi}_1 \tilde{\gamma}^\mu \Psi_2 \sqrt{-g} d\tilde{\sigma}^\mu ,$$

$$g = \det \left\| g_{\mu\nu} \right\| , \quad g_{\mu\nu} = \frac{\partial x^\alpha}{\partial u^\mu} \frac{\partial x_\alpha}{\partial u^\nu} . \tag{2.67}$$

Here $d\tilde{\sigma}^\mu = (du^1du^2du^3, du^0du^2du^3, du^0du^1du^3, du^0du^1du^2)$, (see [292]) and $\tilde{\gamma}^\mu = (\partial u^\mu / \partial x^\alpha)\gamma^\alpha$ are the Dirac matrices transformed to the curvilinear coordinate system.

If we now choose the coordinates u^μ in such a way that the equation of the hypersurface σ has the form $u^0 = u^0(x) = \text{const}$, so that $\mathbf{u} = (u^1, u^2, u^3)$ are coordinates on the hypersurface, we come to the following form for the scalar product of the Dirac wave functions on an arbitrary hypersurface σ:

$$(\Psi_1, \Psi_2) = \int_\sigma \overline{\Psi}_1 \gamma^\mu \Psi_2 \frac{\partial u^0}{\partial x^\mu} \sqrt{-g} d\mathbf{u} \,. \tag{2.68}$$

Now, the independence of the scalar product (2.68) of u^0 becomes a consequence of the corresponding decrease of the wave functions as $\mathbf{u} \longrightarrow \infty$.

With the light-cone coordinates (2.37) the scalar product is defined on the light-cone plane $u^0 = \text{const}$ [77, 99, 99, 229, 275, 276]. Such a scalar product for Dirac wave functions can be obtained from (2.68) and has the form

$$(\Psi_1, \Psi_2)_{u^0} = \int \Psi^+_{1(-)} \Psi_{2(-)} d\mathbf{u} \,, \tag{2.69}$$

where

$$\Psi_{(-)} = P_{(-)} \Psi, \quad P_{(-)} = \frac{1}{2} [1 - (\mathbf{n}\boldsymbol{\alpha})] \,. \tag{2.70}$$

It follows from the properties of the matrices $\boldsymbol{\alpha}$ that $P_{(-)}$ and $P_{(+)} = 1/2[1 + (\mathbf{n}\boldsymbol{\alpha})]$ are orthoprojectors

$$\begin{aligned} &P_{(-)} P_{(+)} = P_{(+)} P_{(-)} = 0, \quad P_{(-)} + P_{(+)} = \mathbb{I}, \\ &P^2_{(-)} = P_{(-)}, \quad P^2_{(+)} = P_{(+)}, \quad P^+_{(-)} = P_{(-)}, \quad P^+_{(+)} = P_{(+)} \,. \end{aligned} \tag{2.71}$$

Although the light-cone plane is not a space-like hypersurface it can be considered as a limit of the one $u^0(\epsilon) = x^0 - (1 - \epsilon)(\mathbf{n}\mathbf{r}) = \text{const}$ as $\epsilon \longrightarrow +0$. This regularization is used for justifying the results obtained within the light-cone formalisms [322].

Transformation laws of bilinear combinations of Dirac wave functions Ψ under Lorentz transformations are well known, see for example, [100, 292]. The combinations $\overline{\Psi}\Psi$ and $\overline{\Psi}\gamma^5\Psi$ are a scalar and a pseudoscalar, respectively. The combinations

$$\overline{\Psi}\gamma^\mu\Psi, \quad \overline{\Psi}\hat{P}^\mu\Psi, \quad \overline{\Psi}\gamma^5\gamma^\mu\Psi, \quad \overline{\Psi}\gamma^5\hat{P}^\mu\Psi \tag{2.72}$$

and their like are contravariant vectors and pseudovectors, while

$$\overline{\Psi}\sigma^{\mu\nu}\Psi, \quad \overline{\Psi}F_{\mu\nu}\Psi, \quad \overline{\Psi}\gamma^5\sigma^{\mu\nu}\Psi, \quad \overline{\Psi}\gamma^5 F_{\mu\nu}\Psi \tag{2.73}$$

and similar combinations are second-rank tensors and pseudotensors of the corresponding covariance (definitions of γ^5 and $\sigma^{\mu\nu}$ are given below).

By convention, we say that an operator $Q = (Q^{\alpha_1 \cdots \alpha_r}_{\beta_1 \cdots \beta_s})$ is a tensor of rank $r + s$, where r is the number of its contravariant indices, and s is the number of covariant ones, provided that the quantity

$$\overline{\Psi}_1 Q \Psi_2 \tag{2.74}$$

is a tensor of rank $r + s$ with the same covariance. Here it is assumed that Ψ_1 and Ψ_2 are solutions of the Dirac equation. It is straightforward to verify that the quantity

$$Q^\mu = \overline{\Psi}_1 \gamma^\mu Q \Psi_2 \tag{2.75}$$

is an $r + 1$-times contravariant and s-times covariant tensor. One can see that matrix elements of the operator Q with the scalar product defined by (2.64) have the same tensor dimensionality as the operator Q itself. The same tensor dimensionality is inherent in matrix elements of Q relative to the scalar product (2.65).

Recall that an operator is called an integral of motion if its average value calculated using any wave function satisfying the wave equation does not depend on time (if the scalar product defined on an arbitrary space-time hypersurface σ is used for calculating the matrix element, the latter should not depend on the choice of σ). For an operator (2.75) to be an integral of motion it is sufficient that the condition $\partial_\mu Q^\mu = 0$ be fulfilled, the operator Q has to commute with the Dirac operator $\hat{\mathfrak{D}}$ (2.55). If the operator is an integral of motion, one can impose the condition that, apart from satisfying the Dirac equation, the Dirac wave function should be an eigenfunction of this operator.

2.3.2 Evolution function and completeness relation

By convention, we shall refer to an orthogonal system of solutions $\psi_n(x)$ as complete, provided that an arbitrary solution $\Psi(x)$ of the Dirac equation can be expanded into the Fourier series

$$\Psi(x) = \sum_n c_n \psi_n(x), \quad c_n = \frac{(\psi_n, \Psi)}{(\psi_n, \psi_n)}, \quad (\psi_n, \psi_m) = 0, \quad n \neq m. \tag{2.76}$$

Let $\psi_n(x)$ be an orthogonal and complete system of solutions of the Dirac equation relative to the scalar product (2.65) on the plane $x^0 = \text{const}$. Since this scalar product does not depend on time, one can calculate (ψ_n, Ψ) in equation (2.76) for an arbitrary time moment y^0, other than x^0. In this way, we come to a relation between solutions of the Dirac equation for different time moments,

$$\Psi(x) = \int G(x, y)\, \Psi(y) d\mathbf{y}. \tag{2.77}$$

The function $G(x, y)$ in (2.77), henceforth to be called *the evolution function of the Dirac equation*, has the form

$$G(x, y) = \sum_n \frac{\psi_n(x)\psi_n^+(y)}{(\psi_n, \psi_n)}. \tag{2.78}$$

The evolution function can be understood as a matrix element in the $\mathbf{x}$-representation of the evolution operator of the Dirac equation in the Schrödinger form (2.62). It satisfies the Dirac equation and the following conditions, which is easy to check using (2.77) and (2.78):

$$G(x, y)\big|_{x^0=y^0} = \delta(\mathbf{x} - \mathbf{y}),$$
$$\int G(x, y) G(y, z) d\mathbf{y} = G(x, z),$$
$$G^+(x, y) = G(y, x). \tag{2.79}$$

The first condition in (2.79), if written in terms of the system $\psi_n(x)$, has the form

$$\sum_n \frac{\psi_{n\alpha}(x)\psi^+_{n\beta}(y)}{(\psi_n, \psi_n)} = \delta_{\alpha,\beta}\delta(\mathbf{x}-\mathbf{y}) \tag{2.80}$$

(here, the spinor indices are written explicitly). Equation (2.80) means that a complete system of solutions of the Dirac equation is, at each time moment, also complete in the space of bispinors that depend on x.

It is easy to verify that the condition (2.80) is equivalent to completeness of the orthogonal system of solutions $\psi_n(x)$ in the space of all solutions of the the Dirac equation.

Note, finally, that the solution of the *Cauchy problem* for the Dirac equation with the initial data $\Psi(x)|_{x^0=0} = \Phi(\mathbf{x})$ can be written using the function $G(x,y)$ as follows:

$$\Psi(x) = \int G(x,y)|_{y^0=0}\, \Phi(\mathbf{y})\, d\mathbf{y}. \tag{2.81}$$

Let us consider now how the completeness conditions should look relative to the scalar product (2.69) on the light-cone plane. To this end, we rewrite the Dirac equation for a bispinor $\Psi(x)$ as a set of two equations in the light-cone variables (2.37),

$$\begin{aligned} i\frac{\partial}{\partial u^0}\Psi_{(-)}(u) &= \tilde{\mathcal{H}}\Psi_{(-)}(u), \\ 2\widehat{\widetilde{P}}_3\Psi_{(+)}(u) &= \left(\gamma^\zeta \mathcal{P}_\zeta + m\right)\tilde{\gamma}^0\Psi_{(-)}(u), \quad \zeta = 1,2\,, \end{aligned} \tag{2.82}$$

for the bispinors $\Psi_{(\pm)} = P_{(\pm)}\Psi$ where the projectors $P_{(\pm)}$ are given by equations (2.71),

$$\tilde{\mathcal{H}} = \left(\gamma^\zeta \mathcal{P}_\zeta + m\right)\tilde{\gamma}^3\left(4\widehat{\widetilde{P}}_3\right)^{-1}\left(\gamma^\zeta \mathcal{P}_\zeta + m\right)\tilde{\gamma}^0 + e\tilde{A}_0$$

and the tilde distinguishes the quantities referring to the coordinate system u^μ. It is seen that the evolution in u^0 is determined by an equation for the component $\Psi_{(-)}$ alone. Once the latter is known, the component $\Psi_{(+)}$ can be reconstructed[3] with its help using the second relation (2.82), at the same time instant u^0. It is, consequently, sufficient to define completeness of a system of solutions $\psi_n(u)$ in the space of the solutions $\Psi_{(-)}(u)$ of the Dirac equation. Therefore, a system of solutions $\psi_n(u)$, orthogonal on the null plane, is complete relative to the scalar product (2.69), provided that the system $\psi_{n(-)}(u)$ is complete in the space of solutions $\Psi_{(-)}(u)$. If the system $\psi_n(u)$ is complete on the null plane, then one can construct the evolution function $G_{(-)}(u,v)$ that relates solutions $\Psi_{(-)}(u)$ at different moments of time u^0,

$$\begin{aligned} \Psi_{(-)}(u) &= \int G_{(-)}(u,v)\,\Psi_{(-)}(v)d\mathbf{v}\,, \\ G_{(-)}(u,v) &= \sum_n \frac{\psi_{n(-)}(u)\psi^+_{n(-)}(v)}{(\psi_n,\psi_n)_{u^0}}\,. \end{aligned} \tag{2.83}$$

3 This treatment is, of course, accurate under the assumption that every function decreases sufficiently fast as $\mathbf{u} \to \infty$.

The function $G_{(-)}(u, v)$ can be understood as an $\mathbf{u}$-representation matrix element of the evolution operator in the first equation of the set (2.82). It satisfies this equation and the conditions

$$G_{(-)}(u, v)\big|_{u^0=v^0} = P_{(-)}\delta(\mathbf{u} - \mathbf{v}),$$
$$\int G_{(-)}(u, v)\, G_{(-)}(v, s)\, d\mathbf{v} = G_{(-)}(u, s),$$
$$G^{+}_{(-)}(u, v) = G_{(-)}(v, u)\,. \qquad (2.84)$$

The first condition in (2.84), when written in terms of the solutions $\psi_{n(-)}$, has the form

$$\sum_n \frac{\psi_{n(-)}(\mathbf{u}, u^0)\, \psi^{+}_{n(-)}(\mathbf{v}, u^0)}{(\psi_n, \psi_n)_{u^0}} = P_{(-)}\delta(\mathbf{u} - \mathbf{v}) \qquad (2.85)$$

and implies that a system of solutions of the Dirac equation, complete relative to the scalar product (2.69), is also complete, in each time instant u^0, in the space of bispinors $\Psi_{(-)}(\mathbf{u})$ that depend only on the spacial $\mathbf{u}$-variables of the light-cone coordinates.

Condition (2.85) is equivalent to the completeness condition for the orthogonal system $\psi_{n(-)}(u)$ relative to the scalar product (2.69) in the space of solutions of the Dirac equation.

Henceforth, we shall refer to equation (2.85) simply as the completeness condition on the null plane.

It is worth noting that to formulate the Cauchy problem with respect to the time u^0 it is sufficient to fix only the component $\Psi_{(-)}(u)$ at the initial moment of time u^0.

2.3.3 Reducing Dirac equation into two independent sets of second-order equations for spinors

Let us represent the Dirac bispinor ψ in the following form:

$$\psi = (\gamma^\mu P_\mu + m)\phi\,, \qquad (2.86)$$

where ϕ is a bispinor. Substituting (2.86) into (2.55), we obtain an equation for ϕ,

$$(\gamma^\mu \gamma^\nu P_\mu P_\nu - m^2)\phi = 0\,. \qquad (2.87)$$

Taking into account the commutator (2.29), we transform equation (2.87) as

$$\left(\hat{\mathcal{K}} - \frac{e}{2}\sigma^{\mu\nu}F_{\mu\nu}\right)\phi = 0\,, \qquad (2.88)$$

where the K–G operator $\hat{\mathcal{K}}$ is given by (2.28) and

$$\frac{e}{2}\sigma^{\mu\nu}F_{\mu\nu} = i(\boldsymbol{\alpha}\mathbf{E}) - (\boldsymbol{\Sigma}\mathbf{H}) = i\rho_1(\boldsymbol{\Sigma}\mathbf{E}) - (\boldsymbol{\Sigma}\mathbf{H})\,. \qquad (2.89)$$

In the standard representation of Dirac matrices, we have

$$\boldsymbol{\Sigma} = \begin{pmatrix} \boldsymbol{\sigma} & 0 \\ 0 & \boldsymbol{\sigma} \end{pmatrix}, \quad \rho_1 = \begin{pmatrix} 0 & I \\ I & 0 \end{pmatrix}. \tag{2.90}$$

Bispinor ϕ can be always represented as

$$\phi = \begin{pmatrix} v + u \\ v - u \end{pmatrix}, \quad u = \begin{pmatrix} u_1 \\ u_2 \end{pmatrix}, \quad v = \begin{pmatrix} v_1 \\ v_2 \end{pmatrix}, \tag{2.91}$$

where u and v are certain spinors. Substituting (2.91) into equation (2.88), we obtain two independent sets of second order equations for the spinors u and v,

$$\left(\hat{\mathcal{K}} + Q\right) u = 0; \quad \left(\hat{\mathcal{K}} + Q^{+}\right) v = 0; \quad Q = (\boldsymbol{\sigma}\mathbf{q}), \quad \mathbf{q} = e\,(\mathbf{H} + i\mathbf{E}). \tag{2.92}$$

Thus, the Dirac equation with an arbitrary electromagnetic field can be reduced to two independent sets of second order equations for two spinors. We note that differential parts of the both equations (2.92) are given by the K–G operator.

2.3.4 Reducing Dirac equation into two independent sets of fourth-order equations for scalar functions

One can rewrite the first equation (2.92) in the following form:

$$\left(\hat{\mathcal{K}} + a\right) u_1 + b u_2 = 0, \quad c u_1 + \left(\hat{\mathcal{K}} - a\right) u_2 = 0;$$
$$a = e\left[H_z + iE_z\right], \quad b = e\left[H_x + E_y + i(E_x - H_y)\right], \quad c = e\left[H_x - E_y + i(E_x + H_y)\right]. \tag{2.93}$$

Let, for example, $b \neq 0$, then

$$u_2 = -b^{-1}\left(\hat{\mathcal{K}} + a\right) u_1, \tag{2.94}$$

and we obtain a fourth-order differential equation for the function u_1

$$\left[\left(\hat{\mathcal{K}} - a\right) b^{-1}\left(\hat{\mathcal{K}} + a\right) - c\right] u_1 = 0. \tag{2.95}$$

In the same way, we obtain from the second equation in (2.92):

$$\left(\hat{\mathcal{K}} + a^*\right) v_1 + c^* v_2 = 0, \quad b^* v_1 + \left(\hat{\mathcal{K}} - a^*\right) v_2 = 0. \tag{2.96}$$

For $b \neq 0$, we have

$$v_1 = -(b^*)^{-1}\left(\hat{\mathcal{K}} - a^*\right) v_2, \tag{2.97}$$

and

$$\left[\left(\hat{\mathcal{K}} + a^*\right)(b^*)^{-1}\left(\hat{\mathcal{K}} - a^*\right) - c^*\right] v_2 = 0. \tag{2.98}$$

Thus, for $b \neq 0$ the components u_1 and v_2 obey fourth-order differential equations (2.95) and (2.98), respectively, whereas the components u_2 and v_1 can be obtained with the help of equations (2.94) and (2.97). Such a procedure determines completely the Dirac bispinor[4].

If $b = 0$ but $c \neq 0$, one can obtain a similar result.

We demonstrate below that either b or c is zero, the Dirac equation can be reduced to two independent second-order equations for scalar functions.

2.3.5 Squaring the Dirac equation

As we have noted above, the Dirac equation is usually more complicated in its structure than the K–G equation. Indeed, the set of four first-order equations comprising the Dirac equation is generally equivalent to two fourth-order equations, see Section 2.3.4, whereas the K–G equation is second-order. Nevertheless, except for the three cases considered in Sections 4.5, one succeeds in finding solutions of the Dirac equation for the same external fields for which solutions of the K–G equation are known. This indicates a profound connection between solutions of the two equations. Finding this connection in the general case has not so far been possible. However, for external fields of a certain special structure it was demonstrated in Refs. [31, 32], how the Dirac equation can be reduced to a set of independent second-order equations for scalar functions whose differential parts coincide with the K–G operator $\hat{\mathcal{K}}$. Moreover, for the so-called crossed fields (to be specified below) the solutions of the Dirac equation can be directly constructed using solutions of the K–G equation. In what follows, we call such a reduction of the Dirac equation *squaring the Dirac equation.*

Now we are in a position to describe the general structure of the electromagnetic fields that allow the squaring of the Dirac equation.

We begin with defining the *crossed electromagnetic fields* as those subject to the conditions

$$\mathbf{H} = [\mathbf{n} \times \mathbf{E}]\,, \quad (\mathbf{nE}) = 0, \mathbf{n}^2 = 1\,, \tag{2.99}$$

where $\mathbf{n}$ is a constant unit three-dimensional vector. The following properties of the crossed fields are consequences of (2.99):

$$\mathbf{E} = -[\mathbf{n} \times \mathbf{H}]\,, \quad \mathbf{E}^2 = \mathbf{H}^2, \quad (\mathbf{nH}) = (\mathbf{nE}) = (\mathbf{EH}) = 0\,. \tag{2.100}$$

This means that the two field invariants (2.8) are zero, $I_1 = I_2 = 0$.

Thus, the crossed fields include electric $\mathbf{E}$ and magnetic $\mathbf{H}$ fields equal to each other in magnitude, and orthogonal to one another and to a constant vector $\mathbf{n}$. A special case of the crossed field is the plane-wave field of the general form.

---

4 It should be noted that the idea of such a reduction was first published in [2].

Let us define now the *longitudinal electromagnetic fields* that obey the conditions

$$\mathbf{E} = E\mathbf{n}, \quad \mathbf{H} = H\mathbf{n}\,, \tag{2.101}$$

where $E = E(x)$ and $H = H(x)$ are some functions of the space-time coordinates. If we agree, for definiteness, that the vector $\mathbf{n}$ is directed along the x^3 axis, and the electromagnetic field is longitudinal, it follows from the Maxwell equations that these functions cannot be arbitrary. One can verify that in the case under consideration the function $E(x)$ may depend only on x^0, x^3, whereas the function $H(x)$ may depend on x^1, x^2 only,

$$E = E(x^0, x^3), \quad H = H(x^1, x^2)\,, \tag{2.102}$$

i.e. that the magnetic field is stationary, while the electric field may be nonstationary.

Let us consider now an arbitrary linear superposition of crossed and longitudinal electromagnetic fields. Evidently, for such a superposition, the magnetic and electric fields are related as

$$\mathbf{H} = [\mathbf{n} \times \mathbf{E}] + (\mathbf{n}\mathbf{H})\mathbf{n}\,. \tag{2.103}$$

The relation (2.103) implies as well that

$$\mathbf{E} = -[\mathbf{n} \times \mathbf{H}] + (\mathbf{n}\mathbf{E})\mathbf{n}\,. \tag{2.104}$$

Fields subjected to the relation (2.103) constitute the general class of fields for which the squaring of the Dirac equation can be performed.

Let $\mathbf{n} = (0, 0, 1)$, then it follows from (2.103) or (2.104) that $H_x = -E_y$, $H_y = E_x$, which corresponds to the condition $b = 0$ according to (2.93). Vice versa, the condition $b = 0$ implies that electromagnetic fields have the structure (2.103) and (2.104) for $\mathbf{n} = (0, 0, 1)$. One can easily verify that the condition $c = 0$ is equivalent to relations (2.103) and (2.104) for $\mathbf{n} = (0, 0, -1)$.

A method for the above-mentioned reduction was, in the general form, first pointed out in [31] for the crossed fields, and in [32] for the general class (2.103). Finally, it was shown in [246] that the class (2.103) exhausts all the fields that allow such a reduction.

The condition (2.103) is not covariant. However, one can easily find its covariant generalization. Such a covariant condition reads

$$F_{\mu\nu} n^{\nu} = (\mathbf{n}\mathbf{E}) n_{\mu}\,, \tag{2.105}$$

where n^{μ} is a constant isotropic four-vector, $n^{\mu} = (1, \mathbf{n})$, $\mathbf{n}^2 = 1$, $n^2 = 0$. This is a consequence of the following identity:

$$F_{\mu\nu} a^{\nu} = \left((\mathbf{a}\mathbf{E})\,, -a^0\mathbf{E} - [\mathbf{a} \times \mathbf{H}]\right)\,, \tag{2.106}$$

which is easy to verify. The equation (2.105) means that n^{μ} is an eigenvector of the matrix $F_{\mu\nu}$ with the eigenvalue equal to $(\mathbf{n}\mathbf{E})$. The condition (2.105) just provides a covariant characterization to the fields of the class (2.103), i.e. once (2.105) is fulfilled, an inertial reference frame exists where the fields obey equation (2.103).

Let us seek solutions of the Dirac equation (2.55) for the fields subject to (2.103) in the following box form [32]:

$$\Psi = N\mathcal{R}\left[\frac{1+(\mathbf{n}\sigma)}{2}\Phi_1(x) + \frac{1-(\mathbf{n}\sigma)}{2}\Phi_{-1}(x)\right]v\,,$$
$$\mathcal{R} = \begin{pmatrix} m + (\sigma\mathbf{F})(\sigma\mathbf{n}) \\ m\,(\sigma\mathbf{n}) - (\sigma\mathbf{F}) \end{pmatrix}, \quad \mathbf{F} = \mathbf{n}\hat{P}^0 - \hat{\mathbf{P}}\,. \tag{2.107}$$

Here N is a normalizing factor, $\hat{P}^\mu$ are components of the kinetic momentum operator, $\Phi_\zeta(x)$, $\zeta = \pm 1$ are some functions of x, and v is an arbitrary constant spinor, henceforth normalized as $v^+v = 1$. By substituting (2.107) into the Dirac equation (it is more convenient to deal with the Dirac equation in its Hamiltonian form (2.62)) we find that (2.107) turns the Dirac equation into an identity, provided that the functions Φ_ζ obey the equations

$$\left[\hat{\mathfrak{K}} - ie\,(\mathbf{n}\mathbf{E}) + \zeta e\,(\mathbf{n}\mathbf{H})\right]\Phi_\zeta = 0\,, \tag{2.108}$$

where $\hat{\mathfrak{K}}$ is the K–G operator.

Therefore, for the fields of the class (2.103), solving the Dirac equation is, indeed, reduced to solving two independent second-order equations (2.108), the differential part of these equations being determined by the K–G operator.

In the special case $(\mathbf{n}\mathbf{H}) = 0$, the form of equations (2.108) does not depend on ζ. Thus, $\Phi_\zeta = \Phi$, where Φ is a solution of the second-order equation

$$\left[\hat{\mathfrak{K}} - ie\,(\mathbf{n}\mathbf{E})\right]\Phi = 0\,. \tag{2.109}$$

The corresponding solutions (2.107) of the Dirac equation have the form

$$\Psi = N\mathcal{R}\Phi v\,. \tag{2.110}$$

In the special case $(\mathbf{n}\mathbf{E}) = (\mathbf{n}\mathbf{H}) = 0$, which means that the electromagnetic fields are crossed, Φ is a solution of the K–G equation, and solutions of the Dirac equation are directly constructed with the help of Φ according to equation (2.110).

In what follows, when representing solutions of the Dirac equation in the form (2.107), we always write down Cartesian coordinates of the vector $\mathbf{F}$, unless otherwise specified.

Solutions (2.107) of the Dirac equation are characterized by the presence of an arbitrary constant spinor v. If the spin operator-valued integral of motion is known, we can require that the wave function (2.107) be an eigenfunction for this integral of motion. This will result in fixing the spinor v. For some external electromagnetic fields, the spin operators can be found. However, the spin operators are not known for every field from the class (2.103). Nevertheless, the presence of the spinor v in (2.107) enables us to classify solutions of the Dirac equation by spin quantum numbers, even when the spin integral is not known explicitly.

One can find from representation (2.107), with account taken of (2.108), that

$$\Psi^+\Psi = N^2 \sum_\zeta [1 + \zeta\,(\overline{\boldsymbol{\sigma}}\mathbf{n})][(\hat{P}^0\Phi_\zeta)^*(n\hat{P})\Phi_\zeta + \Phi_\zeta^*\hat{P}^0(n\hat{P})\Phi_\zeta]\,, \tag{2.111}$$

where the designation $\overline{\boldsymbol{\sigma}} = \upsilon^+\boldsymbol{\sigma}\upsilon$ is introduced.

For the case $(\mathbf{nH}) = 0$, equations (2.111) simplify to become

$$\Psi^+\Psi = 2N^2[(\hat{P}^0\Phi)^*(n\hat{P})\Phi + \Phi^*\hat{P}^0(n\hat{P})\Phi]\,. \tag{2.112}$$

The case $(\mathbf{nE}) = 0$ is especially simple (the condition $(\mathbf{nH}) = 0$ is not even necessary). The Maxwell equations (2.1) for such fields imply

$$\left[(n\hat{P}), F_{\mu\nu}\right]_- = (n\hat{P})F_{\mu\nu} - F_{\mu\nu}(n\hat{P}) = 0\,, \tag{2.113}$$

whence it follows immediately that in both classical and quantum cases the quantity nP is an integral of motion in such fields. Accordingly, one can chose the wave functions to be eigenvectors of the operator $n\hat{P}$,

$$(n\hat{P})\Psi = \lambda\Psi\,. \tag{2.114}$$

By using (2.114) for such fields, we obtain from (2.111) that

$$\Psi^+\Psi = \lambda N^2 \sum_\zeta [1 + \zeta\,(\overline{\boldsymbol{\sigma}}\mathbf{n})][\Phi_\zeta^*\hat{P}^0\Phi_\zeta + (\hat{P}^0\Phi_\zeta)^*\Phi_\zeta]\,. \tag{2.115}$$

Finally, for the crossed fields $(\mathbf{nE}) = (\mathbf{nH}) = 0$, we find

$$\Psi^+\Psi = 2\lambda N^2[\Phi^*\hat{P}^0\Phi + (\hat{P}^0\Phi)^*\Phi]\,. \tag{2.116}$$

The function $\Psi_{(-)}$ defined in (2.70) has an especially simple form for the fields from the class (2.103),

$$\Psi_{(-)} = N(n\hat{P})\begin{pmatrix} -I \\ (\boldsymbol{\sigma}\mathbf{n}) \end{pmatrix}\left[\frac{1 + (\boldsymbol{\sigma}\mathbf{n})}{2}\Phi_1 + \frac{1 - (\boldsymbol{\sigma}\mathbf{n})}{2}\Phi_{-1}\right]\upsilon\,. \tag{2.117}$$

From (2.117) we find

$$\Psi_{(-)}^+\Psi_{(-)} = N^2 \sum_\zeta |(n\hat{P})\Phi_\zeta|^2\,[1 + \zeta\,(\overline{\boldsymbol{\sigma}}\mathbf{n})]\,. \tag{2.118}$$

Here, too, evident simplifications are possible. For instance, for the crossed fields, one has

$$\Psi_{(-)}^+\Psi_{(-)} = 2\lambda^2N^2|\Phi|^2\,. \tag{2.119}$$

2.4 Spin operators

As noted above, the complete set of operator-valued integrals of motion for the Dirac equation consists of four functionally independent operators, at least one of which has no classical counterpart. This operator corresponds to the spin of the electron.

The problem of finding spin operators that are integrals of motion for the Dirac equation has been discussed in the literature for a long time. Whereas for the free Dirac equation this problem may be thought of as solved, the determination of *spin operators* when external fields are present encounters certain difficulties.

The first results in this field were fragmentary. For the Coulomb field, the spin integral of motion was already known to Dirac [121]. As a matter of fact the author of [307] used an operator that was a component of the spin pseudovector. In Ref. [78] the spin pseudovector was introduced, probably for the first time. An interesting approach to finding spin operators was suggested in Ref. [306] and applied, specifically, to reobtain the operator found in Ref. [78]. A noncovariant spin three-vector was constructed in Ref. [315], and a tensor spin operator was introduced in Refs. [79, 308]. These operators were studied in every detail in [96, 106, 148, 186, 205, 287, 305, 309, 320, 342]. The motion of the spin described by these operators in the presence of external fields was considered in [96, 106, 287, 320, 342]. In particular, for the vector and tensor spin operators, the general equations of motion were obtained for arbitrary external fields in Ref. [320]. For the very important case of a homogeneous magnetic field the complete presentation of the corresponding results may be found in Ref. [319].

All the hitherto known spin operators are linear in the electron momentum; in other words, they are first-order differential operators. To justify the consideration of only first-order operators it is usual to refer to the fact that the Dirac equation is itself a set of first-order equations. Hence the search for higher-order operators seems unnatural, although there are, certainly, no restrictions of any profound character. On the other hand, not every first-order differential operator can be called a spin operator (for instance, this is not the case for the angular momentum operator). Another peculiarity of the spin operators is that the momenta in them are multiplied by matrices that are not proportional to the unit matrix, but, to be more precise, contain the spin matrices.

The final results for first-order (i.e. linear in momentum) spin operators are obtained in Refs. [127, 297]. More precisely, the authors have formulated and solved the following problem, which consists of two parts. The first part is to find the general form of the first-order linear differential operator which is an integral of motion for the Dirac equation; the second one is to list all the external fields that allow this operator-valued integral of motion. It was shown in the general case that this first-order operator-valued integral of motion is a scalar formed by contracting constant vectors and tensors with the operators of linear momentum, the angular momentum and with those spin operators that have the Lorentz transformation properties of a vector, pseudovector, second-rank tensor or a scalar.

Let us consider these spin operators. We begin by studying the *spin pseudovector*, first introduced in Ref. [78]. Consider the object

$$T^\mu = \gamma^5(m\gamma^\mu - \hat{P}^\mu)\,, \tag{2.120}$$

which is a pseudovector. Its components have the form

$$T^0 = im\rho_2 + \rho_1\hat{P}^0, \quad \mathbf{T} = m\rho_3\boldsymbol{\Sigma} + \rho_1\hat{\mathbf{P}}\,. \tag{2.121}$$

Another pseudovector T'^μ is also used:

$$T'^\mu = T^\mu + \gamma^\mu\hat{\mathfrak{D}} = -\,i\gamma^5\sigma^{\mu\nu}\hat{P}_\nu\,. \tag{2.122}$$

It is evident that for solutions Ψ of the Dirac equation one has

$$T'^\mu\Psi = T^\mu\Psi\,. \tag{2.123}$$

For the components of T'^μ one finds

$$T'^0 = \left(\boldsymbol{\Sigma}\hat{\mathbf{P}}\right), \quad \mathbf{T}' = \boldsymbol{\Sigma}\hat{P}^0 + i\rho_1[\boldsymbol{\Sigma}\times\hat{\mathbf{P}}]\,. \tag{2.124}$$

We represent below the simplest properties of the operators T^μ and T'^μ:

$$\begin{aligned}
&\gamma_\mu T'^\mu = -T'^\mu\gamma_\mu = \gamma^5(\hat{\mathfrak{D}} - 3m),\\
&T^\nu\gamma^\mu + \gamma^\mu T^\nu = 2im\gamma^5\sigma^{\mu\nu},\\
&T^2 = \hat{\mathfrak{D}}^2 + 2m\hat{\mathfrak{D}} - 3m^2 + \frac{1}{2}\sigma^{\mu\nu}F_{\mu\nu},\\
&T'^\nu\gamma^\mu + \gamma^\mu T'^\nu + T^\mu\gamma^\nu + \gamma^\nu T^\mu = 0\,.
\end{aligned} \tag{2.125}$$

In agreement with (2.75), we construct the tensor

$$G^{\nu\mu} = \overline{\Psi}_1\gamma^\nu T^\mu\Psi_2\,. \tag{2.126}$$

It is straightforward to show that

$$\partial_\nu G^{\nu\mu} = F^{\mu\nu}\overline{\Psi}_1\gamma_\nu\gamma^5\Psi_2\,. \tag{2.127}$$

Using a constant pseudovector n^μ, we form the scalar operator

$$T = -n_\mu T^\mu = (\mathbf{nT}) - n^0T^0\,. \tag{2.128}$$

Let us require that (2.128) be an integral of motion. Then it is seen from (2.127) that this is possible provided that

$$F^{\mu\nu}n_\nu = 0\,. \tag{2.129}$$

From (2.106) there arise the relations for the fields $\mathbf{E}$ and $\mathbf{H}$,

$$(\mathbf{nE}) = 0, \quad n^0\mathbf{E} = -[\mathbf{n}\times\mathbf{H}]\,. \tag{2.130}$$

Equation (2.130) implies, in particular, that the operator T^0 is an integral of motion for an arbitrary magnetic field under the condition $\mathbf{E} = \mathbf{n} = 0$. If $\mathbf{n} \neq 0$, then, without loss of generality, we may assume that $\mathbf{n}^2 = 1$. Then it follows that if

$$n^0\mathbf{E} = -[\mathbf{n} \times \mathbf{H}] \tag{2.131}$$

(and, consequently, $(\mathbf{nE}) = 0$) operator (2.128) is an integral of motion. For instance, (2.128) is an integral of motion for the crossed fields (2.99), and also for the fields of the class (2.103), with $(\mathbf{nE}) = 0$.

The fields subject to the condition (2.131) may be described in a simpler way. If $|n^0| > 1$, the electric field may be nullified by an appropriate Lorentz transformation. Hence, this case is equivalent to that of a pure magnetic field. At the same time, the Maxwell equations (2.1) imply that the magnetic field be constant, and one may solve a stationary problem. If $|n^0| < 1$, an appropriate Lorentz transformation reduces the fields to those subject to the conditions

$$(\mathbf{nE}) = 0, \quad \mathbf{H} = (\mathbf{nH})\,\mathbf{n}\,. \tag{2.132}$$

This means that there exists a longitudinal (i.e. parallel to the vector $\mathbf{n}$) magnetic field $\mathbf{H}$ and an arbitrary electric field $\mathbf{E}$ orthogonal to $\mathbf{H}$. In this case, it follows from equation (2.128) that the operator $(\mathbf{nT})$ is an integral of motion.

Finally, for the case $|n^0| = 1$, there are no simplifications. Setting $n^0 = 1$, we find that the operator

$$T = (\mathbf{nT}) - T^0 \tag{2.133}$$

is the integral of motion for the fields of the class (2.103) providing $(\mathbf{nE}) = 0$. This holds true, in particular, for the plane-wave field, at first pointed out in Ref. [327], and also for the Redmond field [273] (see [37]). In this case the operator $(n\hat{P})$ is also an integral of motion. Taking equation (2.114) into account, we may display the operator (2.133) in the following box form:

$$T = \begin{pmatrix} m\,(\boldsymbol{\sigma}\mathbf{n}) - m - \lambda \\ m - \lambda - m(\boldsymbol{\sigma}\mathbf{n}) \end{pmatrix}. \tag{2.134}$$

It is established by direct verification that, if the spinor in (2.107) obeys the equation

$$(\boldsymbol{\sigma}\mathbf{n})v = \zeta v, \quad \zeta = \pm 1\,, \tag{2.135}$$

then (2.107) is an eigenfunction of the operator (2.134)

$$T\Psi = \zeta\lambda\Psi\,, \tag{2.136}$$

the spin quantum number ζ being the same as in (2.107). Therefore, to classify electron states by their spin we do not need to know, in this case, the function Φ_ζ in (2.107) at all.

Consider next the *tensor spin operator* studied in [79, 308] and [127, 297, 319, 320]. Let us introduce the tensors

$$\Pi^{\mu\nu} = i(\gamma^\mu \hat{P}^\nu - \gamma^\nu \hat{P}^\mu) - m\sigma^{\mu\nu} = F(-\boldsymbol{\epsilon}, \boldsymbol{\mu}),$$
$$\boldsymbol{\epsilon} = im\rho_1\Sigma - i\rho_3\hat{\mathbf{P}} - \rho_2\Sigma\hat{P}^0, \quad \boldsymbol{\mu} = m\Sigma + \rho_2[\Sigma \times \hat{\mathbf{P}}]\,, \tag{2.137}$$

and

$$\Pi'^{\mu\nu} = \Pi^{\mu\nu} - \sigma^{\mu\nu}\hat{\mathfrak{D}} = i\left(\gamma^\mu \hat{P}^\nu - \gamma^\nu \hat{P}^\mu\right) - \sigma^{\mu\nu}\gamma^\gamma \hat{P}_\lambda = F(-\boldsymbol{\epsilon}', \boldsymbol{\mu}'),$$
$$\boldsymbol{\epsilon}' = \rho_3\left[\Sigma \times \hat{\mathbf{P}}\right], \quad \boldsymbol{\mu}' = \rho_3\Sigma\hat{P}^0 - i\rho_2\hat{\mathbf{P}}\,. \tag{2.138}$$

For solutions Ψ of the Dirac equation it obviously holds that $\Pi^{\mu\nu}\Psi = \Pi'^{\mu\nu}\Psi$, and hence it makes no difference which of the two operators is used. We list some simple properties of the operators $\Pi^{\mu\nu}$ and $\Pi'^{\mu\nu}$:

$$\Pi'^{\mu\nu} = \gamma^5 \epsilon^{\mu\nu\alpha\beta}\hat{P}_\alpha\gamma_\beta, \quad \Pi^{*\mu\nu} = -i\gamma^5(\Pi'^{\mu\nu} + m\sigma^{\mu\nu}),$$
$$\Pi'^{*\mu\nu} = -i\gamma^5\left(\Pi^{\mu\nu} + m\sigma^{\mu\nu}\right),$$
$$\Pi_{\mu\nu}\Pi^{\mu\nu} = 12m^2 - 6\hat{P}^2 = 6m^2 - 3\sigma^{\mu\nu}F_{\mu\nu} - 6\hat{\mathfrak{D}}^2 - 12m\hat{\mathfrak{D}}\,. \tag{2.139}$$

The operators $\Pi^{\mu\nu}$ and T^μ are related as follows:

$$2T^\mu = i\gamma^5\gamma_\nu\Pi^{\nu\mu} + \gamma^5\gamma^\mu\hat{\mathfrak{D}}\,,$$
$$2T'^\mu = i\gamma^5\gamma_\nu\Pi'^{\nu\mu}\,,$$
$$4\Pi^{\mu\nu} = i\gamma^5[3(\gamma^\mu T^\nu - \gamma^\nu T^\mu) + T^\mu\gamma^\nu - T^\nu\gamma^\mu]\,. \tag{2.140}$$

Now consider the quantity

$$\Pi^{\lambda\mu\nu} = \overline{\Psi}_1\gamma^\lambda\Pi^{\mu\nu}\Psi_2\,, \tag{2.141}$$

which is a tensor in accordance with (2.75). Its divergence is

$$\partial_\lambda\Pi^{\lambda\mu\nu} = -\overline{\Psi}_1 W^{\mu\nu}\Psi_2\,, \tag{2.142}$$

where the tensor $W^{\mu\nu}$ is defined as in (A.53). It follows from (A.54) that the scalar operator

$$\Pi = \frac{1}{2}\omega_{\mu\nu}\Pi^{\mu\nu}, \quad \omega_{\mu\nu} = aF_{\mu\nu} + bF^*_{\mu\nu} \tag{2.143}$$

is the integral of motion, provided that one can so choose the functions a and b in (2.142) that the tensor $\omega_{\mu\nu}$ be constant. Using (2.142), we find that the operator Π takes the form

$$\Pi = a\left[(\boldsymbol{\mu}\mathbf{H}) - (\boldsymbol{\epsilon}\mathbf{E})\right] - b\left[(\boldsymbol{\epsilon}\mathbf{H}) + (\boldsymbol{\mu}\mathbf{E})\right]\,. \tag{2.144}$$

For the tensor $\omega_{\mu\nu}$ we have

$$\omega_{\mu\nu} = F(\mathbf{t}_1, \mathbf{t}_2), \quad \mathbf{t}_1 = a\mathbf{E} + b\mathbf{H}, \quad \mathbf{t}_2 = a\mathbf{H} - b\mathbf{E} \tag{2.145}$$

and the fields **E** and **H** should be such that $\mathbf{t}_1$ and $\mathbf{t}_2$ be constant vectors. With the use of (2.145) one can express the fields in terms of the constant vectors $\mathbf{t}_1$ and $\mathbf{t}_2$:

$$\mathbf{E} = \frac{a\mathbf{t}_1 - b\mathbf{t}_2}{a^2 + b^2}, \quad \mathbf{H} = \frac{b\mathbf{t}_1 + a\mathbf{t}_2}{a^2 + b^2}. \tag{2.146}$$

If $(\mathbf{t}_1\mathbf{t}_2) \neq 0$, the vectors $\mathbf{t}_1$ and $\mathbf{t}_2$ can be made parallel to one another by an appropriate Lorentz transformation. In this case, the fields (2.146) are also parallel with one another and with the constant unit vector **n**, i.e. we have come to the longitudinal fields (2.101). If $(\mathbf{t}_1\mathbf{t}_2) = 0$, there always exists a vector **n**, such that

$$\mathbf{t}_1 = -[\mathbf{n} \times \mathbf{t}_2], \quad (\mathbf{n}\mathbf{t}_1) = (\mathbf{n}\mathbf{t}_2) = 0. \tag{2.147}$$

From (2.147), and employing (2.145), we obtain

$$\mathbf{t}_1 = a\mathbf{E} + b\mathbf{H} = b[\mathbf{n} \times \mathbf{E}] - a[\mathbf{n} \times \mathbf{H}]. \tag{2.148}$$

Taking into account the constancy of $\mathbf{t}_1$, we conclude that equation (2.148) can be fulfilled for varying fields only under the conditions

$$\mathbf{H} = [\mathbf{n} \times \mathbf{E}], \quad \mathbf{E} = -[\mathbf{n} \times \mathbf{H}], \quad \mathbf{n}^2 = 1,$$

which define the crossed fields (2.99). Thus, the fields (2.146) reduce either to longitudinal or to crossed fields.

Let us dwell upon the crossed fields in more detail. We choose the vector $\mathbf{t}_2$ in the form

$$\mathbf{t}_2 = \mathbf{l} - (\mathbf{n}\mathbf{l})\,\mathbf{n}, \quad \mathbf{l}^2 = 1, \tag{2.149}$$

where **l** is an arbitrary constant unit vector. From (2.147) we find for $\mathbf{t}_1$

$$\mathbf{t}_1 = -[\mathbf{n} \times \mathbf{l}]. \tag{2.150}$$

As the operator (2.133) is also an integral of motion for the crossed fields, we see that they have the following general spin operator [31, 54, 327]:

$$Y = (\mathbf{l}\boldsymbol{\mu}) - (\mathbf{n}\mathbf{l})(\mathbf{n}\boldsymbol{\mu}) + (\boldsymbol{\epsilon}\,[\mathbf{n} \times \mathbf{l}]) + (\mathbf{n}\mathbf{l})\left[(\mathbf{n}\mathbf{T}) - T^0\right]. \tag{2.151}$$

Making the function (2.110) obey the additional equations

$$Y\Psi = \zeta\lambda\Psi, \quad \zeta = \pm 1; \quad (n\hat{P})\Psi = \lambda\Psi, \tag{2.152}$$

we obtain that the first equation in (2.152) is equivalent to the following condition imposed on the spinor v:

$$(\boldsymbol{\sigma}\mathbf{l})v = \zeta v, \quad \zeta = \pm 1. \tag{2.153}$$

Classification of Dirac wave functions according to the spin is often accomplished by solutions of equations such as equation (2.153). For instance, equation (2.135) is its special form. Usually, as far as special calculations are concerned, one can manage without an explicit solution of this equation, although sometimes it is useful to know it. It is not difficult to obtain the general solution of equation (2.153) subject to the orthonormality and completeness conditions

$$v_\zeta^+ v_{\zeta'} = \delta_{\zeta,\zeta'}, \quad \sum_\zeta v_\zeta v_\zeta^+ = \mathbf{I}\,. \tag{2.154}$$

Upon setting $\mathbf{l} = (\sin\theta\cos\varphi, \sin\theta\sin\varphi, \cos\theta)$ in (2.153), it is a matter of simple algebra to determine the required solution as

$$v_\zeta = \frac{1}{\sqrt{2}}\begin{pmatrix} \zeta\sqrt{1+\zeta\cos\theta}\exp\left(-\frac{i}{2}\varphi\right) \\ \sqrt{1-\zeta\cos\theta}\exp\left(\frac{i}{2}\varphi\right) \end{pmatrix}, \quad \zeta = \pm 1\,. \tag{2.155}$$

Now consider a *pseudoscalar spin operator*. Define the pseudoscalar R in the following way [127, 297]:

$$R = -\gamma^5\left(x^\mu \hat{P}_\mu - m\gamma_\mu\gamma^\mu + \frac{3}{2}i\right) = x^\mu T_\mu - \frac{3}{2}i\gamma^5\,. \tag{2.156}$$

One may also define the operator

$$R' = R + \gamma^5(x^\mu\gamma_\mu)\hat{\mathfrak{D}} = x^\mu T'_\mu + \frac{3}{2}i\gamma^5 = -\frac{1}{2}\Lambda^{\mu\nu}\Lambda^*_{\mu\nu}\,. \tag{2.157}$$

The introduced operators are equivalent when acting on solutions Ψ of the Dirac equation, $R'\Psi = R\Psi$. There is a relation between the operators R and T_μ,

$$T_\mu = i[R, \hat{P}_\mu] + \gamma^5 F_{\mu\nu}x^\nu\,. \tag{2.158}$$

By constructing a vector R^ν in accordance with the rule (2.75) we find that

$$\partial_\nu R^\nu = x^\mu F_{\mu\nu}\overline{\Psi}_1\gamma^\nu\gamma^5\Psi_2\,, \tag{2.159}$$

which implies that R be an integral of motion providing

$$x^\mu F_{\mu\nu} = 0\,. \tag{2.160}$$

Using equation (2.106) we may see that (2.160) holds if there exists the following relation between the magnetic and electric fields:

$$x^0\mathbf{E} = -\,[\mathbf{r}\times\mathbf{H}]\,. \tag{2.161}$$

In particular, the pseudoscalar operator R is an integral of motion for the magnetic monopole field (or, more generally, for every central-symmetric field).

Consider an operator-valued vector [121, 127, 297] N^μ,

$$\begin{aligned}
N^\mu &= \gamma^\mu + x_\nu \Pi'^{\nu\mu} = \gamma^\mu\left(1 + \frac{1}{2}\sigma^{\alpha\beta} Z_{\alpha\beta}\right) - iZ^{\mu\nu}\gamma_\nu\,,\\
Z^{\mu\nu} &= x^\mu \hat{P}^\nu - x^\nu \hat{P}^\mu\,,\\
N^0 &= \rho_3\left\{1 + \left(\boldsymbol{\Sigma}\left[\mathbf{r}\times\hat{\mathbf{P}}\right]\right)\right\} = \rho_3\left[\left(\boldsymbol{\Sigma}\hat{\mathbf{J}}\right) - \frac{1}{2}\right],\\
\mathbf{N} &= i\rho_2\left\{\Sigma + \left[\mathbf{r}\times\hat{\mathbf{P}}\right]\right\} - \rho_3 x^0\left[\boldsymbol{\Sigma}\times\hat{\mathbf{P}}\right] + \rho_3\left[\boldsymbol{\Sigma}\times\mathbf{r}\right]\hat{P}^0\,,\\
\hat{\mathbf{J}} &= \left[\mathbf{r}\times\hat{\mathbf{P}}\right] + \frac{1}{2}\boldsymbol{\Sigma}\,.
\end{aligned} \tag{2.162}$$

The spin integral of motion N^0 was known to Dirac [121].

Let us list some simple properties of the operator N^μ:

$$\begin{aligned}
N^\mu &= \gamma^\mu + Z^{*\mu\nu}\gamma_\nu\gamma^5, \quad N_0^2 = \frac{1}{4} + \hat{\mathbf{J}}^2 + (\boldsymbol{\Sigma}\mathbf{r})(\mathbf{r}\mathbf{H})\,,\\
(\mathbf{n}\mathbf{N})^2 &= \hat{\digamma}^2 - (\mathbf{n}\hat{\digamma})^2 - \left(\mathbf{n}\hat{\mathbf{J}}\right)^2 + x_0^2\,(\boldsymbol{\Sigma}\mathbf{n})\,(\mathbf{n}\mathbf{H}) + i\rho_1 x_0(\mathbf{n}\mathbf{H})(\mathbf{n}\,[\boldsymbol{\Sigma}\times\mathbf{r}]) - \frac{1}{4}\,,\\
N^2 &= 2\left(\hat{\digamma}^2 - \hat{\mathbf{L}}^2\right) - x_0^2\,(\boldsymbol{\Sigma}\mathbf{H}) - i\rho_1 x_0(\mathbf{H}\,[\boldsymbol{\Sigma}\times\mathbf{r}]) + (\boldsymbol{\Sigma}\mathbf{r})\,(\mathbf{r}\mathbf{H}) + 1\\
&= \Lambda_{\mu\nu}\Lambda^{\mu\nu} - x_0^2\,(\boldsymbol{\Sigma}\mathbf{H}) - i\rho_1 x_0(\mathbf{H}\,[\boldsymbol{\Sigma}\times\mathbf{r}]) + (\boldsymbol{\Sigma}\mathbf{r})\,(\mathbf{r}\mathbf{H}) + 1\,,\\
\Pi'_{\mu\nu} &= i\left[N_\nu, \hat{P}_\mu\right] + \eta_{\mu\eta}F^{\eta\lambda}\epsilon_{\nu\lambda\alpha\beta}x^\alpha\gamma^\beta\gamma^5\,.
\end{aligned} \tag{2.163}$$

Here

$$\hat{\digamma} = \mathbf{r}\hat{P}^0 - x^0\hat{\mathbf{P}} - \frac{i}{2}\boldsymbol{\alpha}$$

and $\mathbf{n}$ is a unit constant vector.

The tensor $N^{\nu\mu}$ formed following the rule (2.75) has the divergence

$$\partial_\nu N^{\nu\mu} = W^{\mu\nu}x_\nu\,, \tag{2.164}$$

where the tensor $W^{\mu\nu}$ is defined by (A.53). Let us contract the vector N^μ with the constant vector n^μ and require that the operator

$$N = n_\mu N^\mu \tag{2.165}$$

be an integral of motion. Then, using (2.164), we find that the condition

$$n_\mu W^{\mu\nu}x_\nu = 0 \tag{2.166}$$

is necessary and sufficient for this.

It follows from (A.53) that the relations (2.166) are equivalent to

$$\begin{aligned}
n^0\,[\mathbf{H}\times\mathbf{r}] - x^0\,[\mathbf{H}\times\mathbf{n}] + [\mathbf{E}\times[\mathbf{n}\times\mathbf{r}]] &= 0\,,\\
n^0[\mathbf{E}\times\mathbf{r}] - x^0[\mathbf{E}\times\mathbf{n}] + [\mathbf{H}\times[\mathbf{r}\times\mathbf{n}]] &= 0\,.
\end{aligned} \tag{2.167}$$

The constants n^0 and $\mathbf{n}$ are to be found from (2.167). Simultaneously, equations (2.167) determine admissible fields. A simple study shows that equation (2.167) admits three nonequivalent simplest types of fields [127, 297],

$$n^0 = 1, \qquad \mathbf{n} = 0, \qquad \mathbf{E} = a\mathbf{r}, \qquad \mathbf{H} = b\mathbf{r}\,, \tag{2.168}$$

$$n^0 = 0, \qquad \mathbf{n}^2 = 1, \qquad \mathbf{E} = ra\mathbf{e}_2 - x^0 b\mathbf{n}, \qquad \mathbf{H} = rb\mathbf{e}_2 + x^0 a\mathbf{n}\,, \tag{2.169}$$

$$\begin{aligned} n^0 = 1, \qquad \mathbf{n}^2 = 1, \qquad & \mathbf{E} = r\left(a\mathbf{e}_1 + b\mathbf{e}_2\right) - u^0 a\mathbf{n}\,, \\ & \mathbf{H} = -r\left(b\mathbf{e}_1 - a\mathbf{e}_2\right) + u^0 b\mathbf{n} = [\mathbf{n} \times \mathbf{E}] + u^0 b\mathbf{n}\,, \end{aligned} \tag{2.170}$$

where $r^2 = (x^1)^2 + (x^2)^2$, and the vectors $\mathbf{e}_1$, $\mathbf{e}_2$ and $\mathbf{n}$ make a unit basis for the cylindrical coordinate frame. The functions a and b are arbitrary, and $u^0 = x^0 - x^3$. In particular, N^0 is an integral of motion for spherically symmetric fields (2.168). The fields (2.170) represent a special case of the fields (2.103). It is clear that the Maxwell equations (2.1) impose some restrictions on the functions a and b. For instance, for item 3, the functions a and b are found to be arbitrary functions of the following arguments:

$$a = a(x^2, u^0), \quad b = r^{-3} f\left(r/u^0, \varphi\right) \tag{2.171}$$

where φ is the cylindrical coordinate.

It seems reasonable to try to find fields that allow combinations of various contractions of the above operators as integrals of motion. Such fields do exist and correspond to certain combinations of operators N^μ and $\Pi^{\mu\nu}$ as pointed out in [127, 297]. Some of these cases will be mentioned in the discussion of special solutions.

It is natural to seek admissible fields by contracting the spin vectors or tensors with nonconstant vectors and tensors. However, it was shown in [127, 297] that no new nonzero fields exist.

To conclude, we state that the spin operators that we have considered exhaust all possible spin operators linear in the momentum operators. As for spin operators that are differential second- or higher- order operators, the problem of their determination remains completely unsolved. No serious investigations in the field have yet been done.

3 Basic exact solutions

In this chapter we consider exact solutions of the Dirac equation (and related relativistic equations of motion) for a free particle, for a charge in a magnetic field, in a plane-wave field, and in a spherically symmetric field. These solutions have a special importance in relativistic quantum theory. They were the first problems to be solved exactly in relativistic quantum mechanics and they have gained exceptionally wide applications in various physical problems. It is sufficient to note that solutions for free particles provide a basis for perturbation expansions in quantum field theory [100, 292]. Solutions in the magnetic field allow one to calculate quantum corrections to synchrotron radiation [69, 311, 319]. The problem of an electron in the spherically symmetric field is fundamental for relativistic theory of atom and molecule spectra [82]. Solutions for a charge in the plane-wave field (Volkov's solutions [336, 338]) describe quantum motion of charged particles in intense laser fields and have wide physical applications.

3.1 Free particle motion

We consider here solutions of the classical and quantum relativistic equations of motion for the *free relativistic particle*. We shall describe solutions of K–G and Dirac equations with a given linear and angular momentum, solutions in light-cone variables and coherent states. States of the free particle in spherical coordinates will be considered in Section 3.5 as a limit of zero external fields.

3.1.1 Classical motion

It follows from the Lorentz equations (2.18) that the four-momentum p^μ of a free particle ($F^{\mu\nu} = 0$) is conserved,

$$m\dot{x}^\mu = p^\mu = P^\mu = \text{const}\,, \tag{3.1}$$

and satisfies the relation

$$p_\mu p^\mu = p^2 = m^2\,. \tag{3.2}$$

This yields the expression for the energy $\mathcal{E}$ of the free relativistic particle

$$\mathcal{E} = p_0 = \sqrt{\mathbf{p}^2 + m^2}\,. \tag{3.3}$$

In classical theory we consider only the positive value for the energy, i.e. the only positive value for the square root in equation (3.3).

Equations (3.1) are readily integrated to give

$$mx^\mu = p^\mu\tau + x^\mu_{(0)}, \quad (x^\mu_{(0)} = \text{const})\,, \tag{3.4}$$

which corresponds to the rectilinear uniform motion,

$$\mathbf{r} = \boldsymbol{\beta} x^0 + \mathbf{r}_{(0)}, \quad \boldsymbol{\beta} = \mathbf{p}/\mathcal{E} . \tag{3.5}$$

The angular momentum of the free particle

$$\mathbf{L} = [\mathbf{r} \times \mathbf{p}] = [\mathbf{r}_{(0)} \times \mathbf{P}] . \tag{3.6}$$

is conserved as well. One can easily find the geometrical meaning of the vector $\mathbf{L}$ and its components. Calculating the minimal distance $R_{\min} = \min |\mathbf{r}(t)|$ between the particle trajectory and the origin, we find

$$R_{\min}^2 = \mathbf{L}^2/\beta^2\mathcal{E}^2 . \tag{3.7}$$

Therefore, for a fixed particle energy, such a distance is proportional to the magnitude of the angular momentum.

Similarly, for example for $z_{\min} = \min |z(t)|$, we find

$$z_{\min}^2 = L_z^2/\beta^2\mathcal{E}^2 . \tag{3.8}$$

It is useful to obtain expressions for the coordinates $x = x^1$ and $y = x^2$ as functions of the light-cone variable $u^0 = x^0 - x^3$. These expressions follow from (3.5) and, with the use of $t = 2\lambda u^0$, $\lambda = p^0 - p^3$, have the form

$$x^k = \frac{p^k}{2\lambda^2} t + x_{(0)}^k = \frac{p^k}{\lambda} u^0 + x_{(0)}^k, \quad k = 1, 2 , \tag{3.9}$$

where $x_{(0)}^k$ are arbitrary constants (initial values of x^k).

For the free particle case under consideration, the solution of the Hamilton–Jacobi equation (2.27) has the form

$$S = -p_\mu x^\mu . \tag{3.10}$$

3.1.2 States with a given momentum

For a free quantum particle both spinless and spinning, the operator of the generalized momentum $\hat{p}_\mu = i\partial_\mu$ is an integral of motion. Hence, solutions of the K–G and Dirac equations can be looked for as eigenfunctions of this operator with the eigenvalues p_μ.

For the K–G equation, such solutions have the form

$$\begin{aligned}
&\varphi_{\mathbf{p},\epsilon}(x) = \frac{1}{\sqrt{2(2\pi)^3\mathcal{E}}} e^{-ipx} = \frac{1}{\sqrt{2(2\pi)^3\mathcal{E}}} e^{iS} , \\
&\hat{p}_\mu \varphi_{\mathbf{p},\epsilon}(x) = p_\mu \varphi_{\mathbf{p},\epsilon}(x), \quad p^2 = m^2 , \\
&p_0 = \epsilon\mathcal{E}, \quad \mathcal{E} = \epsilon\sqrt{\mathbf{p}^2 + m^2}, \quad \epsilon = \operatorname{sgn} p_0 = \pm 1 .
\end{aligned} \tag{3.11}$$

Thus, the scalar wave functions are labeled by four independent quantum numbers: three linear momentum components $\mathbf{p}$ and the signature ϵ of the particle energy p_0.

The functions (3.11) are normalized with respect to the scalar product (2.32),

$$\left(\varphi_{\mathbf{p},\epsilon}, \varphi_{\mathbf{p}',\epsilon'}\right) = \epsilon\delta_{\epsilon,\epsilon'}\delta(\mathbf{p} - \mathbf{p}') , \tag{3.12}$$

and form a complete system of functions, with the completeness relations (2.45) fulfilled,

$$\begin{aligned}
&\sum_{\epsilon} \epsilon \int \varphi_{\mathbf{p},\epsilon}(x^0, \mathbf{r})\varphi^*_{\mathbf{p},\epsilon}(x^0, \mathbf{r}')d\mathbf{p} = 0 , \\
&\sum_{\epsilon} \int \mathcal{E}\varphi_{\mathbf{p},\epsilon}(x^0, \mathbf{r})\varphi^*_{\mathbf{p},\epsilon}(x^0, \mathbf{r}')d\mathbf{p} = \delta(\mathbf{r} - \mathbf{r}') .
\end{aligned} \tag{3.13}$$

One-particle interpretation of the solutions (3.11) implies that the function

$$\varphi_{\mathbf{p},+}(x) = \frac{1}{\sqrt{2(2\pi)^3\mathcal{E}}} e^{-ipx}, \quad p_0 = \mathcal{E}$$

should be understood as the wave function of a particle having energy $\mathcal{E}$ and the linear three-momentum $\boldsymbol{p}$, while the function

$$\varphi_{-\mathbf{p},-}(x) = \frac{1}{\sqrt{2(2\pi)^3\mathcal{E}}} e^{ipx}, \quad p_0 = \mathcal{E}$$

should be understood as the antiparticle wave function with the energy $\mathcal{E}$ and the linear three-momentum $\mathbf{p}$.

The evolution function $G(x, y)$ (2.43) for the free K–G equation has the form

$$G(x, y) = \frac{1}{2(2\pi)^3} \sum_{\epsilon} \int \frac{e^{-ip(x-y)}}{p_0} d\mathbf{p} = -i\mathcal{D}(x - y) , \tag{3.14}$$

where $\mathcal{D}(x)$ is the Pauli–Jordan function [100],

$$\begin{aligned}
\mathcal{D}(x) &= \frac{1}{(2\pi)^3} \int \frac{\sin(\sqrt{\mathbf{p}^2 + m^2}x^0)}{\sqrt{\mathbf{p}^2 + m^2}} e^{ipx} d\mathbf{p} \\
&= \frac{i}{(2\pi)^3} \int \epsilon e^{-ipx}\delta(p^2 - m^2)dp = \mathcal{D}^+(x) + \mathcal{D}^-(x) , \\
\mathcal{D}^{\epsilon}(x) &= \frac{i}{2(2\pi)^3} \int e^{-ipx} \frac{d\mathbf{p}}{p^0} .
\end{aligned} \tag{3.15}$$

Let us look for solutions of the Dirac equation that are eigenvectors for the momentum operators $\hat{p}_\mu = i\partial_\mu$ with the eigenvalues p_μ. In the case under consideration, all spinning operators considered in Section 2.4 are integrals of motion. Let us chose the helicity operator as $(\boldsymbol{\Sigma}\hat{\boldsymbol{p}})$. It commutes with the momentum operators and is obviously an integral of motion for the free Dirac equation. Solutions of the Dirac equation that are eigenvectors for the momentum and helicity operators have the form

$$\Psi_{\mathbf{p},\epsilon,\zeta}(x) = \frac{e^{-ipx}}{\sqrt{8(2\pi)^3\mathcal{E}(m+\mathcal{E})}}\begin{pmatrix}(1+\epsilon)(m+p_0)-(1-\epsilon)(\boldsymbol{\sigma}\boldsymbol{p})\\ (1+\epsilon)(\boldsymbol{\sigma}\boldsymbol{p})+(1-\epsilon)(m-p_0)\end{pmatrix} v_\zeta(\mathbf{p})\,, \quad (3.16)$$

$$\hat{p}_\mu\Psi_{\mathbf{p},\epsilon,\zeta}(x) = p_\mu\Psi_{\mathbf{p},\epsilon,\zeta}(x), \quad p^2 = m^2, \quad p_0 = \epsilon\mathcal{E}\,,$$
$$(\boldsymbol{\Sigma}\hat{\boldsymbol{p}})\Psi_{\mathbf{p},\epsilon,\zeta}(x) = \zeta\left|\hat{\boldsymbol{p}}\right|\Psi_{\mathbf{p},\epsilon,\zeta}(x), \quad \zeta = \pm 1\,. \quad (3.17)$$

Equation (3.17) implies that the two-component spinors $v_\zeta(\mathbf{p})$ obey the equation

$$(\boldsymbol{\sigma}\boldsymbol{l})\, v_\zeta(\mathbf{p}) = \zeta v_\zeta(\mathbf{p}), \quad \boldsymbol{l} = \mathbf{p}|\mathbf{p}|^{-1}\,.$$

Its general solution has the form (2.155). For every given value $\mathbf{p}$, the spinors $v_\zeta(\mathbf{p})$ obey the following orthogonality and completeness relations (2.154),

$$v_\zeta^+(\mathbf{p})v_{\zeta'}(\mathbf{p}) = \delta_{\zeta,\zeta'}, \quad \sum_\zeta v_\zeta(\mathbf{p})v_\zeta^+(\mathbf{p}) = I\,, \quad (3.18)$$

where I is the unit 2×2 matrix.

Once the spinors $v_\zeta(\mathbf{p})$ are normalized according to (3.18), the solutions (3.16) form an orthonormalized and complete system with respect to the scalar product (2.65),

$$\left(\Psi_{\mathbf{p},\epsilon,\zeta}, \Psi_{\mathbf{p}',\epsilon,\zeta'}\right) = \delta_{\epsilon,\epsilon'}\delta_{\zeta,\zeta'}\delta\left(\mathbf{p}-\mathbf{p}'\right),$$
$$\sum_{\epsilon,\zeta}\int \Psi_{\mathbf{p},\epsilon,\zeta}\left(x^0,\mathbf{r}\right)\Psi^+_{\mathbf{p},\epsilon,\zeta}\left(x^0,\mathbf{r}'\right)d\mathbf{p} = \mathbb{I}\delta\left(\mathbf{r}-\mathbf{r}'\right)\,, \quad (3.19)$$

where $\mathbb{I}$ is the unit 4×4 matrix.

A consistent one-particle interpretation of solutions (3.16) can be obtained in the course of the second quantization. It implies that the function

$$\psi_{\mathbf{p},+,\zeta}(x) = \sqrt{\frac{m}{(2\pi)^3\mathcal{E}}}e^{-ipx}U_\zeta(\mathbf{p}), \quad p_0 = \mathcal{E} = \sqrt{\mathbf{p}^2+m^2}\,,$$
$$U_\zeta(\mathbf{p}) = \sqrt{\frac{m+\mathcal{E}}{2m}}\begin{pmatrix} v_\zeta(\mathbf{p})\\ \frac{(\sigma\mathbf{p})}{m+\mathcal{E}}v_\zeta(\mathbf{p})\end{pmatrix} = \frac{(\gamma p+m)}{\sqrt{2m(m+\mathcal{E})}}\begin{pmatrix} v_\zeta(\mathbf{p})\\ 0\end{pmatrix}$$

should be understood as the wave function of an electron having the energy $\mathcal{E}$, the linear momentum $\mathbf{p}$, and the helicity ζ, while the function

$$\psi_{-\mathbf{p},-,-\zeta}(x) = \sqrt{\frac{m}{(2\pi)^3\mathcal{E}}}e^{ipx}V_\zeta(\mathbf{p}), \quad p_0 = \mathcal{E} = \sqrt{\mathbf{p}^2+m^2}\,,$$
$$V_\zeta(-\mathbf{p}) = \sqrt{\frac{m+\mathcal{E}}{2m}}\begin{pmatrix} \frac{(\sigma\mathbf{p})}{m+\mathcal{E}}v_\zeta(-\mathbf{p})\\ v_\zeta(-\mathbf{p})\end{pmatrix} = \frac{(-\gamma p+m)}{\sqrt{2m(m+\mathcal{E})}}\begin{pmatrix} 0\\ v_\zeta(-\mathbf{p})\end{pmatrix}$$

should be understood as the wave function of a positron having the energy $\mathcal{E}$, the linear momentum $\mathbf{p}$, and the helicity ζ.

The bispinors $U_\zeta(\mathbf{p})$ and $V_\zeta(\mathbf{p})$ obey the equations

$$(\gamma p - m)\, U_\zeta(\mathbf{p}) = \overline{U_\zeta}(\mathbf{p})\,(\gamma p - m) = 0,$$
$$(\gamma p + m)\, V_\zeta(\mathbf{p}) = \overline{V_\zeta}(\mathbf{p})\,(\gamma p + m) = 0 \qquad (3.20)$$

and the completeness and orthonormality relations

$$U_\zeta^+(\boldsymbol{p})U_{\zeta'}(\boldsymbol{p}) = V_\zeta^+(\boldsymbol{p})V_{\zeta'}(\boldsymbol{p}) = \delta_{\zeta\zeta'}\frac{\mathcal{E}}{m},$$
$$U_\zeta^+(\boldsymbol{p})V_{\zeta'}(\boldsymbol{p}) = V_\zeta^+(\boldsymbol{p})U_{\zeta'}(\boldsymbol{p}) = 0,$$
$$\sum_\zeta \left[U_\zeta(\boldsymbol{p})U_\zeta^+(\boldsymbol{p}) + V_\zeta(\boldsymbol{p})V_\zeta^+(\boldsymbol{p})\right] = \mathbb{I}\frac{\mathcal{E}}{m}\,. \qquad (3.21)$$

An alternative form of equations (3.21) reads

$$\overline{U_\zeta}(\boldsymbol{p})U_{\zeta'}(\boldsymbol{p}) = -\overline{V_\zeta}(\boldsymbol{p})V_{\zeta'}(\boldsymbol{p}) = 1,$$
$$\overline{U_\zeta}(\boldsymbol{p})V_{\zeta'}(\boldsymbol{p}) = \overline{V_{\zeta'}}(\boldsymbol{p})U_\zeta(\boldsymbol{p}) = 0,$$
$$\sum_\zeta \left[U_\zeta(\boldsymbol{p})\overline{U_\zeta}(\boldsymbol{p}) - V_\zeta(\boldsymbol{p})\overline{V_\zeta}(\boldsymbol{p})\right] = I\,. \qquad (3.22)$$

Evolution function $G(x, x')$ (2.78) of the free Dirac equation has the form

$$G(x,x') = -i(\gamma\hat{p} + m)\gamma^0\mathcal{D}(\Delta x)$$
$$= \frac{1}{(2\pi)^3}\int\left[\cos\left(\mathcal{E}\Delta x^0\right) - i(\boldsymbol{\alpha}\boldsymbol{p} + m\beta)\mathcal{E}^{-1}\sin\left(\mathcal{E}\Delta x^0\right)\right]e^{i\boldsymbol{p}\Delta\boldsymbol{r}}d\boldsymbol{p}\,, \qquad (3.23)$$

where $\Delta x = x - x'$.

3.1.3 Positive and negative frequency solutions

Let us define *positive and negative frequency solutions* of the K–G and Dirac equations as those which, when decomposed in the complete sets (3.11) or (3.16) contain only solutions with energies $p_0 = \epsilon\mathcal{E}, \epsilon = \pm$. Evolution functions of the K–G and Dirac equations allow one to represent any solution via its initial conditions, see equations (2.47) and (2.81). Let us describe initial conditions for the positive ($\epsilon = +$) and negative ($\epsilon = -$) frequency solution of the K–G and Dirac equations.

Consider first the spinless particle case. Then, for any positive or negative frequency solution $\varphi_\epsilon(x)$, there exists restrictions on the corresponding initial conditions (2.46). One can see that both $u(\mathbf{r})$ and $v(\boldsymbol{r})$ cannot be arbitrary. If $v(\mathbf{r}) = \partial_0\varphi(x)|_{x^0=0}$ is arbitrary, then $u(\boldsymbol{r}) = u_\epsilon(\boldsymbol{r})$ has the form

$$u_\epsilon(\boldsymbol{r}) = \frac{i\epsilon m}{2\pi^2}\int v(\boldsymbol{r}')|\boldsymbol{R}|^{-1}K_1(m|\boldsymbol{R}|)d\boldsymbol{x}', \quad \boldsymbol{R} = \boldsymbol{r} - \boldsymbol{r}'\,, \qquad (3.24)$$

such that

$$\varphi_\epsilon(x) = 2\partial_0 \int \mathcal{D}^{-\epsilon}(x - x')u_\epsilon(\boldsymbol{r}')d\boldsymbol{r}'|_{x_0'=0}$$

where $K_1(x)$ is the MacDonald function, [191].

For spinning particle case, the restriction on the corresponding initial bispinor $\Phi_\epsilon(\boldsymbol{r})$ in (2.81) has the form

$$(\gamma\hat{p} + m)\int \mathcal{D}^\epsilon(x - x')\gamma^0\Phi_\epsilon\left(\boldsymbol{r}'\right)d\boldsymbol{r}' = 0, \quad x_0' = 0\,. \tag{3.25}$$

The solution of equation (3.25) for $\Phi_\epsilon(\boldsymbol{r})$ reads

$$\Phi_\epsilon(\boldsymbol{r}) = N\int u_\epsilon(\boldsymbol{p})e^{i(\boldsymbol{pr})}d\boldsymbol{p},$$

$$u_\epsilon(\boldsymbol{p}) = \begin{pmatrix}(1-\epsilon)(m - p_0) - (1+\epsilon)(\boldsymbol{\sigma p})\\ (1-\epsilon)(\boldsymbol{\sigma p}) + (1+\epsilon)(m + p_0)\end{pmatrix}v(\boldsymbol{p})\,. \tag{3.26}$$

Here N is the normalization factor, $v(\boldsymbol{p})$ is an arbitrary $\boldsymbol{p}$-dependent spinor.

There are other representations for the bispinor $\Phi_\epsilon(\boldsymbol{r})$. For example,

$$\Phi_\epsilon(\boldsymbol{r}) = \Psi_\epsilon(x)|_{x^0=0}, \quad \Psi_\epsilon(x) = \gamma^0(\gamma\hat{p} - m)\psi_\epsilon(x)\,, \tag{3.27}$$

where each component of the bispinor $\psi_\epsilon(x)$ is a ϵ-frequency solution of the K–G equation with an arbitrary $u(\boldsymbol{r})$.

Finally, using equation (3.24), one can find

$$\Phi_\epsilon(\boldsymbol{r}) = 2\pi^2\epsilon u(\boldsymbol{r}) + m[i\rho_1(\boldsymbol{\sigma}\nabla) - m\rho_3]\int u(\boldsymbol{r}')|\boldsymbol{R}|^{-1}K_1(m|\boldsymbol{R}|)d\boldsymbol{r}'\,, \tag{3.28}$$

where $u(\boldsymbol{r})$ is an arbitrary $\boldsymbol{r}$-dependent bispinor and $\boldsymbol{R} = \boldsymbol{r} - \boldsymbol{r}'$.

3.1.4 Light-cone variables and coherent states

K–G equation in light-cone variables

The light-cone variables (2.37) are very convenient for analysis of the relativistic wave equations. In these variables the K–G operator $\hat{\mathcal{K}} = \hat{P}^2 - m^2$ (2.28) takes the form

$$\hat{\mathcal{K}} = 4\hat{\bar{p}}_3\hat{\bar{p}}_0 - \hat{p}_1^2 - \hat{p}_2^2 - m^2\,, \tag{3.29}$$

where

$$\hat{\bar{p}}_3 = i\partial_{u^3}, \quad \hat{\bar{p}}_0 = i\partial_{u^0}\,. \tag{3.30}$$

It is evident that the operators (3.30) are integrals of motion.

Let us look for solutions $\varphi(x)$ of the K–G equation that are eigenfunctions of the operator $\hat{\bar{p}}_3$ with the eigenvalue $\lambda/2$,

$$\hat{\bar{p}}_3\varphi_\lambda(x) = \frac{\lambda}{2}\varphi_\lambda(x)\,. \tag{3.31}$$

The eigenvalue λ is related to the linear momentum components as

$$\lambda = p^0 - p^3 \,. \tag{3.32}$$

We have $\mathrm{sgn}\lambda = \mathrm{sgn}\, p_0$ because $|p^3| < |p^0|$ as it follows from (3.3). In classical theory one has $\lambda > 0$, which corresponds to particle solutions in the quantum case; $\lambda < 0$ corresponds to antiparticles and is forbidden in classical theory. Thus, in the light-cone variables, we do not have the quantum number ϵ (the energy sign), but instead the definition domain of λ becomes larger than in the classical case, $-\infty < \lambda < \infty$, $\lambda \neq 0$.

Equation (3.31) allows us to separate the variable u^3. Representing the scalar wave function as

$$\varphi_\lambda(x) = \exp\left(-i\frac{\lambda}{2}u^3 - i\frac{m^2}{2\lambda}u^0 \right)\chi(x^1, x^2, u^0)\,, \tag{3.33}$$

we obtain that the function χ obeys the *two-dimensional Schrödinger equation* for a free particle of the mass λ,

$$i\partial_{u^0}\chi = \tilde{\mathcal{H}}\chi, \quad \tilde{\mathcal{H}} = \frac{1}{2\lambda}(\hat{p}_1^2 + \hat{p}_2^2)\,. \tag{3.34}$$

Finally, introducing dimensionless coordinates x, y, and time t,

$$x = 2\lambda x^1, \quad y = 2\lambda x^2, \quad t = 2\lambda u^0, \quad \psi(x, y; t) = \chi(x^1, x^2, u^0)\,, \tag{3.35}$$

we reduce equation (3.34) to an universal form that does not contain any dimensional constants,

$$\bar{\mathcal{K}}\psi(x, y; t) = 0, \quad \bar{\mathcal{K}} = i\partial_t + \Delta_2, \quad \Delta_2 = \frac{\partial^2}{\partial x^2} + \frac{\partial^2}{\partial y^2}\,. \tag{3.36}$$

One can easily find the solution of the Cauchy problem for equation (3.36). If $\psi_0(x, y) = \psi(x, y; 0)$, then for any $t \geq 0$, we have

$$\psi(x, y; t) = \int_{-\infty}^{\infty} dx' \int_{-\infty}^{\infty} dy' G(x - x', y - y'; t)\psi_0(x', y')\,, \tag{3.37}$$

where

$$G(x, y; t) = -\frac{i}{4\pi t}\exp\left[\frac{i}{4t}(x^2 + y^2)\right]. \tag{3.38}$$

Coherent states

Let us construct *coherent states* for equation (3.36). To this end, we define creation and annihilation operators a_k^+ and a_k, $k = 1, 2$, by the relations

$$a_1 = \frac{x + \partial_x}{\sqrt{2}}, \quad a_1^+ = \frac{x - \partial_x}{\sqrt{2}}, \quad a_2 = \frac{y + \partial_y}{\sqrt{2}}, \quad a_2^+ = \frac{y - \partial_y}{\sqrt{2}},$$
$$[a_k, a_s] = [a_k^+, a_s^+] = 0, \quad [a_k, a_s^+] = \delta_{k,s}, \quad k, s = 1, 2\,. \tag{3.39}$$

The operator $\bar{\hat{\mathcal{K}}}$ (3.36), when written in terms of these operators, takes the form

$$\bar{\mathcal{K}} = i\partial_t - \bar{\mathcal{H}}, \quad \bar{\mathcal{H}} = \bar{\mathcal{H}}_1 + \bar{\mathcal{H}}_2 ;$$
$$2\bar{\mathcal{H}}_k = a_k a_k^+ + a_k^+ a_k - a_k^2 - a_k^{+2}, \quad k = 1, 2 . \tag{3.40}$$

Let us construct the operators A_k and A_k^+, where

$$A_k = f_k(t)a_k + g_k(t)a_k^+, \quad [A_k, A_s] = [A_k^+, A_s^+] = 0,$$
$$[A_k, A_s^+] = \Delta_k \delta_{k,s}, \quad \Delta_k = |f_k|^2 - |g_k|^2 . \tag{3.41}$$

We chose the functions f_k and g_k such that A_k and A_k^+ commute with the operator $\bar{\mathcal{K}}$,

$$[\bar{\mathcal{K}}, A_k] = [\bar{\mathcal{K}}, A_k^+] = 0 . \tag{3.42}$$

In this case A_k and A_k^+ are integrals of motion for the initial K–G equation (3.29). Equations (3.42) are reduced to the following set of differential equations for the functions f_k and g_k:

$$i\dot{f}_k + f_k + g_k = 0, \quad i\dot{g}_k - f_k - g_k = 0 . \tag{3.43}$$

Its general solution is

$$f_k = c_1^{(k)} + i(c_1^{(k)} + c_2^{(k)})t, \quad g_k = c_2^{(k)} - i(c_1^{(k)} + c_2^{(k)})t , \tag{3.44}$$

where $c_j^{(k)}$ are arbitrary constants. In this case $\Delta_k = |c_1^{(k)}|^2 - |c_2^{(k)}|^2$. Consider the equation

$$A_k \psi = Z_k \psi , \tag{3.45}$$

where ψ are solutions of equation (3.36), and Z_k are some complex numbers[1].

If $\Delta_k > 0, k = 1, 2$, then, without loss of generality, we can set $\Delta_k = 1$, which is equivalent to multiplying A_k by a complex number. In this case, A_k and A_k^+ are some annihilation and creation operators and solutions of equations (3.45) can be constructed as coherent states having a finite norm in a Fock space.

If for a given k, we have $\Delta_k = 0$, then, without loss of generality, we can consider the corresponding A_k as a self-adjoint operator and its eigenvalues Z_k are real numbers,

$$\Delta_k = 0, A_k = A_k^+, \quad Z_k = Z_k^*, \quad g_k = f_k^* . \tag{3.46}$$

In this case, the eigenfunctions of equations (3.45) can be normalized to $\delta(Z_k - Z_k')$ (generalized eigenfunctions).

Finally, if $\Delta_k < 0$ for a given k, then there do not exist eigenfunctions of equations (3.45) that would belong to the Hilbert space or be generalized eigenfunctions (distributions).

1 A complete study of such kind of equations can be found in [24].

In what follows, we consider the case $\Delta_k \geq 0$ only.

It follows from equations (3.40) and (3.45) that the variables x and y can be separated,

$$\psi(x, y; t) = \psi_1(x, t)\psi_2(y, t),$$
$$(i\partial_t - \mathcal{H}_k)\psi_k = 0, \quad A_k\psi_k = Z_k\psi_k, \quad k = 1, 2\,. \tag{3.47}$$

Both equations (3.47) have the same structure. Thus, it is sufficient to solve only the one-dimensional Schrödinger equation of the form

$$i\partial_t\psi(x, t) = \mathcal{H}\psi(x, t), \quad 2\mathcal{H} = aa^+ + a^+a - a^2 - a^{+2};$$
$$a = \frac{x + \partial_x}{\sqrt{2}}, \quad a^+ = \frac{x - \partial_x}{\sqrt{2}}, \quad [a, a^+] = 1 \tag{3.48}$$

under the condition

$$A\psi(x, t) = Z\psi(x, t), \quad A = fa + ga^+\,,$$
$$f = c_1 + i(c_1 + c_2)t, \quad g = c_2 - i(c_1 + c_2)t\,. \tag{3.49}$$

We consider separately two cases $\Delta = |f|^2 - |g|^2 = |c_1|^2 - |c_2|^2 = 0, 1$.

Let $\Delta = 0$. In this case

$$A = A^+ = fa + f^*a^+, \quad f = c + i(c + c^*)t, \quad Z = Z^*, \quad c \neq 0\,, \tag{3.50}$$

and a common solution of equations (3.48) and (3.49) has the form

$$\psi_Z^c(x, t) = \left[\sqrt{2}\pi(f - f^*)\right]^{-1/2} \exp\left[\frac{f + f^*}{2(f^* - f)}\left(x - \frac{\sqrt{2}Z}{f + f^*}\right)^2\right]. \tag{3.51}$$

The functions (3.51) obey the following orthonormality and completeness relations:

$$(\psi_{Z'}^c, \psi_Z^c) = \int_{-\infty}^{\infty} \psi_{Z'}^{*c}(x, t)\psi_Z^c(x, t)dx = \delta(Z - Z')\,,$$
$$\int_{-\infty}^{\infty} \psi_Z^{*c}(x', t)\psi_Z^c(x, t)dZ = \delta(x - x')\,. \tag{3.52}$$

The one-dimensional scalar product in (3.52) will be used throughout this subsection.

It is easy to calculate the overlapping $R^{c',c}(Z', Z)$,

$$R^{c',c}(Z', Z) = (\psi_{Z'}^{c'}, \psi_Z^c) = \frac{(1 + i)\exp Q}{2\sqrt{\pi|c'c^* - cc'^*|}}\,,$$
$$Q = \frac{[Z(c' + c'^*) - Z'(c + c^*)]^2}{2(c'c^* - cc'^*)(c + c^*)(c' + c'^*)}\,,$$
$$\psi_Z^c(x, t) = \int_{-\infty}^{\infty} \psi_{Z'}^{c'}(x, t)R^{c',c}(Z', Z)dZ'\,. \tag{3.53}$$

Let $\Delta = 1$. In this case a common solution of equations (3.48) and (3.49) is defined by Z and by two complex numbers c_1 and c_2 that are constrained by the relation $|c_1|^2 - |c_2|^2 = 1$. Such a solution has the form

$$\psi_Z^{c_1,c_2}(x,t) = \frac{1}{\sqrt{(f-g)\sqrt{\pi}}} \exp\left(Q^* - Q - \bar{p}^2/2\right),$$
$$Q = \frac{(f+g)x^2 - 2\sqrt{2}Zx + (f^* - g^*)Z^2}{4(f-g)},$$
$$\bar{p} = \frac{\sqrt{2}x - Z(f^* - g^*) - Z^*(f-g)}{\sqrt{2}|f-g|}. \tag{3.54}$$

Semicoherent states

It should be noted that in the general case the operator A^+ does not commute with A and is also an integral of motion. Moreover, it is a symmetry operator of equation (3.48). That is why by acting on the functions (3.54) with the operators $(A^+)^n, n \in \mathbb{N}$, we obtain some solutions of equation (3.48), but these solutions will not obey equation (3.49). They are not the coherent states. However, such kinds of states can be useful in different applications, see [36]. We call these states the *semicoherent states*. Here we construct semiclassical states as follows:

$$\psi_{n,Z}^{c_1,c_2}(x,t) = \frac{(A^+ - Z^*)^n}{\sqrt{n!}} \psi_Z^{c_1,c_2}(x,t), \quad n \in \mathbb{N}. \tag{3.55}$$

If we define $n!|_{n=0} = 1$, then $\psi_{0,Z}^{c_1,c_2}(x,t) = \psi_Z^{c_1,c_2}(x,t)$.

The set $\psi_{n,Z}^{c_1,c_2}(x,t), n \in \mathbb{Z}_+$ has the following properties:

$$(A - Z)\psi_{n,Z}^{c_1,c_2} = \sqrt{n}\psi_{n-1,Z}^{c_1,c_2}, \quad (A^+ - Z^*)\psi_{n,Z}^{c_1,c_2} = \sqrt{n+1}\psi_{n+1,Z}^{c_1,c_2},$$
$$\hat{N}_Z \psi_{n,Z}^{c_1,c_2} = n\psi_{n,Z}^{c_1,c_2}, \quad \hat{N}_Z = (A^+ - Z^*)(A - Z),$$
$$(\psi_{n',Z}^{c_1,c_2}, \psi_{n,Z}^{c_1,c_2}) = (\psi_Z^{c_1,c_2}, \psi_Z^{c_1,c_2})\delta_{n,n'}. \tag{3.56}$$

In the coordinate representation, the semicoherent states can be expressed via the Hermite functions $U_n(x)$

$$\psi_{n,Z}^{c_1,c_2}(x,t) = (f-g)^{-1/2} \left(\frac{f^* - g^*}{f-g}\right)^{\frac{n}{2}} e^Q U_n(\bar{p}), \tag{3.57}$$

where Q and $\bar{p}$ are given by equations (3.54).

At any fixed Z, the set $\psi_{n,Z}^{c_1,c_2}(x,t), n \in \mathbb{Z}_+$ is complete:

$$\sum_{n=0}^{\infty} \psi_{n,Z}^{*c_1,c_2}(x',t)\psi_{n,Z}^{c_1,c_2}(x,t) = \delta(x - x'). \tag{3.58}$$

At any fixed $n \in \mathbb{Z}_+$, the functions $\psi_{n,Z}^{c_1,c_2}(x,t)$ are not orthogonal for different Z, the set is overcomplete. The completeness relation has the form

$$\int d^2 Z \psi_{n',Z}^{*c_1,c_2}(x',t)\psi_{n,Z}^{c_1,c_2}(x,t) = \pi\delta_{n',n}\delta(x-x'), \quad d^2 Z = d\,(\operatorname{Re} Z)\, d\,(\operatorname{Im} Z)\ . \quad (3.59)$$

Relations (3.35) and (3.39) allow us to express initial coordinates and momenta in terms of the creation and annihilation operators,

$$\hat{x}^k = \frac{1}{2\sqrt{2}\lambda}(a_k + a_k^+), \quad \hat{p}^k = i\sqrt{2}\lambda(a_k^+ - a_k), \quad k = 1,2\,. \quad (3.60)$$

For $\Delta_k = 1$, we obtain with the use of (3.41):

$$\begin{aligned}\hat{x}^k &= \frac{1}{2\sqrt{2}\lambda}[(f_k^* - g_k^*)A_k + (f_k - g_k)A_k^+]\,,\\ \hat{p}^k &= i\sqrt{2}\lambda[(f_k + g_k)A_k^+ - (f_k^* + g_k^*)A_k]\,.\end{aligned} \quad (3.61)$$

Taking into account equations (3.56) and (3.44), we calculate the mean values of the operators (3.61) in the semicoherent states (3.57)

$$\begin{aligned}\overline{p^k} &= i\sqrt{2}\lambda[(c_1^{(k)} + c_2^{(k)})Z_k^* - (c_1^{*(k)} + c_2^{*(k)})Z_k]\,,\\ \overline{x^k} &= \frac{(c_1^{*(k)} - c_2^{*(k)})Z_k + (c_1^{(k)} - c_2^{(k)})Z_k^*}{2\sqrt{2}\lambda} + \frac{\overline{p^k}}{2\lambda^2}t\,.\end{aligned} \quad (3.62)$$

The mean trajectories (3.62) and the classical ones (3.9) can be identified if

$$\begin{aligned}p^k &= i\sqrt{2}\lambda[(c_1^{(k)} + c_2^{(k)})Z_k^* - (c_1^{*(k)} + c_2^{*(k)})Z_k]\,,\\ x_{(0)}^k &= \frac{(c_1^{*(k)} - c_2^{*(k)})Z_k + (c_1^{(k)} - c_2^{(k)})Z_k^*}{2\sqrt{2}\lambda}\,.\end{aligned} \quad (3.63)$$

For $\Delta_k = 1$, it follows from (3.63) that

$$Z_k = \sqrt{2}\lambda x_{(0)}^k(c_1^{(k)} + c_2^{(k)}) + \frac{ip^k}{2\sqrt{2}\lambda}(c_1^{(k)} - c_2^{(k)})\,. \quad (3.64)$$

Thus, all the mean trajectories $\overline{x^k}$ and $\overline{p^k}$ in semicoherent states $\psi_{n,Z}^{c_1,c_2}$ with arbitrary n correspond to a classical trajectory with given p^k and $x_{(0)}^k$ if Z_k and $c_{1,2}^{(k)}$ are defined by (3.64).

Let us calculate the following mean values:

$$\begin{aligned}\sigma_{1k} &= \overline{(\Delta x^k)^2}, \quad \sigma_{2k} = \overline{(\Delta p^k)^2}\,,\\ \sigma_{3k} &= \overline{(\Delta x^k)(\Delta p^k) + (\Delta p^k)(\Delta x^k)}\,,\end{aligned} \quad (3.65)$$

where $\Delta x^k = \hat{x}^k - \overline{x^k}$, $\Delta p^k = \hat{p}^k - \overline{p^k}$. Taking into account equations (3.62), we obtain

$$\begin{aligned}
\sigma_{sk} &= \sigma_{sk}^{(0)}(2n_k + 1),\ s, k = 1, 2, \\
\sigma_{1k}^{(0)} &= \frac{|c_1^{(k)} - c_2^{(k)} + 2i(c_1^{(k)} + c_2^{(k)})t|^2}{8\lambda^2}, \\
\sigma_{2k}^{(0)} &= 2\lambda^2 |c_1^{(k)} + c_2^{(k)}|^2, \\
\sigma_{3k}^{(0)} &= i\left(c_1^{*(k)} c_2^{(k)} - c_1^{(k)} c_2^{*(k)}\right) + 2|c_1^{(k)} + c_2^{(k)}|^2 t .
\end{aligned} \tag{3.66}$$

Then it follows from (3.66) that

$$J_k = \sigma_{1k}\sigma_{2k} - \sigma_{3k}^2/4 = (n_k + 1/2)^2 . \tag{3.67}$$

One can see that coherent states ($n_k = 0$) provide a minimum to the quantities J_k. The dispersions (3.66) do not depend on the parameters Z_k. In the general case these dispersions change with the time t.

Semicoherent states of a free spinning particle can be constructed with the help of the semicoherent states of spinless particles. To this end one has to use the decomposition $\Psi = \Psi_{(-)} + \Psi_{(+)}$ (see Section 2.3.1),

$$\Psi_{(-)} = \begin{pmatrix} u \\ -(\boldsymbol{\sigma n})u \end{pmatrix}, \quad \Psi_{(+)} = \begin{pmatrix} v \\ (\boldsymbol{\sigma n})v \end{pmatrix}, \tag{3.68}$$

where $\boldsymbol{n} = (0, 0, 1)$. Each component of the spinor u obeys the K–G equation with the operator (3.29), or, with the account of (3.33), obeys also equation (3.36), while the spinor v has the form

$$v = \lambda^{-1}[m - (\boldsymbol{\sigma p}_\perp)(\boldsymbol{\sigma n})]u, \quad \boldsymbol{p}_\perp = \boldsymbol{p} - \boldsymbol{n}(\boldsymbol{pn}) . \tag{3.69}$$

One can see that all the mean values and dispersions of differential operators in the semicoherent states coincide with the ones for spinless cases.

3.1.5 States with given angular momentum projection

Below, we construct free particle states that are eigenvectors for the operator $\hat{L}_z$, the-z-projection of the angular momentum operator $\hat{\mathbf{L}} = [\mathbf{r} \times \hat{\mathbf{P}}]$. At the same time, we demand such states to also be eigenvectors of the operator $\widehat{\overline{p}}_3$ (which commutes with $\hat{L}_z$) with the eigenvalue $\lambda/2$, as in equation (3.31).

We first consider the corresponding K–G wave functions. In the cylindric coordinates, $x = r\cos\varphi$ and $y = r\sin\varphi$, the operator $\hat{L}_z$ reads $\hat{L}_z = -i\partial_\varphi$, so that

$$\hat{L}_z \psi_l = l\psi_l, \quad \psi_l(r, \varphi; t) = \frac{\exp(il\varphi)}{\sqrt{2\pi}} \Phi_l(r, t), \quad l \in \mathbb{Z} . \tag{3.70}$$

States with different l are orthogonal,

$$(\psi_{l'}, \psi_l) = \int_0^{2\pi} d\varphi \int_0^{\infty} r dr \psi_{l'}^*(r, \varphi; t)\psi_l(r, \varphi; t) = \delta_{l,l'} \int_0^{\infty} \Phi_l^*(r, t)\Phi_l(r, t) r dr \,. \quad (3.71)$$

For ψ_l to be a solution of equation (3.36), the functions $\Phi_l(r, t)$ have to satisfy the following equation:

$$\tilde{\mathcal{K}}\Phi_l(r, t) = 0, \tilde{\mathcal{K}} = i\partial_t + \partial_r^2 + r^{-1}\partial_r - r^{-2}l^2, \quad r > 0 \,. \quad (3.72)$$

Stationary states represent a particular case of solutions of equation (3.72),

$$i\partial_t \Phi_{l,q}(r, t) = q\Phi_{l,q}(r, t), \quad q > 0 \,,$$
$$\Phi_{l,q}(r, t) = \exp(-iqt) J_l\left(\sqrt{q}r\right) \,, \quad (3.73)$$

where $J_l(x)$ are Bessel functions, see Ref. [191]. For the functions $\Phi_{l,q}$, we have the orthonormality and completeness relations

$$\int_0^{\infty} \Phi_{l,q'}^*(r, t)\Phi_{l,q}(r, t) r dr = 2\delta(q - q') \,,$$
$$\frac{r'}{2} \int_0^{\infty} \Phi_{l,q}^*(r', t)\Phi_{l,q}(r, t) dq = \delta(r - r'), \quad r, r' > 0 \,. \quad (3.74)$$

The general solution of equation (3.72) with the initial condition $\Phi_l(r, 0) = \varphi(r)$ has the form

$$\Phi_l(r, t) = \int_0^{\infty} G_l(r, r'; t)\varphi(r') dr' \,,$$
$$G_l(r, r'; t) = \frac{(-i)^{l+1} r'}{2t} J_l\left(\frac{rr'}{2t}\right) \exp\left(i\frac{r^2 + r'^2}{4t}\right) . \quad (3.75)$$

At any fixed r', the function $G_l(r, r'; t)$ obeys equation (3.72) and has the property

$$\lim_{t \to 0} G_l(r, r'; t) = \delta(r - r'), \quad r, r' > 0 \,. \quad (3.76)$$

One can express the operator $\hat{L}_z$ via the creation and annihilation operators (3.39),

$$L_z = i(a_1 a_2^+ - a_1^+ a_2) \,. \quad (3.77)$$

One can easily verify that $\hat{L}_z$ does not commute with any linear combination of the operators a_k and a_k^+. The coherent states introduced above cannot be eigenvectors of the operator $\hat{L}_z$. One can demonstrate that only the following quadratic combination of the operators a_k and a_k^+

$$\hat{R} = \alpha_1(a_1^2 + a_2^2) + \alpha_2(a_1 a_1^+ + a_1^+ a_1 + a_2 a_2^+ + a_2^+ a_2) + \alpha_3(a_1^{+2} + a_2^{+2}), \tag{3.78}$$

where the functions $\alpha_1(t)$, $\alpha_2(t)$, and $\alpha_3(t)$ do not depend on x, y, commutes with $\hat{L}_z$. If the latter functions obey the equations

$$i\dot{\alpha}_1 + 2\alpha_1 + 2\alpha_2 = 0, \quad i\dot{\alpha}_2 - \alpha_1 + \alpha_3 = 0, \quad i\dot{\alpha}_3 - 2\alpha_2 - 2\alpha_3 = 0, \tag{3.79}$$

then the operator $\hat{R}$ is an integral of motion of equation (3.34). The general solution of equations (3.79) has the form

$$\alpha_1 = 2f - i\dot{f} - g^2/2, \alpha_2 = -2f - g^2/2, \alpha_3 = 2f + i\dot{f} - g^2/2, \tag{3.80}$$

where the function $f = f(t)$ obeys the equation $\dddot{f} = 2g^2$, with g^2 being an arbitrary complex constant. It is convenient to write

$$f = g^2t^2 + 2\beta_2 t + \beta_1^2 = g^2(b - it)(\bar{b} + it),$$
$$g^2 b = \beta - i\beta_2, \quad g^2\bar{b} = \beta + i\beta_2, \quad \beta^2 = g^2\beta_1^2 - \beta_2^2, \tag{3.81}$$

where β_1 and β_2 are arbitrary and the constant β is defined up to a sign, this sign is then chosen to obey condition (3.84), see below.

If f is real then $\hat{R}$ is self-adjoint. Now, in addition to equations (3.36) and (3.70), we impose the following condition on the functions ψ_l:

$$(R + 4p)\psi_l = 0. \tag{3.82}$$

It is equivalent to the following equation:

$$\left[\partial_r^2 + \left(\frac{1}{r} - \frac{i\dot{f}}{2f}r\right)\partial_r - \frac{g^2r^2}{4f} - \frac{l^2}{r^2} - \frac{i\dot{f}}{2f} + \frac{p}{f}\right]\Phi_{l,p}(r,t) = 0 \tag{3.83}$$

on the functions $\Phi_{l,p}(r,t)$ (these functions obey equation (2.90)). For $\beta \neq 0$, and[2]

$$\operatorname{Re} b \leq 0, \quad g^2 b = \beta - i\beta_2, \tag{3.84}$$

the equation (3.83) has normalizable solutions,

$$\int_0^\infty \Phi_{l,p}^*(r,t)\Phi_{l,p}(r,t)r\,dr = 1. \tag{3.85}$$

2 In particular, the sign of β is to be determined from condition (3.84).

The functions $\Phi_{l,p}(r,t)$ can be written in terms of the Laguerre functions with the complex index, see Section B.1,

$$\Phi_{l,p}(r,t) = N(b-it)^{\frac{p}{2\beta}-\frac{1}{2}}(\bar{b}+it)^{-\frac{p}{2\beta}-\frac{1}{2}}\exp\left(i\frac{\dot{f}}{8f}r^2\right)I_{|l|+s,s}\left(\frac{\beta r^2}{2f}\right),$$
$$g^2\bar{b} = \beta + i\beta_2, \quad 2s = p\beta^{-1} - 1 - |l|. \qquad (3.86)$$

If (3.84) is a strong inequality, then

$$N = \frac{(\bar{b}+\bar{b}^*)^{\frac{p}{2\beta}}}{(\bar{b}+b)^{\frac{|l|}{2}}(b-\bar{b}^*)^s}\left[\frac{\Gamma(1+s)\Gamma(1+|l|)}{2\Gamma(1+s+|l|)F(-s^*,-s;1+|l|;x)}\right]^{1/2},$$
$$x = \frac{(\bar{b}+b)(\bar{b}^*+b^*)}{(b^*-\bar{b})(b-\bar{b}^*)},$$

where $F(\alpha,\beta;\gamma;x)$ is the hypergeometric function, see [191].

If $\beta = 0$, then $f = \omega^2, \omega = gt + \beta_1$, and solutions $\Phi_{l,p}(r,t)$ can be written in terms of the Bessel functions,

$$\Phi_{l,p}(r,t) = \frac{N_0}{\omega}\exp\left[\frac{i}{4g\omega}(4p + g^2r^2)\right]J_l\left(\omega^{-1}\sqrt{p}r\right). \qquad (3.87)$$

These solutions are normalized to the unity if

$$q \geq 0, \quad q = i(g^*\beta_1 - g\beta_1^*). \qquad (3.88)$$

If (3.88) is a strong inequality, then

$$N_0 = \exp\left(\frac{g^*p}{gq}\right)\sqrt{\frac{q}{2I_l(2|p|q^{-1})}}$$

where $I_l(x)$ are Bessel functions of an imaginary argument.

Finally, if $g = 0$, we are left with the stationary case (2.92) already studied above.

Solutions Ψ of the Dirac equation that are eigenvectors for the operator $\hat{J}_z$,

$$\hat{J}_z\Psi = \left(l - \frac{1}{2}\right)\Psi, \quad \hat{J}_z = \hat{L}_z + \frac{1}{2}\Sigma_3, \qquad (3.89)$$

have the form (3.68), where the spinor u is

$$u = \begin{pmatrix}\psi_{l-1}(r,\varphi;t)\\ \psi_l(r,\varphi;t)\end{pmatrix}, \qquad (3.90)$$

and $\psi_l(r,\varphi;t)$ are the corresponding solutions (2.89) of the K–G equation.

3.2 Particles in plane-wave field

3.2.1 Plane-wave electromagnetic field

We call free electromagnetic fields *plane-waves* propagating along a unit vector $\mathbf{n}$ ($\mathbf{n}^2 = 1$), if their electric and magnetic strengths have the form

$$\begin{aligned} &\mathbf{E} = \mathbf{E}(u_0), \quad \mathbf{H} = \mathbf{H}(u_0)\,, \\ &u_0 = (nx) = x^0 - (\mathbf{nr}), \quad n^\mu = (1, \mathbf{n})\,. \end{aligned} \tag{3.91}$$

If the fields (3.91) are free but not constant and uniform then they are related as follows:

$$\begin{aligned} &(\mathbf{nE}) = (\mathbf{nH}) = (\mathbf{EH}) = 0, \quad |\mathbf{E}| = |\mathbf{H}|\,; \\ &\mathbf{H} = [\mathbf{n} \times \mathbf{E}], \quad \mathbf{E} = -[\mathbf{n} \times \mathbf{H}], \quad \mathbf{n} = \mathbf{E}^{-2}[\mathbf{E} \times \mathbf{H}]\,. \end{aligned} \tag{3.92}$$

It is evident that the plane-wave field is a particular case of the crossed field (2.99).

Electromagnetic potentials for the fields (3.92) can be chosen as $A^\mu = A^\mu(u_0)$ and subjected, for example, to the condition

$$(nA) = 0\,, \tag{3.93}$$

which provides (due to $\partial_\mu u_0 = n_\mu$) the Lorentz gauge $\partial_\mu A^\mu = 0$. In addition, we impose the gauge condition $A_0 = 0$. Along with (3.93) this creates the following representation for the potentials of the plane-wave field to be used in what follows:

$$A^\mu = \left(0, \mathbf{A}(u_0)\right), \quad (\mathbf{nA}) = 0\,. \tag{3.94}$$

Then

$$\mathbf{E}(u_0) = -\mathbf{A}'(u_0), \quad \mathbf{H}(u_0) = -\left[\mathbf{n} \times \mathbf{A}'(u_0)\right]\,. \tag{3.95}$$

Here and in what follows in this Section, we denote by primes derivative with respect to the light-cone time u_0, for example $A'^\mu = dA^\mu/du_0$.

The following useful relations for the strength tensor of the plane-wave field are worth listing:

$$\begin{aligned} F_{\mu\nu} &= n_\mu A'_\nu - n_\nu A'_\mu = F_{\mu\nu}(u_0)\,, \\ n^\mu F_{\mu\nu} &= n^\mu \tilde{F}_{\mu\nu} = 0\,, \\ F_{\mu\alpha} F^{\alpha\nu} &= \tilde{F}_{\mu\alpha} \tilde{F}^{\alpha\nu} = n_\mu n^\nu \mathbf{E}^2\,, \\ F_{\mu\alpha} F^{\alpha\mu} &= F_{\mu\alpha} \tilde{F}^{\alpha\mu} = 0, \quad (I_1 = I_2 = 0)\,. \end{aligned} \tag{3.96}$$

3.2.2 Classical motion in the plane-wave field

The Lorentz equations of motion (2.18) for a charge in the plane-wave field can be written with the use of (3.96) as

$$m\ddot{x}^0 = -e(\dot{\mathbf{r}}\mathbf{A}'), \quad m\ddot{\mathbf{r}} = -e\left[\mathbf{A}'\dot{u}_0 + \mathbf{n}(\dot{\mathbf{r}}\mathbf{A}')\right] ;$$
$$(\dot{x}^0)^2 - \dot{\mathbf{r}}^2 = 1 . \tag{3.97}$$

Here, we denote derivatives with respect to the proper time by dots.

One can easily see that these equations imply three first integrals of motion. Indeed, in the gauge under consideration, we have $m\ddot{u}_0 = 0$ as it follows from equations (3.97). Then

$$m\ddot{u}_0 = 0 \to m\dot{u}_0 = m(n\dot{x}) = (nP)$$
$$= P^0 - (\mathbf{n}\mathbf{P}) = (np) = \lambda = \text{const} . \tag{3.98}$$

Besides,

$$\boldsymbol{\pi} + e\mathbf{A} = \mathbf{k}, \quad \boldsymbol{\pi} = \boldsymbol{\pi}(u_0) = \mathbf{P} - \mathbf{n}(\mathbf{n}\mathbf{P}) = \mathbf{k} - e\mathbf{A}(u_0) ;$$
$$\mathbf{k} = \mathbf{p} - \mathbf{n}(\mathbf{n}\mathbf{p}), \quad \mathbf{p} = \mathbf{P} + e\mathbf{A}, \quad (\mathbf{n}\boldsymbol{\pi}) = (\mathbf{n}\mathbf{k}) = 0 , \tag{3.99}$$

where $\mathbf{k}$ is an arbitrary constant vector orthogonal to the vector $\mathbf{n}$. Thus, any two linearly independent components of $\mathbf{k}$ and λ are the first integrals of motion.

Using these integrals of motion and the last equation in (3.97), we find kinetic momenta $P^\mu = m\dot{x}^\mu = (P^0, \mathbf{P})$ as functions of the light-cone time u^0 and of integrals of motion $\mathbf{k}$ and λ,

$$P^0(\mathbf{k}, \lambda; u_0) = \frac{m^2 + \boldsymbol{\pi}^2(u_0) + \lambda^2}{2\lambda} ,$$
$$\mathbf{P}(\mathbf{k}, \lambda; u_0) = \boldsymbol{\pi}(u_0) + \mathbf{n}\frac{m^2 + \boldsymbol{\pi}^2(u_0) - \lambda^2}{2\lambda} . \tag{3.100}$$

Let us introduce a constant four-vector $q^\mu = (q^0, \mathbf{q})$,

$$q^0 = \frac{m^2 + \mathbf{k}^2 + \lambda^2}{2\lambda}, \quad \mathbf{q} = \mathbf{k} + \mathbf{n}\frac{m^2 + \mathbf{k}^2 - \lambda^2}{2\lambda} . \tag{3.101}$$

It satisfies the following conditions:

$$q^2 = (q^0)^2 - \mathbf{q}^2 = m^2, \quad \mathbf{q} - \mathbf{n}(\mathbf{n}\mathbf{q}) = \boldsymbol{k}, \quad (nq) = q^0 - (\mathbf{n}\mathbf{q}) = \lambda . \tag{3.102}$$

With the help of the introduced vector, one can write equations (3.100) in the following covariant form:

$$P^\mu = P^\mu(u_0) = q^\mu - eA^\mu + n^\mu \frac{2e(qA) - e^2A^2}{2(nq)} ;$$
$$P^2 = q^2 = m^2, \quad (nP) = (nq) = \lambda . \tag{3.103}$$

Finally, introducing another four-vector $Q^\mu(u_0) = q^\mu - eA^\mu(u_0)$, we obtain

$$P = P(u_0) = Q + n\frac{m^2 - Q^2}{2(nq)}, \quad p(u_0) = q + n\frac{m^2 - Q^2}{2(nq)}; \quad m^2 - Q^2 = \pi^2 - \mathbf{k}^2. \quad (3.104)$$

Taking all that into account, by integrating equations (3.100) over u_0, we find the general solution of equations (3.97), parametrized by the parameter u_0, and the relation of the parameter with the proper time τ,

$$\mathbf{r}(u_0) = \frac{1}{\lambda}\int \pi(u_0)du_0 + \frac{\mathbf{n}}{2\lambda^2}\int \left[m^2 + \pi^2(u_0) - \lambda^2\right] du_0 + \mathbf{r}_{(0)},$$
$$x^0(u_0) = \frac{1}{2\lambda^2}\int \left[m^2 + \pi^2(u_0) + \lambda^2\right] du_0 + x^0_{(0)}, u_0 = \frac{\lambda}{m}(\tau - \tau_0). \quad (3.105)$$

Using equations (3.104), we rewrite (3.105) in the following covariant form:

$$x^\mu(u_0) = \frac{1}{(nq)}\int Q^\mu(u_0)du_0 + \frac{n^\mu}{2(nq)^2}\int \left[m^2 - Q^2(u_0)\right] du_0 + x^\mu_{(0)}. \quad (3.106)$$

The solution is given in a parametric form, with u_0 being the parameter, and depends on six constants, i.e. integrals of motion. These are λ, $\mathbf{r}_{(0)}$, and two independent components of the vector $\mathbf{k}$. The constant $t_{(0)}$ is related to the choice of initial time.

Let us analyze the motion of a charge in the plane-wave under the following assumptions about the plane-wave potentials[3]:

$$\overline{\mathbf{A}} = 0, \quad \overline{\mathbf{A}^2} < \infty. \quad (3.107)$$

For example, conditions (3.107) take place for potentials bounded for any u_0. We note that if the first condition (3.107) is not fulfilled, but instead $\overline{\mathbf{A}} = \mathbf{A}_{(0)} \neq 0$, one can obviously achieve its fulfilment by the change $\mathbf{A}(\xi) \to \mathbf{A}(\xi) - \mathbf{A}_{(0)}$, which does not violate the earlier accepted gauge conditions.

Assuming conditions (3.107), we find from (3.105) that the charged particle in plane-wave field has a constant mean velocity,

$$\bar{\mathbf{v}} = \lim_{x^0\to\infty}\frac{\mathbf{r}}{x^0} = \frac{2\lambda\mathbf{k} + \mathbf{n}\left[m^2\left(1+\gamma^2\right) + \mathbf{k}^2 - \lambda^2\right]}{m^2\left(1+\gamma^2\right) + \mathbf{k}^2 + \lambda^2},$$
$$\gamma^2 = \frac{e^2\overline{\mathbf{A}^2}}{m^2}. \quad (3.108)$$

3 The averaging is understood as

$$\overline{f(u_0)} = \lim_{L\to\infty}\frac{1}{L}\int_0^L f(u_0)du_0.$$

Therefore, the charged particle in the plane-wave field undergoes oscillations about a certain average position as it drifts with the constant mean velocity $\bar{\mathbf{v}}$. If

$$\mathbf{k} = 0, \quad \lambda^2 = m^2\left(1 + \gamma^2\right), \tag{3.109}$$

then $\bar{\mathbf{v}} = 0$, which corresponds to the charge at rest on the average.

The mean (in time) energy follows from (3.93) to be

$$\overline{\mathcal{E}} = \overline{P^0} = \frac{m^2\left(1 + \gamma^2\right) + \mathbf{k}^2 + \lambda^2}{2\lambda}. \tag{3.110}$$

It follows from (3.110) that the sign of the mean energy is determined by the sign of the integral of motion λ. Consequently, we must retain only positive $\lambda > 0$ in classical theory, whereas in relativistic quantum theory the sign of λ serves to distinguish states with positive and negative energies.

We can see that for a charge at rest on the average it follows from (3.110) with the use of (3.109) that

$$\left.\overline{\mathcal{E}}\right|_{\bar{\mathbf{v}}=0} = m\sqrt{1 + \gamma^2} = m^*. \tag{3.111}$$

This shows that in the presence of the plane wave the rest mass of the particle becomes effectively $\sqrt{1 + \gamma^2}$ times as much.

The solution of the Hamilton–Jacobi equation (2.27) that depends on the integrals of motion λ and $\mathbf{k}$ has the form [231]

$$\begin{aligned} S(\mathbf{k}, \lambda; x) &= (\mathbf{kr}) - \lambda x^0 - \frac{1}{2\lambda}\int\left[m^2 + \pi^2(u_0) - \lambda^2\right] du_0 \\ &= (\mathbf{kr}) - \lambda\left[x^0 + (\mathbf{nr}(u_0))\right]. \end{aligned} \tag{3.112}$$

The function $S(x)$ can be rewritten in a covariant form

$$S = S(q; x) = -(qx) - \frac{1}{2(nq)}\int\left[m^2 - Q^2(u_0)\right] du_0\,. \tag{3.113}$$

By differentiating (3.113) with respect to spacial coordinates, we obtain an expression for the generalized momentum $p(u_0)$ given by equation (3.104),

$$p_\mu(u_0) = -\partial_\mu S(q; x)\,. \tag{3.114}$$

3.2.3 Quantum motion in plane-wave field

For the first time, exact solutions of the Dirac equation with a plane-wave field were obtained by Volkov in the works [336–338].

In the relativistic quantum theory of a charged particle moving in the plane-wave field the operators

$$\hat{\lambda} = (n\hat{p}), \quad \hat{\mathbf{k}} = \hat{\mathbf{p}} - \mathbf{n}\,(\mathbf{n}\hat{\mathbf{p}}), \tag{3.115}$$

where $\hat{p}_\mu = i\partial_\mu = (i\partial_0, -\hat{\mathbf{p}} = i\nabla)$, commute with the K–G operator $\hat{\mathfrak{K}}$ and with the Dirac operator $\hat{\mathfrak{D}}$, and therefore they are quantum integrals of motion. In his pioneering works Volkov was considering solutions (*Volkov's solutions*) that are eigenvectors for the operators (3.115). Far later, other types of solutions were obtained in [35, 61, 122].

Spinless particle in plane-wave field

The quantum motion of spinless relativistic particle in the plane-wave field is described by scalar wave functions $\Phi(x)$, the latter obeying the K–G equation (2.28) with such a field. Here, we shall be interested in such solutions of the K–G equation with the plane-wave field that are the eigenfunctions $\Phi_{k,\lambda}(x)$ of the operators (3.115) with the eigenvalues λ and $\mathbf{k}$, respectively. Such solutions have to satisfy the following set of equations:

$$\hat{\lambda}\Phi_{k,\lambda}(x) = \lambda\Phi_{k,\lambda}(x), \quad \hat{\mathbf{k}}\Phi_{k,\lambda}(x) = \mathbf{k}\Phi_{k,\lambda}(x), \quad \hat{\mathfrak{K}}\Phi_{k,\lambda}(x) = 0 \,. \tag{3.116}$$

Any solution of such a kind (labeled by the quantum numbers λ and $\mathbf{k}$) has the form

$$\Phi_{k,\lambda}(x) = N \exp iS\left(q; x\right) \,, \tag{3.117}$$

where $S(q; x)$ is the classical action (3.112) or (3.113) and N is a normalization constant. The semiclassical form of the solution is due to the structure of the plane-wave field and to the choice of the quantum integrals of motion.

To verify that (3.117) satisfies equations (3.116), we consider the quantity $\hat{p}_\mu e^{iS}$. Taking into account equation (3.114), we get the first two equations in (3.116)

$$\begin{aligned} &\hat{p}_\mu e^{iS} = -(\partial_\mu S)e^{iS} = p_\mu(u_0)e^{iS} \,, \\ &\hat{P}e^{iS} = P(u_0)e^{iS}, \quad \hat{\lambda}e^{iS} = \lambda e^{iS}, \quad \hat{\mathbf{k}}e^{iS} = \mathbf{k}e^{iS} \,, \end{aligned} \tag{3.118}$$

where $P(u_0)$ and $p(u_0)$ are the classical kinetic and generalized momenta (3.104). Then, using these relations, we verify that the function (3.117) also satisfies the third equation in (3.116),

$$\begin{aligned} \hat{P}^2 e^{iS} &= \eta^{\mu\nu}\hat{P}_\mu P_\nu(u_0)e^{iS} = \eta^{\mu\nu}[\hat{P}_\mu P_\nu(u_0) - P_\nu(u_0)\hat{P}_\mu + P_\nu(u_0)\hat{P}_\mu]e^{iS} \\ &= \eta^{\mu\nu}[\hat{P}_\mu, P_\nu(u_0)]e^{iS} + \eta^{\mu\nu}P_\nu(u_0)\hat{P}_\mu e^{iS} \\ &= ie^{iS}\frac{d}{du_0}\eta^{\mu\nu}n_\mu P_\nu(u_0) + \eta^{\mu\nu}P_\nu(u_0)P_\mu(u_0)e^{iS} \\ &= ie^{iS}\frac{d(nP)}{du_0} + P^2(u_0)e^{iS} = ie^{iS}\frac{d\lambda}{du_0} + P^2(u_0)e^{iS} = m^2 e^{iS} \,. \end{aligned}$$

The states (3.117) with $\lambda > 0$ describe particles, whereas these states with $\lambda < 0$ describe antiparticles.

It is convenient to use the scalar product (2.38) on the hyperplane $u_0 = \text{const}$ to normalize the functions (3.117). If $N = (2(2\pi)^3|\lambda|)^{-1/2}$, then it is easy to make sure that solutions (3.117) are orthonormal with the scalar product (2.38),

$$(\Phi_{\mathbf{k}',\lambda'}, \Phi_{\mathbf{k},\lambda})_{u_0} = \frac{\lambda}{|\lambda|}\delta(\mathbf{k} - \mathbf{k}')\delta(\lambda - \lambda'), \tag{3.119}$$

and form a complete system on the hyperplane $u_0 = \text{const}$,

$$\int_{-\infty}^{+\infty} d\lambda|\lambda| \int d\mathbf{k}\Phi_{\mathbf{k},\lambda}(u_0, \mathbf{u})\Phi^*_{\mathbf{k},\lambda}(u_0, \mathbf{u}') = \delta\left(\mathbf{u} - \mathbf{u}'\right),$$

see (2.54).

The solutions (3.117) form a complete and orthogonal set with respect to the usual scalar product (2.32) on the hyperplane $x^0 = \text{const}$. The proof of this fact is not trivial and was presented for the first time in [170]. Consider the scalar product of two solutions (3.117) labeled by quantum numbers $\lambda, \mathbf{k}$ and $\lambda', \mathbf{k}'$. The orthogonality with respect to the quantum numbers $\mathbf{k}$ and $\mathbf{k}'$ is obvious. Then, taking into account equations (3.118), (3.100), and (3.112), we obtain

$$\begin{aligned} F &= (\hat{P}^0\Phi_{\mathbf{k},\lambda'})^*\Phi_{\mathbf{k},\lambda} + \Phi^*_{\mathbf{k},\lambda'}\hat{P}^0\Phi_{\mathbf{k},\lambda} \\ &= (\lambda + \lambda')\frac{d\varphi(u_0)}{du_0}\exp i[\lambda'^0 + (\lambda - \lambda')\varphi(u_0)], \\ \varphi(u_0) &= \frac{1}{2\lambda\lambda'}\int \left[m^2 + \pi^2(u_0) + \lambda\lambda'\right] du_0, \end{aligned}$$

so that

$$(\Phi_{\mathbf{k}',\lambda'}, \Phi_{\mathbf{k},\lambda}) = (2\pi)^2|N|^2\delta(\mathbf{k} - \mathbf{k}') \int_{-\infty}^{\infty} F dx^3. \tag{3.120}$$

On the other hand, the following relation holds:

$$\begin{aligned} \int_{-\infty}^{\infty} F dx^3 &= \int_{-\infty}^{\infty} F du_0 = (\lambda + \lambda') \times \int_{-\infty}^{\infty} \frac{d\varphi(u_0)}{du_0} \exp i\left[\lambda'^0 + (\lambda - \lambda')\varphi(u_0)\right] du_0 \\ &= (\lambda + \lambda')\exp i\lambda'^0 \int_{-\infty}^{\infty} \exp[i(\lambda - \lambda')\varphi]d\varphi = 4\pi\lambda\delta(\lambda - \lambda'). \end{aligned}$$

Thus, for $N = (2|\lambda|(2\pi)^3)^{-1/2}$, we have

$$(\Phi_{\mathbf{k}',\lambda'}, \Phi_{\mathbf{k},\lambda}) = \frac{\lambda}{|\lambda|}\delta(\mathbf{k} - \mathbf{k}')\delta(\lambda - \lambda'). \tag{3.121}$$

According to equations (3.117) and (3.101), the wave functions can be labeled (parametrized) by three quantum numbers $\mathbf{q}$. In such parametrization and for $N = (2|q_0|(2\pi)^3)^{-1/2}$, we have

$$(\Phi_{\mathbf{q}'}, \Phi_{\mathbf{q}})_{u_0} = (\Phi_{\mathbf{q}'}, \Phi_{\mathbf{q}}) = \frac{q_0}{|q_0|}\delta(\mathbf{q} - \mathbf{q}'). \tag{3.122}$$

Reduction to free two-dimensional Schrödinger equation

It turns out that the K–G equation with the plane-wave field can be reduced to a free Schrödinger equation for two-dimensional motion by a simple change of variables and of the wave function. The latter equation was studied in detail in the Section 3.1.4.

Some hints about such variable changes can be obtained already from the classical theory. Let $\mathbf{r}_\perp$ be a position vector of a particle in the plane that is orthogonal to the vector $\mathbf{n}$,

$$\mathbf{r}_\perp = \mathbf{r} - \mathbf{n}(\mathbf{nr}), \quad (\mathbf{nr}_\perp) = 0 . \tag{3.123}$$

It follows from (3.97) that classical equations of motion for $\mathbf{r}_\perp$ have the form

$$m\ddot{\mathbf{r}}_\perp = -e\mathbf{A}'\dot{u}_0 . \tag{3.124}$$

Let us now introduce instead of the variables $\mathbf{r}_\perp$ two new dimensionless variables $\boldsymbol{\rho}_\perp = (\tilde{x}, \tilde{y})$ according to the relations

$$\boldsymbol{\rho}_\perp = 2\lambda\mathbf{r}_\perp + 2\int e\mathbf{A}(u_0)du_0, \quad (\mathbf{n}\boldsymbol{\rho}_\perp) = 0 . \tag{3.125}$$

With the use of equations (3.98) and (3.124), we obtain free classical equations of motion for $\boldsymbol{\rho}_\perp$,

$$m\ddot{\boldsymbol{\rho}}_\perp = 0 . \tag{3.126}$$

Returning to the K–G equation with the plane-wave field, we will again consider its solutions that are eigenfunctions for the operator $\hat{\lambda}$,

$$\hat{\lambda}\Phi_\lambda(x) = \lambda\Phi_\lambda(x) . \tag{3.127}$$

Then we represent the functions $\Phi_\lambda(x)$ as follows:

$$\Phi_\lambda(x) = \exp\left(- i\lambda x^0 + i\frac{\lambda^2 - m^2}{2\lambda}u_0\right)\overline{\Phi}(u_0; \mathbf{r}_\perp) . \tag{3.128}$$

One can easily check that functions (3.128) satisfy equation (3.127) for any $\overline{\Phi}(u_0; \mathbf{r}_\perp)$, the latter functions can be found from the corresponding K–G equation. Namely, substituting (3.128) into the K–G equation, we get an equation for the functions $\overline{\Phi}(u_0; \mathbf{r}_\perp)$,

$$2i\lambda\partial_{u_0}\overline{\Phi}(u_0; \mathbf{r}_\perp) = \hat{\mathbf{P}}_\perp^2\overline{\Phi}(u_0; \mathbf{r}_\perp), \quad \hat{\mathbf{P}}_\perp = \hat{\mathbf{k}} - e\mathbf{A}(u_0) , \tag{3.129}$$

with operator $\hat{\mathbf{k}}$ given in equation (3.115).

In equation (3.129), we go over from the variables $\mathbf{r}_\perp$ to the variables $\boldsymbol{\rho}_\perp$ defined according to (3.125), also from the light-cone time u_0 to the new dimensionless time $\tilde{t} = 2\lambda u_0$, and to a new function $\psi(\tilde{t}; \boldsymbol{\rho}_\perp)$,

$$\overline{\Phi}(u_0; \mathbf{r}_\perp) = \exp\left(-\frac{i}{2\lambda}\int e^2\mathbf{A}^2(u_0)du_0\right)\psi(\tilde{t}; \boldsymbol{\rho}_\perp) . \tag{3.130}$$

Then equation (3.129) implies the following equation for the function $\psi(\tilde{t};\boldsymbol{\rho}_\perp)$:

$$i\partial_{\tilde{t}}\psi(\tilde{t};\boldsymbol{\rho}_\perp) = -\left(\partial_{\tilde{x}}^2 + \partial_{\tilde{y}}^2\right)\psi(\tilde{t};\boldsymbol{\rho}_\perp), \quad \boldsymbol{\rho}_\perp = (\tilde{x},\tilde{y}), \tag{3.131}$$

which is free Schrödinger equation for two-dimensional motion.

The free Schrödinger equation (3.131) was studied in detail in Section 3.1.4. All results obtained there can be used here in context of the K–G equation with the plane-wave field. In particular, using coherent states obtained in Section 3.1.4, one can form relativistic coherent states for a spinless particle in the plane wave. Such coherent states were first presented in the works [35, 122].

Spinning particle in plane-wave field

Because the plane-wave field is a special case of the crossed field, there exist the relation (2.110) which establishes the connection between solutions of the Dirac and K–G equations in such a field. According to this relation, the Dirac wave function has the form

$$\Psi(x) = N\hat{R}\Phi(x)v\,, \tag{3.132}$$

where N is the normalization factor, $\Phi(x)$ is a solution of the K–G equation with the plane-wave field, v is an arbitrary constant spinor, and then, in the case under consideration, the operator $\hat{R}$ can be represented as

$$\hat{R} = \begin{pmatrix} m + \hat{\lambda} - (\boldsymbol{\sigma}\hat{\mathbf{P}}_\perp)(\boldsymbol{\sigma}\mathbf{n}) \\ (m - \hat{\lambda})(\boldsymbol{\sigma}\mathbf{n}) + (\boldsymbol{\sigma}\hat{\mathbf{P}}_\perp) \end{pmatrix}. \tag{3.133}$$

The operators $\hat{\lambda}$ and $\hat{\mathbf{P}}_\perp$ are defined in (3.115) and (3.129) respectively.

It follows from equations (3.132) and (3.133) that components $\Psi_{(-)}(x)$ (defined by relation (2.70)) that correspond to the bispinors $\Psi(x)$ have the form

$$\Psi_{(-)}(x) = N\hat{\lambda}\Phi(x)\hat{R}_{(-)}v,\, \hat{R}_{(-)} = \begin{pmatrix} 1 \\ -(\boldsymbol{\sigma}\mathbf{n}) \end{pmatrix}. \tag{3.134}$$

If states (3.132) are considered that are eigenvectors for the operators $\hat{\lambda}$ and $\hat{\mathbf{k}}$, then the operator $\hat{R}$ takes a more simple form,

$$\hat{R} = \begin{pmatrix} m + \lambda - (\boldsymbol{\sigma}\boldsymbol{\pi}(u_0))(\boldsymbol{\sigma}\mathbf{n}) \\ (m - \lambda)(\boldsymbol{\sigma}\mathbf{n}) + (\boldsymbol{\sigma}\boldsymbol{\pi}(u_0)) \end{pmatrix}, \tag{3.135}$$

where the vector function $\boldsymbol{\pi}(u_0)$ is given by (3.99).

Specifying the arbitrary constant spinors v in the solutions (3.132) and (3.134), one can classify these solutions as states with a certain spin polarization. As it follows from Section 2.4, for the Dirac equation with the plane-wave field, there exists a spin integral of motion Y (2.151). It was pointed out for the first time in [327]. If we chose the solutions (3.132) and (3.134) to be eigenvectors for Y, then the spinor v takes the form (2.155).

3.3 Particles in BGY field

3.3.1 BGY field

In this section, we study classical and quantum motion of a charged particle in superpositions of the plane-wave field with other electromagnetic fields. Among these additional fields, we consider a constant uniform magnetic field collinear with the plane-wave propagation direction, and an electric field propagating along the same direction. The exact solutions of the Dirac and K–G equations in the superposition of the plane-wave field and the collinear magnetic field were obtained for the first time by Redmond in [273] (classical motion in such fields was discussed in [274]). Since that time this field configuration is called the *Redmond field* and the corresponding solutions *Redmond solutions*. Exact solutions in the superposition of the Redmond field and electric field propagating along the plane-wave were first found by Bagrov, Gitman, and Yushin in [25]. Such a superposition is called *BGY superposition of electromagnetic fields* and the corresponding exact solutions are called *BGY solutions*. The quantum effects of scattering and pair creation in a BGY field were considered in Refs. [153, 175].

The BGY electromagnetic field is described by electromagnetic potentials $A_\mu^{\mathrm{BGY}}(x)$ of the form

$$A_0^{\mathrm{BGY}} = A_3^{\mathrm{BGY}} = A_0(u_0), \quad A_1^{\mathrm{BGY}} = \frac{H}{2}x^2 + A_1(u_0), \quad A_2^{\mathrm{BGY}} = -\frac{H}{2}x^1 + A_2(u_0), \tag{3.136}$$

where[4] $u_0 = x^0 - x^3$, $H = \mathrm{const}$, and $A_\mu(u_0)$, $\mu = 0, 1, 2$ are arbitrary functions of the light-cone time u_0. Cartesian coordinates of the corresponding electromagnetic fields are

$$E^1 = E_x = H^2 = H_y = A_1'(u_0), \quad E^2 = E_y = -H^1 = -H_x = A_2'(u_0), \tag{3.137}$$

$$E^3 = E_z = 2A_0'(u_0), \quad H^3 = H_z = H. \tag{3.138}$$

Thus, the BGY field is a superposition of the plane-wave field, given by equations (3.137), of a constant uniform magnetic field, collinear to it, and of electric field, propagating in the same direction, both given by equations (3.138).

3.3.2 Classical motion in a BGY field

The Lorentz equations of motion (2.18) with the BGY field have the form

$$m\dot{P}^0 = e(\mathbf{EP}), \quad m\dot{\mathbf{P}} = e\mathbf{E}P^0 + e[\mathbf{P}\times\mathbf{H}], \quad P^\mu = m\dot{x}^\mu. \tag{3.139}$$

4 Here, we chose $\boldsymbol{n} = (0, 0, 1)$.

It follows from the second equation (3.139) that

$$m\dot{P}^3 = e(\mathbf{EP}) + 2eA_0'(u_0)(P^0 - P^3) \,. \tag{3.140}$$

Combining equation (3.140) and the first equation (3.139), we find

$$m\dot{Q} + 2eA_0'(u_0)Q = 0, Q = P^0 - P^3 \,. \tag{3.141}$$

With the help of the relations

$$\begin{gathered} Q = P^0 - P^3 = m(\dot{x}^0 - \dot{x}^3) = m\dot{u}_0 \Rightarrow \\ A_0'(u_0)Q = mA_0'(u_0)\dot{u}_0 = m\dot{A}_0(u_0)\,, \end{gathered}$$

we can rewrite equation (3.141) first as

$$\dot{Q} + 2e\dot{A}_0(u_0) = 0 \Rightarrow Q + 2eA_0(u_0) = \lambda, \quad \lambda = \text{const}\,,$$

and then as follows:

$$P^0 - P^3 = m\dot{u}_0 = \rho(u_0), \quad \rho(u_0) = \lambda - 2eA_0(u_0); \quad p^0 - p^3 = (np) = \lambda\,. \tag{3.142}$$

In particular, equations (3.142) imply the relation between the proper time τ and light-cone time u_0,

$$\tau = \tau(u_0) = m\int \frac{du_0}{\rho(u_0)}\,, \tag{3.143}$$

such that the differentiation with respect to the τ can be replaced by the differentiation with respect to u_0,

$$\frac{d}{d\tau} = \frac{\rho(u_0)}{m}\frac{d}{du_0}\,. \tag{3.144}$$

Taking this into account in equations (3.139), we obtain differential equations for the functions $P^\mu = P^\mu(u_0)$,

$$\begin{aligned} &\rho(u_0)P_0'(u_0) = e(\mathbf{E}(u_0)\mathbf{P}(u_0))\,, \\ &\rho(u_0)\mathbf{P}'(u_0) = e\left\{\mathbf{E}(u_0)P_0(u_0) + [\mathbf{P}(u_0) \times \mathbf{H}(u_0)]\right\}\,. \end{aligned} \tag{3.145}$$

These equations have two first integrals of motion

$$P^0(u_0) - P^3(u_0) = \rho(u_0), \quad P_0^2(u_0) - \mathbf{P}^2(u_0) = m^2\,. \tag{3.146}$$

In addition, equations (3.145) have the following consequences for the functions $P^1(u_0)$ and $P^2(u_0)$:

$$\begin{aligned} &\rho(u_0)P'^1(u_0) = e\left\{A_1'(u_0)[P^0(u_0) - P^3(u_0)] + HP^2(u_0)\right\}\,, \\ &\rho(u_0)P'^2(u_0) = e\left\{A_2'(u_0)[P^0(u_0) - P^3(u_0)] - HP^1(u_0)\right\}\,. \end{aligned} \tag{3.147}$$

These equations and the first equation (3.146) can be transformed into a separate set of equations for the functions $P^1(u_0)$ and $P^2(u_0)$,

$$\rho(u_0)P'^1(u_0) = eA'_1(u_0)\rho(u_0) + \gamma P^2(u_0)\,,$$
$$\rho(u_0)P'^2(u_0) = eA'_2(u_0)\rho(u_0) - \gamma P^1(u_0), \quad \gamma = eH\,. \tag{3.148}$$

This set can be integrated as follows. A linear combination of equations (3.148) has the form

$$\rho(u_0)\overline{P}\,'(u_0) = \rho(u_0)\overline{A}\,'(u_0) - i\gamma\overline{P}(u_0)\,, \tag{3.149}$$

where complex functions $\overline{P}(u_0)$ and $\overline{A}(u_0)$ are introduced:

$$\overline{P}(u_0) = [P^1(u_0) + iP^2(u_0)], \overline{A}(u_0) = e[A_1(u_0) + iA_2(u_0)]\,. \tag{3.150}$$

One can easily write the general solution of equation (3.149)

$$\overline{P}(u_0) = \overline{A}(u_0) + \overline{k}\exp\left[-i\frac{\gamma}{m}s(u_0)\right] - i\gamma\phi(u_0), \quad \overline{k} = k^1 + ik^2 = \text{const.}\,,$$
$$\phi(u_0) = \exp\left[-i\frac{\gamma}{m}\tau(u_0)\right]\int\frac{\overline{A}(u_0)}{\rho(u_0)}\exp\left[i\frac{\gamma}{m}\tau(u_0)\right]du_0\,, \tag{3.151}$$

where k^1 and k^2 are arbitrary real constants, and the function $\tau(u_0)$ is given by equation (3.143).

We note that the complex function $\phi(u_0)$ satisfies the following equations:

$$\rho(u_0)\phi'(u_0) + i\gamma\phi(u_0) - \overline{A}(u_0) = 0 \Rightarrow$$
$$m\dot{\phi}(u_0) + i\gamma\phi(u_0) - \overline{A}(u_0) = 0 \Rightarrow$$
$$m\ddot{\phi}(u_0) + i\gamma\dot{\phi}(u_0) - \frac{\rho(u_0)}{m}\overline{A}\,'(u_0) = 0\,. \tag{3.152}$$

Finally, the functions $P^\mu(u_0)$ can be found from equations (3.151) and (3.146) to be

$$P^0(u_0) = \frac{m^2 + |\overline{P}(u_0)|^2 + \rho^2(u_0)}{2\rho(u_0)}, \quad P^1(u_0) = \frac{\overline{P}(u_0) + \overline{P}^*(u_0)}{2}\,,$$
$$P^2(u_0) = \frac{\overline{P}(u_0) - \overline{P}^*(u_0)}{2i}, \quad P^3(u_0) = \frac{m^2 + |\overline{P}(u_0)|^2 - \rho^2(u_0)}{2\rho(u_0)}\,. \tag{3.153}$$

Then taking into account equation (3.144), one can easily integrate equations (3.153) and find classical trajectories parametrized by the light-cone time,

$$P^\mu(u_0) = \hbar\rho(u_0)\frac{dx^\mu(u_0)}{du_0} \Rightarrow x^\mu(u_0) = \int\frac{P^\mu(u_0)du_0}{\hbar\rho(u_0)} + x_0^\mu\,, \tag{3.154}$$

where the constants x_0^μ represent initial data in the parametrization under consideration.

Finally, let us return to equations (3.139) for the variables x^1 and x^2. With the use of equations (3.146) these equations can be written in the form

$$m\ddot{x}^1 = eA_1'(u_0)\frac{\rho(u_0)}{m} + \gamma\dot{x}^2, \quad m\ddot{x}^2 = eA_2'(u_0)\frac{\rho(u_0)}{m} - \gamma\dot{x}^1 . \tag{3.155}$$

Introducing new variables $\tilde{x}$ and $\tilde{y}$,

$$\tilde{x} = x^1 - \phi_1(u_0), \quad \tilde{y} = x^2 - \phi_2(u_0) , \tag{3.156}$$

instead of the Cartesian x^1 and x^2, into equations (3.155), we obtain the following equations for the new variables:

$$m\ddot{\tilde{x}} - \gamma\dot{\tilde{y}} = 0, \quad m\ddot{\tilde{y}} + \gamma\dot{\tilde{x}} = 0 . \tag{3.157}$$

We see that the change of variables (3.156) removes the plane-wave field and the electric field from the equations of motion, the motion of the variables $\tilde{x}$ and $\tilde{y}$ being in fact governed by the magnetic field alone.

3.3.3 Quantum motion in a BGY field

Here we represent solutions of the K–G and Dirac equations with a BGY field.

In the case under consideration (as in the plane-wave case), the operator $\hat{\lambda} = \hat{p}_0 + \hat{p}_3$ commutes both with K–G operator $\hat{\mathfrak{K}}$ and Dirac operator $\hat{\mathfrak{D}}$. Therefore, the same as in classical theory, it is an integral of motion. Let us look for solutions of both equations that are eigenvectors for the operator $\hat{\lambda}$, i.e. for solutions $\Phi_\lambda(x)$ that obey the equation (3.127).

Considering the K–G equation, we pass to the variables $\tilde{x}$ and $\tilde{y}$ according to equations (3.156), and we replace the differentiation with respect to u_0 by the differentiation with respect to the proper time τ according to (3.143) and (3.144). Then we represent the scalar wave function in the following form:

$$\Phi_\lambda(x) = \frac{N}{\sqrt{\rho(u_0)}} \exp\left(-i\Gamma\right) \psi(\tau(u_0), \tilde{x}, \tilde{y}) , \tag{3.158}$$

where

$$\Gamma = \lambda x^0 - \frac{\lambda}{2}u_0 + i\frac{\gamma}{4}[\phi(u_0)(\tilde{x} - i\tilde{y}) - \phi^*(u_0)(\tilde{x} + i\tilde{y})] + \int \frac{2m^2 + 2|\overline{A}(u_0)|^2 + i\gamma[\overline{A}(u_0)\phi^*(u_0) - \overline{A}^*(u_0)\phi(u_0)]}{4\rho(u_0)} du_0 , \tag{3.159}$$

and N is a normalization factor, which can be found using the scalar product on the light-cone plane $u_0 = \text{const}$, see [170]. The function $\psi(\tau, x, y)$ satisfies the equation

$$i\partial_\tau\psi(\tau, x, y) = \frac{\hat{\pi}_1^2 + \hat{\pi}_2^2}{2m}\psi(\tau, x, y), \quad \hat{\pi}_1 = i\partial_{\tilde{x}} - \frac{\gamma}{2}\tilde{y}, \quad \hat{\pi}_2 = i\partial_{\tilde{y}} + \frac{\gamma}{2}\tilde{x} . \tag{3.160}$$

This is the two-dimensional Schrödinger equation with a constant and uniform magnetic field.

Thus, we see that the plane-wave field and the collinear running electric field can be excluded from the K–G equation by a change of variables and of the wave function.

Solutions of the Dirac equation with a BGY field can be constructed by the formulas (2.107) and (2.108). In the special case under consideration, these expressions can be simplified in such a way that the Dirac wave function $\Psi_\lambda(x)$ has the form

$$\Psi_\lambda(x) = \hat{R}\,\frac{V(u_0)v}{\sqrt{\rho(u_0)}}\,\Phi_\lambda(x), \quad \hat{\lambda}\Psi_\lambda(x) = \lambda\Psi_\lambda(x)\,,$$
$$\hat{R} = \begin{pmatrix} m + \rho(u_0) - (\boldsymbol{\sigma}\hat{\mathbf{P}}_\perp)(\boldsymbol{\sigma}\mathbf{n}) \\ [m - \rho(u_0)]\,(\boldsymbol{\sigma}\mathbf{n}) + (\boldsymbol{\sigma}\hat{\mathbf{P}}_\perp) \end{pmatrix}, \quad \hat{\mathbf{P}}_\perp = \hat{\mathbf{P}} - \mathbf{n}(\mathbf{n}\hat{\mathbf{P}})\,,$$
$$V(u_0) = \cos\left[\frac{\gamma\tau(u_0)}{2m}\right] + i(\boldsymbol{\sigma}\mathbf{n})\sin\left[\frac{\gamma\tau(u_0)}{2m}\right], \tag{3.161}$$

where v is an arbitrary constant spinor.

It should be noted that solutions in a constant electric field and a colinear plane-wave field were studied in detail in Ref. [248].

3.4 Particles in a constant and uniform magnetic field

3.4.1 Introduction

In this Section, we consider classical and quantum motion of a charge e in a *constant and uniform magnetic field* of the strength $\mathbf{H}$. Such a field can be described, for example, by potentials of the form

$$A_0 = 0, \quad \mathbf{A} = \frac{1}{2}[\mathbf{H} \times \mathbf{r}]\,. \tag{3.162}$$

If $\mathbf{H}$ is directed along the axis x^3, i.e. $\mathbf{H} = (0, 0, H > 0)$, then

$$A_0 = A_3 = 0, \quad A_1 = Hx^2/2, \quad A_2 = -Hx^1/2\,. \tag{3.163}$$

This is the so-called *symmetric gauge* for potentials of the magnetic field. Sometimes, it is convenient to use another gauge (*Landau gauge*),

$$A_0 = A_1 = A_3 = 0, \quad A_2 = -Hx^1\,, \tag{3.164}$$

which differs from (3.163) by a gauge transformation.

In the symmetric gauge for potentials, the Lorentz equations of motion (2.18) take the form

$$m\ddot{x}^0 = 0, \quad m\ddot{x}^1 = eH\dot{x}^2, \quad m\ddot{x}^2 = -eH\dot{x}^1, \quad m\ddot{x}^3 = 0\,,$$
$$\dot{x}_\mu\dot{x}^\mu = 1\,. \tag{3.165}$$

The general solution of these equations reads

$$
\begin{aligned}
x^1 &= R\cos\Omega + x^1_{(0)}, & x^2 &= -R\sin\Omega + x^2_{(0)}\,,\\
x^3 &= -p_3 m^{-1}(\tau - \tau_0), & x^0 &= p_0 m^{-1}\tau\,,\\
\Omega &= \omega_0\tau + \varphi_0, & \omega_0 &= eH/m\,,
\end{aligned}
\tag{3.166}
$$

where the constants $p_0, p_3, R, x^1_{(0)}, x^2_{(0)}, \tau_0$, and φ_0, are integrals of motion. They are related between themselves as follows

$$\dot{x}_\mu \dot{x}^\mu = 1 \Longrightarrow p_0^2 = m^2 + p_3^2 + (eH)^2 R^2\,. \tag{3.167}$$

The kinetic momenta P_μ have the form

$$P_0 = p_0, \quad P_1 = eHR\sin\Omega, \quad P_2 = eHR\cos\Omega, \quad P_3 = p_3\,. \tag{3.168}$$

Because $A_0 = 0$, the particle energy $\mathcal{E}$ is $P_0 = p_0$. In classical theory, the energy is positive,

$$E_{\mathrm{cl}} = p_0 = \sqrt{m^2 + p_3^2 + (eH)^2 R^2}\,. \tag{3.169}$$

The quantity E_r,

$$E_{\mathrm{r}} = P_1^2 + P_2^2 = \gamma^2 R^2\,, \tag{3.170}$$

is traditionally called the radial energy.

The projection L_z of the angular momentum $\mathbf{L} = [\mathbf{r}\times\mathbf{p}]$ on the axis x^3,

$$L_z = \frac{eH}{2}(R_0^2 - R^2), \quad R_0^2 = \left(x^1_{(0)}\right)^2 + \left(x^2_{(0)}\right)^2\,, \tag{3.171}$$

is also an integral of motion in the symmetric gauge. It is easy to see that

$$-2L_z/eH \le R^2 = E_{\mathrm{r}}/(eH)^2\,. \tag{3.172}$$

It follows from equations (3.166) that the trajectory of the particle in the magnetic field is a screw line. Its projection on the $z = 0$ plane is a circle of the radius R with a centre in the point $(x^1_{(0)}, x^2_{(0)})$,

$$R^2 = \left(x^1 - x^1_{(0)}\right)^2 + \left(x^2 - x^2_{(0)}\right)^2\,. \tag{3.173}$$

The particle rotates along a circle with a constant angular frequency ω,

$$\omega = \frac{d\Omega}{dx^0} = \frac{\dot{\Omega}}{\dot{x}^0} = \frac{m\omega_0}{E_{\mathrm{cl}}} = \frac{eH}{E_{\mathrm{cl}}}, \quad \Omega = \omega x^0 + \varphi_0\,, \tag{3.174}$$

and, at the same time, moves along the axis z with a constant velocity β_3,

$$\beta_3 = \frac{dx^3}{dx^0} = \frac{\dot{x}^3}{\dot{x}^0} = -\frac{p_3}{E_{\mathrm{cl}}}, \quad |\beta_3| < 1\,. \tag{3.175}$$

One can obtain trajectories parametrized with the light-cone time $u^0 = x^0 - x^3$ by replacing $\Omega = \omega_0\tau + \varphi_0$ in (3.166) and (3.168) with

$$\Omega = \tilde{\omega}u^0 + \varphi_0, \qquad u^0 = \frac{\lambda}{m}\tau + u^0_{(0)},$$
$$\tilde{\omega} = eH/\lambda = \frac{\omega}{(1-\beta_3)}, \qquad \lambda = p_0 + p_3 = E_{\rm cl}(1-\beta_3), \tag{3.176}$$

where $u^0_{(0)}$ is a constant.

The solution of the Hamilton–Jacobi equation (2.27) for the case under consideration has the form

$$S = -p_0x^0 - p_3x^3 + L_z \arctan\left(\frac{x^2}{x^1}\right) + \frac{1}{2}\int^{\rho} \sqrt{R_1(q)}\frac{dq}{q}, \tag{3.177}$$

where $\rho = (|eH|/2)[(x^1)^2 + (x^2)^2]$, and

$$R_1(q) = 2q(p_0^2 - p_3^2 - m^2)/|eH| - [L_z + q\nu]^2$$
$$= 2q|eH|R^2 - [L_z + q\nu]^2 \geq 0, \tag{3.178}$$

where $\nu = \mathrm{sgn}(eH)$.

For the first time solutions of the Dirac equation with the constant and uniform magnetic field were found by Rabi [268], just after the appearance of the pioneering works [120] of Dirac. Solutions of the K–G equations in this case were obtained by Page [260] (some relevant remarks can be found even earlier in a work of Brillouin [105]). Then, such solutions were studied in detail [208, 266]. Rabi´s and Page´s solutions were widely used in theoretical analysis of a number of physical phenomena, for example in synchrotron radiation theory, in the theory of dia- and paramagnetism, in the theory of magneto-optic phenomena, and so on. Below we consider these solutions.

3.4.2 Page´s and Rabi´s solutions

The quantum motion of a spinless relativistic particle in the constant and uniform magnetic field is described by the scalar wave function $\varphi(x)$, which obeys the K–G equation (2.28) with such a field. We are going to use the symmetric gauge (3.163) for the potentials. In this case the corresponding K–G operator $\hat{\mathcal{K}}$ reads

$$\hat{\mathcal{K}} = \hat{p}_0^2 - \hat{p}_3^2 - \hat{E}_{\rm r} - m^2,$$
$$\hat{p}_0 = i\partial_0, \quad \hat{p}_3 = i\partial_3, \quad \hat{E}_{\rm r} = \hat{P}_1 + \hat{P}_2,$$
$$\hat{P}_1 = i\partial_1 - eHx^2/2, \quad \hat{P}_2 = i\partial_2 + eHx^1/2. \tag{3.179}$$

The set of mutually commuting operators $\hat{p}_0$, $\hat{p}_3$, and $\hat{E}_r$ commutes also with the operator $\hat{\mathcal{K}}$ and represents a complete set of integrals of motion. We shall here be interested in such solutions $\varphi(x)$ of the K–G equation that are eigenfunctions of these operators (with the eigenvalues p_0, p_3, and E_r),

$$\hat{p}_0\varphi = p_0\varphi, \quad \hat{p}_3\varphi = p_3\varphi, \quad \hat{E}_r\varphi = E_r\varphi . \tag{3.180}$$

In this case the K–G equation $\hat{\mathcal{K}}\varphi = 0$ implies classical relations (3.167) and (3.170) between the quantum numbers p_0, p_3, and E_r.

$$p_0^2 = m^2 + p_3^2 + E_r . \tag{3.181}$$

The variables x^0 and x^3 can be separated in the K–G equation by the substitution

$$\varphi(x) = \exp(-ip_0x^0 - ip_3x^3)\Phi(x^1, x^2) . \tag{3.182}$$

Then the function $\Phi(x^1, x^2)$ has to satisfy the equation

$$\hat{E}_r\Phi(x^1, x^2) = E_r\Phi(x^1, x^2) . \tag{3.183}$$

In the cylindric coordinates r, φ, ($x^1 = r\cos\varphi$, $x^2 = r\sin\varphi$), the equation (3.183) takes the form

$$\left[\partial_r^2 + r^{-1}\partial_r - r^{-2}\hat{L}_z^2 + eH\hat{L}_z - (eH)^2r^2/4 + \hat{E}_r\right]\Phi = 0 , \tag{3.184}$$

where $\hat{L}_z = -i\partial_\varphi$. The operator $\hat{L}_z$ commutes with the K–G operator $\hat{\mathcal{K}}$ and is, therefore, an integral of motion. One can chose solutions $\varphi(x)$ to be eigenfunctions for $\hat{L}_z$,

$$\hat{L}_z\varphi(x) = l\varphi(x), \; l \in \mathbb{Z} . \tag{3.185}$$

Now, the variables r and φ can be also separated,

$$\begin{aligned} &\Phi(x^1, x^2) = \exp(il\varphi)f(\rho), \quad \rho = |eH|r^2/2 , \\ &4\rho^2 f''(\rho) + 4\rho f'(\rho) - [\rho^2 - 2\rho(E_r + eHl)/|eH| + l^2]f(\rho) = 0 . \end{aligned} \tag{3.186}$$

Solutions of this equation are Laguerre functions $I_{n,s}(\rho)$ (see Section B.1), such that

$$f(\rho) = I_{n,s}(\rho), \quad l = (s-n)\,\nu, \quad E_r = |eH|(2n+1), \quad n, s \in \mathbb{Z}_+ . \tag{3.187}$$

It follows from (3.187) that there exists a restriction on 1,

$$-\infty < -l\nu \leq n . \tag{3.188}$$

The restriction (3.188) is a quantum analog of the classical inequality (3.172).

With the use of (3.187) and (3.181), we finally obtain Page's solutions of the K–G equation with a constant uniform magnetic field [260],

$$\begin{aligned} &\varphi_{n,s,p_3,\epsilon}(x) = N\exp(il\varphi - ip_0x^0 - ip_3x^3)I_{n,s}(\rho) , \\ &l = (s-n)\,\nu, \quad s, n \in \mathbb{Z}_+ , \\ &p_0 = \epsilon E, \quad E = \sqrt{m^2 + p_3^2 + |eH|(2n+1)}, \quad \epsilon = \pm 1 . \end{aligned} \tag{3.189}$$

For $N = (2\pi)^{-1}\sqrt{|eH|/2E}$ the wave functions (3.189) obey the following orthonormality relations:

$$(\varphi_{n',s',p_3',\epsilon'}, \varphi_{n,s,p_3,\epsilon}) = \epsilon\delta_{n,n'}\,\delta_{s,s'}\,\delta(p_3 - p_3')\,\delta_{\epsilon,\epsilon'} \tag{3.190}$$

with respect to the scalar product (2.32).

Rabi´s solution of the Dirac equation

Turning to the spinning particle case, we, following Rabi [268], chose Dirac wave functions Ψ to be eigenvectors of the commuting operators $\hat{p}_0 = i\partial_0$, $\hat{p}_3 = i\partial_3$, and $\hat{J}_z$ that are integrals of motion in the case under consideration. We recall that $\hat{J}_z$ is the z-projection of the total angular momentum operator $\hat{J}$ (A.57), so that

$$\hat{J}_z = \hat{L}_z + \frac{1}{2}\Sigma_3 . \tag{3.191}$$

Eigenvalues of the operator $\hat{J}_z$ are half-integers and will be written as

$$J_z = l + \nu/2 , \quad l \in \mathbb{Z} , \tag{3.192}$$

for convenience. The *Rabi solutions* have the form

$$\begin{aligned}
&\Psi_{n,s,p_3,\epsilon}(x) = N\exp(iJ_z\varphi - ip_0x^0 - ip_3x^3)\psi(\rho,\varphi) , \\
&l = \nu(s-n) , \quad -\infty < -l\nu \le n, \quad n, s \in \mathbb{Z}_+ ,
\end{aligned} \tag{3.193}$$

where the bispinor $\psi(\rho,\varphi)$ has the form

$$\psi(\rho,\varphi) = \begin{pmatrix} (1+i\nu)c_1\exp(-i\varphi/2)I_{n-1/2+\nu/2,s}(\rho) \\ (1-i\nu)c_2\exp(i\varphi/2)I_{n-1/2-\nu/2,s}(\rho) \\ (1+i\nu)c_3\exp(-i\varphi/2)I_{n-1/2+\nu/2,s}(\rho) \\ (1-i\nu)c_4\exp(i\varphi/2)I_{n-1/2-\nu/2,s}(\rho) \end{pmatrix} , \tag{3.194}$$

and the constant bispinor C with elements $c_a, a = 1,2,3,4$, has to obey a uniform algebraic set of equations

$$AC = 0, \quad A = \gamma^0 p_0 + \gamma^3 p_3 - \gamma^1\sqrt{2|eH|n} - m . \tag{3.195}$$

A nontrivial solution of this set does exist provided that

$$\det A = (p_0^2 - p_3^2 - 2|eH|n - m^2)^2 = 0 . \tag{3.196}$$

The energy spectrum follows from this condition,

$$p_0 = \epsilon E_D, \quad E_D = \sqrt{p_3^2 + 2|eH|n + m^2}, \quad \epsilon = \pm 1 . \tag{3.197}$$

One can see that under condition (3.196), the rank of the matrix A is equal to 2, thus, the bispinor C is defined by equation (3.195) up to an arbitrary spinor ϑ. For example, the spinor $C = C_\epsilon$ can be written in the following form:

$$C_\epsilon = \frac{1}{2}\begin{pmatrix} [(1+\epsilon)(E_D+m) - (1-\epsilon)(\sigma_1\sqrt{2|eH|n} - \sigma_3 p_3)]\vartheta \\ [(1+\epsilon)(\sigma_1\sqrt{2|eH|n} - \sigma_3 p_3) + (1-\epsilon)(E_D+m)]\vartheta \end{pmatrix},$$
$$C^+_{\epsilon'}C_\epsilon = 2E_D(E_D+m)\vartheta^+\vartheta\,\delta_{\epsilon,\epsilon'}\,. \tag{3.198}$$

The state $n = 0$ is very special for the Dirac wave function (3.194). In this case ϑ is not arbitrary, but has the form

$$\vartheta = \frac{1}{2}\begin{pmatrix} 1+\nu \\ 1-\nu \end{pmatrix}, \tag{3.199}$$

because for $n = 0$ and $\nu = 1$ one has to set $c_2 = c_4 = 0$ in (3.194), whereas for $n = 0$ and $\nu = -1$ one has to set $c_1 = c_3 = 0$ there. One can verify that in this case the Dirac bispinor is an eigenvector for the operator Σ_3 with the eigenvalue ν,

$$\Sigma_3\Psi(x) = \nu\Psi(x)\,. \tag{3.200}$$

That means that only electrons with spin directed against the magnetic field can exist on the lowest level $n = 0$. Respectively, only positrons with spin directed along the magnetic field can exist on the lowest level $n = 0$.

If we chose

$$N = (4\pi)^{-1}\sqrt{\frac{|eH|}{E_D(E_D+m)\vartheta^+\vartheta}},$$

then the wave functions (3.193) obey the following orthonormality relations:

$$(\Psi_{n',s',p_3',\epsilon'}, \Psi_{n,s,p_3,\epsilon}) = \delta_{n,n'}\,\delta_{s,s'}\,\delta(p_3 - p_3')\,\delta_{\epsilon,\epsilon'}$$

with respect to the scalar product (2.65).

The arbitrary spinor ϑ (for $n \neq 0$) can be further specified if we chose the spin integral of motion[5] in a way adequate to the physical problem under consideration, see Section 2.4 and Ref. [11, 310, 311, 319]. We note that in all such cases, equations on the spinor ϑ acquire the form

$$(\boldsymbol{\sigma}\mathbf{l})\vartheta = \zeta\vartheta, \quad \zeta = \pm 1\,, \tag{3.201}$$

where $\mathbf{l}$ is a real unit vector ($\mathbf{l}^2 = 1$). The general solution of equation (3.201) is studied in detail in Chapter 8. The spin quantum number ζ represents the orientation of the particle spin with respect to the direction $\mathbf{l}$. The spinor $\vartheta = \vartheta_\zeta$ depends now on ζ so that the Dirac wave function (3.193) adopts the spin quantum number ζ.

---

5 In his work [268] Rubi did not discuss spin integrals for the Dirac equation with the magnetic field.

The Rabi solutions were widely used in calculating various physical effects with electrons in the constant and uniform magnetic field, for example characteristics of synchrotron radiation, see for example [310], pair creation by a photon in the magnetic field [72, 331], and so on.

Another type of solutions of the K–G and Dirac equations with the constant and uniform magnetic field may be found in [34, 35, 59, 211].

We show below that all known exact solutions of these equations can be described in a unique way.

3.4.3 Creation and annihilation operators

The same as in the nonrelativistic quantum problem with a constant and uniform magnetic field [109], in the relativistic case it is convenient to introduce some creation and annihilation operators and study solutions of the K–G and Dirac equations in the corresponding Fock spaces, see [59]. All known and some new solutions of these equations can be described in the framework of such techniques considered below.

In what follows, when studying solutions of the K–G and Dirac equations with the magnetic field, we select $\nu = -1$, which correspond to the magnetic field directed in the positive direction of the z-axis and introduce the notation $\gamma = |eH|$. Results for $\nu = 1$ can be easily restored in final formulas. We are going to use the light-cone variables (2.37), dimensionless Cartesian coordinates x, y and modified cylindrical coordinates ρ, φ that are defined by the relations

$$
\begin{aligned}
& x = \sqrt{\gamma/2}x^1 = \sqrt{\rho}\cos\varphi, \quad y = \sqrt{\gamma/2}x^2 = \sqrt{\rho}\sin\varphi, \quad x, y \in \mathbb{R}, \\
& \rho = x^2 + y^2, \quad x + iy = \sqrt{\rho}\exp i\varphi, \quad \rho \in \mathbb{R}_+, \quad 0 \le \varphi < 2\pi, \\
& dx^1 dx^2 = (2/\gamma)dx\,dy = (1/\gamma)d\rho\,d\varphi, \quad \gamma = |eH| .
\end{aligned} \tag{3.202}
$$

Let us introduce dimensionless operators $\hat{a}_1, \hat{a}_1^+, \hat{a}_2$, and $\hat{a}_2^+$ as follows:

$$
\begin{aligned}
\hat{a}_1 &= -\left(i\hat{P}_1 + \hat{P}_2\right)/\sqrt{2\gamma}, \quad \hat{a}_2 = \left[\hat{P}_2 - i\hat{P}_1 + \gamma\left(x^1 + ix^2\right)\right]/\sqrt{2\gamma}, \\
\hat{a}_1^+ &= \left(i\hat{P}_1 - \hat{P}_2\right)/\sqrt{2\gamma}, \quad \hat{a}_2^+ = \left[\hat{P}_2 + i\hat{P}_1 + \gamma\left(x^1 - ix^2\right)\right]/\sqrt{2\gamma},
\end{aligned} \tag{3.203}
$$

where $\hat{P}_k, k = 1, 2$, are operators of the kinetic momenta (2.29), which have the following form in the case under consideration:

$$
\hat{P}_1 = i\partial_1 + \frac{\gamma}{2}x^2, \quad \hat{P}_2 = i\partial_2 - \frac{\gamma}{2}x^1 .
$$

The introduced operators $\hat{a}_k^+, \hat{a}_k, k = 1, 2$, are Bose creation and annihilation operators, they obey the following commutation relations:

$$
[\hat{a}_k, \hat{a}_s^+] = \delta_{k,s}, \quad [\hat{a}_k, \hat{a}_s] = [\hat{a}_k^+, \hat{a}_s^+] = 0, \quad k, s = 1, 2 . \tag{3.204}
$$

In the dimensionless coordinates (3.202) they have the form:

$$
\begin{aligned}
\hat{a}_1 &= \frac{\left(x - iy + \partial_x - i\partial_y\right)}{2} = \frac{e^{-i\varphi}\left(\rho - i\partial_\varphi + 2\rho\partial_\rho\right)}{2\sqrt{\rho}}, \\
\hat{a}_1^+ &= \frac{\left(x + iy - \partial_x - i\partial_y\right)}{2} = \frac{e^{i\varphi}\left(\rho - i\partial_\varphi - 2\rho\partial_\rho\right)}{2\sqrt{\rho}}, \\
\hat{a}_2 &= \frac{\left(x + iy + \partial_x + i\partial_y\right)}{2} = \frac{e^{i\varphi}\left(\rho + i\partial_\varphi + 2\rho\partial_\rho\right)}{2\sqrt{\rho}}, \\
\hat{a}_2^+ &= \frac{\left(x - iy - \partial_x + i\partial_y\right)}{2} = \frac{e^{-i\varphi}\left(\rho + i\partial_\varphi - 2\rho\partial_\rho\right)}{2\sqrt{\rho}}.
\end{aligned}
\tag{3.205}
$$

Below, we list expressions for all the relevant operators in terms of the creation and annihilation operators.

Operators of coordinates x^k and momenta $\hat{P}_k$, are

$$
\begin{aligned}
x^1 &= \sqrt{2\gamma}\,[\hat{a}_1 + \hat{a}_1^+ + \hat{a}_2 + \hat{a}_2^+], \\
x^2 &= i\sqrt{2\gamma}\,(\hat{a}_1 - \hat{a}_1^+ - \hat{a}_2 + \hat{a}_2^+), \\
\hat{P}_k &= i\sqrt{\gamma/2}\left[(-1)^{k-1}\hat{a}_1 - \hat{a}_1^+\right].
\end{aligned}
\tag{3.206}
$$

Operators of radial energy $\hat{E}_r$ and of the z-projection of the angular momentum $\hat{L}_z$, are

$$
\hat{E}_r = \hat{P}_1^2 + \hat{P}_2^2 = 2\gamma\hat{N}_1 + \gamma\,, \tag{3.207}
$$

$$
\hat{L}_z = \hat{N}_1 - \hat{N}_2\,, \tag{3.208}
$$

where

$$
\hat{N}_k = \hat{a}_k^+\hat{a}_k,\, k = 1, 2\,. \tag{3.209}
$$

K–G and Dirac operators, are

$$
\begin{aligned}
\hat{\mathfrak{K}} &= \hat{p}_0^2 - \hat{p}_3^2 - 2\gamma\hat{N}_1 - m_{\text{eff}}^2, \quad m_{\text{eff}}^2 = m^2 + \gamma\,, \\
\hat{\mathfrak{D}} &= \gamma^0\hat{p}_0 + \gamma^3\hat{p}_3 - \sqrt{\gamma/2}\left[\left(\gamma^2 - i\gamma^1\right)\hat{a}_1 + \left(\gamma^2 + i\gamma^1\right)\hat{a}_1^+\right] - m\,.
\end{aligned}
\tag{3.210}
$$

The quantity m_{eff} is called the effective mass of a spinless particle.

It should be noted that the operators $\hat{\mathfrak{K}}$ and $\hat{\mathfrak{D}}$ do not contain operators $\hat{a}_2^+$ and $\hat{a}_2$. Moreover, the latter operators commute with the K–G and Dirac operators. Thus, $\hat{a}_2^+$, $\hat{a}_2$, and $\hat{N}_2$ are integrals of motion. One can easily see that the operators $\hat{p}_0$ and $\hat{p}_3$ are integrals of motion as well.

The operator $\hat{N}_1$ is an integral of motion for the K–G equation, it commutes with the set of operators $\hat{\mathfrak{K}}$, $\hat{p}_0$, $\hat{p}_3$, $\hat{a}_2^+$, and $\hat{a}_2$. Its analog for the Dirac equation is the operator

$$
\hat{\mathfrak{N}}_{\text{D}} = \hat{N}_1 + \frac{1}{2}\Sigma_3\,. \tag{3.211}
$$

The latter operator also commutes with the set of operators $\hat{\mathfrak{D}}$, $\hat{p}_0$, $\hat{p}_3$, $\hat{a}_2^+$, and $\hat{a}_2$.

Relations (3.208) and (3.210) allow one to see that the operator $\hat{L}_z$, which commutes with $\hat{\mathfrak{K}}$, $\hat{N}_1$, $\hat{p}_0$, and $\hat{p}_3$, is an integral of motion for the K–G equation, whereas in the Dirac case the z-projection of the total angular momentum operator $\hat{\mathbf{J}}$ is an integral of motion,

$$\hat{J}_z = \hat{L}_z + \frac{1}{2}\Sigma_3 = \hat{a}_1^+\hat{a}_1 - \hat{a}_2^+\hat{a}_2 + \frac{1}{2}\Sigma_3 = \hat{\mathfrak{N}}_D - \hat{a}_2^+\hat{a}_2 \,. \tag{3.212}$$

The operator $\hat{J}_z$ commutes with the operators $\hat{\mathfrak{D}}$, $\hat{\mathfrak{N}}_D$, $\hat{p}_0$, and $\hat{p}_3$, but does not commute with the operators $\hat{a}_2^+$ and $\hat{a}_2$.

The K–G operator, as written in the light cone variables and in terms of the creation and annihilation operators, has the following form:

$$\widehat{\mathfrak{K}} = 4\widehat{\widehat{P_3}}\,\widehat{\widehat{P_0}} - 2\gamma\widehat{N}_1 - m_{\text{eff}}^2, \quad \widehat{\widehat{P_0}} = i\partial_{u^0}, \quad \widehat{\widehat{P_3}} = i\partial_{u^3} \,. \tag{3.213}$$

In the same terms the Dirac equation reads (see (2.82))

$$4\widehat{\widehat{P_3}}\,\widehat{\widehat{P_0}}\Psi_{(-)} = \left(2\gamma\widehat{\mathfrak{N}}_D + m_{\text{eff}}^2\right)\Psi_{(-)} \,, \tag{3.214}$$

$$2\widehat{\widehat{P_3}}\Psi_{(+)} = \left[\left(\boldsymbol{\alpha}\hat{\mathbf{P}}_\perp\right) + \rho_3 m\right]\Psi_{(-)} \,, \tag{3.215}$$

where $\hat{\mathbf{P}}_\perp = -(\hat{P}_1, \hat{P}_2, 0)$. In the case under consideration, bispinors Ψ can be expressed via solutions of the corresponding K–G equation as follows:

$$\Psi_{(-)} = \left[i\left(1+\rho_1\right)\left(1-\Sigma_3\right) + \left(1-\rho_1\right)\left(1+\Sigma_3\right)\hat{a}_1\right]C\Phi \,, \tag{3.216}$$

where C is an arbitrary constant bispinor, and functions Φ satisfy the K–G equation with the K–G operator of the form,

$$\widehat{\widehat{K}} = 4\widehat{\widehat{P_3}}\,\widehat{\widehat{P_0}} - 2\gamma\widehat{N}_1 - m^2 \,, \tag{3.217}$$

compare with the K–G operator (3.213).Then the bispinor $\Psi_{(+)}$ can be found from equation (3.215).

Since the creation and annihilation operators with different indices commute, one can try to find a coordinate representation for them, in which this fact would be seen explicitly. To this end, we pass to a partial Fourier transform of a K–G wave function φ (the same can be done for the Dirac wave function Ψ),

$$\varphi(x^0, x^3, x, y) = \frac{1}{\sqrt{2\pi}}\int_{-\infty}^{\infty} e^{i\kappa y}\chi(x^0, x^3, x, \kappa)d\kappa \,. \tag{3.218}$$

It is easy to see that the property

$$\int_{-\infty}^{\infty} dx \int_{-\infty}^{\infty} dy\, |\varphi(x^0, x^3, x, y)|^2 = \int_{-\infty}^{\infty} dx \int_{-\infty}^{\infty} d\kappa |\chi(x^0, x^3, x, \kappa)|^2 \tag{3.219}$$

takes place. The operations of multiplication by y and differentiation ∂_y, as they act on φ, induce the following operations $i\partial_\kappa$ and $-\kappa$ in the space of the functions $\chi(x^0, x^3, x, \kappa)$, i.e. $y \to i\partial_\kappa$, $\quad i\partial_y \to -\kappa$. In this space the creation and annihilation operators have the form

$$\hat{a}_1 = (x + \kappa + \partial_x + \partial_\kappa)/2, \quad \hat{a}_1^+ = (x + \kappa - \partial_x - \partial_\kappa)/2,$$
$$\hat{a}_2 = (x - \kappa + \partial_x - \partial_\kappa)/2, \quad \hat{a}_2^+ = (x - \kappa - \partial_x + \partial_\kappa)/2. \tag{3.220}$$

Introducing new coordinates ξ and η by the relations

$$\sqrt{2}\xi = x + \kappa, \quad \sqrt{2}\eta = x - \kappa, \quad \sqrt{2}x = \xi + \eta, \quad \sqrt{2}\kappa = \xi - \eta,$$

we find from (3.220) that

$$\hat{a}_1 = (\xi + \partial_\xi)/\sqrt{2}, \quad \hat{a}_1^+ = (\xi - \partial_\xi)/\sqrt{2},$$
$$\hat{a}_2 = (\eta + \partial_\eta)/\sqrt{2}, \quad \hat{a}_2^+ = (\eta - \partial_\eta)/\sqrt{2}. \tag{3.221}$$

Thus, in the space of functions $\chi(x^0, x^3, \eta, \xi)$, the creation and annihilation operators have canonical coordinate representation (3.221) and the operators with different indices act on different variables.

In the space of functions $\chi(x^0, x^3, \eta, \xi)$, the K–G and the Dirac operators (3.210), the operator $\hat{N}_1$ (3.209), and the operator $\hat{L}_z$ (3.208) take the form

$$\hat{\mathfrak{K}} = \hat{p}_0^2 - \hat{p}_3^2 + \gamma\left(\partial_\xi^2 - \xi^2\right) - m^2, \tag{3.222}$$

$$\hat{\mathfrak{D}} = \gamma^0 \hat{p}_0 + \gamma^3 \hat{p}_3 - \sqrt{\gamma}\left(\gamma^2 \xi - i\gamma^1 \partial_\xi\right) - m,$$
$$\hat{N}_1 = \left(\xi^2 - \partial_\xi^2 - 1\right)/2, \quad \hat{L}_z = \left(\xi^2 - \partial_\xi^2 - \eta^2 + \partial_\eta^2\right)/2. \tag{3.223}$$

One can see that both the K–G operator and the Dirac operator do not contain the variable η, and therefore the K–G and Dirac equations do not determine η-dependence of wave functions. Thus, solving these equations in the space of functions $\chi(x^0, x^3, \eta, \xi)$, we can explicitly control all the functional arbitrariness in their solutions. This functional arbitrariness will be used in what follows to construct different sets of exact solutions of these equations.

3.4.4 Stationary states

Below we consider stationary states of a charged spinless and spinning particle in the magnetic field under consideration. Such states are eigenvectors of the energy operator $\hat{p}_0 = i\partial_0$. We chose the stationary states to be also eigenvectors for the operator $\hat{p}_3 = i\partial_3$ and for the operator $\hat{N}_1$. We denote eigenvalues of the operators $\hat{p}_0$ and $\hat{p}_3$ as p_0 and p_3, respectively. Representatives $\chi(x^0, x^3, \eta, \xi)$ (defined by (3.218)) of K–G stationary wave functions φ have thus the form

$$\chi(x^0, x^3, \xi, \eta) = \exp\left(-ip_0 x^0 - ip_3 x^3\right) \Phi_n(\xi, \eta), \tag{3.224}$$

where $\Phi_n(\eta, \xi)$ are eigenfunctions of the operator $\hat{N}_1$ (3.223) with the eigenvalues n,

$$\hat{N}_1 \Phi_n = n\Phi_n \Longrightarrow$$
$$\left(\xi^2 - \partial_\xi^2 - 1\right) \Phi_n = 2n\Phi_n, \quad n \in \mathbb{Z}_+ . \tag{3.225}$$

The latter equation determines only the ξ-dependence of the functions Φ_n,

$$\Phi_n(\xi, \eta) = U_n(\xi)\phi(\eta) . \tag{3.226}$$

Here $U_n(\xi)$ are Hermite functions (see Section B.2) and $\phi(\eta)$ are arbitrary functions of η.

It follows from the K–G equation (with the use of (3.222)) that the eigenvalues p_0, p_3, and n are related as follows:

$$p_0^2 = p_3^2 + 2\gamma n + m_{\text{eff}}^2 . \tag{3.227}$$

Thus, the energy spectrum of a spinless quantum particle in the magnetic field under consideration has the form

$$p_0 = \epsilon E, \quad E = \sqrt{p_3^2 + 2\gamma n + m_{\text{eff}}^2}, \quad \epsilon = \pm 1 , \tag{3.228}$$

which obviously coincides with the one given by equation (3.189).

Finally, as it follows from (3.218), the stationary solutions of the K–G equation have the form

$$\varphi_{\epsilon,p_3,n} = N \exp(-ip_0 x^0 - ip_3 x^3)\Phi_n(x, y) ,$$

where

$$\Phi_n(x, y) = \frac{1}{\sqrt{2\pi}} \int_{-\infty}^{\infty} d\kappa \exp(i\kappa y)\, U_n\left(\frac{x+\kappa}{\sqrt{2}}\right) \phi\left(\frac{x-\kappa}{\sqrt{2}}\right)$$
$$= \frac{\exp(ixy)}{\sqrt{\pi}} \int_{-\infty}^{\infty} d\eta \exp\left(-i\sqrt{2}y\eta\right) U_n\left(\sqrt{2}x - \eta\right) \phi(\eta) , \tag{3.229}$$

and $\phi(\eta)$ is an arbitrary function of the variable η. The relation (3.219) implies that

$$\int_{-\infty}^{\infty} dx \int_{-\infty}^{\infty} dy |\Phi_n(x, y)|^2 = \int_{-\infty}^{\infty} d\eta |\phi(\eta)|^2 . \tag{3.230}$$

As follows from the properties of the Hermite functions and with the use of (3.226) (under the supposition that $\phi(\eta)$ does not depend on n), the following relations take place:

$$\hat{a}_1 \Phi_n = \sqrt{n}\Phi_{n-1} \quad \Longrightarrow \quad \hat{a}_1 \varphi_{\epsilon,p_3,n} = \sqrt{n}\varphi_{\epsilon,p_3,n-1},$$
$$\hat{a}_1^+ \Phi_n = \sqrt{n+1}\Phi_{n+1} \Longrightarrow \hat{a}_1^+ \varphi_{\epsilon,p_3,n} = \sqrt{n+1}\varphi_{\epsilon,p_3,n+1} , \tag{3.231}$$
$$\hat{N}_1 \Phi_n = n\Phi_n \quad \Longrightarrow \hat{N}_1 \varphi_{\epsilon,p_3,n} = n\varphi_{\epsilon,p_3,n} . \tag{3.232}$$

We stress that the constructed stationary states (3.229) contain an arbitrary function $\phi(\eta)$. Depending on the choice of this function, we derive different sets of stationary states. This function can be specified, for example, if we demand for the complete wave function φ to be an eigenvector of some additional operator $\hat{Q}$ that is an integral of motion and that commutes with $\hat{N}_1$, $\hat{p}_0$, and $\hat{p}_3$. Then we have $\hat{Q}\phi_q(\eta) = q\phi_q(\eta)$, such that $\Phi_n(x,y) \to \Phi_{n,q}(x,y)$ and $\varphi_{\epsilon,p_3,n} \to \varphi_{\epsilon,p_3,n,q}$.

With the use of the orthogonality of the Hermite functions, we can find

$$\left(\Phi_{n',q'},\Phi_{n,q}\right) = \int_{-\infty}^{\infty} dx \int_{-\infty}^{\infty} dy \Phi^*_{n',q'}(x,y)\Phi_{n,q}(x,y) = \delta_{n',n}\int_{-\infty}^{\infty} d\eta \phi^*_{q'}(\eta)\phi_q(\eta)\,. \tag{3.233}$$

The bilinear form in the first line of equation (3.233) is considered in this section as a scalar product in the space of functions of x, y.

The relations (3.231) mean that any state $\Phi_n(x,y)$ can be created from a vacuum state $\Phi_0(x,y)$,

$$\Phi_n(x,y) = \frac{(\hat{a}_1^+)^n}{\sqrt{n!}}\Phi_0(x,y), \quad \hat{a}_1\Phi_0(x,y) = 0\,, \tag{3.234}$$

$$\Phi_0(x,y) = \pi^{-3/4}\exp\left(-x^2+ixy\right)\int_{-\infty}^{\infty} d\eta \exp\left[-\frac{\eta^2}{2}+\sqrt{2}\eta(x-iy)\right]\phi(\eta)\,.$$

Going over to the spinning particle case, we select the stationary Dirac wave functions Ψ to be eigenvectors of the operators $\hat{p}_0 = i\partial_0$, $\hat{p}_3 = i\partial_3$, and $\hat{\mathrm{N}}_\mathrm{D}$, with the eigenvalues p_0, p_3, and n respectively. Such solutions have the form

$$\Psi = N\exp(-ip_0x^0 - ip_3x^3)\psi_{n,\epsilon}(x,y)\,, \tag{3.235}$$

where

$$\psi_{n,1}(x,y) = \begin{pmatrix} c_1\Phi_{n-1}(x,y) \\ ic_2\Phi_n(x,y) \\ c_3\Phi_{n-1}(x,y) \\ ic_4\Phi_n(x,y) \end{pmatrix},$$

$$\psi_{n,-1}(x,y) = \begin{pmatrix} c_1\Phi_n(x,y) \\ ic_2\Phi_{n-1}(x,y) \\ c_3\Phi_n(x,y) \\ ic_4\Phi_{n-1}(x,y) \end{pmatrix}, \tag{3.236}$$

and functions $\Phi_n(x,y)$ are given by (3.229). The constant bispinor C with elements c_a, $a = 1,2,3,4$ has to obey a uniform algebraic set of equations

$$AC = 0, \quad A = \gamma^0 p_0 + \gamma^3 p_3 - \sqrt{2\gamma n}\gamma^1 - m\,. \tag{3.237}$$

The condition of the existence of a nontrivial solutions of this system

$$\det A = 0 = (p_0^2 - p_3^2 - 2\gamma n - m^2)^2 \tag{3.238}$$

leads us to the energy spectrum

$$p_0 = \epsilon E, \quad \epsilon = \pm 1, \quad E = \sqrt{p_3^2 + 2\gamma n + m^2} \,. \tag{3.239}$$

It differs from the K–G case (3.227) by the substitution $m_{\text{eff}} \rightarrow m$.

One can see that the rank of the matrix A is 2, thus the bispinor C is determined by the system (3.237) up to an arbitrary spinor v. For example, the bispinor can be written in the following form:

$$C = \begin{pmatrix} (p_0 + m)v \\ (\sqrt{2\gamma n}\sigma_1 - p_3\sigma_3)\, v \end{pmatrix}, \quad C^+C = 2p_0(p_0 + m)v^+v \,. \tag{3.240}$$

The state $n = 0$ for the Dirac wave functions is special. As was already discussed above, in this case, one has to set $c_1 = c_3 = 0$ in (3.236). The spinor v from (3.240), which provides such a structure, has the form

$$v = \begin{pmatrix} 0 \\ 1 \end{pmatrix} \,. \tag{3.241}$$

One can verify that in this case the complete Dirac wave function is an eigenvector for the spin operator Σ_3 with the eigenvalue -1,

$$\Sigma_3 \Psi(x) = -\Psi(x^\mu) \,. \tag{3.242}$$

This means that only electrons with the spin directed opposite to the magnetic field can exist on the lowest level $n = 0$.

The arbitrary spinor v may be specialized if we select complementary spin operators – integrals of motion. Thus, we get an algebraic set of equations, additional to equation (3.237) (and consistent with it), which allows one to determine v (up to a normalization). Note that all (consistent with the Dirac equation) spin operators, which were considered in Section 2.4, lead always to an uniform set of equations for the spinor v. The choice of the spin operators is essentially related to peculiarities of the physical problem, see for example the relevant works [11, 311, 319]. The above-mentioned equations for the spinor v can be always written in the following form:

$$(\boldsymbol{\sigma}\mathbf{e})\, v = \zeta v, \quad \zeta = \pm 1 \,, \tag{3.243}$$

where $\mathbf{e}$ is a three-dimensional real unit vector ($\mathbf{e}^2 = 1$), and ζ is a spin quantum number, which defines the orientation of the spin along ($\zeta = 1$) or opposite ($\zeta = -1$) the vector $\mathbf{e}$. Thus, the stationary Dirac wave function always adopts a spin quantum number.

It follows from the above consideration that both K–G and Dirac stationary wave functions are defined up to a functional arbitrariness, which is described explicitly by the equations (3.229). That arbitrariness corresponds to an infinite degeneracy of each energy level n. Physically, such a degeneracy is related, for example, to the independence of the radial energy of the coordinate of the center of the orbit. To lift this degeneracy one has to select additional operators – integrals of motion. Some important possibilities will be discussed below.

Let us fix z-projections of the orbital momentum. These are L_z in the K–G case or J_z in the Dirac case. According to equations (3.208) and (3.212) this means that the functions $\Phi_n(x, y)$ have to be eigenvectors of the operator $\hat{N}_2 = \hat{a}_2^+ \hat{a}_2$. The corresponding eigenvalues (quantum numbers) are integers and will be denoted by s,

$$\hat{N}_2 \Phi_{n,s}(x, y) = s\Phi_{n,s}(x, y), \quad s \in \mathbb{Z}_+ . \tag{3.244}$$

The quantum number s is called the *radial quantum number*.

According to equations (3.208) and (3.232), eigenvalues of the operator L_z are integer as well, they are called *orbital quantum numbers*,

$$\hat{L}_z \Phi_{n,s}(x, y) = l\Phi_{n,s}(x, y), \quad l = n - s, \quad -\infty < l \leq n . \tag{3.245}$$

In accordance with classical relation (3.171), we then obtain

$$2s = \gamma R_0^2, \quad R_0^2 = \left(x_{(0)}^1\right)^2 + \left(x_{(0)}^2\right)^2 . \tag{3.246}$$

Thus, the quantum number s characterizes the distance between the origin and the center of the particle orbit.

The Dirac wave functions (3.235) are eigenvectors of the operator J_z,

$$\hat{J}_z \Psi_{n,s}(x, y) = (l - 1/2)\Psi_{n,s}(x, y), \quad l = n - s , \tag{3.247}$$

if we set

$$\Phi_n(x, y) = \Phi_{n,s}(x, y), \quad \Phi_{n-1}(x, y) = \Phi_{n-1,s}(x, y) ,$$

in equations (3.236). At the same time the constant bispinor C is given by the same expression (3.240).

Returning to equation (3.244), we can verify (using (3.221) and (3.223)) that it results in the following equation for the function $\Phi(\eta)$ from (3.226):

$$\left(\eta^2 - \partial_\eta^2 - 1\right) \Phi_s(\eta) = 2s\Phi_s(\eta) . \tag{3.248}$$

Thus, we find

$$\Phi_s(\eta) = U_s(\eta), \quad s \in \mathbb{Z}_+ , \tag{3.249}$$

where $U_s(\eta)$ are Hermite functions. Substituting equation (3.249) into (3.229), and using the formula (B.153), we find the functions $\Phi_{n,s}$ written in the coordinates ρ, φ,

$$\Phi_{n,s}(\rho, \varphi) = \frac{(-1)^n}{\sqrt{\pi}} \exp\left[i(n - s)\varphi\right] I_{s,n}(\rho) , \tag{3.250}$$

where $I_{s,n}(x)$ are Laguerre functions (B.1). When written in Cartesian coordinates the functions $\Phi_{n,s}$ have the form

$$\Phi_{n,s}(x, y) = \frac{(-1)^n}{\sqrt{\pi}} \sqrt{\frac{n!}{s!}} (x - iy)^{s-n} \exp\left(-\frac{x^2 + y^2}{2}\right) L_n^{s-n}(x^2 + y^2) . \tag{3.251}$$

In addition to equation (3.231), the functions $\Phi_{n,s}$ satisfy the relations

$$\hat{a}_2\Phi_{n,s} = \sqrt{s}\Phi_{n,s-1}, \quad \hat{a}_2^+\Phi_{n,s} = \sqrt{s+1}\Phi_{n,s+1} . \tag{3.252}$$

Thus, they can be obtained from a unique vacuum state $\Phi_{0,0}$ by means of the action of the creation operators,

$$\begin{aligned} &\Phi_{n,s}(x,y) = \frac{(\hat{a}_1^+)^n(\hat{a}_2^+)^s}{\sqrt{\Gamma(n+1)\Gamma(s+1)}}\Phi_{0,0}(x,y), \\ &\hat{a}_1\Phi_{0,0} = \hat{a}_2\Phi_{0,0} = 0, \quad \Phi_{0,0}(x,y) = (1/\sqrt{\pi})\exp[-(x^2+y^2)/2] . \end{aligned} \tag{3.253}$$

Below we consider a different way to construct stationary states.

Let us chose a linear combination of the creation and annihilation operators $\hat{a}_2^+$ and $\hat{a}_2$ as an additional integral of motion,

$$\hat{A}_2(\alpha,\beta) = \alpha\hat{a}_2 + \beta\hat{a}_2^+ . \tag{3.254}$$

Here α and β are arbitrary complex constants. One can verify (see [24] and Section 7.8) that eigenvectors of the operator $\hat{A}_2(\alpha,\beta)$ have finite norms for $|\alpha| > |\beta|$ and can be normalized to δ-function for $|\alpha| = |\beta|$. For $|\alpha| < |\beta|$ they do not belong to a Hilbert space, and are not normalizable. We suppose now that the condition $|\alpha| \geq |\beta|$ holds.

Consider first the limiting case $|\alpha| = |\beta|$. In this case one can always set

$$\alpha = \alpha_0 e^{i\varphi}/2, \quad \beta = \alpha_0 e^{i\psi}/2, \quad \alpha_0 \neq 0 ,$$

where α_0, φ, and ψ are arbitrary real numbers. Introducing new real angles ν, μ, as $\varphi = \nu + \mu$, $\psi = \nu - \mu$, we reduce the operators $\hat{A}_2(\alpha,\beta)$ to the following form: $\hat{A}_2(\alpha,\beta) = \alpha_0\, e^{i\nu}\hat{A}(\mu)$, where $\hat{A}(\mu)$ is a self-adjoint operator,

$$\hat{A}(\mu) = \frac{1}{2}\left(e^{i\mu}a_2 + e^{-i\mu}a_2^+\right), \quad A^+(\mu) = A(\mu) . \tag{3.255}$$

To study the eigenvalue problem for $\hat{A}_2$ it is sufficient to consider the self-adjoint operator $A(\mu)$.

The function $\Phi(\eta)$ from (3.229) has to be an eigenvector for the operator (3.255)

$$\hat{A}(\mu)\Phi_z^\mu(\eta) = z\Phi_z^\mu(\eta), z = z^* .$$

Taking into account the explicit form (3.221) of the operators a_2^+ and a_2, we easily find

$$\Phi_z^\mu(\eta) = \left(2\sqrt{2}\pi\sin\mu\right)^{-1/2}\exp\left(\frac{i\cos\mu}{2\sin\mu}\eta^2 - i\frac{z}{\sqrt{2}\sin\mu}\eta\right) . \tag{3.256}$$

The functions (3.256) satisfy the orthonormality relation

$$\int_{-\infty}^{\infty}\Phi_{z'}^{*\mu}(\eta)\Phi_z^\mu(\eta)d\eta = \delta(z-z') . \tag{3.257}$$

Substituting (3.256) into (3.229) and integrating over η, we get

$$\Phi^{\mu}_{n,z}(x,y) = \sqrt{2/\pi}\exp i(\theta - n\mu)U_n(Q),$$
$$\theta = (x\sin\mu + y\cos\mu)(x\cos\mu - y\sin\mu - 2z)\,,$$
$$Q = \sqrt{2}(x\cos\mu - y\sin\mu - z)\,. \tag{3.258}$$

The functions $\Phi^{\mu}_{n,z}(x,y)$ are eigenvectors of the operator (3.255) (with representation (3.205) taken for a_2^+ and a_2). They satisfy relations (3.231) and (3.232). For the values $\mu = 0, \pi/2$ they are known in the literature [311].

Consider now the case $|\alpha| > |\beta|$. Multiplying operator (3.254) by a constant factor, we can always provide the following relation:

$$|\alpha|^2 - |\beta|^2 = 1\,, \tag{3.259}$$

for the new operator. Without loss of generality we accept that this relation holds. Then $\hat{A}_2(\alpha,\beta)$ is an annihilation operator,

$$\left[\hat{A}_2(\alpha,\beta), \hat{A}_2^+(\alpha,\beta)\right] = 1\,, \tag{3.260}$$

and their eigenvectors are coherent states.

Function $\Phi(\eta)$ from (3.229) can be now specified as

$$\hat{A}_2(\alpha,\beta)\Phi^{\alpha,\beta}_z(\eta) = z\Phi^{\alpha,\beta}_z(\eta)\,,$$
$$\Phi^{\alpha,\beta}_z(\eta) = (\alpha-\beta)^{-1/2}U_0(P)\exp\theta\,,$$
$$P = \frac{2\eta - \sqrt{2}z\,(\alpha^* - \beta^*) - \sqrt{2}z^*\,(\alpha-\beta)}{2|\alpha-\beta|}\,,$$
$$4\theta = \left[2\,(\alpha\beta^* - \alpha^*\beta)\,\eta^2 + 2\sqrt{2}\left[z\,(\alpha^*-\beta^*) - z^*\,(\alpha-\beta)\right]\eta\right.$$
$$\left. + \left[z^*(\alpha-\beta)\right]^2 - \left[z(\alpha^*-\beta^*)\right]^2\right]|\alpha-\beta|^{-2}\,. \tag{3.261}$$

In particular, for $\beta = 0$ and $\alpha = 1$ we have

$$\hat{a}_2\Phi_z(\eta) = z\Phi_z(\eta),$$
$$\Phi_z(\eta) = U_0\left(\eta - \frac{z+z^*}{\sqrt{2}}\right)\exp\left(\frac{z-z^*}{\sqrt{2}}\eta + \frac{z^{*2}-z^2}{4}\right)\,. \tag{3.262}$$

Substituting (3.261) into (3.229), we find a coordinate representation for stationary coherent states

$$\Phi^{\alpha,\beta}_{n,z}(x,y) = (-1)^n\pi^{-1/4}\alpha^{-1/2}\,(\beta/\alpha)^{n/2}\,U_n\!\left(\frac{q}{\sqrt{2\alpha\beta}}\right)\exp(Q/4\alpha\beta)\,,$$
$$Q = (1+2|\beta|^2)z^2 - 2\alpha\beta|z|^2 + (z+q)\left[\beta(x-iy) - \alpha(x+iy)\right],$$
$$q = z - \alpha(x+iy) - \beta(x-iy)\,. \tag{3.263}$$

For $\beta = 0$ and $\alpha = 1$, these states have an especially simple form. Substituting (3.262) into (3.229) and integrating over η, we obtain (such states were first found in the work [211])

$$\Phi_{n,z}(x,y) = \varphi_{n,z}(x,y)\exp\left(-|z|^2/2\right) ,$$
$$\varphi_{n,z}(x,y) = \frac{(x+iy-z)^n}{\sqrt{\pi\Gamma(n+1)}}\exp\left[z(x-iy) - \left(x^2+y^2\right)/2\right] . \tag{3.264}$$

The functions $\Phi_{n,z}(x,y)$ and $\varphi_{n,z}(x,y)$ are eigenvectors of the operator $\hat{a}_2$ (3.205). The action of the creation operator $\hat{a}_2^+$ on these functions reads:

$$\hat{a}_2^+\Phi_{n,z}(x,y) = (x-iy)\Phi_{n,z}(x,y) - \sqrt{n}\Phi_{n-1,z}(x,y) ,$$
$$\hat{a}_2^+\varphi_{n,z}(x,y) = (x-iy)\varphi_{n,z}(x,y) - \sqrt{n}\varphi_{n-1,z}(x,y) . \tag{3.265}$$

However, the second equation (3.265) can be also written as

$$a_2^+\varphi_{n,z}(x,y) = \frac{\partial}{\partial z}\varphi_{n,z}(x,y) . \tag{3.266}$$

As it is an integral of motion, the operator $\hat{a}_2^+$ is at the same time a symmetry operator. This means that we can get new stationary solutions by acting with this operator on the functions $\varphi_{n,z}(x)$. Applying successive differentiation to the functions $\varphi_{n,z}(x)$, we get the following family of stationary states:

$$\begin{aligned}\bar{\Phi}_{n,s,z}(x,y) &= \frac{\partial^s}{\partial z^s}\varphi_{n,z}(x,y)\\ &= (-1)^n(N/\sqrt{\pi})\exp\left[i(n-s)\varphi + (\rho-\tilde{q})/2\right](\tilde{q}/\rho)^{(n-s)/2}I_{s,n}(\tilde{q})\\ &= (-1)^n(N/\sqrt{\pi})\exp\left[z(x-iy)/2\right]\left(\frac{x+iy-z}{x-iy}\right)^{(n-s)/2}I_{s,n}(\tilde{q}),\\ \tilde{q} &= \rho - z\sqrt{\rho}\exp(-i\varphi) = (x-iy)(x+iy-z) .\end{aligned} \tag{3.267}$$

The functions (3.267) are written in the coordinates x, y, ρ, φ, and N is a normalization factor to be specified below. By a straightforward calculation, one can verify that the states (3.267) obey the relations

$$\hat{a}_2\bar{\Phi}_{n,s,z} = z\bar{\Phi}_{n,s,z} + \sqrt{s}\bar{\Phi}_{n,s-1,z} ,$$
$$\hat{a}_2^+\bar{\Phi}_{n,s,z} = \frac{\partial}{\partial z}\bar{\Phi}_{n,s,z} = \sqrt{s+1}\bar{\Phi}_{n,s+1,z} . \tag{3.268}$$

This implies that the states $\bar{\Phi}_{n,s,z}$ can be created from a "vacuum" by the action of the operators $\hat{a}_1^+$ and $\hat{a}_2^+$,

$$\bar{\Phi}_{n,s,z} = \frac{\left(a_1^+\right)^n\left(a_2^+\right)^s}{\sqrt{\Gamma(n+1)\Gamma(s+1)}}\bar{\Phi}_{0,0,z} ,$$
$$\bar{\Phi}_{0,0,z} = (1/\sqrt{\pi})\exp\left[z(x-iy) - \left(x^2+y^2\right)/2\right] . \tag{3.269}$$

The states (3.250) present a special case of the states (3.267) with $z = 0$. At the same time relations (3.268) and (3.269) generalize relations (3.252) and (3.253).

It is easy to see that an orthogonal system of generalized coherent states can be constructed acting by the operator $[A_2^+(\alpha,\beta) - z^*]^s$ on the set (3.263). First, let us act by this operator on the functions $\Phi_z^{\alpha,\beta}(\eta)$ given by equation (3.261). Thus, we obtain

$$\Phi_{s,z}^{\alpha,\beta}(\eta) = (\alpha-\beta)^{-1/2}\left(\frac{\alpha^*-\beta^*}{\alpha-\beta}\right)^{s/2} U_s(P)\exp\theta\,, \tag{3.270}$$

where P and θ are defined in (3.261). Substituting (3.270) into (3.229), we obtain the most general form of stationary solutions

$$\Phi_{n,s,z}^{\alpha,\beta}(x,y) = \frac{\exp(ixy)}{\sqrt{\pi}}\int_{-\infty}^{\infty} d\eta \exp\left(-iy\eta\sqrt{2}\right) U_n\left(x\sqrt{2}-\eta\right)\Phi_{s,z}^{\alpha,\beta}(\eta)\,. \tag{3.271}$$

The integral over η can be expressed in terms of some cumbersome finite sums of products of the Hermite functions. However, the integral representation (3.271) itself may be effectively used in physical applications.

In the special case $\alpha = 1$ and $\beta = 0$ one can get from (3.270) or from (3.262) the following result:

$$\Phi_{s,z}(\eta) = U_s\left(\eta - \frac{z+z^*}{\sqrt{2}}\right)\exp\left(\frac{z-z^*}{\sqrt{2}}\eta + \frac{z^{*2}-z^2}{4}\right), \tag{3.272}$$

and the integral in (3.229) can be calculated to give

$$\begin{aligned}\Phi_{n,s,z}(x,y) &= \frac{(-1)^n}{\sqrt{\pi}}\left(\frac{x+iy-z}{x-iy-z^*}\right)^{\frac{n-s}{2}}\\ &\times \exp[z(x-iy) - z^*(x+iy)]I_{s,n}\left(|x+iy-z|^2\right)\,.\end{aligned} \tag{3.273}$$

The functions $\Phi_{n,s,z}(x,y)$ satisfy relations (3.231) and (3.232) and the operators $\hat{a}_2$ and $\hat{a}_2^+$ act on these functions as follows:

$$\begin{aligned} a_2\Phi_{n,s,z} &= z\Phi_{n,s,z} + \sqrt{s}\Phi_{n,s-1,z}\,,\\ a_2^+\Phi_{n,s,z} &= z^*\Phi_{n,s,z} + \sqrt{s+1}\Phi_{n,s+1,z}\,.\end{aligned} \tag{3.274}$$

Thus, we have an analog of the representation (3.269)

$$\begin{aligned}\Phi_{n,s,z} &= \frac{(a_2^+-z^*)^s(a_1^+)^n}{\sqrt{\Gamma(n+1)(s+1)}}\Phi_{0,0,z},\\ \Phi_{0,0,z} &= \frac{1}{\sqrt{\pi}}\exp\left[z(x-iy) - \frac{1}{2}\left(x^2+y^2\right) - \frac{|z|^2}{2} + \frac{z^{*2}-z^2}{4}\right].\end{aligned} \tag{3.275}$$

The states (3.250) represent a special case of the states (3.273) for $z = 0$, and the states (3.263) are a special case of the states (3.273) for $s = 0$. Thus, states (3.273) represent the most general form of stationary solutions in a constant uniform magnetic field.

3.4.5 Orthonormality and completeness of stationary states

Here we are going to discuss properties of orthonormality and completeness of stationary states presented above. To this end we use the inner product (3.233), which is defined in the two-dimensional space x, y or ρ, φ. As eigenvectors of the self-adjoint operator (3.232) all the solutions under consideration are orthogonal with respect to the index n in virtue of general relations (3.233). However, a straightforward check of this fact can encounter technical difficulties and needs the use of different properties of Laguerre and Hermite functions listed in Chapter B.

For example, one can verify that the set of functions (3.273) obeys the following orthonormality relation:

$$\left(\Phi_{n',s',z}, \Phi_{n,s,z}\right) = \delta_{n,n'}\delta_{s,s'} \,. \tag{3.276}$$

As a consequence of that fact the set (3.250) obeys the same relation. For different values of z the orthogonality is absent. Using, for example, the formula (B.1), we get

$$\begin{aligned}\left(\Phi_{n',s',z'}, \Phi_{n,s,z}\right) &= \delta_{n,n'} R_{s',s}(z', z), R_{s',s}(z', z) \\ &= \left(\frac{z - z'}{z^* - z'^*}\right)^{\frac{s'-s}{2}} I_{s',s}(|z - z'|^2) \exp\left(\frac{z'^* z - z' z^*}{2}\right) .\end{aligned} \tag{3.277}$$

For the functions (3.270) the inner product does not depend on α, β and coincides with (3.277).

Only the set (3.259) is orthogonal with respect to the index z,

$$\left(\Phi^{\mu}_{n',z'}, \Phi^{\mu}_{n,z}\right) = \delta_{n,n'}\delta(z - z') \,. \tag{3.278}$$

All other sets of functions (excluding (3.270), which is orthogonal with respect to two indices n, s) are orthogonal only with respect to the index n. Consider the following cases.

For the set (3.263) we have

$$\left(\Phi^{\alpha,\beta}_{n',z'}, \Phi^{\alpha,\beta}_{n,z}\right) = \delta_{n,n'} \exp S/2, \quad S = zz'^* - z^* z' - |z - z'|^2 \,. \tag{3.279}$$

The same relation holds for the set (3.264), which is a special case (for $\alpha = 1, \beta = 0$) of the set (3.263). Taking into account (3.264), one can derive from (3.279) that

$$(\varphi_{n',z'}, \varphi_{n,z}) = \delta_{n,n'} \exp\left(zz'^*\right) , \tag{3.280}$$

where $\varphi_{n,z}$ are defined in (3.264).

For the set (3.267) the orthonormality relations can be derived in the following way. First, we find

$$\begin{aligned}\left(\bar{\Phi}_{n',s',z'}, \bar{\Phi}_{n,s,z}\right) &= N'^* N \delta_{n,n'} J^n_{s,s'}(z, z') , \\ J^n_{s,s'}(z, z') &= \left[\Gamma(s+1)\Gamma(s'+1)\right]^{-1/2} \frac{\partial^{s'}}{\partial z'^*} \frac{\partial^s}{\partial z^s} (\varphi_{n,z'}, \varphi_{n,z}) \,.\end{aligned}$$

Now, using equations (3.280), (3.233), (3.262) and (B.21), we find

$$J^n_{s,s'} = \sqrt{\frac{\Gamma(s+1)}{\Gamma(s'+1)}} z^{s'-s} e^{zz'^*} L_s^{s'-s}(-zz'^*),\ s \le s' ,$$
$$J^n_{s,s'} = \sqrt{\frac{\Gamma(s'+1)}{\Gamma(s+1)}} (z'^*)^{s-s'} e^{zz'^*} L_{s'}^{s-s'}(-zz'^*),\ s' \le s , \tag{3.281}$$

where $L_s^{\alpha}(x)$ are Laguerre polynomials (B.5). Thus, one can see that the functions $\bar{\Phi}_{n,s,z}$ are normalized to the unity if the normalization factor in (3.267) has the form

$$N = N_s(z) = \exp\left(-|z|^2/2\right)\left[L_s(-|z|^2)\right]^{-1/2} . \tag{3.282}$$

The normalization (3.282) always exists since $L_s(x) > 0$ for $x < 0$ (all roots of the polynomials $L_s(x)$ are positive). It should be noted that relations (3.268) and (3.269) hold only for $N = 1$.

Let us consider now completeness relations for some sets of the solutions under consideration.

First we start with solutions (3.250). Consider a sum

$$F(x, y; x', y'; z) = \sum_{n=0}^{\infty}\sum_{s=0}^{\infty} \Phi^*_{n,s}(x', y')\Phi_{n,s}(x, y)z^n, \quad |z| < 1 . \tag{3.283}$$

It can be calculated by means of the corresponding formulas (B.108) and (B.64) to be

$$F = \frac{1}{\pi(1-z)} \exp\left\{i(xy' - x'y) + \frac{(1-3z)[(x-x')^2 + (y-y')^2]}{2(1-z)}\right\} . \tag{3.284}$$

Then we consider the integral

$$\int_{-\infty}^{\infty} dx' \int_{-\infty}^{\infty} dy' F(x, y; x', y'; z) f(x', y') = J(x, y; z) , \tag{3.285}$$

which exists in the domain

$$|z - 2/3| < 1/3 , \tag{3.286}$$

if, for example, the function $f(x, y)$ is a function of a limited growth. The restriction (3.286) is stronger than $|z| < 1$. By the change of variables

$$x' = p\sqrt{2\mu} + x - iy\mu, \quad y' = q\sqrt{2\mu} + y + ix\mu ,$$
$$dx'dy' = 2\mu dp dq, \qquad \mu = (1-z)/(3z-1) ,$$

in (3.285), J is reduced to the following form:

$$J = \int_{-\infty}^{\infty} dp \int_{-\infty}^{\infty} dq f\left(x + p\sqrt{2\mu} - iy\mu, y + q\sqrt{2\mu} + ix\mu\right) G ,$$
$$G = 2\mu F\left(x, y; x + p\sqrt{2\mu} - iy\mu, y + q\sqrt{2\mu} + ix\mu; z\right) . \tag{3.287}$$

Let z tends to $1-0$, remaining in the area (3.286). In this case $\mu \to 0$. Then it follows from (3.284) and (3.287) that

$$\lim_{z \to 1-0} G = \frac{1}{\pi} \exp(-p^2 - q^2), \quad \lim_{z \to 1-0} J(x, y; z) = f(x, y).$$

Thus, due to the arbitrariness of the function $f(x, y)$ we obtain the completeness relation for the functions $\Phi_{n,s}(x, y)$,

$$\sum_{n=0}^{\infty} \sum_{s=0}^{\infty} \Phi^*_{n,s}(x', y')\Phi_{n,s}(x, y) = \delta(x - x')\delta(y - y'). \tag{3.288}$$

For the functions $\Phi^{\mu}_{n,s}(x, y)$ such a relation can be obtained in a similar way by means of the formula (B.1),

$$\int_{-\infty}^{\infty} dz \sum_{n=0}^{\infty} \Phi^{*\mu}_{n,z}(x', y')\Phi^{\mu}_{n,z}(x, y) = \delta(x - x')\delta(y - y'). \tag{3.289}$$

For the nonorthogonal set (3.264) the completeness relation

$$\int d^2z \sum_{n=0}^{\infty} \Phi^*_{n,z}(x', y')\Phi_{n,z}(x, y) = \pi\delta(x - x')\delta(y - y'),$$
$$d^2z = d\,\mathrm{Re}\, z \;\, d\,\mathrm{Im}\, z, \tag{3.290}$$

can be derived as follows. By a straightforward calculation we find

$$\sum_{n=0}^{\infty} \Phi^*_{n,z}(x', y')\Phi_{n,z}(x, y) = \frac{1}{\pi} \exp\left[iw - \frac{1}{2}(x - x')^2 - \frac{1}{2}(y - y')^2\right],$$
$$w = yx' - xy' + (x - x')(z + z^*) + i(y - y')(z^* - z).$$

Then by taking into account the simple relations $z + z^* = 2\,\mathrm{Re}\, z$, and $i(z^* - z) = 2\,\mathrm{Im}\, z$ we arrive at (3.290).

More complicated calculations with the help of equation (B.1) allow us to establish a completeness relation in the same form as (3.290) for the functions (3.263).

The set (3.273) is complete for any given z,

$$\sum_{n=0}^{\infty} \sum_{s=0}^{\infty} \Phi^*_{n,s,z}(x', y')\Phi_{n,s,z}(x, y) = \delta(x - x')\delta(y - y'), \tag{3.291}$$

and for any given s, s',

$$\int d^2z \sum_{n=0}^{\infty} \Phi^*_{n,s',z}(x', y')\Phi_{n,s,z}(x, y) = \pi\delta_{s',s}\delta(x - x')\delta(y - y'). \tag{3.292}$$

The proof of (3.291) is similar to that of (3.288); whereas the proof of (3.292) is similar to that of (3.290) if the formula (B.1) is used.

It is easy to obtain a simple relation

$$\int d^2z\Phi^{*\alpha,\beta}_{s',z}(\eta')\Phi^{\alpha,\beta}_{s,z}(\eta) = \pi\delta_{s',s}\delta(\eta-\eta')\,, \tag{3.293}$$

for the functions $\Phi^{\alpha,\beta}_{s,z}(\eta)$. This equation and the representation (3.270) allow us to obtain completeness relations that literally coincide with (3.291) and (3.292).

All the functions under consideration are eigenvectors of the operator (3.223). That is why they have mutual decompositions for any fixed n. Consider below such decompositions.

The functions (3.258) have the following mutual decomposition:

$$\Phi^{\mu'}_{n,z'}(x,y) = \int_{-\infty}^{\infty} C(z,z';\mu-\mu')\Phi^{\mu}_{n,z}(x,y)dz,$$

$$C(z,z';\gamma) = \frac{(1+i)\exp(i\delta)}{\sqrt{2\pi\sin\gamma}}, \quad \delta = \frac{2zz' - z^2\cos\gamma}{\sin\gamma} - \frac{\gamma}{2}, \tag{3.294}$$

which can be checked by using (B.1). The same functions can be decomposed in the set (3.250),

$$\Phi^{\mu}_{n,z}(x,y) = \sum_{s=0}^{\infty} e^{-is\mu}U_s\left(\sqrt{2}z\right)\Phi_{n,s}(x,y)\,. \tag{3.295}$$

The validity of (3.295) can be checked with the help of equation (B.1). Taking into account the orthogonality of the Hermite functions, one can derive from (3.295) an inverse decomposition

$$\Phi_{n,s}(x,y) = \sqrt{2}e^{is\mu}\int_{-\infty}^{\infty} dzU_s\left(\sqrt{2}z\right)\Phi^{\mu}_{n,z}(x,y)\,. \tag{3.296}$$

For $N = 1$ the functions (3.267) have the following mutual decompositions:

$$\bar{\Phi}_{n,s+k,z'}(x,y) = \sqrt{\frac{\Gamma(k+1)}{\Gamma(s+k+1)}}\int \frac{d^2z}{\pi} z^{*s} e^{(z'z^*-|z|^2)}\bar{\Phi}_{n,k,z}(x,y)\,, \tag{3.297}$$

$$\bar{\Phi}_{n,s,z'}(x,y) = \sum_{k=0}^{\infty}\sqrt{\frac{\Gamma(k+s+1)}{\Gamma(s+1)}}\frac{(z'-z)^k}{k!}\bar{\Phi}_{n,s+k,z}(x,y)\,. \tag{3.298}$$

In particular, setting $z = 0$ in (3.298), we get a decomposition of the functions into the set (3.250), and setting $k = 0$ in (3.297) we get a decomposition into the set (3.264).

The set (3.270) has the mutual decomposition of the form

$$\Phi^{\alpha,\beta}_{n,s,z}(x,y) = \frac{1}{\pi}\int d^2z'\Phi^{\alpha',\beta'}_{n,s',z'}(x,y)R_{s',s}(\alpha',\beta';\alpha,\beta;z',z),$$

$$\Phi^{\alpha,\beta}_{n,s,z}(x,y) = \sum_{s'=0}^{\infty}\Phi^{\alpha',\beta'}_{n,s',z'}(x,y)R_{s',s}(\alpha',\beta';\alpha,\beta;z',z)\,, \tag{3.299}$$

where the coefficient function R is one and the same in the both expressions (this follows from (3.293)), and has the form

$$R_{s',s}(\alpha',\beta';\alpha,\beta;z',z) = \int_{-\infty}^{\infty} \Phi^{*\alpha',\beta'}_{s',z'}(\eta)\Phi^{\alpha,\beta}_{s,z}(\eta)\,d\eta\,, \tag{3.300}$$

where $\Phi^{\alpha,\beta}_{s,z}(\eta)$ is defined by equation (3.270).

Unfortunately, a compact form of the functions R is unknown in the general case $\alpha\prime \neq \alpha$ and $\beta\prime \neq \beta$. However, in some special cases there exist exclusions. For example, the set (3.263) has a relatively simple decomposition with respect to the set (3.250),

$$\Phi^{\alpha,\beta}_{n,z}(x,y) = \pi^{1/4}\alpha^{-1/2}\exp\left[\frac{(1+2|\beta|^2)z^2 - 2\alpha\beta|z|^2}{4\alpha\beta}\right] \times \sum_{s=0}^{\infty}\left(\frac{\beta}{\alpha}\right)^{\frac{s}{2}} U_s\left(\frac{z}{\sqrt{2\alpha\beta}}\right)\Phi_{n,s}(x,y)\,. \tag{3.301}$$

One can also find explicitly the coefficient function $R_{0,0}$. Thus, we know the decomposition of the functions $\Phi^{\alpha,\beta}_{n,z}$ with respect to $\Phi^{\alpha\prime,\beta\prime}_{n,z\prime}$,

$$R_{0,0}(\alpha',\beta';\alpha,\beta;z',z) = (\alpha\alpha'^{*} - \beta\beta'^{*})^{-1/2}\exp(Q/2)\,,$$
$$Q = \frac{z^2(\alpha'^{*}\beta^{*} - \alpha^{*}\beta'^{*}) + (z'^{*})^2(\alpha\beta' - \alpha'\beta) + 2zz'^{*}}{\alpha\alpha'^{*} - \beta\beta'^{*}} - z'z^{*} - zz'^{*} - |z-z'|^2\,. \tag{3.302}$$

In the case $\alpha\prime = \alpha$ and $\beta\prime = \beta$, the coefficient functions do not depend on α and β, they can be found in the general case and coincide with the functions $R_{s\prime,s}(z',z)$ given by equation (3.277).

The especially simple form has a decomposition of the functions (3.264) over the set (3.250),

$$\Phi_{n,z}(x,y) = \sum_{s=0}^{\infty}\frac{z^s}{\sqrt{s!}}\Phi_{n,s}(x,y)\exp(-|z|^2/2)\,. \tag{3.303}$$

The coefficient functions that correspond to this case are a special case of (3.277) for $s = 0$ and $z\prime = 0$.

3.4.6 Coherent states

One of the most interesting examples of nonstationary states are *coherent states*. Below we are going to discuss such states for spinless and spinning particles moving in a constant and uniform magnetic field. To this end light-cone variables (2.37) are widely used. As was already remarked, in these variables solutions of the Dirac equation can be found via solutions of the K–G equation (3.213) by means of equations (3.216) and (3.217). In the case under consideration the operators $\widehat{\widehat{P_0}}$ and $\widehat{\widehat{P_3}}$, are integrals of motion. Let us search the K–G wave functions as eigenvectors for the operator $\widehat{\widehat{P_3}}$,

$$\widehat{\widehat{P_3}}\,\Phi(x) = \frac{\lambda}{2}\varphi(x)\,. \tag{3.304}$$

The eigenvalue λ can be related to the classical momenta p_0 and p_3,

$$\lambda = p_0 + p_3\,. \tag{3.305}$$

The signs of λ and p_0 coincide, so that states with $\lambda > 0$ describe particles, whereas ones with $\lambda < 0$ describe antiparticles. It follows from (3.304) that the variable u^3 can be separated. If we write the scalar wave function in the following form:

$$\varphi(x) = N\exp\left(-i\lambda u^3/2 - im_{\text{eff}}^2 u^0/2\lambda\right)\Phi(u^0, x, y)\,, \tag{3.306}$$

where dimensionless variables x and y are defined in (3.202), then the function $\Phi(u^0, x, y)$ satisfies an equation of the Schrödinger type,

$$\bar{\hat{K}}\Phi(u^0, x, y) = 0, \quad \bar{\hat{K}} = i\partial_{u^0} - \tilde{\omega}\hat{a}_1^+\hat{a}_1, \quad \tilde{\omega} = \gamma/\lambda\,, \tag{3.307}$$

with the Hamiltonian quadratic in creation and annihilation operators.

If one separates also the variable u^0 (this is possible, since $\widehat{\widehat{P_0}}$ is an integral of motion), then we get stationary states which were already studied above.

Let us write the Dirac wave function $\Psi_{(-)}(x)$ from (3.214) in the following form:

$$\Psi_{(-)}(x) = N\exp(-i\lambda u^3/2 - im^2u^0/2\lambda)W\,C\,\Phi(u^0, x, y)\,, \tag{3.308}$$

where C is a constant bispinor, W a unitary matrix (ψ_0 is an arbitrary constant phase),

$$W = \cos\Omega_0 - i\Sigma_3\sin\Omega_0, \quad \Omega_0 = (\tilde{\omega}u^0 + \psi_0)/2\,, \tag{3.309}$$

and the scalar wave function $\Phi_0(u^0, x, y)$ is a solution of the equation (3.307). If necessary, the component $\Psi_{(+)}$ can be restored according to $\Psi_{(-)}$ from the equation (3.215),

$$\Psi_{(+)} = \lambda^{-1}\left[(\boldsymbol{\alpha}\hat{\mathbf{P}}_\perp) + m\rho_3\right]\Psi_{(-)}\,. \tag{3.310}$$

Let us construct an operator-integral of motion that is a linear combination of the creation and annihilation operators a_1^+, a_1,

$$\hat{A}_1 = f(u^0)\hat{a}_1 + g(u^0)\hat{a}_1^+\,. \tag{3.311}$$

To be an integral of motion, such an operator has to commute with the operator $\bar{\hat{K}}$. Taking into account the commutators,

$$[\hat{a}_1^+\hat{a}_1, \hat{a}_1] = -\hat{a}_1, \quad [\hat{a}_1^+\hat{a}_1, \hat{a}_1^+] = \hat{a}_1^+\,,$$

we find the following equations for the coefficient functions f and g:

$$i\dot{f} + \tilde{\omega} f = 0, \quad i\dot{g} - \tilde{\omega} g = 0\,, \tag{3.312}$$

where the dot above means a derivative with respect to the light-cone time u^0. These equations have simple solutions

$$f = f_0 \exp\left(i\tilde{\omega} u^0\right), \quad g = g_0 \exp\left(-i\tilde{\omega} u^0\right)\,, \tag{3.313}$$

where f_0 and g_0 are arbitrary complex numbers. If we choose these numbers to obey the relation

$$|f_0|^2 - |g_0|^2 = 1\,, \tag{3.314}$$

then A_1 and A_1^+ are some creation and annihilation operators. Equation (3.314) has the general solution

$$f_0 = e^{i\mu} \cosh\alpha_0, \quad g_0 = e^{-i\nu} \sinh\alpha_0\,, \tag{3.315}$$

with α_0, μ and ν being arbitrary real numbers. Thus, we get a set of creation and annihilation operators

$$\begin{aligned} \hat{A}^+ &= \hat{a}_1^+ \cosh\alpha_0 \exp\left[-i\left(\tilde{\omega}u^0 + \mu\right)\right] + \hat{a}_1 \sinh\alpha_0 \exp\left[i\left(\tilde{\omega}u^0 + \nu\right)\right], \\ \hat{A} &= \hat{a}_1 \cosh\alpha_0 \exp\left[i\left(\tilde{\omega}u^0 + \mu\right)\right] + \hat{a}_1^+ \sinh\alpha_0 \exp\left[-i\left(\tilde{\omega}u^0 + \nu\right)\right]\,, \end{aligned} \tag{3.316}$$

which are at the same time integrals of motion for equation (3.307). The inverse relations read

$$\hat{a}_1 = f^* \hat{A}_1 - g\hat{A}_1^+, \quad \hat{a}_1^+ = f\hat{A}_1^+ - g^*\hat{A}_1\,. \tag{3.317}$$

Thus, we can express all basic operators of the problem under consideration in terms of the integrals of motion, which are A_1, A_1^+, and $A_2(\alpha,\beta)$, $A_2^+(\alpha,\beta)$,

$$\begin{aligned} \hat{P}_1 &= i\sqrt{\gamma/2}\,[(g^* + f^*)\hat{A}_1 - (g + f)\hat{A}_1^+], \\ \hat{P}_2 &= \sqrt{\gamma/2}\,[(g^* - f^*)\hat{A}_1 + (g - f)\hat{A}_1^+]\,, \end{aligned} \tag{3.318}$$

$$\begin{aligned} x^1 &= [(\alpha^* - \beta^*)\hat{A}_2 + (\alpha - \beta)\hat{A}_2^+]/\sqrt{2\gamma} - \hat{P}_2/\gamma, \\ x^2 &= i[(\alpha + \beta)\hat{A}_2^+ - (\alpha^* + \beta^*)\hat{A}_2]/\sqrt{2\gamma} + \hat{P}_1/\gamma\,, \end{aligned} \tag{3.319}$$

$$\hat{N}_1 = \left(|f_0|^2 + |g_0|^2\right)\hat{A}_1^+\hat{A}_1 - f_0 g_0 \hat{A}_1^+\hat{A}_1^+ - g_0^* f_0^* \hat{A}_1\hat{A}_1 + |g_0|^2\,. \tag{3.320}$$

The operators $\hat{A}_2 = \hat{A}_2(\alpha,\beta)$ are defined by equations (3.254) and (3.259).

Let us consider states that are eigenvectors for the commuting operators $\hat{A}_1$ and $\hat{A}_2$,

$$\begin{aligned} \hat{A}_1 \Phi_{z_1,z_2}(u^0, x, y) &= z_1 \Phi_{z_1,z_2}(u^0, x, y), \\ \hat{A}_2 \Phi_{z_1,z_2}(u^0, x, y) &= z_2 \Phi_{z_1,z_2}(u^0, x, y)\,. \end{aligned} \tag{3.321}$$

These are the so-called coherent states. Acting on these states by the symmetry operators $\hat{A}_1^+ - z_1^*$, and $\hat{A}_2^+ - z_2^*$, we can construct a set of *semicoherent states*,

$$\Phi_{n,s,z_1,z_2}(u^0, x, y) = \frac{\left(\hat{A}_1^+ - z_1^*\right)^n \left(\hat{A}_2^+ - z_2^*\right)^s}{\sqrt{\Gamma(n+1)\Gamma(s+1)}} \Phi_{z_1,z_2}(u^0, x, y) \,, \tag{3.322}$$

see [36]. Functions from this set are labeled by integral quantum numbers n, s, and by complex numbers z_1, z_2, and depend on four complex parameters f_0, g_0, α, and β, related by equations (3.314) and (3.259). The set (3.322) can be represented in the following integral form:

$$\Phi_{n,s,z_1,z_2}(u^0, x, y) = (1/\sqrt{\pi}) \int_{-\infty}^{\infty} \exp\left(ixy - iy\eta\sqrt{2}\right) \Phi^{f,g}_{n,z_1}\left(x\sqrt{2} - \eta\right) \Phi^{\alpha,\beta}_{s,z_2}(\eta) d\eta \,, \tag{3.323}$$

where the functions $\Phi^{\alpha,\beta}_{s,z}(\eta)$ are defined by equation (3.270).

The integral in (3.323) can be calculated in some particular cases. For example, for $n = s = 0$ we obtain

$$\begin{aligned}
\Phi_{z_1,z_2}(u^0, x, y) &= \left[\pi(\alpha f - \beta g)|\alpha f|/\alpha f\right]^{-1/2} \exp(Q/2), \\
Q &= (q + Q_1)/(\alpha f - \beta g) - |z_1|^2 - |z_2|^2, \\
q &= 2i(\beta f - \alpha g)xy - (\alpha + \beta)(f + g)x^2 - (\alpha - \beta)(f - g)y^2 \\
&\quad + 2x[(\alpha + \beta)z_1 + (f + g)z_2] + 2iy[(\alpha - \beta)z_1 - (f - g)z_2], \\
Q_1 &= (\alpha g^* - \beta f^*)z_1^2 + (\beta^* f - \alpha^* g)z_2^2 - 2z_1 z_2 \,.
\end{aligned} \tag{3.324}$$

The set (3.324) with $\beta = g = 0$ was known [34, 35]. In the same particular case one can find an explicit coordinate representation for the semicoherent states,

$$\begin{aligned}
\Phi_{n,s,z_1,z_2}(u^0, x, y) &= \frac{(-1)^n}{\sqrt{\pi}} \left(\frac{x + iy - \bar{z}_1^* - z_2}{x - iy - \bar{z}_1 - z_2^*}\right) \exp[\bar{q}/2 - iu^0 n\tilde{\omega}] I_{s,n}(P), \\
P &= |x + iy - \bar{z}_1^* - z_2|^2, \bar{z}_1 = z_1 \exp(-i\tilde{\omega}u^0), \\
\bar{q} &= (\bar{z}_1 - z_2^*)(x + iy) - (\bar{z}_1^* - z_2)(x - iy) + \bar{z}_1^* z_2^* - \bar{z}_1 z_2 \,.
\end{aligned} \tag{3.325}$$

In the general case, when $\beta \neq 0$, $g \neq 0$, the mutual decompositions take place

$$\begin{aligned}
\Phi_{n,s,z_1,z_2}(u^0, x, y) &= \sum_{n'=0}^{\infty} R_{n',n}(z_1', z_1)\Phi_{n',s,z_1',z_2}(u^0, x, y) \,, \\
\Phi_{n,s,z_1,z_2}(u^0, x, y) &= \sum_{s'=0}^{\infty} R_{s',s}(z_2', z_2)\Phi_{n,s',z_1,z_2'}(u^0, x, y) \,,
\end{aligned} \tag{3.326}$$

where the coefficient functions R are given by equations (3.277). Thus, one can get a double decomposition

$$\Phi_{n,s,z_1,z_2}(u^0,x,y) = \sum_{n'=0}^{\infty}\sum_{s'=0}^{\infty} R_{n',n}(z_1',z_1)R_{s',s}(z_2',z_2) \times \Phi_{n',s',z_1',z_2'}(u^0,x,y)\,. \tag{3.327}$$

All these decompositions take an especially simple form if $n = s = 0$ and $z_1' = z_2' = 0$. In this case

$$\Phi_{z_1,z_2}(u^0,x,y) = \exp(-|z_1|^2/2 - |z_2|^2/2)\sum_{n=0}^{\infty}\sum_{s=0}^{\infty}\frac{\bar{z}_1^n}{\sqrt{n!}}\frac{z_2^2}{\sqrt{s!}}\Phi_{n,s}(x,y)\,, \tag{3.328}$$

where $\Phi_{n,s}(x,y)$ is the set of stationary states (3.250).

Functions (3.323) satisfy the equations

$$\begin{aligned}
(\hat{A}_1 - z_1)\Phi_{n,s,z_1,z_2} &= \sqrt{n}\Phi_{n-1,s,z_1,z_2}\,,\\
(\hat{A}_1^+ - z_1^*)\Phi_{n,s,z_1,z_2} &= \sqrt{n+1}\Phi_{n+1,s,z_1,z_2}\,,\\
(\hat{A}_2 - z_2)\Phi_{n,s,z_1,z_2} &= \sqrt{s}\Phi_{n,s-1,z_1,z_2}\,,\\
(\hat{A}_2^+ - z_2^*)\Phi_{n,s,z_1,z_2} &= \sqrt{s+1}\Phi_{n,s+1,z_1,z_2}\,.
\end{aligned} \tag{3.329}$$

Using representations (3.318), one can easily calculate mean values of the momentum operator and of the operator $\hat{N}_1$ in the states (3.323),

$$\begin{aligned}
\bar{P}_1 &= i\sqrt{\gamma/2}\,[(g^* + f^*)z_1 - (g+f)z_1^*]\,,\\
\bar{P}_2 &= \sqrt{\gamma/2}\,[(g^* - f^*)z_1 + (g-f)z_1^*]\,,\\
\bar{N}_1 &= (|f_0|^2 + |g_0|^2)(|z_1|^2 + n) - g_0^* f_0^* z_1^2 - g_0 f_0 z_1^{*2} + |g_0|^2\,.
\end{aligned} \tag{3.330}$$

If z_1 is taken in the form

$$z_1 = \sqrt{\gamma/2}R\left(f_0 e^{-i\varphi_0} + g_0 e^{i\varphi_0}\right)\,, \tag{3.331}$$

where the classical quantities R and φ_0 were introduced in (3.166), then, with the account of (3.176), the mean values $\bar{P}_1$ and $\bar{P}_2$ coincide with the classical expressions (3.168). It should be stressed that the mean values $\bar{P}_1$ and $\bar{P}_2$ do not depend on quantum numbers n and s.

The mean values of the coordinates x^1, x^2 can be found from (3.319). They also do not depend on n and s and coincide with their classical expressions (3.166), if

$$\begin{aligned}
(\alpha^* - \beta^*)\,z_2 + (\alpha - \beta)\,z_2^* &= \sqrt{2\gamma}x^1_{(0)},\\
i\,[(\alpha + \beta)z_2^* - (\alpha^* + \beta^*)\,z_2] &= \sqrt{2\gamma}x^2_{(0)}\,.
\end{aligned} \tag{3.332}$$

Setting z_2 in the form

$$z_2 = \sqrt{\gamma/2}\left[(\alpha+\beta)x^1_{(0)} + i(\alpha-\beta)x^2_{(0)}\right], \tag{3.333}$$

we can satisfy (3.332) for any α and β.

Quadratic dispersions calculated in the states (3.323) are

$$\begin{aligned}
\overline{(\Delta P_1)^2} &= \gamma|f+g|^2(n+1/2), \quad \overline{(\Delta P_2)^2} = \gamma|f-g|^2(n+1/2),\\
\gamma\overline{(\Delta x^1)^2} &= |f-g|^2(n+1/2) + |\alpha-\beta|^2(s+1/2),\\
\gamma\overline{(\Delta x^2)^2} &= |f+g|^2(n+1/2) + |\alpha+\beta|^2(s+1/2),\\
\overline{[(\Delta x^1)(\Delta P_1)+(\Delta P_1)(\Delta x^1)]} &= -\overline{[(\Delta x^2)(\Delta P_2)+(\Delta P_2)(\Delta x^2)]}\\
&= i\left(fg^* - gf^*\right)(2n+1).
\end{aligned} \tag{3.334}$$

These do not depend on z_1 and z_2, but do depend on f_0, g_0, α, β, on the light-cone time u^0, and on quantum numbers n, s. Introducing the quantities

$$J_k = \left(\overline{\Delta x^k}\right)^2\left(\overline{\Delta P_k}\right)^2 - \left[\overline{(\Delta x^k)(\Delta P_k)+(\Delta P_k)(\Delta x^k)}\right]^2/4, \quad k=1,2,$$

one can find from (3.334) that

$$\begin{aligned}
J_1 &= (n+1/2)[n+1/2+|(\alpha-\beta)(f+g)|^2(s+1/2)],\\
J_2 &= (n+1/2)[n+1/2+|(\alpha+\beta)(f-g)|^2(s+1/2)].
\end{aligned} \tag{3.335}$$

By an appropriate choice of parameters f_0, g_0, α, β, we can construct a state (at a fixed u^0) with given dispersions $\overline{(\Delta x^k)^2}$ or $\overline{(\Delta P_k)^2}$. Such states are called squeezed states. The evolution of the dispersions in the time u^0 is controlled by equations (3.334).

Similar results hold for the Dirac wave functions, if we use the corresponding inner product on the null-plane (2.69).

3.4.7 Zero magnetic field limit

It is well known that usual stationary sates in the magnetic field do not admit a direct limiting transition to zero magnetic field $\gamma \to 0$. However, such a limiting transition can be fulfilled in the coherent states considered above. Namely, they pass to free particle solutions as $\gamma \to 0$. Consider below such a limit in the solutions (3.324) of the K–G equation.

First of all, we represent the quantities q and Q_1 as follows:

$$\begin{aligned}
q &= 2i\gamma(\beta f - \alpha g)x^1x^2 - \gamma(\alpha+\beta)(f+g)(x^1)^2 - \gamma(\alpha-\beta)(f-g)(x^2)^2\\
&\quad + 2\sqrt{2\gamma}\,x^1[(\alpha+\beta)z_1 + (f+g)z_2] + 2i\sqrt{2\gamma}\,x^2[(\alpha-\beta)z_1 - (f-g)z_2],\\
Q_1 &= \bar{Q}_1 + Q_1^0, \quad Q_1^0 = Q_1(f=f_0, g=g_0).
\end{aligned} \tag{3.336}$$

One can easily see that

$$\bar{Q}_1 = Q_1 - Q_1^0 = 2i(\alpha f_0 - \beta g_0)^{-1}(\alpha z_1 + g_0 z_2)(\beta z_1 + f_0 z_2)\sin\tilde{\omega}u^0 \\ - 2z_1 z_2(1 - \cos\tilde{\omega}u^0)\,.$$

The quantity Q_1^0 is a constant, thus it has to be included in the normalization factor in (3.324). Now we replace $\sqrt{2\gamma}z_1$ and $\sqrt{2\gamma}z_2$ in all the expressions as follows:

$$\sqrt{2\gamma}z_1 = -(f_0 + g_0)p_2 + i(f_0 - g_0)p_1,$$
$$\sqrt{2\gamma}z_2 = (\alpha + \beta)p_2 - i(\alpha - \beta)p_1\,, \tag{3.337}$$

where p_1 and p_2 are arbitrary real numbers. Thus, in fact, we restrict possible values of z_1 and z_2 and the set of states (3.324). Now we can perform the limit $\gamma \to 0$. Taking into account that

$$\lim_{\gamma\to 0}\frac{\sin\tilde{\omega}u^0}{\gamma} = \frac{u^0}{\lambda}, \quad \lim_{\gamma\to 0}\frac{1-\cos\tilde{\omega}u^0}{\gamma} = 0\,, \tag{3.338}$$

and returning to the coordinates x^1, x^2, we obtain

$$\varphi(x) = N\exp(-iS), \quad S = \frac{\lambda^2 + p_1^2 + p_2^2 + m^2}{2\lambda}x^0 \\ + p_1 x^1 + p_2 x^2 + \frac{\lambda^2 - p_1^2 - p_2^2 - m^2}{2\lambda}x^3\,. \tag{3.339}$$

Now one can see (taking into account (3.305)) that

$$p_0 = \frac{\lambda^2 + p_1^2 + p_2^2 + m^2}{2\lambda}, \quad p_3 = \frac{\lambda^2 - p_1^2 - p_2^2 - m^2}{2\lambda}, \tag{3.340}$$

and the following relations take place:

$$S = p_\mu x^\mu, \quad p_\mu p^\mu = m^2\,. \tag{3.341}$$

The wave functions (3.339) corresponds to the free particle states with given four momenta.

3.4.8 Some other types of nonstationary states

Let us recall that the relativistic quantum problem of particle motion in the constant and uniform magnetic field was reduced in Section 3.4.6 to solving the Schrödinger equation (3.307) with a quadratic Hamiltonian. All integrals of motion of this equation can be constructed as (functional) combinations of the set of the operators

$$f\hat{a}_1, \quad g\hat{a}_1^+, \quad \hat{a}_2, \quad \hat{a}_2^+\,, \tag{3.342}$$

which are also integrals of motion if the functions f and g obey the relations (3.312) and (3.313). Coherent states are eigenfunctions for a linear combinations of these operators. Semicoherent states (3.322) are eigenvectors of the operator-integrals of motion $\hat{N}_1$ and $\hat{N}_2$,

$$\hat{N}_1 = (\hat{A}_1^+ - z_1^*)(\hat{A}_1 - z_1), \quad \hat{N}_2 = (\hat{A}_2^+ - z_2^*)(\hat{A}_2 - z_2), \tag{3.343}$$

where $\hat{A}_1$ and A_2 are given by equations (3.311) and (3.254) respectively.

However, besides operators $\hat{N}_1$, $\hat{N}_2$, $\overset{+}{a_1}a_1$, and $\overset{+}{a_2}a_2$, one can find another quadratic operator integral of motion, which commutes with L_z. Such an operator has the form

$$\bar{\hat{A}} = f\hat{a}_1\hat{a}_2 + g\hat{a}_1^+\hat{a}_2^+ . \tag{3.344}$$

It commutes also with the operator J_z. We are going to consider only the case $|g| \le |f|$, because for $|g| > |f|$ normalized (even to δ-function) eigenvectors for the operator (3.344) do not exist. In the case $|g| \le |f|$, we can, in fact, consider the operator

$$\hat{A}^{(p)} = e^{i\Omega}(\hat{a}_1\hat{a}_2 - \bar{p}^2\hat{a}_1^+\hat{a}_2^+), \quad \bar{p} = p\exp(-i\Omega), \tag{3.345}$$

with Ω defined in (3.176) and p being a real number, $-1 \le p \le 1$. Indeed, the operator (3.344) differs only by a complex factor from (3.345).

Let us consider wave functions $\Phi(u^0, \rho, \varphi)$ which obey equation (3.307) and the following equations:

$$\hat{A}^{(p)}\Phi_{q,l}^p = -q\Phi_{q,l}^p, \quad L_z\Phi_{q,l}^p = l\Phi_{q,l}^p, \quad l \in \mathbb{Z}, \tag{3.346}$$

(for further convenience we have denoted the eigenvalue of the operator $\hat{A}^p$ by $-q$). Such a wave function has the form

$$\begin{aligned}
&\Phi_{q,l}^p(u^0, \rho, \varphi) = N\exp(il\varphi - \Gamma/2)(1+\bar{p})^{-\nu/2}(1-\bar{p})^{-\mu/2}I_{|l|+s,s}(x), \\
&\Gamma = ilu^0\gamma/\lambda + \rho(1+\bar{p}^2)/(1-\bar{p}^2), \quad x = 2\bar{p}\rho/(1-\bar{p}^2), \\
&\nu = 1 - q/p, \quad \mu = 1 + q/p, \quad s = (q/p - |l| - 1)/2,
\end{aligned} \tag{3.347}$$

where $I_{m,n}(x)$ are the Laguerre functions (B.1), see Section B.1. For $p^2 = 1$, the operator (3.345) is anti-self-adjoint, such that $q^* = -q$. For $p = 0$ the corresponding solutions have a simple form

$$\begin{aligned}
&\Phi_{q,l}^0(u^0, \rho, \varphi) = N_0\exp(il\varphi + \bar{q} - \Gamma_0/2)J_{|l|}\left(2\sqrt{\bar{q}\rho}\right), \\
&\Gamma_0 = ilu_0\gamma/\lambda + \rho, \quad \bar{q} = q\exp(-i\Omega),
\end{aligned} \tag{3.348}$$

where $J_\nu(x)$ are the Bessel functions. Using equation (B.35), one can show that these solutions can be obtained from the functions (3.347) in the limit $p \to 0$.

The functions (3.347) and (3.348) are orthogonal with respect to the quantum number l, and are not orthogonal with respect to q,

$$(\Phi^p_{q',l'}, \Phi^p_{q,l}) = \delta_{l,l'} Q F(-s, -s'^*; 1 + |l|; y), \quad y = \left(\frac{2p}{1+p^2}\right)^2 ,$$

$$Q = \left[\frac{\Gamma(1+|l|+s)\Gamma(1+|l|+s'^*)}{\Gamma(1+s)\Gamma(1+s'^*)}\right]^{1/2} \frac{\pi N N'^* y^{(1+|l|)/2}(1-y)^{-(q+q'^*)/4p}}{p\Gamma(1+|l|)} ,$$

$$(\Phi^0_{q',l'}, \Phi^0_{q,l}) = \delta_{l,l'} 2\pi N_0 N_0'^* I_{|l|}\left(2\sqrt{qq'^*}\right) . \tag{3.349}$$

Here $F(\alpha, \beta; \gamma; x)$ is the hypergeometric function, and $I_\alpha(x)$ is the Bessel function of an imaginary argument. In deriving (3.349) we have used equation (B.70).

The set of functions (3.347) is complete for $p| = 1$ and overcomplete for $|p| < 1$.

In spite of the fact that the states (3.347) are not coherent ones, they have quite similar properties. Indeed, as it follows from (3.171) and (3.176) the following classical equations hold:

$$\rho = \rho(u^0) = \sqrt{L_z^2 + 4|a_1 a_2|^2} - a_1 a_2 - \overset{+}{a}_1 \overset{+}{a}_2 ,$$

$$a_1 = \sqrt{\gamma/2}\, R \exp(-i\Omega), \quad a_2 = \sqrt{\gamma/2}\,(x^1_{(0)} + i x^2_{(0)}) . \tag{3.350}$$

It follows from (3.345) and (3.346) for $p = 0$ that $a_1 a_2 = -\bar{q}, \quad L_z = l$. This allows one to rewrite (3.350) in the form

$$\rho(u^0) = \rho_0^{\text{cl}} + \bar{q} + \bar{q}^*, \quad \rho_0^{\text{cl}} = \sqrt{l^2 + 4|q|^2} . \tag{3.351}$$

Calculating the mean value of ρ in the states (3.348), we find

$$\bar{\rho} = \rho_0 + \bar{q} + \bar{q}^*, \quad \rho_0 = |l| - 1 - 2|q| I_{|l|-1}(2|q|) I^{-1}_{|l|}(2|q|) . \tag{3.352}$$

Thus, the time dependence of $\bar{\rho}$ (which is determined by $\bar{q}$) has the classical form. Only the constant ρ_0 can differ from its classical value.

3.5 Particles in spherically symmetric fields

3.5.1 General

The problem of a charged particle moving in a central field is among the most important problems in quantum mechanics. In the particular case, when the field is the Coulomb field of a point like charge, this is the problem of the hydrogen atom.

In relativistic quantum mechanics such a problem is completely solved only for the Coulomb field [116, 120, 188, 247, 291, 304, 339], for the field of a magnetic monopole [196], for the combination of the Coulomb field and the field of a magnetic monopole [199, 200, 321], and for the combination of the latter fields with the Aharonov–Bohm (AB) field of an infinitesimally thin and infinitely long solenoid [118, 119, 203, 295, 312, 334]. No other combinations that include spherically symmetric fields and allow exact solutions in an explicit form have yet been found. Below we study all the above cases.

In this Section, we use the spherical coordinates r, φ, and θ,

$$x^1 = r\sin\theta\cos\varphi, \quad x^2 = r\sin\theta\sin\varphi, \quad x^3 = r\cos\theta\,,$$
$$0 \le r < \infty, \quad 0 \le \varphi < 2\pi, \quad 0 \le \theta \le \pi\,. \tag{3.353}$$

In this coordinate system, we specify the components of the electromagnetic potentials that describe all the above-mentioned configurations as follows:

$$eA_0 = -A(r), \quad A_1 = A_3 = 0, \quad eA_2 = -B(\theta)\,. \tag{3.354}$$

Here, $A(r)$ and $B(\theta)$ are arbitrary functions of their arguments, which are assumed to be smooth (or piecewise-smooth). The vector-potential $\mathbf{A}$ has the form

$$e\mathbf{A} = A_\varphi \mathbf{e}_\varphi, \quad A_\varphi = \frac{B(\theta)}{r\sin\theta}\,, \tag{3.355}$$

where $\mathbf{e}_\varphi = -\mathbf{i}\sin\varphi + \mathbf{j}\cos\varphi$ is the unit basis vector corresponding to the φ-axis of the spherical reference frame. It follows from (3.355) that for $\theta = 0, \pi$ the function $B(\theta)$ cannot be arbitrary.

The fields $\mathbf{E}$ and $\mathbf{H}$ that correspond to the potentials (3.354) are

$$e\mathbf{E} = E\mathbf{e}_r, \quad e\mathbf{H} = H\mathbf{e}_r, \quad E = A'(r), \quad H = \frac{B'(\theta)}{r^2\sin\theta}\,, \tag{3.356}$$

where $\mathbf{e}_r = \mathbf{r}/r$ is the unit basis vector corresponding to the r-axis of the spherical reference frame. The primes stand for derivatives with respect to the corresponding variables.

Some remarks should be made regarding the expressions (3.355) and (3.356). It is easy to see that the case $B(\theta) = \text{const}$ corresponds to the AB field (see equations (3.465) and (3.467)), although a formal differentiation in equation (3.356) produces a vanishing field. Calculating the flux Φ of the magnetic field through the sphere with the radius R and the center being the point of origin, one finds

$$\frac{e\Phi}{2\pi} = \oint_{r=R} \mathbf{H}ds = B(\theta = \pi) - B(\theta = 0)\,. \tag{3.357}$$

On the other hand, the flux of the magnetic field through any closed surface (in the Maxwell electrodynamics) must be zero. This implies the condition for $B(\theta)$

$$B(\theta = 0) = B(\theta = \pi)\,. \tag{3.358}$$

If, for a specific choice of $B(\theta)$, this condition does not hold (we shall encounter this case later on), its fulfillment can be provided, for instance, by making a transition to the function

$$B_\alpha(\theta) = \frac{B(\theta)\sin\theta}{\sin\theta + \alpha^2}, \quad B_\alpha(0) = B_\alpha(\pi) = 0, \quad \alpha = \text{const} \neq 0\,. \tag{3.359}$$

The function $B_\alpha(\theta)$ satisfies conditions (3.358) for any bounded function $B(\theta)$, and as $\alpha \to 0$, we have

$$\lim_{\alpha\to 0} B_\alpha(\theta) = B(\theta), \quad \theta \neq 0, \pi\,. \tag{3.360}$$

The function $B_\alpha(\theta)$ gives rise to a magnetic field with zero magnetic flux through any closed surface. For $\theta \neq 0\,, \pi$, this field tends to (3.356) as $\alpha \to 0$. For $\theta = 0, \pi$ (i.e. on the x^3-axis), one can encounter peculiarities, as in the AB field case.

The Lorentz equations of motion (2.18) with the fields (3.356), as being written in the coordinates (3.353), have the form

$$\begin{aligned}
m\ddot{x}^0 - A'(r)\dot{r} &= 0\,,\\
m\left(\ddot{r} - r\dot{\varphi}^2\sin^2\theta - r\dot{\theta}^2\right) - A'(r)\dot{x}^0 &= 0\,,\\
mr(\ddot{\varphi}r\sin\theta + 2\dot{r}\dot{\varphi}\sin\theta + 2r\dot{\varphi}\dot{\theta}\cos\theta)\sin\theta + B'(\theta)\dot{\theta} &= 0\,,\\
mr\left(\ddot{\theta}r + 2\dot{r}\dot{\theta} - r\dot{\varphi}^2\sin\theta\cos\theta\right) - B'(\theta)\dot{\varphi} &= 0\,.
\end{aligned} \tag{3.361}$$

We remind that derivatives with respect to the proper time are denoted by dots above.

One easily finds three first integrals of motion k_s, $s = 0, 1, 2$ for equations (3.361) and the relation between them,

$$\begin{aligned}
&m\dot{x}^0 - P_0 = 0, \quad P_0 = k_0 + A(r)\,,\\
&mr^2\dot{\varphi}\sin^2\theta + P_2 = 0, \quad P_2 = k_2 + B(\theta)\,,\\
&m^2r^4\dot{\theta}^2 + P_2^2\sin^{-2}\theta - k_1^2 = 0\,,\\
&m^2\dot{r}^2 + m^2 + k_1^2r^{-2} - P_0^2 = 0\,.
\end{aligned} \tag{3.362}$$

The physical meaning of these integrals is clear: k_0 is the total energy of the particle and

$$k_2 = -\left[\mathbf{r}\times\mathbf{P}\right]_z = -L_z, \quad k_1^2 = \left[\mathbf{r}\times\mathbf{P}\right]^2\,. \tag{3.363}$$

The solution of equations (3.362) with arbitrary functions $A(r)$ and $B(\theta)$ is reduced to the calculation of quadratures

$$\begin{aligned}
&x^0 = \int \frac{P_0}{\sqrt{\chi(r)}}dr, \qquad \chi(r) = P_0^2 - m^2 - k_1^2r^{-2}\,,\\
&\varphi = -\int \frac{P_2 d\theta}{\sqrt{\pi(\theta)}\sin^2\theta}, \quad \pi(\theta) = k_1^2 - P_2^2\sin^{-2}\theta\,,\\
&\int \frac{d\theta}{\sqrt{\pi(\theta)}} = \int \frac{dr}{r^2\sqrt{\chi(r)}}, \qquad \tau = m\int \frac{dr}{\sqrt{\chi(r)}}\,.
\end{aligned} \tag{3.364}$$

It is obvious that the classically admissible domains of the variables r and θ are determined by $\chi(r) \geq 0, \pi(\theta) \geq 0$.

Equations (3.362) and (3.364) imply that the radial motion is determined only by the electric field (i.e. by the component A_0 of the potential). This fact is obvious, since the magnetic field does not affect the motion along its direction. Thus, the radial and angular motions can be analyzed separately.

In the case under consideration, the solution of the Hamilton–Jacobi equation (2.27) has the form

$$S = -k_0 x^0 - \varphi k_2 + \int \sqrt{\chi(r)} dr + \int \sqrt{\pi(\theta)} d\theta \,. \tag{3.365}$$

3.5.2 Separation of variables in K–G and Dirac equations

The scalar wave function (a solution of the K–G equation) corresponding to the given integrals of motion k_s, $s = 0, 1, 2$, we represent as

$$\Phi = \exp[-i(k_0 x^0 + k_2 \varphi)] f(r) g(\theta) \,, \tag{3.366}$$

where k_2 is an integer, $k_2 \in \mathbb{Z}$, and functions $f(r)$ and $g(\theta)$ are solutions of a set of ordinary differential equations,

$$f''(r) + 2r^{-1} f'(r) + \chi(r) f(r) = 0 \,, \tag{3.367}$$

$$g''(\theta) + g'(\theta) \cot\theta + \pi(\theta) g(\theta) = 0 \,. \tag{3.368}$$

It follows from equations (3.367) and (3.368) that, as in the classical case, the radial and angular motions can also be analyzed separately.

Let us construct solutions of the Dirac equation with the fields under consideration.

The stationary Dirac wave function can be chosen as an eigenvector of the operator $\hat{J}_z$ (the projection of the total angular momentum to the z-axis):

$$\hat{J}_z \Psi = -\left(k_2 + \frac{1}{2}\right)\Psi, \quad \hat{J}_z = [\mathbf{r} \times \hat{\mathbf{p}}]_z + \frac{1}{2}\Sigma_3 \,. \tag{3.369}$$

As follows from equation (2.169), the operator $\hat{N}^0$ (2.162) is a spin integral of motion for the fields (3.356), and we shall consider its eigenvectors:

$$\hat{N}^0 \Psi = \zeta n_0 \Psi, \quad \hat{N}^0 = \rho_3 \{1 + (\boldsymbol{\Sigma}[\mathbf{r} \times \hat{\mathbf{P}}])\} \,. \tag{3.370}$$

Here the quantum number $\zeta = \pm 1$ describes the spin orientation, while n_0 is a subject for determination. Owing to the properties (2.163), n_0^2 is an analog to k_1^2.

To find the solution of the Dirac equation we use the Ansatz

$$\Psi = \exp(-ik_0 x^0) \begin{pmatrix} if_1(r) \\ (\boldsymbol{\sigma}\mathbf{e}_r) f_2(r) \end{pmatrix} G(\varphi, \theta) \,, \tag{3.371}$$

where the bispinor $G(\varphi, \theta)$ has the form

$$G(\varphi, \theta) = \begin{pmatrix} \exp[-i(k_2 + 1)\varphi]g_1(\theta) \\ \exp(-ik_2\varphi)g_2(\theta) \end{pmatrix} . \tag{3.372}$$

Substituting (3.371) into the Dirac equation with the field (3.356), we obtain

$$\begin{aligned} &i(P_0 - m)f_1 G - (\boldsymbol{\sigma}\hat{\mathbf{P}})\frac{(\boldsymbol{\sigma}\mathbf{r})}{r} f_2 G = 0 , \\ &i(\boldsymbol{\sigma}\hat{\mathbf{P}})f_1 G - (P_0 + m)\frac{(\boldsymbol{\sigma}\mathbf{r})}{r} f_2 G = 0 . \end{aligned} \tag{3.373}$$

Upon the substitution of (3.371), equation (3.370) takes the form

$$\begin{aligned} &(1 + \hat{R})f_1 G - \zeta n_0 f_1 G = 0, \quad \hat{R} = (\boldsymbol{\sigma}[\mathbf{r} \times \hat{\mathbf{P}}]) , \\ &(1 + \hat{R})\frac{(\boldsymbol{\sigma}\mathbf{r})}{r} f_2 G + \zeta n_0 \frac{(\boldsymbol{\sigma}\mathbf{r})}{r} f_2 G = 0 . \end{aligned} \tag{3.374}$$

It is straightforward to check the following relations:

$$\begin{aligned} &[\mathbf{r} \times \hat{\mathbf{P}}]\frac{(\boldsymbol{\sigma}\mathbf{r})}{r}6 = i\frac{[\boldsymbol{\sigma} \times \mathbf{r}]}{r} + \frac{(\boldsymbol{\sigma}\mathbf{r})}{r}[\mathbf{r} \times \hat{\mathbf{P}}] , \\ &\hat{\mathbf{P}}\frac{(\boldsymbol{\sigma}\mathbf{r})}{r} = i\frac{(\boldsymbol{\sigma}\mathbf{r})\mathbf{r}}{r^3} - i\frac{\boldsymbol{\sigma}}{r} + \frac{(\boldsymbol{\sigma}\mathbf{r})}{r}\hat{\mathbf{P}} . \end{aligned} \tag{3.375}$$

Hence, with allowance for the properties of the Pauli matrices, the following identities hold true:

$$\begin{aligned} &\hat{R}\frac{(\boldsymbol{\sigma}\mathbf{r})}{r} = -2\frac{(\boldsymbol{\sigma}\mathbf{r})}{r} - i\frac{(\boldsymbol{\sigma}\mathbf{r})}{r}(\mathbf{r}\hat{\mathbf{P}}) + ir(\boldsymbol{\sigma}\hat{\mathbf{P}}) , \\ &\frac{(\boldsymbol{\sigma}\mathbf{r})}{r}\hat{R} = i\frac{(\boldsymbol{\sigma}\mathbf{r})}{r}(\mathbf{r}\hat{\mathbf{P}}) - ir(\boldsymbol{\sigma}\hat{\mathbf{P}}) , \end{aligned}$$

$$(1 + \hat{R})\frac{(\boldsymbol{\sigma}\mathbf{r})}{r} = -\frac{(\boldsymbol{\sigma}\mathbf{r})}{r}(1 + \hat{R}) , \tag{3.376}$$

$$(\boldsymbol{\sigma}\hat{\mathbf{P}})\frac{(\boldsymbol{\sigma}\mathbf{r})}{r} = -\frac{2i}{r} + \frac{1}{r}(\mathbf{r}\hat{\mathbf{P}}) - \frac{i}{r}\hat{R} , \tag{3.377}$$

$$\frac{(\boldsymbol{\sigma}\mathbf{r})}{r}(\boldsymbol{\sigma}\hat{\mathbf{P}}) = \frac{1}{r}(\mathbf{r}\hat{\mathbf{P}}) + \frac{i}{r}\hat{R} . \tag{3.378}$$

One easily finds the explicit form of the operators $\hat{R}$ and $(\mathbf{r}\hat{\mathbf{P}})$ for the potentials (3.354),

$$\begin{aligned} 2\hat{R} = &(\sigma_1 + i\sigma_2)e^{-i\varphi}(\hat{P}_2 \cot\theta - \partial_\theta) + (\sigma_1 - i\sigma_2)e^{i\varphi}(\hat{P}_2 \cot\theta + \partial_\theta) \\ &- 2\sigma_3\hat{P}_2, \quad \hat{P}_2 = i\partial_\varphi + B(\theta); \quad (\mathbf{r}\hat{\mathbf{P}}) = -ir\partial_r . \end{aligned} \tag{3.379}$$

This implies the commutation properties

$$\hat{R}f_k(r) = f_k(r)\hat{R}, \quad k = 1, 2; \quad (\mathbf{r}\hat{\mathbf{P}})G = G(\mathbf{r}\hat{\mathbf{P}}) . \tag{3.380}$$

Hence, the set of equations (3.374), with allowance for (3.376), is equivalent to the following equation for the spinor G:

$$\hat{R}G = (\zeta n_0 - 1)G\,. \tag{3.381}$$

Substituting (3.372) into equation (3.381), and taking into account the explicit form (3.379) of the operator $\hat{R}$, we find the following set of ordinary first-order differential equations for the functions $g_1(\theta)$ and $g_2(\theta)$:

$$\begin{aligned} &g_1'(\theta) + (1 + P_2)g_1(\theta)\cot\theta + (1 + P_2 - \zeta n_0)g_2(\theta) = 0\,,\\ &\qquad g_2'(\theta) - P_2 g_2(\theta)\cot\theta + (P_2 + \zeta n_0)g_1(\theta) = 0, \quad P_2 = k_2 + B(\theta)\,. \end{aligned} \tag{3.382}$$

The system (3.382) can be rewritten in the matrix form

$$\begin{aligned} &\tilde{g}'(\theta) + \left[\left(\frac{1+\sigma_3}{2} + P_2\sigma_3\right)\cot\theta + \left(\tfrac{1}{2} + P_2\right)\sigma_1 + i\left(\tfrac{1}{2} - \zeta n_0\right)\sigma_2\right]\tilde{g}(\theta) - 0,\\ &\qquad \tilde{g}(\theta) = \begin{pmatrix} g_1(\theta) \\ g_2(\theta) \end{pmatrix}. \end{aligned} \tag{3.383}$$

The Dirac equation (3.373), with allowance for (3.377), (3.378) and (3.381), is now equivalent to a set of two ordinary differential equations for the functions $f_1(r)$ and $f_2(r)$:

$$\begin{aligned} &f_1'(r) + \frac{1 - \zeta n_0}{r} f_1(r) - (P_0 + m)f_2(r) = 0, \quad P_0 = k_0 + A_0(r)\,.\\ &f_2'(r) + \frac{1 + \zeta n_0}{r} f_2(r) + (P_0 - m)f_1(r) = 0\,. \end{aligned} \tag{3.384}$$

This set can be also rewritten in the matrix form

$$\tilde{f}'(r) + \left(\frac{1 - \zeta n_0\sigma_3}{r} - m\sigma_1 - iP_0\sigma_2\right)\tilde{f}(r) = 0, \quad \tilde{f}(r) = \begin{pmatrix} f_1(r) \\ f_2(r) \end{pmatrix}. \tag{3.385}$$

Thus, the Dirac equation has been reduced to independent sets of fist-order differential equations (3.383) and (3.385), describing the angular and radial motions of the Dirac particle (analogous to equations (3.367) and (3.368) for the scalar particle).

3.5.3 Specification of potentials and complete classical solution

Explicit solutions of relativistic wave equations with spherically symmetric fields are known only in the case

$$A(r) = -\frac{q_1}{r}, \quad B(\theta) = -q_2\cos\theta - F\,, \tag{3.386}$$

where q_1, q_2 and F are dimensionless constants.

The field corresponding to the potential (3.386) is a combination of the Coulomb electric field of a point-like charge q_1 with the field of a magnetic monopole with the coupling constant q_2 (with Dirac's threads on the z-axis) and with the AB field along the z-axis with magnetic flux $F\Phi_0$, where Φ_0 is the *quantum of magnetic flux* (3.474). In quantum theory we shall assume, in accordance with (3.473), that

$$F = l_0 + \mu, \quad 0 \leq \mu < 1\,, \tag{3.387}$$

where l_0 is an integer, $l_0 \in \mathbb{Z}$, and μ is the (nonnegative) fractional part of magnetic flux. We choose k_2 in the form

$$k_2 = l_z - l_0\,, \tag{3.388}$$

and introduce the notation

$$\bar{k}_2 = k_2 + F = l_z + \mu, \quad l_z \in \mathbb{Z}\,. \tag{3.389}$$

A simple calculation shows that, in classical theory, for the fields (3.386), there exists the conserved vector

$$\boldsymbol{\Upsilon} = [\mathbf{r} \times \mathbf{P}] - q_2\frac{\mathbf{r}}{r}\,. \tag{3.390}$$

In particular, equation (3.390) implies

$$\Upsilon_z = -\bar{k}_2, \quad \Upsilon^2 = k_1^2 + q_2^2\,. \tag{3.391}$$

Classical motion has the simplest form in the coordinate system, where the constant vector Υ has a single component, for instance, the z-component, $\Upsilon = (0, 0, -\bar{k}_2)$. In this coordinate system, the relations

$$q_2 = \bar{k}_2 \cos\theta, \quad k_1^2 = \bar{k}_2^2 \sin^2\theta, \quad P_2 = \bar{k}_2 \sin^2\theta, \tag{3.392}$$

hold. It follows from (3.392) that classical trajectories belong to the conical surface $\theta = \theta_0 = \text{const}$, and the classical first integrals of motion (3.362), with allowance for (3.392), can be written as

$$m\dot{x}^0 = P_0 = k_0 - \frac{q_1}{r}, \quad mr^2\dot{\varphi} + \bar{k}_2 = 0, \quad m^2\dot{r}^2 = \chi(r)\,, \tag{3.393}$$

where,

$$\chi(r) = k_0^2 - m^2 - \frac{2k_0 q_1}{r} + \frac{q_1^2 - k_1^2}{r^2} \geq 0\,. \tag{3.394}$$

It is remarkable that equations (3.393) do not contain any characteristics of the magnetic field.

If $q_1 > 0$ (repulsion), the motion is possible only for $k_0^2 > m^2$, and is not restricted to any finite domain of r (scattering). If $q_1 < 0$ (attraction), for $k_0^2 \geq m^2$ there is either scattering or a collapse to the central point; however, in this case the motion is also possible for $k_0^2 < m^2$, being restricted to a finite domain of r (a possible collapse to the central point). In all the cases, the following inequality holds true:

$$m^2 q_1^2 + k_1^2(k_0^2 - m^2) \geq 0 \Longrightarrow a = [m^2 q_1^2 + k_1^2(k_0^2 - m^2)]^{1/2} \geq 0\,. \tag{3.395}$$

Classical motion is most easily described in parametric form (t is the parameter). If $k_0^2 \neq m^2$, the following relations take place:

$$\begin{aligned} r(t) &= \frac{r_0(t)}{k_0^2 - m^2}, \quad \tau(t) = \frac{m\tau_0(t)}{|k_0^2 - m^2|^{3/2}}\,, \\ x^0(t) &= \frac{k_0}{m}\tau(t) - \frac{q_1 t}{\sqrt{|k_0^2 - m^2|}}\,. \end{aligned} \tag{3.396}$$

If $k_0^2 > m^2$, one should set

$$r_0(t) = a\,\cosh t + b, \quad \tau_0(t) = a\,\sinh t + bt, \quad b = k_0 q_1\,. \tag{3.397}$$

For $b < 0$, it is possible that $k_0^2 < m^2$. In this case

$$r_0(t) = -a\cos t - b, \quad \tau_0(t) = -a\sin t - bt\,. \tag{3.398}$$

The function $\varphi(t)$ has the following form:

For $k_0^2 > m^2$,

$$\begin{aligned} \varphi(t) &= 2c\,\arctan\psi_1(t), \quad k_1^2 > q_1^2; \qquad c = -\frac{\bar{k}_2}{\sqrt{|k_1^2 - q_1^2|}}\,, \\ \varphi(t) &= c\,\ln\left|\frac{1+\psi_1(t)}{1-\psi_1(t)}\right|, \quad k_1^2 < q_1^2; \quad \psi_1(t) = \sqrt{\left|\frac{b-a}{b+a}\right|}\tanh\frac{t}{2}\,, \\ \varphi(t) &= d\frac{\cosh t - b/|b|}{\sinh t}, \quad k_1^2 = q_1^2; \qquad d = -\frac{\bar{k}_2}{b}\sqrt{|k_0^2 - m^2|}\,. \end{aligned} \tag{3.399}$$

For $k_0^2 < m^2 (b < 0)$,

$$\varphi(t) = c\left[t - 2\arctan\left(\frac{a\sin t}{b - \sqrt{b^2 - a^2} - a\cos t}\right)\right], \quad k_1^2 > q_1^2\,. \tag{3.400}$$

This case corresponds to a motion in a restricted domain of r without collapsing to the central point. In this case the inequality $m^2 q_1^2 > k_0^2 k_1^2$ holds.

In the two other cases the particle collapses to the central point:

$$\begin{aligned} \varphi(t) &= -c\,\ln\left|\frac{1+\psi_2(t)}{1-\psi_2(t)}\right|, \quad \psi_2(t) = \sqrt{\frac{a-b}{a+b}}\tan\frac{t}{2}, \quad k_1^2 < q_1^2\,; \\ \varphi(t) &= d\,\cot\frac{t}{2}, \quad k_1^2 = q_1^2\,. \end{aligned} \tag{3.401}$$

Finally, if $q_1 < 0$ ($b < 0$), a motion is also possible for $k_0 = m$ (scattering or collapse to the central point). This motion is described by the following parametric expressions:

$$
\begin{aligned}
&r(t) = \frac{t^2 + k_1^2 - q_1^2}{2m|q_1|}, \qquad \tau(t) = \frac{t^3 + 3(k_1^2 - q_1^2)t}{6mq_1^2}, \qquad mx^0(t) = m\tau(t) + t\,;\\
&\varphi(t) = 2c \arctan \psi_3(t), \qquad k_1^2 > q_1^2; \quad \psi_3(t) = \frac{t}{\sqrt{|k_1^2 - q_1^2|}},\\
&\varphi(t) = -c \ln\left|\frac{1 + \psi_3(t)}{1 - \psi_3(t)}\right|, \qquad k_1^2 < q_1^2,\\
&\varphi(t) = 2\bar{k}_2 t^{-1}, \qquad k_1^2 = q_1^2\,.
\end{aligned}
\tag{3.402}
$$

Note that the collapse to the central point in all the cases occupies a finite interval of both proper and laboratory time. Equations (3.396)–(3.402) describe the complete solution of the classical problem for the fields (3.386).

3.5.4 Azimuthal motion

Let us study exact solutions of equations (3.368) and (3.382) for the azimuthal motion with the potentials (3.386). It is convenient to use the variable

$$
x = \cos\theta, \quad -1 \le x \le 1\,. \tag{3.403}
$$

Note that in this Section, we consider only the segment $x \in [-1, 1]$. Equation (3.368) for a scalar particle with the potential (3.386) reads

$$
\begin{aligned}
&(1 - x^2)g''(x) - 2xg'(x) + \pi(x)g(x) = 0\,,\\
&\pi(x) = k_1^2 + q_2^2 - \frac{\alpha^2}{2(1 - x)} - \frac{\beta^2}{2(1 + x)}\,,\\
&\alpha^2 = (\bar{k}_2 - q_2)^2, \quad \beta^2 = (\bar{k}_2 + q_2)^2, \quad \bar{k}_2 = l_z + \mu, \quad l_z \in \mathbb{Z}\,.
\end{aligned}
\tag{3.404}
$$

For $q_2 = \mu = 0$ (free azimuthal motion), solutions of equation (3.404), which are square-integrable on the segment $x \in [-1, 1]$, are given by the associated Legendre functions (see [191, equation 8.810])

$$
g(x) = P_l^{|l_z|}(x), \quad |l_z| \le l, \quad l \in \mathbb{Z}_+\,. \tag{3.405}
$$

The quantity k_1^2 is quantized:

$$
k_1^2 = l(l + 1)\,. \tag{3.406}
$$

The functions (3.405) are bounded on $x \in [-1, 1]$ and form an orthogonal with respect to l system.

If at least one of the quantities q_2, μ is nonzero, then every integrable solution of equation (3.404) has the form

$$g(x) = (1-x)^{\frac{\alpha}{2}}(1+x)^{\frac{\beta}{2}}P_n^{(\alpha,\beta)}(x), \quad n \in \mathbb{Z}_+ . \tag{3.407}$$

Here, $P_n^{(\alpha,\beta)}(x)$ are Jacobi polynomials (see [191, equation 8.960]).

If

$$\alpha, \beta > -1 , \tag{3.408}$$

the functions (3.407) are integrable and form an orthogonal system with respect to n.

The quantity k_1^2 is quantized,

$$\begin{aligned} k_1^2 &= n^2 + n(\alpha+\beta+1) - q_2^2 + \frac{(\alpha+\beta)(\alpha+\beta+2)}{4} \\ &= \left(n + \frac{\alpha+\beta+1}{2}\right)^2 - q_2^2 - \frac{1}{4} . \end{aligned} \tag{3.409}$$

In accordance with (3.404), there always exists a choice of nonnegative α and β,

$$\alpha = |\bar{k}_2 - q_2|, \quad \beta = |\bar{k}_2 + q_2| , \tag{3.410}$$

which ensures that (3.408) holds. In this case, the functions (3.407) are bounded and equation (3.409) implies that

$$k_1^2 \geq |q_2| , \tag{3.411}$$

where the equality takes place when $n = 0$ and $|q_2| \geq |\bar{k}_2|$. If at least one of the numbers α and β is negative, but (3.408) is fulfilled, then the functions (3.407) are unbounded,

$$-1 < \alpha < 0, \quad g(x=1) = \infty , \tag{3.412}$$

$$-1 < \beta < 0, \quad g(x=-1) = \infty , \tag{3.413}$$

but still integrable. As far as inequalities (3.408) are more restrictive, one can easily see that for $\mu = 0$ (q_2 is an integer) and for $\mu = 1/2$ (q_2 is a half-integer) only nonnegative α and β are possible (3.410). Inequality (3.412) takes place only if $q_2 - \mu$ is not an integer. Indeed, in this case we have

$$q_2 - \mu = \bar{l}_0 + \nu, \quad 0 < \nu < 1 , \tag{3.414}$$

where $\bar{l}_0$ is an integer. Then there exist two (and only two) values l_z

$$l_z = \bar{l}_0, \quad l_z = \bar{l}_0 + 1 , \tag{3.415}$$

that allow one to choose negative α,

$$\alpha = -\nu, \quad \alpha = \nu - 1 , \tag{3.416}$$

which obey the condition (3.412).

It follows from (3.409) that at least for one of the two α's from (3.416) we have

$$k_1^2(n=0) < 0, \quad q_2 \neq 0\,, \tag{3.417}$$

which means that the K–G operator is not self-adjoint.

Analogously, the realization (3.413) is possible only if $q_2 + \mu$ is not an integer. In this case

$$q_2 + \mu = \tilde{l}_0 + \tilde{\nu}, \quad 0 < \tilde{\nu} < 1\,, \tag{3.418}$$

where $\tilde{l}_0$ is an integer (it is easy to see that $\bar{l}_0 \leq \tilde{l}_0 \leq \bar{l}_0 + 2$). There exist two (and only two) values of l_z

$$l_z = -\tilde{l}_0, \quad l_z = -\tilde{l}_0 - 1\,, \tag{3.419}$$

that admit negative values of β,

$$\beta = -\tilde{\nu}, \quad \beta = \tilde{\nu} - 1\,, \tag{3.420}$$

obeying inequality (3.413). Here, (3.417) holds for at least one of the values (3.420).

Finally, the simultaneous fulfillment of both (3.412) and (3.413) is possible only in two cases:

$$l_z = 0, \quad |q_2| < 1 - \mu; \quad l_z = -1, \quad |q_2| < \mu\,. \tag{3.421}$$

Both values of l_z in (3.421) are simultaneously admissible for

$$|q_2| < \min(\mu, 1-\mu)\,. \tag{3.422}$$

In the case

$$\min(\mu, 1-\mu) \leq |q_2| < \max(\mu, 1-\mu)\,, \tag{3.423}$$

one of l_z from (3.421) is possible. For

$$\max(\mu, 1-\mu) \leq |q_2|\,,$$

inequalities (3.412) and (3.413) cannot be realized simultaneously. It follows from (3.409) that if both inequalities (3.412) and (3.413) hold then (3.417) holds as well. We note that in the case (3.408) and for $n \geq 1$ equation (3.409) always implies $k_1^2 > 0$, although the wave functions (3.407) can have integrable singularities. We have the same situation in the Aharonov–Bohm field combined with the magnetic field (in the magnetic-solenoid field).

Finishing the analysis of the azimuthal motion of a scalar particle, we note that for $q_2 = 0$ and a nonzero μ there exist two (and only two) normalized states

$$l_z = 0, \quad g_0(x) = \sqrt{\frac{\Gamma(3/2-\mu)}{\sqrt{\pi}\Gamma(1-\mu)}}(1-x^2)^{-\frac{\mu}{2}}, \quad \alpha = \beta = -\mu\,;$$

$$l_z = -1, \quad g_{-1}(x) = \sqrt{\frac{\Gamma(1-\mu)(1/2+\mu)\sin\mu\pi}{\pi\sqrt{\pi}}}(1-x^2)^{-\frac{1-\mu}{2}}, \quad \alpha = \beta = \mu - 1\,;$$

$$\int_{-1}^{1} g_s^2(x)dx = 1, \quad s = 0, -1\,, \tag{3.424}$$

corresponding to negative values of k_1^2 (coinciding for both the states) $k_1^2 = -\mu(1-\mu)$. Both functions (3.424) tend to ∞ as $x \to \pm 1$. The function $g_0(x)$ for $l = 0$ is reduced to (3.405) as $\mu \to 0$. The function $g_{-1}(x)$ vanishes as $\mu \to 0$.

The set of azimuthal Dirac equations (3.382) with the potential (3.386) has the form

$$\begin{aligned}(1-x^2)g_1'(x) - (1+\bar{k}_2 - q_2x)xg_1(x) - (1 - \zeta n_0 + \bar{k}_2 - q_2x)\sqrt{1-x^2}g_2(x) = 0\,,\\ (1-x^2)g_2'(x) + (\bar{k}_2 - q_2x)xg_2(x) - (\zeta n_0 + \bar{k}_2 - q_2x)\sqrt{1-x^2}g_1(x) = 0\,.\end{aligned} \tag{3.425}$$

For $q_2 = \mu = 0$ (free azimuthal motion), integrable solutions of this system can be expressed in terms of associated Legendre functions (see [191], 8.810)

$$\begin{aligned} g_1(x) &= [2\epsilon + (|l_z| + \zeta n_0)(1-\epsilon)]P_l^{|1+l_z|}(x), \quad l \in \mathbb{Z}_+\,,\\ g_2(x) &= [1 - \epsilon - (l_z + \zeta n_0)(1+\epsilon)]P_l^{|l_z|}(x), \quad -1 - l \le l_z \le l\,,\\ \epsilon &= \begin{cases} 1, & l_z \ge 0 \\ -1, & l_z < 0 \end{cases}, \quad n_0 = l + \frac{1+\zeta}{2}\,. \end{aligned} \tag{3.426}$$

For $l_z = l$ and $l_z = -1 - l$, nonzero functions $g_1(x)$ and $g_2(x)$ are possible only if $\zeta = 1$. Hence, for $q_2 = \mu = 0$ there exist no states with $n_0 = 0$ (i.e. always $n_0 \ge 1$), which is evidently implied by (2.163) at $\mathbf{H} = 0$. The functions (3.426) are bounded and form an orthogonal (in l) system.

For nonvanishing q_2 or μ, integrable solutions of the system (3.425) for $n_0 \neq 0$ are expressed in terms of the Jacobi polynomials (see [191], 8.960) and have the form

$$\begin{aligned} g_1(x) &= (1-x)^{\frac{\alpha-1}{2}}(1+x)^{\frac{\bar\beta}{2}}\Big\{(\bar k_2 - q_2 - \alpha)P_n^{(\alpha,\bar\beta)}(x) \\ &\quad + (1-x)\left[(n+\bar\beta)P_{n-1}^{(\alpha+1,\bar\beta)}(x) + aP_n^{(\alpha,\bar\beta)}(x)\right]\Big\}, \\ g_2(x) &= (1-x)^{\frac{\alpha}{2}}(1+x)^{\frac{\bar\beta-1}{2}}\Big\{(1+\bar k_2 + q_2 + \bar\beta)P_n^{(\alpha,\bar\beta)}(x) \\ &\quad + (1+x)\left[(n+\alpha)P_{n-1}^{(\alpha,\bar\beta)}(x) - aP_n^{(\alpha,\bar\beta)}(x)\right]\Big\}; \\ \alpha^2 &= (\bar k_2 - q_2)^2, \quad \bar\beta^2 = (1+\bar k_2 + q_2)^2, \quad a = n + q_2 - \zeta n_0 + (1+\alpha+\bar\beta)/2, \\ 2n_0 &= \sqrt{(2n+\alpha+\bar\beta+1)^2 - 4q_2^2}, \quad n \in \mathbb{Z}_+ . \end{aligned} \tag{3.427}$$

For $\mu = 0$ and q_2 integer, or $\mu = 1/2$ and q_2 half-integer, integrable bounded solutions (3.427) correspond to nonnegative α and $\bar\beta$ (which is always possible).

For non-integer $(q_2 - \mu)$, equation (3.414) holds true and there exists a unique, in this case, as distinct from (3.415), number $l_z = \bar l_0$ allowing both positive and negative $\alpha = \pm\nu$. Here, the functions (3.427) have an integrable singularity at the point $x = 1$. In other cases, integrable (and bounded) functions correspond to $\alpha > 0$.

For non-integer $(q_2 + \mu)$, equation (3.418) holds true and there exists a unique, in contrast to (3.419), number $l_z = -\tilde l_0 - 1$ allowing $\bar\beta = \pm\tilde\nu$. Here, the functions (3.427) have an integrable singularity at $x = -1$. For other values of l_z, one should choose $\bar\beta > 0$, and then functions (3.427) are bounded. For nonnegative α and $\bar\beta$, the system (3.427) is orthogonal in n. Finally, only if $l_z = -1$ and (3.422) holds, one can set

$$\alpha = \pm(1 - \mu + q_2), \quad \bar\beta = \pm(\mu + q_2) . \tag{3.428}$$

In this case, functions (3.427) have integrable singularities in both points $x = \pm 1$. In all these cases, the value of n_0 is nonnegative, and its nonphysical values do not exist.

If $n_0 = 0$, the general solution of the system (3.425) has the form

$$\begin{aligned} g_1(x) &= (1-x^2)^{-1/2}\left[B(1-x)^{\tilde\alpha}(1+x)^{\tilde\beta+1} + A(1-x)^{-\tilde\alpha+1}(1+x)^{-\tilde\beta}\right], \\ g_2(x) &= B(1-x)^{\tilde\alpha}(1+x)^{\tilde\beta} - A(1-x)^{-\tilde\alpha}(1+x)^{-\tilde\beta}, \\ 2\tilde\alpha &= q_2 - \bar k_2, \quad 2\tilde\beta = q_2 + \bar k_2 , \end{aligned} \tag{3.429}$$

where A and B are arbitrary functions. An elementary analysis shows that these solutions are integrable, free of singularities, and orthogonal to the solutions with $n_0 \neq 0$, provided that

$$\begin{aligned} &B \neq 0, \quad A = 0, \quad q_2 - \mu - 1 \geq l_z \geq -q_2 - \mu, \quad 2q_2 \geq 1 , \\ &B = 0, \quad A \neq 0, \quad -1 - q_2 - \mu \geq l_z \geq q_2 - \mu, \quad 2q_2 \leq -1 . \end{aligned} \tag{3.430}$$

If (3.414) takes place, inequalities (3.430) are violated for $l_z = \bar{l}_0$; however, for $B \neq 0, A = 0, 2q_2 > -1$ solution (3.429) remains integrable, but contains singularities. If (3.418) takes place, inequalities (3.430) are violated for $l_z = -\tilde{l}_0 - 1$; however, solution (3.429) has integrable singularities for $B = 0, A \neq 0, 2q_2 < 1$. Finally, in case (3.422) takes place, only for $l_z = -1$ solution (3.429) is integrable for arbitrary A, B, and is equal to ∞ at the points $x = \pm 1$.

In the general case, solutions (3.429) which have integrable singularities are not orthogonal to solutions without singularities, nor to solutions with $n_0 \neq 0$. The above consideration also implies that for $q_2 = 0$ and for $\mu \neq 0$, there exists an integrable Dirac particle state (which does not exist for $\mu = 0$), corresponding to $n_0 = 0$ (again only for $l_z = -1$). This state contains an arbitrary complex parameter (or two arbitrary real parameters). The explicit form of such a state follows from (3.429) and reads

$$g_1(x) = \sqrt{\frac{\sin \mu\pi}{2\pi}} \left(Q^{\frac{\mu}{2}} e^{i\gamma} \sin\delta + Q^{-\frac{\mu}{2}} e^{-i\gamma} \cos\delta \right),$$

$$g_2(x) = \sqrt{\frac{\sin \mu\pi}{2\pi}} \left(Q^{\frac{\mu-1}{2}} e^{i\gamma} \sin\delta - Q^{\frac{1-\mu}{2}} e^{-i\gamma} \cos\delta \right),$$

$$Q = \frac{1+x}{1-x}, \quad \int_{-1}^{1} \left(|g_1(x)|^2 + |g_2(x)|^2 \right) dx = 1 . \tag{3.431}$$

Here γ and δ are arbitrary real parameters. The functions $g_1(x)$ and $g_2(x)$ tend to ∞ as $x \to \pm 1$. For $\mu \to 0$, this state vanishes.

3.5.5 Radial motion

In the scalar particle case, the equation for the radial motion (3.367) with the potential (3.386), with the use of (3.394), takes the form

$$f''(r) + \frac{2}{r} f(r) + \left(k_0^2 - m^2 - \frac{2k_0 q_1}{r} - \frac{k_1^2 - q_1^2}{r^2} \right) f(r) = 0 . \tag{3.432}$$

Let us first examine solutions of this equation in the absence of the Coulomb field, i.e. for $q_1 = 0$.

For $q_2 = \mu = 0$ (free scalar particle), relation (3.406) takes place and the motion is possible only if $k_0^2 \geq m^2$, whereas solutions of equation (3.432) bounded at $r \geq 0$ can be expressed in terms of the Bessel functions of half-integer index (see [191], 8.461)

$$f(r) = f_{k_0}^l(r) = r^{-1/2} J_{l+1/2}\left(r\sqrt{k_0^2 - m^2} \right),$$

$$k_0^2 > m^2, \quad k_1^2 = l(l+1), \quad |l_z| \leq l, \quad l \in \mathbb{Z}_+ . \tag{3.433}$$

With the aid of equation (B.80), one can easily verify that the functions (3.433) are orthogonal,

$$\int_0^\infty r^2 f^l_{k_0}(r) f^l_{k_0'}(r) dr = \frac{1}{k_0}\delta(k_0 - k_0'), \tag{3.434}$$

and form a complete system,

$$\int_{m^2}^\infty f^l_{k_0}(r') f^l_{k_0}(r) dk_0^2 = \frac{2}{r^2}\delta(r - r'), \quad r > 0, \quad r' > 0. \tag{3.435}$$

For $k_0^2 = m^2$ and $q_1 = 0$, the general solution of equation (3.432) has the form

$$k_0^2 = m^2, \quad f^{\bar{l}}_m(r) = ar^{\bar{l}-\frac{1}{2}} + br^{-\bar{l}-\frac{1}{2}}, \quad 2\bar{l} = \sqrt{4k_1^2 + 1}, \tag{3.436}$$

where a and b are arbitrary constants. If $q_1 = 0$, but q_2 and μ are generally nonzero, we have the following solutions:

$$f^{\bar{l}}_{k_0}(r) = r^{-1/2} J_{\bar{l}}\left(r\sqrt{k_0^2 - m^2}\right),$$
$$k_0^2 > m^2, \quad \bar{l}^2 = k_1^2 + 1/4, \quad \operatorname{Re}\bar{l} > -1, \tag{3.437}$$

instead of (3.433).

In accordance with equation (3.409), the quantity $\bar{l}$ is not half-integer in the general case, and is determined by the relation

$$\bar{l}^2 = \left(n + \frac{\alpha + \beta + 1}{2}\right)^2 - q_2^2, \quad n \in \mathbb{Z}_+. \tag{3.438}$$

Hence it follows that $\bar{l}^2$ is always real, thereby $\bar{l}$ is either real or purely imaginary. For nonnegative α and β, in accordance with (3.411), one can always choose $\bar{l} > 0$. For real $\bar{l} > -1$, orthogonality and completeness relations (3.434) and (3.435) remain valid.

If $\bar{l}^2$ obtained from (3.438) satisfies the inequality

$$0 < \bar{l}^2 < 1, \tag{3.439}$$

then we can choose solutions of equation (3.432) with $q_1 = 0$ as arbitrary linear combinations of the form

$$f^{\bar{l}}_{k_0}(r) = r^{-1/2}\left[aJ_{\bar{l}}\left(r\sqrt{k_0^2 - m^2}\right) + bJ_{-\bar{l}}\left(r\sqrt{k_0^2 - m^2}\right)\right], \tag{3.440}$$

where a and b are arbitrary constants.

Two sets of functions (3.440), one corresponding to $a = 1, b = 0$ and the other one corresponding to $a = 0, b = 1$, each obey the conditions of orthogonality and completeness (3.434) and (3.435), however, they are not orthogonal between themselves. For these values of $\bar{l}$, the K–G operator fails to be self-adjoint. In particular, both the states (3.424) admit solutions (3.440) with $2\bar{l} = 1 - 2\mu$.

However, if (3.439) holds, the fact that the K–G operator is not self-adjoint appreciably complicates the situation. For $q_1 = 0$, equation (3.432) allows integrable solutions in nonphysical domains of k_0^2. One easily checks that for any (complex!) k_0, with the only condition

$$\operatorname{Re}\sqrt{m^2 - k_0^2} > 0\,, \tag{3.441}$$

there exist integrable solutions of the form

$$f_{k_0}^{\bar{l}}(r) = r^{-1/2} K_{\bar{l}}\left(r\sqrt{m^2 - k_0^2}\right), \quad 0 < \bar{l} < 1\,, \tag{3.442}$$

where $K_\alpha(x)$ are MacDonald functions (see [191], 8.407). In particular, condition (3.441) allows any $k_0^2 < m^2$ (including $k_0^2 < 0$). The norm of the function (3.442) is finite,

$$\int_0^\infty r^2 \left| f_{k_0}^{\bar{l}}(r) \right|^2 dr = \frac{\bar{l}\pi}{2|m^2 - k_0^2| \sin \bar{l}\pi}, \quad |\bar{l}| < 1\,. \tag{3.443}$$

The solutions (3.442) make sense in the case $q_2 = \mu = 0$, $\bar{l} = 1/2$, which corresponds to $l = 0$ in equation (3.433), and implies that the set of functions (3.433) can be complemented by the integrable solution

$$f(r) = r^{-1/2} K_{1/2}\left(r\sqrt{m^2 - k_0^2}\right), \quad l = 0\,, \tag{3.444}$$

in the case when (3.441) holds true. The functions (3.442) with different k_0's are not orthogonal.

Let us turn to the analysis of solutions of equation (3.432) for $q_1 \neq 0$. If $k_0^2 \neq m^2$, one easily checks that, in accordance with equation (B.17), these solutions can be expressed in terms of the Laguerre functions (see Section B.1):

$$f(r) = f_{\bar{s}}^{\bar{\nu}}(r) = z^{-1/2} I_{\bar{\nu}+\bar{s},\bar{s}}(z), \quad z = 2r\sqrt{m^2 - k_0^2}\,. \tag{3.445}$$

One can see that for physically admissible k_0^2 the argument of the Laguerre functions can be either a nonnegative real or a purely imaginary number. The indices of the Laguerre functions are determined by the relations

$$\begin{aligned} \bar{\nu}^2 &= 1 + 4(k_1^2 - q_1^2) = (2n + \alpha + \beta + 1)^2 - 4(q_1^2 + q_2^2)\,, \\ \bar{s} &= -\frac{1 + \bar{\nu}}{2} - \frac{q_1 k_0}{\sqrt{m^2 - k_0^2}}\,. \end{aligned} \tag{3.446}$$

From (3.446) it follows that $\bar{\nu}^2$ and $(2\bar{s} + 1 + \bar{\nu})^2$ are always real in physically admissible domains of parameters; therefore, these quantities can be either real or purely imaginary. Thus, in general, the indices of the Laguerre functions in (3.445) are complex.

If $k_0^2 = m^2$, a solution of equation (3.432) exists and has the form

$$f(r) = r^{-1/2} J_{\bar{\nu}} \left(2\sqrt{2m|q_1|r} \right) . \tag{3.447}$$

It can be obtained from (3.445) as $k_0^2 \to m^2$, by using the limit (B.38).

As it is known from classical theory, for $q_1 < 0$ there exists a domain of parameters corresponding to the restricted classical motion (3.400). According to equations (3.446), if the classical restriction $k_1^2 \geq q_1^2$ holds true, one can always choose $\bar{\nu} \geq 1$. In quantum theory, a restricted motion corresponds to solutions with half-integer values of $\bar{s}$. In this case, the functions (3.445) can be expressed in terms of the Laguerre polynomials (B.5), and have the form

$$f_s^{\bar{\nu}}(r) = N z^{\frac{\bar{\nu}-1}{2}} e^{-\frac{z}{2}} L_{\bar{s}}^{\bar{\nu}}(z), \quad \bar{s} \in \mathbb{Z}_+ . \tag{3.448}$$

The energy is quantized,

$$k_0^2 = m^2 \left[1 - \frac{4q_1^2}{4q_1^2 + (1 + \bar{\nu} + 2\bar{s})^2} \right] . \tag{3.449}$$

The classical restriction $k_1^2 > q_1^2$ implies $\bar{\nu} \geq 1$, and the functions (3.448) have no singularities for $r \geq 0 (z \geq 0)$, whereas their norm

$$\int_0^\infty x^2 |f_{\bar{s}}^{\bar{\nu}}(x)|^2 dx = \int_0^\infty x^{1+\bar{\nu}} e^{-x} [L_{\bar{s}}^{\bar{\nu}}(x)]^2 dx = \frac{(2\bar{s} + \bar{\nu} + 1)\Gamma(\bar{s} + \bar{\nu} + 1)}{\Gamma(\bar{s} + 1)} \tag{3.450}$$

remains finite at $\bar{\nu} > -2$. Therefore, in the domain $1 > \bar{\nu} > -2$ the functions (3.448) tend to ∞ as $x = 0$; however, this singularity is integrable, in the sense of (3.450). Note that in accordance with (3.446) the domain $-1 > \bar{\nu} > -2$ does not contradict the classical restriction $k_1^2 > q_1^2$, whereas the domain $1 > \bar{\nu} \geq -1$ does.

In the domain $|\bar{\nu}| < 2$, the situation is essentially more complicated. Namely in this domain, for complex k_0^2 to satisfy the condition (3.441), equation (3.432) has integrable, in the sense of (3.450), solutions of the form

$$f_{k_0}(r) = z^{-1/2} \psi_{\lambda,\bar{\nu}}(z), \quad \lambda = -\frac{k_0 q_1}{\sqrt{m^2 - k_0^2}} , \tag{3.451}$$

where $\psi_{\lambda,\bar{\nu}}$ are determined by equations (B.126). The functions (3.451) tend to ∞ as $z = 0$.

The existence of integrable solutions in nonphysical domains of parameters indicates that the K–G operator is not self-adjoint in these domains.

Let us turn to the set of radial equations (3.384) with potentials (3.386),

$$\begin{aligned} f_1'(r) + \frac{1 - \zeta n_0}{r} f_1(r) - \left(k_0 + m - \frac{q_1}{r} \right) f_2(r) = 0 , \\ f_2'(r) + \frac{1 + \zeta n_0}{r} f_2(r) + \left(k_0 - m - \frac{q_1}{r} \right) f_1(r) = 0 . \end{aligned} \tag{3.452}$$

The quantity n_0 is given by (3.427).

Let us first investigate solutions of the set (3.452) at $q_1 = 0$.

One easily verify that the functions

$$f_1^{k_0}(r) = \sqrt{\frac{k_0 + m}{x}} J_{n_0 - \frac{\zeta}{2}}(x), \quad f_2^{k_0}(r) = -\zeta \sqrt{\frac{k_0 - m}{x}} J_{n_0 + \frac{\zeta}{2}}(x),$$
$$x = r\sqrt{k_0^2 - m^2}, \tag{3.453}$$

are solutions of equations (3.452) when $q_1 = 0$. Here $J_\alpha(x)$ are Bessel functions (see [191], 8.402). For $q_2 = \mu = 0$ (a free Dirac particle), n_0 is given by (3.426), and the Bessel functions have half-integer indices,

$$f_1^{k_0}(r) = \sqrt{\frac{k_0 + m}{x}} J_{l+\frac{1}{2}}(x), \quad f_2^{k_0}(r) = -\zeta \sqrt{\frac{k_0 - m}{x}} J_{l+\frac{1}{2}+\zeta}(x), \tag{3.454}$$

where for $\zeta = -1$, we have $l \in \mathbb{N}$, whereas for $\zeta = 1$, we have $l \in \mathbb{Z}_+$. The solutions (3.454) are bounded as $x \geq 0$.

In the case $n_0 < 1$, which is possible for nonvanishing q_2 or μ (we remind that $n_0 \geq 0$ in the general case), the functions (3.453) have singularities in the point $x = 0$. However, for every $n_0 \geq 0$ the functions (3.453) satisfy the orthogonality condition

$$\int_0^\infty r^2 (f_1^{*k_0'} f_1^{k_0} + f_2^{*k_0'} f_2^{k_0}) dr = 2\delta(k_0 - k_0'). \tag{3.455}$$

For $1 > 2n_0 \geq 0$, the K–G operator is not self-adjoint and the set (3.452) allows, besides (3.453), the following solutions:

$$f_1 = \sqrt{\frac{m + k_0}{r}} K_\gamma \left(r\sqrt{m^0 - k_0^2} \right), \quad \gamma = \frac{1}{2} - \zeta n_0,$$
$$f_2 = -\sqrt{\frac{m - k_0}{r}} K_{1-\gamma} \left(r\sqrt{m^2 - k_0^2} \right), \quad 0 < \gamma < 1. \tag{3.456}$$

These solutions tend to ∞ as $r = 0$ but remain integrable,

$$\int_0^\infty r^2 \left(|f_1|^2 + |f_2|^2 \right) dr = N^2 < \infty, \tag{3.457}$$

as long as condition (3.441) holds true. For instance, if k_0 is real and (3.441) is fulfilled, we have

$$N^2 = \frac{(m - 2\zeta n_0 k_0)\pi}{2(m^2 - k_0^2)\cos n_0 \pi}, \quad m^2 > k_0^2 > 0,$$
$$N^2 = \frac{\pi}{2\sqrt{m^2 - k_0^2} \cos n_0 \pi}, \quad k_0^2 \leq 0. \tag{3.458}$$

The functions (3.456) are not orthogonal for different k_0.

Finally, we consider solutions of the set (3.452) of the radial equations for $q_1 \neq 0$.

For $k_0^2 \neq m^2$, one easily checks, using (B.15) and (B.16), that solutions of this set can be expressed in terms of the Laguerre functions,

$$f_j^{k_0}(r) = \frac{1}{r}\sqrt{m-(-1)^j k_0}\Big[\sqrt{\omega+\zeta n_0}I_{\nu+s\,,s}(z)$$
$$-(-1)^j\sqrt{\omega-\zeta n_0}I_{\nu+s+1\,,s+1}(z)\Big], \quad z = 2r\sqrt{m^2-k_0^2}\,,$$
$$\nu^2 = 4(n_0^2-q_1^2) = (2n+\alpha+\bar{\beta}+1)^2 - 4(q_1^2+q_2^2)\,,$$
$$s = -1-\frac{\nu}{2}-\frac{k_0 q_1}{\sqrt{m^2-k_0^2}}, \quad \omega = \frac{m q_1}{\sqrt{m^2-k_0^2}}, \quad j = 1,2\,. \tag{3.459}$$

These functions are bounded in the domain $z \geq 0$ for $\nu \geq 2$, and for $2 > \nu > -1$ have an integrable (in the sense of (3.457)) singularity at the origin.

For $q_1 < 0$, there exists a region of a restricted motion. In this region, $s \in \mathbb{Z}_+$, and the functions (3.459) can be expressed in terms of the Laguerre polynomials,

$$f_j^{k_0}(r) = N z^{\frac{\nu-2}{2}} e^{\frac{-z}{2}}\sqrt{m-(-1)^j k_0}\Big[\sqrt{(\omega+\zeta n_0)(s+\nu+1)}L_s^{\nu}(z)$$
$$-(-1)^j\sqrt{(\omega-\zeta n_0)(s+1)}L_{s+1}^{\nu}(z)\Big], \quad j = 1,2;\ s \in \mathbb{Z}_+\,. \tag{3.460}$$

Here the notation from (3.459) is used. As in (3.449), the energy is quantized,

$$k_0^2 = m^2\left[1-\frac{4q_1^2}{4q_1^2+(2s+2+\nu)^2}\right]. \tag{3.461}$$

As in the scalar case, the Dirac operator is not self-adjoint for $-1 < \nu < 1$. For such ν, the set (3.452) has (besides (3.459)) integrable, in the sense of (3.457), solutions (with an integrable singularity at $r = 0$), which are expressed in terms of the functions $\psi_{\lambda,\nu}(z)$ given in (B.126),

$$f_j(r) = \frac{1}{r}\sqrt{m-(-1)^j k_0}\left[(\omega+\zeta n_0)\psi_{\lambda-1/2\,,\nu}(z) + (-1)^j\psi_{\lambda+1/2\,,\nu}(z)\right],$$
$$\lambda = -\frac{k_0 q_1}{\sqrt{m^2-k_0^2}}, \quad j = 1,2\,. \tag{3.462}$$

Using equation (B.127), one easily checks that functions (3.462) are solutions of the set (3.452). The fact that these solutions have a finite norm (3.457), with $\nu^2 < 1$ and any complex k_0^2 subject to (3.441), follows from (B.132) and (B.15).

For $q_1 < 0$, the motion is possible for $k_0 = m$ (for $q_1 > 0$, the motion is possible for $k_0 = -m$). In this case, the wave functions are expressed in terms of the Bessel functions,

$$
\begin{aligned}
rf_1(r) &= yJ_{\nu-1}(y) - (\nu - 2\zeta n_0)J_\nu(y)\,, \\
rf_2(r) &= -2|q_1|J_\nu(y), \quad k_0 = m, \quad q_1 < 0\,; \qquad (3.463) \\
rf_1(r) &= -2|q_1|J_\nu(y)\,, \\
rf_2(r) &= yJ_{\nu-1}(y) - (\nu + 2\zeta n_0)J_\nu(y), \quad k_0 = -m, \quad q_1 > 0\,; \qquad (3.464) \\
y &= 2\sqrt{2m|q_1|r}, \quad \nu^2 = 4(n_0^2 - q_1^2)\,.
\end{aligned}
$$

For $\nu \geq 2$, these functions are bounded for any $r \geq 0$, whereas in the domain $-1 < \nu < 2$ they tend to ∞ as $r = 0$, but remain integrable (in the sense of (3.457)). The solutions (3.463) and (3.464) can be derived from the solutions (3.459) as $k_0 \to \pm m$, with the help of the limit (III.40).

3.6 Particles in the Aharonov–Bohm field and in its superpositions with other fields

3.6.1 Introduction

The *Aharonov–Bohm (AB) effect* plays an important role in quantum theory revealing a peculiar status of electromagnetic potentials in the theory [195, 258, 264]. This effect was discussed in [7] when studying the scattering of a nonrelativistic charged spinless particle by a magnetic field of finite magnetic flux produced by an infinitely long and infinitely thin solenoid, the *AB field* in what follows (a similar effect was discussed earlier by Ehrenberg and Siday [128]). There exists a nontrivial problem with the self-adjointness of Dirac Hamiltonians with AB fields (see [177, 332]), which we do not touch here.

A splitting of Landau levels in a superposition of the AB field with a parallel uniform magnetic field gives an example of the AB effect for bound states. In what follows, we call such a superposition the *magnetic-solenoid field* (*MSF*). Solutions of the nonrelativistic stationary Schrödinger equation with the MSF were first studied in [235]. Solutions of the relativistic wave equations (K–G and Dirac ones) with the MSF were first obtained in [56] and studied in [155–157]. Then these solutions were used in [57, 58, 60] to study the AB effect in cyclotron and synchrotron radiations. Coherent states in the MSF were constructed in [63]. A complete spectral analysis for all the s.a. nonrelativistic and relativistic Hamiltonians with the MSF was performed in [176]. Exact solutions of relativistic wave equations with an external field that represents a combination of the AB field and additional electromagnetic fields were completely described in [65, 66].

The AB field is a field of an infinitely long and infinitesimally thin solenoid with a finite constant internal magnetic flux Φ along the axis z that can be described by the following nonzero electromagnetic potentials A_μ^{AB}:

$$A_x^{\mathrm{AB}} = -\frac{\Phi}{2\pi}\frac{y}{r^2} = \frac{\Phi}{2\pi}\frac{\partial\varphi}{\partial x}, \qquad A_y^{\mathrm{AB}} = \frac{\Phi}{2\pi}\frac{x}{r^2} = \frac{\Phi}{2\pi}\frac{\partial\varphi}{\partial y},$$
$$\mathbf{A}^{\mathrm{AB}} = \left(A_x^{(0)}, A_y^{(0)}, 0\right) = \frac{\Phi}{2\pi r}\mathbf{e}_\varphi, \qquad \mathbf{e}_\varphi = -\mathbf{i}\sin\varphi + \mathbf{j}\cos\varphi\,, \tag{3.465}$$

where r, φ are cylindrical coordinates,

$$x = r\cos\varphi, \quad y = r\sin\varphi, \quad r^2 = x^2 + y^2.$$

The magnetic field of an AB solenoid has the form $\mathbf{B}^{\mathrm{AB}} = (0, 0, B^{\mathrm{AB}})$. It is easy to see that outside the z axis the magnetic field $\mathbf{B}^{\mathrm{AB}} = \operatorname{rot}\mathbf{A}^{\mathrm{AB}}$ is equal to zero. Nevertheless, for any surface Σ with a boundary L being any contour (even an infinitely small one) around the z axis, the circulation of the vector potential along L does not vanish and reads $\oint_L \mathbf{A}^{\mathrm{AB}} d\mathbf{l} = \Phi$. If we interpret this circulation as the flux of the magnetic field $\mathbf{B}^{\mathrm{AB}}$ through the surface Σ,

$$\int_\Sigma \mathbf{B}^{\mathrm{AB}} d\sigma = \oint_L \mathbf{A}^{\mathrm{AB}} d\mathbf{l} = \Phi = \text{const}\,, \tag{3.466}$$

then we obtain an expression for the magnetic field,

$$B^{\mathrm{AB}} = \Phi\delta(x)\delta(y)\,,$$

where the term “infinitely thin solenoid” comes from.

One can see that $\mathbf{A}^{\mathrm{AB}} = -\operatorname{rot}\Psi$, $\Psi = (0, 0, \Phi/(2\pi)\ln r)$, so that $\operatorname{div}\mathbf{A}^{\mathrm{AB}} = 0$, and again

$$\mathbf{B}^{\mathrm{AB}} = \operatorname{rot}\mathbf{A}^{\mathrm{AB}} = (0, 0, B^{\mathrm{AB}}), \quad B^{\mathrm{AB}} = \frac{\Phi}{2\pi}\Delta\ln r = \Phi\delta(x)\delta(y)\,. \tag{3.467}$$

The singular AB field (2.86) and (2.87) can be considered as the limit of a smooth magnetic field. Let us consider the vector potential

$$\mathbf{A}_a = \Phi\frac{1 - f(q)}{2\pi r}\mathbf{e}_\varphi, \quad q = \frac{r}{a}, \quad 0 < a = \text{const}\,, \tag{3.468}$$

where a smooth function $f(q)$ is defined for $q \geq 0$ and satisfies the conditions

$$f(0)=1, \quad f'(0)=0, \quad \lim_{q\to\infty} f(q) = \lim_{q\to\infty} qf'(q) = 0, \quad \lim_{q\to+0} f'(q)/q = \alpha, \quad |\alpha| < \infty\,. \tag{3.469}$$

For example, the function $f(q) = (1 + q^2)^{-s}$, $s > 0$ satisfies the conditions (2.89). The corresponding magnetic field is

$$\mathbf{B}_a(r) = (0, 0, B(a, r))\,, \quad B(a, r) = -\frac{\Phi}{2\pi a^2}\frac{f'(q)}{q}\,. \tag{3.470}$$

The potentials (2.90) and the field (2.91) are nonsingular for $r \geq 0$ and the function $B(a, r)$ vanishes as $r \to \infty$.

Let Σ in equation (2.88) be the circle of the radius R with the centrum in the point $x = 0, y = 0$. Then the magnetic flux $\Phi_a(R) = \int_\Sigma \mathbf{B}_a d\sigma$ has the form

$$\Phi_a(R) = \Phi\left[1 - f(R/a)\right], \quad \lim_{R\to\infty} \Phi_a(R) = \lim_{a\to+0} \Phi_a(R) = \Phi\,. \tag{3.471}$$

The potential (2.90) is reduced to (2.87) and the field (2.91) is reduced to (2.86) as $a \to +0$,

$$\mathbf{A}^{\mathrm{AB}} = \lim_{a\to+0} \mathbf{A}_a, \quad \mathbf{B}^{\mathrm{AB}} = \lim_{a\to+0} \mathbf{B}_a(r)\,. \tag{3.472}$$

In quantum theory, it is convenient to represent the magnetic flux Φ of the AB solenoid via the Dirac's fundamental magnetic flux as follows:

$$\Phi = -\frac{e}{|e|}\left(l_0 + \mu\right)\Phi_0, \quad l_0 \in \mathbb{Z}, \quad 0 \leq \mu < 1\,, \tag{3.473}$$

where Φ_0 is *a fundamental unit of magnetic flux*. In Gaussian units it is

$$\Phi_0 = 2\pi c\hbar/|e| = 4,135 \times 10^{-7}\,\text{Gauss}\cdot\text{cm}^2\,, \tag{3.474}$$

l_0 is an integer and the quantity μ is called the mantissa of the magnetic flux.

Using the introduced quantity, we obtain for potentials of the AB field

$$\begin{aligned} eA_\nu^{\mathrm{AB}} &= -\frac{e}{|e|}\frac{\Phi}{\Phi_0}\partial_\nu\varphi = \left(l_0 + \mu\right)\partial_\nu\varphi = \left(0, -e\mathbf{A}^{\mathrm{AB}}\right); \\ e\mathbf{A}^{\mathrm{AB}} &= -\frac{l_0 + \mu}{r}\mathbf{e}_\varphi\,. \end{aligned} \tag{3.475}$$

Sometimes, it is more convenient to use another representation,

$$\Phi = -\frac{e}{|e|}(\bar{l}_0 - \bar{\mu})\Phi_0, \quad \bar{l}_0 \in \mathbb{Z}, \quad 0 \leq \bar{\mu} < 1\,. \tag{3.476}$$

Then

$$eA_\nu^{\mathrm{AB}} = (\bar{l}_0 - \bar{\mu})\partial_\nu\varphi\,. \tag{3.477}$$

Comparing (3.473) and (3.476), we find $\bar{l}_0 = l_0 + 1$, $\bar{\mu} = 1 - \mu$, so that we have two possible definitions of the mantissa of the magnetic flux.

As was already said, the MSF is defined as a superposition of a constant uniform magnetic field of strength B directed along the axis z and the AB field with the flux Φ in the same direction. The MSF can be described by electromagnetic potentials of the form $A^\mu = (0, \mathbf{A})$, $\mathbf{A} = (A^k, k = 1, 2, 3)$,

$$A_x = A_x^{\mathrm{AB}} - \frac{By}{2}, \quad A_y = A_y^{\mathrm{AB}} + \frac{Bx}{2}, \quad A_z = 0\,. \tag{3.478}$$

The potentials (3.478) define the magnetic field $\mathbf{H} = (0, 0, B + B^{\mathrm{AB}})$. Such a magnetic field is called the MSF.

Some arguments that stress the importance of the mantissa μ are given below.

Let us consider a charged particle moving in a superposition of the AB field and of an additional electromagnetic field. Such a superposition will be called the combined electromagnetic field. Electromagnetic potentials $\mathcal{A}_\nu(x)$ of the combined field are chosen as follows:

$$\mathcal{A}_\nu = (\mathcal{A}_0, \boldsymbol{\mathcal{A}}) = A_\nu + A_\nu^{\mathrm{AB}} = \left(A_0, -\mathbf{A} - \mathbf{A}^{\mathrm{AB}}\right),$$
$$\mathcal{A}_0 = A_0, \quad \boldsymbol{\mathcal{A}} = \mathbf{A} + \mathbf{A}^{\mathrm{AB}}, \tag{3.479}$$

where A_ν are potentials of the additional electromagnetic field.

Electromagnetic potentials of the combined field enter quantum wave equations via the operators of the kinetic momenta $\hat{P}_\nu = \hat{p}_\nu - e\mathcal{A}_\nu$,

$$\hat{P}_0 = \hat{p}_0 - e\mathcal{A}_0, \quad \hat{\mathbf{P}} = \hat{\mathbf{p}} - e\boldsymbol{\mathcal{A}}.$$

Representing the corresponding wave functions Ψ in the form

$$\Psi = \exp(-il_0\varphi)\widetilde{\Psi}, \tag{3.480}$$

one can easily see that the resulting equations for the function $\widetilde{\Psi}$ are obtained from the initial equations for Ψ by the replacement of $\hat{P}_\nu$ by $\widetilde{\hat{P}}_\nu$, where

$$\widetilde{\hat{P}}_\nu = \exp(il_0\varphi)\hat{P}_\nu \exp(-il_0\varphi) = \hat{P}_\nu + l_0\,\partial_\nu\varphi = \hat{p}_\nu - eA_\nu - e\widetilde{A}_\nu^{\mathrm{AB}},$$
$$e\widetilde{A}_\nu^{\mathrm{AB}} = \left(0, -e\widetilde{\mathbf{A}}^{\mathrm{AB}}\right) = \mu\partial_\nu\varphi, \quad e\widetilde{\mathbf{A}}^{\mathrm{AB}} = -\frac{\mu}{r}\mathbf{e}_\varphi. \tag{3.481}$$

Thus, equations for the function $\widetilde{\Psi}$ contain the mantissa of the AB field alone and do not contain the characteristic l_0 of the magnetic flux. This implies that $|\Psi|^2$ and mean values of operators that do not contain derivatives with respect to φ depend only on μ.

In what follows, we are going to consider the AB field only with nonzero mantissa. In this case, cylindrical and spherical coordinates are physically preferable.

3.6.2 Aharonov–Bohm field

Let us consider solutions of relativistic wave equations with a pure AB field. It should be noted that the AB field can be treated as a special case of longitudinal stationary fields. In this case both the particle energy (operator $i\partial_0$) and z-component of the particle momentum (operator $i\partial_z$) are integrals of motion. The operators $\hat{L}_z = -i\partial_\varphi$ and $\hat{J}_z = \hat{L}_z + 1/2\,\Sigma_3$ are integrals of motion for the K–G and Dirac equations respectively. All these integrals of motion mutually commute, so that we choose solutions Ψ_K of the K–G equation and solutions Ψ^e of the Dirac equation as eigenfunctions for these operators with the eigenvalues $k_0, k_3, L_z = l - l_0$, and $J_z = l - l_0 - 1/2, l \in \mathbb{Z}$, respectively. Then

$$\Psi_K(x) = \exp(i\Gamma)\, v_0^l(r), \quad \Gamma = (l - l_0)\varphi - k_0x^0 - k_3z, \tag{3.482}$$

where $v_0^l(r)$ is a solution of an ordinary linear differential equation of the second order,

$$\left[\frac{d^2}{dr^2}+\frac{1}{r}\frac{d}{dr}-\frac{(l+\mu)^2}{r^2}+k^2\right]v_0^l(r)=0,\quad k_0^2=m^2+k_3^2+k^2\,. \tag{3.483}$$

The Dirac wave function has the form

$$\Psi^e(x)=e^{i\Gamma}\begin{pmatrix} e^{-i\varphi}\psi_1(r)\\ i\,\psi_2(r)\\ e^{-i\varphi}\,\psi_3(r)\\ i\,\psi_4(r)\end{pmatrix}. \tag{3.484}$$

The functions $\psi_s(r)$, $s=1,2,3,4$ obey the following set of differential equations:

$$\overline{D}\,\overline{\Phi}=0,\quad \overline{\Phi}=\overline{\Phi}(r)=\begin{pmatrix}\psi_1(r)\\ \psi_2(r)\\ \psi_3(r)\\ \psi_4(r)\end{pmatrix}, \tag{3.485}$$

where the matrix operator $\overline{D}$ has the form

$$\overline{D}=\rho_3k_0+i\rho_2\Sigma_3k_3+\rho_2\Sigma_2\left(\frac{d}{dr}+\frac{1}{2r}\right)-i\rho_2\Sigma_1\frac{l+\mu-1/2}{r}-mI\,. \tag{3.486}$$

Here ρ_k and Σ_k, $k=1,2,3$ are Dirac matrices in the standard representation, see Section A.2.1. Solutions of the set (3.485) have the form

$$\overline{\Phi}(r)=\begin{pmatrix}\left[m+k_0+k_3-\hat{Q}\sigma_3\right]V^l(r)\\ \left[(m-k_0-k_3)\sigma_3+\hat{Q}\right]V^l(r)\end{pmatrix},$$
$$V^l(r)=\begin{pmatrix}v_1^l(r)\\ v_2^l(r)\end{pmatrix},\quad \hat{Q}=\sigma_1\frac{l+\mu-1/2}{r}+i\sigma_2\left(\frac{d}{dr}+\frac{1}{2r}\right), \tag{3.487}$$

where the functions $v_1^l(r)$ and $v_2^l(r)$ are expressed via the functions $v_0^l(r)$ from (3.483) as follows:

$$v_1^l(r)=v_0^{l-1}(r),\quad v_2^l(r)=v_0^l(r)\,.$$

Thus, the problem is reduced to solving equation (3.483). Its general solution has the form

$$v_0^l(r)=AJ_\nu(x)+BJ_{-\nu}(x),\quad \nu=l+\mu,\,x=kr\,,$$

where A and B are arbitrary constants and $J_\nu(x)$ are the Bessel functions. Because the index $\nu=l+\mu$ is not an integer for nonzero μ, one has to choose the constants A and B to provide square integrability of the corresponding wave functions.

3.6.3 Magnetic-solenoid field

Classical motion in the MSF

In this section we present a brief description of classical motion of a charge e with a mass M in the MSF. Trajectories $x^\mu(\tau)$ are parametrized by the proper time and obey the Lorentz equations of motion (2.18). Trajectories that do not intersect the axis z, do not sense the presence of the AB solenoid and are identical to those in a pure constant and uniform magnetic field, see Section 3.4.1.

In what follows we need trajectories parametrized by the light-cone time $u_0 = x^0 - z$. They have the form

$$\tau(u_0) = \frac{u_0}{\lambda}, \quad x^0(u_0) = \frac{P^0}{m\lambda}u_0, \quad x(u_0) = x_{(0)} + R\cos\Omega, \quad y(u_0) = y_{(0)} - \epsilon R\sin\Omega\,,$$
$$z(u_0) = \frac{P^3}{m\lambda}u_0 + z_{(0)}; \quad \Omega = \frac{\gamma}{m\lambda}u_0 + \varphi_{(0)}, \quad \gamma = |eH|, \quad \epsilon = \operatorname{sgn}(eH)\,. \tag{3.488}$$

Here $x_{(0)}, y_{(0)}, z_{(0)}, R, \varphi_{(0)}$, and λ are integrals of motion. In addition

$$P_0 = p_0 = \frac{m^2 + \gamma^2R^2 + m^2\lambda^2}{2m\lambda}, \quad P_3 = p_3 = -\frac{m^2 + \gamma^2R^2 - m^2\lambda^2}{2m\lambda}, \quad \frac{P_0 + P_3}{m} = \lambda\,,$$
$$P_1 = \epsilon\gamma\left[y_{(0)} - y(u_0)\right], \quad P_2 = \epsilon\gamma\left[x(u_0) - x_{(0)}\right], \quad P_\perp^2 = P_1^2 + P_2^2 = \gamma^2R^2\,,$$
$$p_1 = \epsilon\gamma\left[y_{(0)} - \frac{1}{2}y(u_0)\right] + \frac{e\Phi}{2\pi}\frac{y(u_0)}{r^2(u_0)}, \quad p_2 = \epsilon\gamma\left[\frac{1}{2}x(u_0) - x_{(0)}\right] - \frac{e\Phi}{2\pi}\frac{x(u_0)}{r^2(u_0)}\,. \tag{3.489}$$

Using equations (3.488) and (3.489), we obtain

$$L_z = yp_1 - xp_2 = \frac{\epsilon\gamma}{2}(R_0^2 - R^2) + \frac{e\Phi}{2\pi}, \quad R_0^2 = x_{(0)}^2 + y_{(0)}^2\,. \tag{3.490}$$

Already in classical theory, it is convenient to introduce dimensionless complex quantities a_1 and a_2 as follows:

$$a_1 = \frac{iP_1 - \epsilon P_2}{\sqrt{2\gamma}}, \quad a_2 = \frac{\gamma(x - i\epsilon y) - iP_1 - \epsilon P_2}{\sqrt{2\gamma}}\,. \tag{3.491}$$

On the classical trajectories (3.488) and (3.489), we have

$$a_1 = -\sqrt{\frac{\gamma}{2}}R\exp(-i\Omega), \quad a_2 = \sqrt{\frac{\gamma}{2}}R_0\exp(-i\epsilon\alpha), \quad x_{(0)} = R_0\cos\alpha, \quad y_{(0)} = R_0\sin\alpha\,. \tag{3.492}$$

One can see that $a_1\exp((i\gamma/m\lambda)u_0)$ and a_2 are complex (dependent) integrals of motion. Finally, it follows from (3.490) and (3.492) that

$$L_z = \epsilon(a_2^*a_2 - a_1^*a_1) + \frac{e\Phi}{2\pi}\,. \tag{3.493}$$

Quantum consideration

Let us introduce operators $\hat{a}_1, \hat{a}_2$ (and operators $\hat{a}_1^+, \hat{a}_2^+$ formally adjoint to them) that correspond to the classical quantities (3.491),

$$\hat{a}_1 = \frac{i\hat{P}_1 - \epsilon\hat{P}_2}{\sqrt{2\gamma}}, \quad \hat{a}_2 = \frac{\gamma(x - i\epsilon y) - i\hat{P}_1 - \epsilon\hat{P}_2}{\sqrt{2\gamma}},$$
$$\hat{a}_1^+ = -\frac{i\hat{P}_1 + \epsilon\hat{P}_2}{\sqrt{2\gamma}}, \quad \hat{a}_2^+ = \frac{\gamma(x + i\epsilon y) + i\hat{P}_1 - \epsilon\hat{P}_2}{\sqrt{2\gamma}}. \tag{3.494}$$

To find commutator between the introduced operators, we take into account that

$$\left[\hat{P}_\nu, x^\alpha\right] = i\delta_\nu^\alpha, \quad \left[\hat{P}_\nu, \hat{P}_\alpha\right] = -ieF_{\nu\alpha}, \tag{3.495}$$

and the fact that for the MSF under consideration the only nonzero commutators $[\hat{P}_\nu, \hat{P}_\alpha]$ are

$$\left[\hat{P}_1, \hat{P}_2\right] = -\left[\hat{P}_2, \hat{P}_1\right] = i\epsilon\gamma(1 + f), \quad f = \frac{\Phi}{\pi H}\frac{\delta(r)}{r}. \tag{3.496}$$

Thus, using (3.495) and (3.496), we derive the following relations:

$$[\hat{a}_1, \hat{a}_1^+] = 1 + f, \quad [\hat{a}_2, \hat{a}_2^+] = 1 - f, \quad [\hat{a}_1, \hat{a}_2] = f, \quad [\hat{a}_1, \hat{a}_2^+] = 0. \tag{3.497}$$

It follows from equations (3.497) that on functions that decrease sufficiently fast as $r \to 0$, the introduced operators behave as creation and annihilation operators.

Relations (3.494) can be solved with respect to the coordinate and momentum operators,

$$\hat{P}_1 = i\frac{m\sqrt{2b}}{2}(\hat{a}_1^+ - \hat{a}_1), \quad \hat{P}_2 = -\epsilon\frac{m\sqrt{2b}}{2}(\hat{a}_1^+ + \hat{a}_1);$$
$$x = \frac{\hat{a}_2^+ + \hat{a}_2 - \hat{a}_1^+ - \hat{a}_1}{m\sqrt{2b}}, \quad y = i\epsilon\frac{\hat{a}_2 - \hat{a}_2^+ + \hat{a}_1 - \hat{a}_1^+}{m\sqrt{2b}},$$
$$x_{(0)} = \frac{\hat{a}_2^+ + \hat{a}_2}{m\sqrt{2b}}, \quad y_{(0)} = i\epsilon\frac{\hat{a}_2 - \hat{a}_2^+}{m\sqrt{2b}}, \tag{3.498}$$

where

$$b = \frac{\gamma}{m^2} = \frac{|H|}{H_0} > 0, \quad H_0 = \frac{m^2}{|e|}\left(H_0 = \frac{m^2c^3}{|e|\hbar}\right), \tag{3.499}$$

and H_0 is the *Schwinger critical field.*

It follows from equations (3.498) that

$$\hat{P}_\perp^2 = \hat{P}_1^2 + \hat{P}_2^2 = m^2b(\hat{a}_1^+\hat{a}_1 + \hat{a}_1\hat{a}_1^+) = m^2b(2\hat{a}_1^+\hat{a}_1 + 1 + f). \tag{3.500}$$

In the MSF under consideration, the operator $\hat{L}_z$ can be written as follows:

$$\hat{L}_z = y\hat{p}_1 - x\hat{p}_2 = y\hat{P}_1 - x\hat{P}_2 + \frac{\epsilon bm^2}{2}(x^2 + y^2) + \frac{e\Phi}{2\pi}. \tag{3.501}$$

Then using equations (3.499) and (3.497), we obtain its representation in terms of the creation and annihilation operators

$$\hat{L}_z = \frac{e\Phi}{2\pi} + \epsilon(\hat{a}_2^+\hat{a}_2 - \hat{a}_1^+\hat{a}_1 - f)\,, \tag{3.502}$$

which is an analog of the classical relation (3.493).

Let us introduce dimensionless variables ρ by the relation

$$\rho = \frac{\gamma}{2}r^2 = \frac{bm^2r^2}{2}\,, \tag{3.503}$$

then

$$x = \frac{1}{m}\sqrt{\frac{2\rho}{b}}\cos\varphi, \quad y = \frac{1}{m}\sqrt{\frac{2\rho}{b}}\sin\varphi\,.$$

In the case under consideration, it is convenient to define the mantissa μ of the magnetic flux Φ as follows:

$$\begin{aligned} \frac{e\Phi}{2\pi} &= \frac{e}{|e|}\frac{\Phi}{\Phi_0} = \epsilon(l_0+\mu) \to \\ \frac{H\Phi}{|H|\Phi_0} &= l_0+\mu, \quad 0 \le \mu < 1, \quad l_0 \in \mathbb{Z}\,. \end{aligned} \tag{3.504}$$

In fact, we use the definition (3.473) for $\epsilon = -1$ and the definition (3.476) for $\epsilon = 1$.

Relation (3.504) implies the following representation for the singular function f from (3.496):

$$f = \frac{\Phi}{\pi H}\frac{\delta(r)}{r} = 2(l_0+\mu)\delta(\rho)\,. \tag{3.505}$$

Using the well-known formulas

$$\begin{aligned} \partial_x r &= \partial_r x = \cos\varphi, \qquad \partial_y r = \partial_r y = \sin\varphi\,, \\ \partial_x\varphi &= \frac{\partial_\varphi x}{r^2} = -\frac{\sin\varphi}{r}, \quad \partial_y\varphi = \frac{\partial_\varphi y}{r^2} = \frac{\cos\varphi}{r}\,, \end{aligned}$$

we obtain the so-called coordinate representation for the creation and annihilation operators

$$\begin{aligned} \hat{a}_1 &= -\frac{\exp(i\epsilon\varphi)}{2\sqrt{\rho}}(2\rho\partial_\rho + \rho + \hat{A})\,, \\ \hat{a}_1^+ &= \frac{\exp(-i\epsilon\varphi)}{2\sqrt{\rho}}(2\rho\partial_\rho - \rho - \hat{A})\,, \\ \hat{a}_2 &= \frac{\exp(-i\epsilon\varphi)}{2\sqrt{\rho}}(2\rho\partial_\rho + \rho - \hat{A})\,, \\ \hat{a}_2^+ &= -\frac{\exp(i\epsilon\varphi)}{2\sqrt{\rho}}(2\rho\partial_\rho - \rho + \hat{A})\,, \end{aligned}$$

where

$$\hat{A} = i\epsilon\partial_\varphi + l_0 + \mu = l_0 + \mu - \epsilon\hat{L}_z \,. \qquad (3.506)$$

In classical theory, the quantity L_z given by (3.490) is an integral of motion. In quantum mechanics of spinless particles, the corresponding operator $\hat{L}_z$ is also an integral of motion. Consider eigenfunctions of this operator with eigenvalues L_z which are integers. Let us represent L_z as follows:

$$L_z = \epsilon(l_0 - l), \quad l \in \mathbb{Z} \,. \qquad (3.507)$$

Then, with account taken of (3.504), it follows from (3.490) that

$$2(l + \mu) = bm^2(R^2 - R_0^2) \,. \qquad (3.508)$$

The presence of the AB solenoid breaks the translational symmetry in the xy-plane. There appear two types of trajectories: those with $l \geq 0$ embrace the solenoid and those with $l < 0$ do not embrace the solenoid.

One obtains expressions for particle number operators from (3.506)

$$\begin{aligned} \hat{N}_1 &= \hat{a}_1^+\hat{a}_1 = -(4\rho)^{-1}\left[4\rho^2\partial_{\rho^2}^2 + 4\rho\partial_\rho - \rho^2 + 2\rho(1 - \hat{A}) - \hat{A}^2\right] , \\ \hat{N}_2 &= \hat{a}_2^+\hat{a}_2 = -(4\rho)^{-1}\left[4\rho^2\partial_{\rho^2}^2 + 4\rho\partial_\rho - \rho^2 + 2\rho(1 + \hat{A}) - \hat{A}^2\right] . \end{aligned} \qquad (3.509)$$

Then it follows from (3.502) and (3.500) that the following relations hold:

$$\hat{L}_z = \epsilon(l_0 + \mu + \hat{N}_2 - \hat{N}_1 - f), \quad \hat{P}_\perp^2 = \hat{P}_1^2 + \hat{P}_2^2 = m^2b(2\hat{N}_1 + 1 + f) \,. \qquad (3.510)$$

Let us find common eigenfunctions for the operators: $\hat{L}_z$, given by equation (3.502), and operator $\hat{N}_1$, given by equation (3.509). As it follows from (3.510) such functions will be also eigenfunctions for the operator $\hat{N}_2$, given by equation (3.509). Equations that have to be solved are

$$\begin{aligned} &\hat{L}_z\Psi = -i\partial_\varphi\Psi = \epsilon(l_0 - l)\Psi \Rightarrow \hat{A}\Psi = (l + \mu)\Psi; \\ &\hat{N}_1\Psi = n_1\Psi; \quad \hat{N}_2\Psi = n_2\Psi; \quad n_1 \geq 0, n_2 \geq 0, n_1 - n_2 = l + \mu \,. \end{aligned} \qquad (3.511)$$

As follows from the last equation (3.511), the numbers n_1 and n_2 cannot at the same time be both integer if $\mu \neq 0$.

Equations (3.511) define only φ and ρ dependence of the complete wave function Ψ. Thus, we have

$$\Psi = \aleph \exp[i\epsilon(l_0 - l)\varphi]\psi(\rho), \quad l \in \mathbb{Z} \,, \qquad (3.512)$$

and according to (3.509) and (3.510) $\psi(\rho)$ satisfies the following differential equation:

$$4\rho^2\psi'' + 4\rho\psi'^2 - 2\rho(1 + 2n_1 - l - \mu) + (l + \mu)^2]\psi = 0 \,. \qquad (3.513)$$

The factor $\aleph$ can be any function of the rest of the variables. It can be specified from additional considerations.

Equation (3.513) is a particular case of the Laguerre equation

$$4x^2 I''_{k,s}(x) + 4xI'^2_{k,s} - 2x(1+k+s) + (k-s)^2]I_{k,s}(x) = 0 \quad (3.514)$$

for the Laguerre functions $I_{k,s}(x)$, see Section B.1. The indices k, s of these functions can be arbitrary complex numbers.

Equation (3.514) has a square-integrable solution only for $s \in \mathbb{Z}_+$, $\operatorname{Re}(k-s) > -1$. In this case we have

$$I_{k,s}(x) = \sqrt{\frac{\Gamma(1+s)}{\Gamma(1+k)}} \exp\left(-\frac{x}{2}\right) x^{\frac{k-s}{2}} L_s^{k-s}(x), \quad s \in \mathbb{Z}_+ , \quad (3.515)$$

where $L_s^\alpha(x)$ are the Laguerre polynomials

$$L_s^\alpha(x) = \frac{e^x x^{-\alpha}}{s!} \frac{d^s}{dx^s} e^{-x} x^{s+\alpha}$$
$$= \sum_{n=0}^{s} \binom{s+\alpha}{s-n} \frac{(-x)^n}{n!} = \binom{s+\alpha}{s} \Phi(-s, 1+\alpha; x) ;$$

where binomial coefficients are given by equation (B.6) and $\Phi(\alpha, \beta; x)$ is the confluent hypergeometric function.

For any real $\alpha > -1$, the functions $I_{\alpha+s,s}(x)$ form a complete orthonormalized system that satisfies the corresponding relations

$$\int_0^\infty I_{\alpha+k,k}(x) I_{\alpha+s,s}(x) dx = \delta_{k,s}, \quad \sum_{s=0}^{\infty} I_{\alpha+s,s}(y) I_{\alpha+s,s}(x) = \delta(x-y), \quad x, y \in \mathbb{R}_+ . \quad (3.516)$$

It should be noted that equation (3.514) has a square-integrable and bounded at $x \in \mathbb{R}_+$ solution for

$$s \in \mathbb{Z}_+, \quad \operatorname{Re}(k-s) \geq 0 . \quad (3.517)$$

Equation (3.513) is a particular case of equation (3.514) for the indices k and s that satisfy the restrictions

$$(k-s)^2 = (l+\mu)^2, \quad k+s = 2n_1 - l - \mu . \quad (3.518)$$

These restrictions imply an equation for n_1,

$$(n_1 - s)(n_1 - s - l - \mu) = 0 . \quad (3.519)$$

It follows from (3.519) and (3.518) that two values of n_1 and correspondingly of k are possible. These k's have to be matched with inequalities (3.517).

Taking all the said into account, we obtain the following form for the wave functions (3.512):

$$\Psi = \aleph e^{i\epsilon(l_0 - l)\varphi} \begin{cases} I_{n+\mu,n-l}(\rho), & 0 \leq l \leq n, n_1 = n+\mu, \quad n_2 = n-l \\ I_{n-l-\mu,n}(\rho), & -\infty < l \leq -1, n_1 = n, \quad n_2 = n-l-\mu \end{cases} , \quad (3.520)$$

where $n \in \mathbb{Z}_+$.

As long as functions (3.520) vanish as $x = 0$, the operators $\hat{L}_z$, $\hat{N}_1$, and $\hat{N}_2$ commute on solutions of the wave equations. However eigenvalues of the operators $\hat{N}_1$ and $\hat{N}_2$ may be not integer, the operators $\hat{a}_1^+, \hat{a}_1; \hat{a}_2^+, \hat{a}_2$ not being creation and annihilation operators in the strict sense.

The two types of solutions (3.520) will be labeled by an additional index $j = 1, 2$,

$$j = \begin{array}{ll} 1, & 0 \le l \le n - \text{states of the first type} \\ 0, & -\infty < l \le -1 - \text{states of zeroth type} \end{array} . \tag{3.521}$$

Then the wave functions (3.512) can be written as

$$\begin{aligned} &\Psi = \Psi^{(j)}_{n_1,n_2}(\varphi, \rho) = \aleph \exp\{i\epsilon[(l_0 - l)\varphi + j\pi l]\}[jI_{n_1,n_2}(\rho) + (1 - j)I_{n_2,n_1}(\rho)] , \\ &n_1 = n_1^{(j)} = n + j\mu, \quad n_2 = n_2^{(j)} = n - l + (j - 1)\mu , \end{aligned} \tag{3.522}$$

or

$$\begin{aligned} \Psi^{(0)}_{n_1,n_2}(\varphi, \rho) &= \aleph \exp[i\epsilon(l_0 - l)\varphi]I_{n-l-\mu,n}(\rho), \quad -\infty < l \le -1 , \\ &n_1 = n, \quad n_2 = n - l - \mu , \\ \Psi^{(1)}_{n_1,n_2}(\varphi, \rho) &= \aleph \exp\{i\epsilon[(l_0 - l)\varphi + \pi l]\}I_{n+\mu,n-l}(\rho), \quad 0 \le l \le n , \\ &n_1 = n + \mu, \quad n_2 = n - l , \end{aligned}$$

where

$$n_1^{(j)} - n_2^{(j)} = l + \mu, \quad n \in \mathbb{Z}_+, \quad -\infty < l \le n .$$

Using the properties of the Laguerre function (see Section B.1), one can find the following relations:

$$\begin{aligned} \hat{a}_1\Psi^{(j)}_{n_1,n_2}(\varphi, \rho) &= \sqrt{n_1^{(j)}}\; \Psi^{(j)}_{n_1-1,n_2}(\varphi, \rho) , \\ \hat{a}_1^+\Psi^{(j)}_{n_1,n_2}(\varphi, \rho) &= \sqrt{n_1^{(j)} + 1}\;\; \Psi^{(j)}_{n_1+1,n_2}(\varphi, \rho) , \\ \hat{a}_2\Psi^{(j)}_{n_1,n_2}(\varphi, \rho) &= \sqrt{n_2^{(j)}}\; \Psi^{(j)}_{n_1,n_2-1}(\varphi, \rho) , \\ \hat{a}_2^+\Psi^{(j)}_{n_1,n_2}(\varphi, \rho) &= \sqrt{n_2^{(j)} + 1}\;\; \Psi^{(j)}_{n_1,n_2+1}(\varphi, \rho) . \end{aligned} \tag{3.523}$$

Acting successively by the operators $\hat{a}_1$ and $\hat{a}_2$ on the functions (3.522), we can find states with minimum possible quantum numbers $n_1^{(j)}(\min) = j\mu$ and $n_2^{(j)}(\min) = (1 - j)(1 - \mu)$. Such states will be called vacuum states and denoted by $\Psi^{(j)}_{\text{vac}}(\varphi, \rho)$. Using equations (3.522) and (3.515), we find an explicit form of such states

$$\begin{aligned} \Psi^{(j)}_{\text{vac}}(\varphi, \rho) &= \aleph \exp[i\epsilon(l_0 + 1 - j)\varphi]\; I_{\alpha,0}(\rho) \\ &= \aleph[\Gamma(1 + \alpha)]^{-1/2} \exp[i\epsilon(l_0 + 1 - j)\varphi - \rho/2]\; \rho^{\alpha/2} , \\ \alpha = \alpha(j) &= 1 - j + (2j - 1)\mu , \end{aligned}$$

or

$$\Psi^{(j)}_{\text{vac}}(\varphi, \rho) = \Psi^{(j)}_{n_1,n_2}(\varphi, \rho) \quad \text{for } n = 0, l = j - 1 \,. \tag{3.524}$$

It follows from equations (3.523) that the states (3.522) can be obtained by successive action of the operators $\hat{a}_1^+$ and $\hat{a}_2^+$ on the vacuum states,

$$\Psi^{(j)}_{n_1,n_2}(\varphi, \rho) = \sqrt{\frac{\Gamma(1+\alpha(j))}{\Gamma(1+n_1^{(j)})\Gamma(1+n_2^{(j)})}}(a_1^+)^n(a_2^+)^{n-l-1+j}\,\Psi^{(j)}_{\text{vac}}(\varphi, \rho) \,. \tag{3.525}$$

If $\mu = 0$, the vacuum states $\Psi^{(0)}_{\text{vac}}(\varphi, \rho)$ and $\Psi^{(1)}_{\text{vac}}(\varphi, \rho)$ can be related as

$$\begin{aligned} &a_2\Psi^{(0)}_{\text{vac}}(\varphi, \rho) = \Psi^{(1)}_{\text{vac}}(\varphi, \rho), \quad a_2^+\Psi^{(1)}_{\text{vac}}(\varphi, \rho) = \Psi^{(0)}_{\text{vac}}(\varphi, \rho)\,; \\ &\Psi^{(1)}_{\text{vac}}(\varphi, \rho) = \aleph \exp(i\epsilon l_0\varphi - \rho/2), \quad \mu = 0\,. \end{aligned} \tag{3.526}$$

If $\mu \neq 0$ this is impossible.

Thus, for $\mu = 0$ all the states (3.522) can be created from one and the same vacuum $\Psi^{(1)}_{\text{vac}}(\varphi, \rho)$. Taking into account the relation (B.1), one can see that the states of the zeroth and the first type are coinciding for $\mu = 0$, they do not depend on the index j and can be written as

$$\begin{aligned} &\Psi_{n_1,n_2}(\varphi, \rho) = \aleph\,\frac{(a_1^+)^n(a_2^+)^{n-l}\exp(i\epsilon l_0\varphi - \rho/2)}{\sqrt{\Gamma(1+n)\Gamma(1+n-l)}}\,, \\ &\mu = 0, \quad n_1 = n, \quad n_2 = n - l, \quad l \leq n\,. \end{aligned} \tag{3.527}$$

For $\mu \neq 0$, definitions (3.522) and relations (3.523) hold true for any values of the parameters $n_1^{(j)}$, $n_2^{(j)}$ (even for negative ones). For example, acting by annihilation operators $\hat{a}_1$ and $\hat{a}_2$ on the vacuum states, we get

$$\begin{aligned} \Phi^{(1)} &= \hat{a}_1\Psi^{(j)}_{\text{vac}}(\varphi, \rho) = -j\aleph[\Gamma(\mu)]^{-1/2}\exp[i\epsilon(l_0+1)\varphi - \rho/2]\rho^{-(1-\mu)/2}\,, \\ \Phi^{(0)} &= \hat{a}_2\Psi^{(j)}_{\text{vac}}(\varphi, \rho) = (1-j)\aleph[\Gamma(1-\mu)]^{-1/2}\exp(i\epsilon l_0\varphi - \rho/2)\rho^{-\mu/2}\,. \end{aligned} \tag{3.528}$$

The functions in the right hand sides of equations (3.528) are unbounded but still square-integrable. However, they do not belong to the domain of a correct defined self-adjoint Hamiltonian, see Refs. ([155, 176]). One can verify that $\Phi^{(1)}$ is an eigenfunction of the operators N_1 and N_2 with the eigenvalues $n_1 = -(1-\mu) < 0$ and $n_2 = 0$, and $\Phi^{(0)}$ is an eigenfunction of the same operators with the eigenvalues $n_1 = 0$ and $n_2 = -\mu < 0$. This is in complete agreement with equations (3.523).

Let us consider the following functions:

$$\begin{aligned} &\Phi_s^{(1)k} = (a_2^+)^k(a_1)^s\Psi^{(1)}_{\text{vac}}(\varphi, \rho), \quad k \in \mathbb{Z}_+, \quad s \in \mathbb{N}\,, \\ &\Phi_s^{(0)k} = (a_1^+)^k(a_2)^s\Psi^{(0)}_{\text{vac}}(\varphi, \rho), \quad \Phi_1^{(j)0} = \Phi^{(j)}\,. \end{aligned} \tag{3.529}$$

The functions $\Phi_s^{(1)k}$ are eigenfunctions of the operators N_1 and N_2 with the eigenvalues $n_1 = \mu - s < 0$, $n_2 = k \geq 0$, and $\Phi_s^{(0)k}$ are eigenfunction of the same operators with the eigenvalues $n_1 = k \geq 0$, $n_2 = 1 - s - \mu < 0$. For $s = 1$ and $k = 0$ the functions (3.529) are normalized, but unbounded (they tend to ∞ as $\rho \to 0$); for $k + s > 1$ the functions (3.529) are unbounded and not square-integrable.

The functions (3.522) satisfy the following orthonormality relations:

$$(\Psi^{(j')}_{n'_1,n'_2}, \Psi^{(j)}_{n_1,n_2}) = \int_0^{2\pi} d\varphi \int_0^{\infty} d\rho\, \Psi^{*(j')}_{n'_1,n'_2}(\varphi,\rho)\, \Psi^{(j)}_{n_1,n_2}(\varphi,\rho) = 2\pi \aleph'^{*} \aleph\, \delta_{n'_1,n_1} \delta_{n'_2,n_2} \delta_{j',j} \,. \tag{3.530}$$

Solutions of the K–G equation

Consider now the K–G equation with the MSF,

$$\hat{\mathfrak{K}}\Psi^B(x) = 0, \hat{\mathfrak{K}} = \left(\frac{\hat{P}_0}{m}\right)^2 - \left(\frac{\hat{P}_3}{m}\right)^2 - \left(\frac{\hat{P}_\perp}{m}\right)^2 - 1\,. \tag{3.531}$$

Here $\Psi^B(x)$ is the scalar wave function of a spinless particle that is moving in the MSF. According to (3.510) the operator $\hat{P}_\perp^2$ can be expressed via the operator $\hat{N}_1$.

In the beginning, we introduce eigenfunctions $\Psi^B = \Psi^B_{p_3,n,l}(x)$ of the operators $\hat{P}_0$, $\hat{P}_3$, $\hat{N}_1$, and $\hat{L}_z$,

$$\hat{P}_0\Psi^B = p_0\Psi^B, \quad \hat{N}_1\Psi^B = n_1\Psi^B, \quad \hat{P}_3\Psi^B = p_3\Psi^B, \quad \hat{L}_z\Psi^B = \epsilon(l_0 - l)\Psi^B\,. \tag{3.532}$$

They have the form

$$\Psi^B_{p_3,n,l}(x) = \Psi^{B(j)}_{p_3,n,l}(x) = \aleph \exp\left[-i(p_0^{(j)}x^0 + p_3 z)\right] \Psi^{(j)}_{n_1,n_2}(\varphi,\rho)\,, \tag{3.533}$$

where the $\Psi^{(j)}_{n_1,n_2}(\varphi,\rho)$ are given in (3.522), and quantum numbers are related by the following equation:

$$\left(p_0^{(j)}/m\right)^2 = 1 + (p_3/m)^2 + 2b\left(n_1^{(j)} + 1/2\right), \quad n_1^{(j)} = n + j\mu\,. \tag{3.534}$$

Solutions of the Dirac equation

In the case under consideration, we write the Dirac equation as

$$\gamma^0\hat{\mathcal{D}}\Psi^e(x) = \hat{\bar{\mathcal{D}}}\Psi^e(x) = 0, \quad \hat{\bar{\mathcal{D}}} = \hat{P}_0 - (\boldsymbol{\alpha}\hat{\mathbf{P}}) - m\rho_3\,. \tag{3.535}$$

Introducing projection operators $P_{(\pm)}$ by equations (2.71) and the state vectors $\Psi_{(\pm)}(x)$ as

$$\Psi_{(\pm)}(x) = P_{(\pm)}\Psi^e(x), \quad \Psi^e(x) = \Psi_{(+)}(x) + \Psi_{(-)}(x)\,,$$

we reduce the Dirac equation to the following set of equivalent equations:

$$(\hat{P}_0 + \hat{P}_3)\Psi_{(+)}(x) = \hat{\mathfrak{Q}}\Psi_{(-)}(x), \quad (\hat{P}_0 - \hat{P}_3)\Psi_{(-)}(x) = \hat{\mathfrak{Q}}\Psi_{(+)}(x);$$
$$\hat{\mathfrak{Q}} = (\boldsymbol{\alpha}_\perp\hat{\mathbf{P}}_\perp) + m\rho_3, \quad \boldsymbol{\alpha}_\perp = (\alpha_1,\alpha_2,0), \quad \hat{\mathbf{P}}_\perp = -(\hat{P}_1,\hat{P}_2,0)\,. \tag{3.536}$$

The operators $\hat{P}_0 + \hat{P}_3$, $\hat{P}_0 - \hat{P}_3$ and $\hat{Q}$ mutually commute in the case under consideration, which allows one to verify that the set (3.536) implies the same equations for the bispinors $\Psi_{(\pm)}(x)$,

$$(\hat{P}_0^2 - \hat{P}_3^2 - \hat{Q}^2)\Psi_{(\pm)}(x) = 0\,,$$
$$\hat{Q}^2 = \hat{P}_\perp^2 + m^2 - \epsilon m^2 b(1+f)\Sigma_3 = m^2\{1 + b[2\hat{N}_1 + (1+f)(1-\epsilon\Sigma_3)]\}\,. \quad (3.537)$$

It is enough to find only one bispinor, say $\Psi_{(-)}(x)$, then the other one $\Psi_{(+)}(x)$ can be restored from the set (3.536).

If the Dirac matrices are chosen in the standard representation and $\Psi_s^{(e)}(x)$ are components of the bispinor $\Psi^{(e)}(x)$, then

$$\Psi_{(-)}(x) = \frac{1}{2}\begin{pmatrix} \Psi_1^{(e)}(x) - \Psi_3^{(e)}(x) \\ \Psi_2^{(e)}(x) + \Psi_4^{(e)}(x) \\ \Psi_3^{(e)}(x) - \Psi_1^{(e)}(x) \\ \Psi_2^{(e)}(x) + \Psi_4^{(e)}(x) \end{pmatrix}. \quad (3.538)$$

This bispinor can be written as

$$\Psi_{(-)}(x) = \frac{1}{2}\begin{pmatrix} \psi(x) \\ -\sigma_3\psi(x) \end{pmatrix}, \quad (3.539)$$

where the spinor $\psi(x)$ reads

$$\psi(x) = \begin{pmatrix} \psi_1(x) = \Psi_1^{(e)}(x) - \Psi_3^{(e)}(x) \\ \psi_2(x) = \Psi_2^{(e)}(x) + \Psi_4^{(e)}(x) \end{pmatrix}.$$

It satisfies the equation

$$\left\{\frac{\hat{P}_0^2 - \hat{P}_3^2}{m^2} - 1 - b[2\hat{N}_1 + (1+f)(1-\epsilon\sigma_3)]\right\}\psi(x) = 0 \quad (3.540)$$

as it follows from equations (3.537).

As it is usual in the light-cone variables, when working in the spaces where the inverse operator $(\hat{P}_0 + \hat{P}_3)^{-1}$ does exist we can calculate the bispinor $\Psi_{(+)}(x)$,

$$\Psi_{(+)}(x) = (\hat{P}_0 + \hat{P}_3)^{-1}\hat{Q}\Psi_{(-)}(x) = \frac{1}{2}\begin{pmatrix} \chi(x) \\ \sigma_3\chi(x) \end{pmatrix}, \quad (3.541)$$

where

$$\chi(x) = m(\hat{P}_0 + \hat{P}_3)^{-1}\left\{1 + \left[(\sigma_2 - i\epsilon\sigma_1)a_1^+ - (\sigma_2 + i\epsilon\sigma_1)a_1\right]\sqrt{b/2}\right\}\psi(x)\,. \quad (3.542)$$

Thus, the Dirac wave function is completely defined via the spinor $\psi(x)$.

The operator $\hat{J}_z = \hat{L}_z + 1/2\,\Sigma_3$ is an integral of motion in the case under consideration. Let us consider eigenvectors of the mutually commuting integrals of motion

$\hat{P}_0$, $\hat{P}_3$, and $\hat{J}_z$ with respective eigenvalues p_0, p_3, and $j_z = \epsilon(l_0 - l + 1/2)$, $l \in \mathbb{Z}$. It is easy to see that the corresponding spinors $\psi(x)$ must satisfy the following equations:

$$P_0\psi(x) = p_0\psi(x), \quad P_3\psi(x) = p_3\psi(x)\,,$$
$$\hat{j}_z\psi(x) = j_z\psi(x); \qquad \hat{j}_z = \hat{L}_z + \frac{1}{2}\sigma_3\,. \tag{3.543}$$

Spinors $\psi(x)$ that satisfy both sets of equations (3.540) and (3.543) have the form $\psi(x) = \psi^{(j)}_{n,l,p_3}(x)$,

$$\psi^{(j)}_{n,l,p_3}(x) = \aleph \exp\left[-i(p_0^{(j)}x^0 + p_3 z)\right] F^{(j)} v_0^{(j)}\,, \tag{3.544}$$

where $v_0^{(j)}$ are arbitrary complex spinors, and $F^{(j)}$ is a 2×2 matrix having the form

$$F^{(j)} = (1 + \epsilon\sigma_3)\Psi^{(j)}_{n_1,n_2}(\varphi,\rho) + (1 - \epsilon\sigma_3)\Psi^{(j)}_{n_1-1,n_2}(\varphi,\rho)\,. \tag{3.545}$$

The quantum numbers are related by the equation

$$\left(p_0^{(j)}/m\right)^2 = 1 + (p_3/m)^2 + 2bn_1^{(j)}, \quad n_1^{(j)} = n + j\mu\,, \tag{3.546}$$

similar to (3.534).

For states with $n = 0$ the spinor $v_0^{(j)}$ is to be chosen as

$$v_0^{(j)} = \frac{1}{2}\begin{pmatrix} 1+\epsilon \\ 1-\epsilon \end{pmatrix}\,. \tag{3.547}$$

This choice provides both square-integrability and boundedness of the bispinors $\Psi_{(-)}(x)$ for any $j = 0, 1$. The corresponding bispinors $\Psi_{(+)}(x)$ are square-integrable, but unbounded for $j = 1$ as $\rho \to 0$. We stress that any different choice of the spinor $v_0^{(j)}$ leads to bispinors $\Psi_{(+)}(x)$ that are not square-integrable.

3.6.4 Quasicoherent states in the magnetic-solenoid field

Instantaneous quasicoherent states

Let us introduce the so-called *instantaneous quasicoherent states* in the MSF as follows:

$$\Psi^{(j)}_{z_1,z_2}(\varphi,\rho) = \sum_{n_1,n_2} \frac{z_1^{n_1^{(j)}} z_2^{n_2^{(j)}} \Psi^{(j)}_{n_1,n_2}(\varphi,\rho)}{\sqrt{\Gamma(1+n_1^{(j)})\,\Gamma(1+n_2^{(j)})}}\,, \tag{3.548}$$

where possible values of the quantum numbers n_1 and n_2 are given in (3.522). The instantaneous quasicoherent states of the type $j = 0$ can be written in two equivalent forms

$$\Psi^{(0)}_{z_1,z_2}(\varphi,\rho) = \exp(i\epsilon l_0\varphi)\sum_{n=0}^{\infty}\sum_{l=-1}^{-\infty}\frac{z_1^n\, z_2^{n-l-\mu}\exp(-i\epsilon l\varphi)\, I_{n-l-\mu,\, n}(\rho)}{\sqrt{\Gamma(1+n)\Gamma(1+n-l-\mu)}}$$
$$= \exp[i\epsilon(l_0+1)\varphi]\sum_{n=0}^{\infty}\sum_{s=0}^{\infty}\frac{z_1^n\, z_2^{n+1-\mu+s}\exp(i\epsilon s\varphi)\, I_{n+1-\mu+s,\, n}(\rho)}{\sqrt{\Gamma(1+n)\Gamma(2+n-\mu+s)}}\,. \tag{3.549}$$

The instantaneous quasicoherent states of the type $j = 1$ have the following two equivalent forms:

$$\Psi^{(1)}_{z_1,z_2}(\varphi,\rho) = \exp(i\epsilon l_0\varphi)\sum_{n=0}^{\infty}\sum_{s=0}^{n}\frac{z_1^{n+\mu}\, z_2^{n-s}\exp[i\epsilon s(\pi-\varphi)]\, I_{n+\mu,\, n-s}(\rho)}{\sqrt{\Gamma(1+n-s)\Gamma(1+n+\mu)}}\,,$$
$$= \exp(i\epsilon l_0\varphi)\sum_{n=0}^{\infty}\sum_{s=0}^{\infty}\frac{z_1^{n+\mu+s}\, z_2^{n}\exp[i\epsilon s(\pi-\varphi)]\, I_{n+\mu+s,\, n}(\rho)}{\sqrt{\Gamma(1+n)\Gamma(1+n+\mu+s)}}\,. \tag{3.550}$$

The latter result is based on the relation

$$\sum_{n=0}^{\infty}\sum_{s=0}^{n} a(n,s) = \sum_{n=0}^{\infty}\sum_{s=0}^{\infty} a(n+s,s)\,, \tag{3.551}$$

which is a consequence (as $k \to \infty$) of the obvious relation

$$\sum_{n=0}^{k}\sum_{s=0}^{n} a(n,s) = \sum_{s=0}^{k}\sum_{n=s}^{k} a(n,s) = \sum_{s=0}^{k}\sum_{n=0}^{k-s} a(n+s,s)\,.$$

Representations (3.549) and (3.550) can be written in a different form via a function $F_\alpha(z_1,z_2;\varphi,\rho)$

$$\Psi^{(0)}_{z_1,z_2}(\varphi,\rho) = \exp[i\epsilon(l_0+1)\varphi]F_{1-\mu}(z_1,z_2;\varphi,\rho)\,,$$
$$\Psi^{(1)}_{z_1,z_2}(\varphi,\rho) = \exp(i\epsilon l_0\varphi)F_{\mu}(z_2,z_1;\pi-\varphi,\rho)\,, \tag{3.552}$$

where

$$F_\alpha(z_1,z_2;\varphi,\rho) = \sum_{n=0}^{\infty}\sum_{s=0}^{\infty}\frac{z_1^n\, z_2^{n+\alpha+s}\exp(i\epsilon s\varphi)\, I_{n+\alpha+s,\, n}(\rho)}{\sqrt{\Gamma(1+n)\Gamma(1+n+\alpha+s)}}$$
$$= \exp\left(z_1z_2 - \frac{\rho}{2}\right)\sum_{s=0}^{\infty}\left(\sqrt{\frac{z_2}{z_1}}\right)^{\alpha+s} e^{i\epsilon s\varphi} J_{\alpha+s}(2\sqrt{z_1z_2\rho})\,. \tag{3.553}$$

We have used the well-known sum

$$\sum_{n=0}^{\infty}\frac{z^n\, I_{\alpha+n,\, n}(x)}{\sqrt{\Gamma(1+n)\Gamma(1+\alpha+n)}} = z^{-\frac{\alpha}{2}}\exp\left(z - \frac{x}{2}\right) J_\alpha(2\sqrt{xz}) \tag{3.554}$$

to get the second equivalent representation for $F_\alpha(z_1,z_2;\varphi,\rho)$ in (3.553).

Let is consider the scalar product of two functions $F_\alpha(z_1, z_2; \varphi, \rho)$ and $F_\alpha(z_1', z_2'; \varphi, \rho)$

$$\left(F_\alpha(z_1, z_2; \varphi, \rho),\, F_\alpha(z_1', z_2'; \varphi, \rho)\right) = \frac{1}{2\pi}\int_0^\infty d\rho \int_0^{2\pi} d\varphi\, F_\alpha^*(z_1, z_2; \varphi, \rho)\, F_\alpha(z_1', z_2'; \varphi, \rho)\,.$$

Using equations (6.615) and (8.445) from [191], one can easily find

$$\left(F_\alpha(z_1, z_2; \varphi, \rho)\,,\, F_\alpha(z_1', z_2'; \varphi, \rho)\right) = \sum_{n=0}^\infty \sum_{s=0}^\infty \frac{(z_1^* z_2^* z_1' z_2')^n (z_2'^* z_2')^{\alpha+s}}{\Gamma(1+n)\Gamma(1+\alpha+s+n)}$$

$$= \sum_{s=0}^\infty \left(\sqrt{\frac{z_2^* z_2'}{z_1^* z_1'}}\right)^{\alpha+s} I_{\alpha+s}\left(2\sqrt{z_1^* z_2^* z_1' z_2'}\right) = Q_\alpha\left(\sqrt{z_1^* z_1'},\, \sqrt{z_2^* z_2'}\right), \tag{3.555}$$

with the function $Q_\alpha(x, y)$ defined as follows:

$$Q_\alpha(x, y) = \sum_{s=0}^\infty \left(\frac{y}{x}\right)^{\alpha+s} I_{\alpha+s}(2xy)\,,$$

where $I_\alpha(x)$ is the Bessel function of the imaginary argument.

In particular, at $z_1 = z_1'$ and $z_2 = z_2'$ we have

$$\left(F_\alpha(z_1, z_2; \varphi, \rho), F_\alpha(z_1, z_2; \varphi, \rho)\right) = Q_\alpha(|z_1|, |z_2|)\,. \tag{3.556}$$

For the instantaneous quasicoherent states (3.552) the scalar product reads

$$\left(\Psi^{(j)}_{z_1, z_2}, \Psi^{(j')}_{z_1', z_2'}\right) = \delta_{j,j'} R^{(j)}, \quad R^{(j)} = \begin{cases} Q_{1-\mu}\left(\sqrt{z_1^* z_1'},\, \sqrt{z_2^* z_2'}\right), & j = 0 \\ Q_\mu\left(\sqrt{z_2^* z_2'},\, \sqrt{z_1^* z_1'}\right), & j = 1 \end{cases}. \tag{3.557}$$

With the help of equations (3.548) and (3.523), one gets

$$
\begin{aligned}
a_1\Psi^{(0)}_{z_1,z_2}(\varphi,\rho) &= z_1\left\{\Psi^{(0)}_{z_1,z_2}(\varphi,\rho) - e^{[i\epsilon(l_0+1)\varphi]}\sum_{n=0}^{\infty}\frac{(z_1z_2)^n z_2^{1-\mu}\, I_{1-\mu+n,n}(\rho)}{\sqrt{\Gamma(1+n)\Gamma(2-\mu+n)}}\right\} \\
&= z_1 e^{[i\epsilon(l_0+1)\varphi]}\left[F_{1-\mu}(z_1,z_2;\varphi,\rho) - e^{(z_1z_2-\frac{\rho}{2})}\left(\sqrt{\tfrac{z_2}{z_1}}\right)^{1-\mu} J_{1-\mu}\left(2\sqrt{z_1z_2\rho}\right)\right], \\
a_2\Psi^{(0)}_{z_1,z_2}(\varphi,\rho) &= z_2\left\{\Psi^{(0)}_{z_1,z_2}(\varphi,\rho) + e^{(i\epsilon l_0\varphi)}\sum_{n=0}^{\infty}\frac{(z_1z_2)^n z_2^{-\mu}\, I_{n-\mu,n}(\rho)}{\sqrt{\Gamma(1+n)\Gamma(1-\mu+n)}}\right\} \\
&= z_2 e^{[i\epsilon(l_0+1)\varphi]}\left[F_{1-\mu}(z_1,z_2;\varphi,\rho) + e^{(z_1z_2-i\epsilon\varphi-\frac{\rho}{2})}\left(\sqrt{\tfrac{z_2}{z_1}}\right)^{-\mu} J_{-\mu}\left(2\sqrt{z_1z_2\rho}\right)\right], \\
a_1\Psi^{(1)}_{z_1,z_2}(\varphi,\rho) &= z_1\left\{\Psi^{(1)}_{z_1,z_2}(\varphi,\rho) - e^{[i\epsilon(l_0+1)\varphi]}\sum_{n=0}^{\infty}\frac{(z_1z_2)^n z_1^{\mu-1}\, I_{\mu-1+n,n}(\rho)}{\sqrt{\Gamma(1+n)\Gamma(\mu+n)}}\right\} \\
&= z_1 e^{(i\epsilon l_0\varphi)}\left[F_{\mu}(z_2,z_1;\pi-\varphi,\rho) - e^{(z_1z_2+i\epsilon\varphi-\frac{\rho}{2})}\left(\sqrt{\tfrac{z_2}{z_1}}\right)^{1-\mu} J_{\mu-1}\left(2\sqrt{z_1z_2\rho}\right)\right], \\
a_2\Psi^{(1)}_{z_1,z_2}(\varphi,\rho) &= z_2\left\{\Psi^{(1)}_{z_1,z_2}(\varphi,\rho) - e^{(i\epsilon l_0\varphi)}\sum_{n=0}^{\infty}\frac{(z_1z_2)^n z_1^{\mu}\, I_{\mu+n,n}(\rho)}{\sqrt{\Gamma(1+n)\Gamma(1+\mu+n)}}\right\} \\
&= z_2 e^{(i\epsilon l_0\varphi)}\left[F_{\mu}(z_2,z_1;\pi-\varphi,\rho) - e^{(z_1z_2-\frac{\rho}{2})}\left(\sqrt{\tfrac{z_1}{z_2}}\right)^{\mu} J_{\mu}\left(2\sqrt{z_1z_2\rho}\right)\right].
\end{aligned}
\tag{3.558}
$$

Relations (3.558) allow one to calculate the matrix elements $(\Psi^{(j)}_{z_1,z_2}, a_s\Psi^{(j')}_{z_1',z_2'}) = (a_s)^{(j,j')}_{z_1,z_2;z_1',z_2'}$, $s = 1,2$

$$
\begin{aligned}
&(a_1)^{(0,0)}_{z_1,z_2;z_1',z_2'} = z_1' Q^{-}_{1-\mu}\left(\sqrt{z_1^* z_1'}, \sqrt{z_2^* z_2'}\right), \quad (a_1)^{(1,1)}_{z_1,z_2;z_1',z_2'} = z_1' Q_{\mu}\left(\sqrt{z_2^* z_2'}, \sqrt{z_1^* z_1'}\right), \\
&(a_1)^{(0,1)}_{z_1,z_2;z_1',z_2'} = -z_1'\left(\sqrt{\frac{z_2^* z_2'}{z_1^* z_1'}}\right)^{1-\mu} \mathfrak{I}_{1-\mu}\left(\sqrt{z_1^* z_2^*}, \sqrt{z_1' z_2'}\right), \quad (a_1)^{(1,0)}_{z_1,z_2;z_1',z_2'} = 0, \\
&(a_2)^{(0,0)}_{z_1,z_2;z_1',z_2'} = z_2' Q_{1-\mu}\left(\sqrt{z_1^* z_1'}, \sqrt{z_2^* z_2'}\right), \quad (a_2)^{(1,1)}_{z_1,z_2;z_1',z_2'} = z_2' Q^{-}_{\mu}\left(\sqrt{z_2^* z_2'}, \sqrt{z_1^* z_1'}\right), \\
&(a_2)^{(1,0)}_{z_1,z_2;z_1',z_2'} = z_2'\left(\sqrt{\frac{z_1^* z_1'}{z_2^* z_2'}}\right)^{\mu} \mathfrak{I}_{\mu}\left(\sqrt{z_1^* z_2^*}, \sqrt{z_1' z_2'}\right), \quad (a_2)^{(0,1)}_{z_1,z_2;z_1',z_2'} = 0,
\end{aligned}
\tag{3.559}
$$

where

$$
\begin{aligned}
Q^{-}_{\alpha}(x,y) &= \sum_{s=1}^{\infty}\left(\frac{y}{x}\right)^{\alpha+s} I_{\alpha+s}(2xy) = Q_{\alpha}(x,y) - \left(\frac{y}{x}\right)^{\alpha} I_{\alpha}(2xy), \\
\mathfrak{I}_{\mu}(x,y) &= \exp(x^2+y^2)\int_0^{\infty} e^{-\rho} J_{\mu}(2x\sqrt{\rho})\, J_{-\mu}(2y\sqrt{\rho})\, d\rho.
\end{aligned}
\tag{3.560}
$$

Time dependent quasicoherent states of a scalar particle in the MSF

Let us search for solutions of equation (3.531) with an integral of motion λ given by (3.489),

$$(\hat{P}^0 - \hat{P}^3)\Psi_\lambda^B(x^i) = (\hat{P}_0 + \hat{P}_3)\Psi_\lambda^B(x^i) = \lambda m \Psi_\lambda^B(x^i)\,, \tag{3.561}$$

by introducing the light-cone variables $u_0 = x^0 - z$ and $u_3 = x^0 + z$. One can easily see that equations (3.561) and (3.531) are obeyed if the function $\Psi_\lambda^B(x^i)$ has the form

$$\Psi_\lambda^B(x^i) = \aleph \exp\left\{-\frac{im}{2\lambda}[\lambda^2 u_3 + (1+b)u_0]\right\}\Phi_\lambda^B(u_0, \varphi, \rho)\,, \tag{3.562}$$

and the function $\Phi_\lambda^B(u_0, \varphi, \rho)$ satisfies the following “Schrödinger equation”:

$$i\frac{\partial\Phi_\lambda^B(u_0, \varphi, \rho)}{\partial u_0} = \mathcal{H}^B\Phi_\lambda^B(u_0, \varphi, \rho), \quad \mathcal{H}^B = \tilde{\omega}\hat{N}_1, \tilde{\omega} = \frac{mb}{\lambda}\,. \tag{3.563}$$

One can see that there exist solutions of the latter equation in the form of instantaneous quasicoherent states (3.548), $\Phi_\lambda^B(u_0, \varphi, \rho) = \Psi_{z_1, z_2}^{(j)}(\varphi, \rho)$ where the parameter z_1 is a complex function of the variable u_0, and z_2 is a complex constant. Indeed, it follows from equation (3.548) that

$$\mathcal{H}^B\Psi_{z_1,z_2}^{(j)}(\varphi, \rho) = \tilde{\omega} z_1 \frac{\partial \Psi_{z_1,z_2}^{(j)}(\varphi, \rho)}{\partial z_1};$$
$$i\frac{\partial \Psi_{z_1,z_2}^{(j)}(\varphi, \rho)}{\partial u_0} = i\dot{z}_1\frac{\partial \Psi_{z_1,z_2}^{(j)}(\varphi, \rho)}{\partial z_1}, \quad \dot{z}_1 = \frac{dz_1}{du_0}\,. \tag{3.564}$$

Then substituting (3.564) into (3.563), we find $i\dot{z}_1 = \tilde{\omega} z_1$, which yields

$$z_1 = z_1^{(\lambda)}(u_0) = z_1^0 \exp(-i\tilde{\omega}u_0)\,, \quad |z_1| = |z_1^0|\,, \tag{3.565}$$

where z_1^0 is an arbitrary complex constant. We chose this constant in the following form:

$$z_1^0 = -|z_1^0|\exp(-i\varphi_{(0)}) = -|z_1|\exp(-i\varphi_{(0)})\,,$$

where $|z_1| \geq 0$ and $\varphi_{(0)}$ are some real constants. Then

$$z_1 = z_1^{(\lambda)}(u_0) = -|z_1|\exp(-i\Omega), \quad \Omega = \tilde{\omega}u_0 + \varphi_{(0)}, \,. \tag{3.566}$$

The quantity Ω coincides with that present in (3.488), if equation (3.499) is taken into account.

Thus, the functions

$$\Psi_{coherent}^{B(j)}(x) = \aleph \exp\left\{-\frac{im}{2\lambda}[\lambda^2 u_3 + (1+b)u_0]\right\}\Psi_{z_1^{(\lambda)}(u_0),\, z_2}^{(j)}(\varphi, \rho) \tag{3.567}$$

are solutions of the K–G equation. We call them quasicoherent states of Bose particles in the MSF.

Any linear combinations of the functions (3.567)

$$\Psi(A_0, A_1; x^i) = A_0 \Psi^{B(0)}_{coherent}(x^i) + A_1 \Psi^{B(1)}_{coherent}(x^i)\,, \tag{3.568}$$

where A_0 and A_1 are arbitrary constants, are solutions of the K–G equation as well. Solutions (3.568) for $A_0 A_1 \neq 0$ represent mixed semicoherent states in the MSF.

Relations (3.559) allow one to find mean values $\overline{(a_s)}_{(j)}$ of the operators a_s, $s = 1, 2$ in the semicoherent states (3.567). These mean values are

$$\begin{aligned} &\overline{(a_1)}_{(0)} = z_1^{(\lambda)}(u_0)\Delta_{1-\mu}(|z_1|, |z_2|), && \overline{(a_2)}_{(0)} = z_2, \\ &\overline{(a_1)}_{(1)} = z_1^{(\lambda)}(u_0), && \overline{(a_2)}_{(1)} = z_2\Delta_\mu(|z_2|, |z_1|)\,, \end{aligned} \tag{3.569}$$

where

$$\Delta_\alpha(x, y) = \frac{Q_\alpha^-(x, y)}{Q_\alpha(x, y)}; \quad 0 < \Delta_\alpha(x, y) < 1, \quad \alpha > 0, \quad x > 0, \quad y > 0\,. \tag{3.570}$$

The given estimation of the functions $\Delta_\alpha(x, y)$ follows from equation (3.560).

Under certain conditions, trajectories of the mean values (3.569) with z_1 given by (3.566) coincide with the classical trajectories (3.492). Then the mean trajectories in the (x, y)-plane coincide with the corresponding classical trajectories due to equations (3.498).

For states with $j = 0$ these conditions are

$$\begin{aligned} &|z_1|\Delta_{1-\mu}(|z_1|, |z_2|) = \sqrt{\frac{\gamma}{2}}R = Rm\sqrt{\frac{b}{2}} \to R = \frac{|z_1|\Delta_{1-\mu}(|z_1|, |z_2|)}{m}\sqrt{\frac{2}{b}}\,, \\ &z_2 = \sqrt{\frac{\gamma}{2}}(x_0 - i\epsilon y_0) = (x_0 - i\epsilon y_0)\, m\sqrt{\frac{b}{2}} \to R_0 = \frac{|z_2|}{m}\sqrt{\frac{2}{b}}\,. \end{aligned} \tag{3.571}$$

It was established (see equation (3.508) and the following discussion) that for the states with $j = 0$, the inequality $R < R_0$ holds. In the case under consideration, this inequality has the form

$$|z_1|\Delta_{1-\mu}(|z_1|, |z_2|) < |z_2| \tag{3.572}$$

in accordance with (3.571). But due to (3.570) this inequality holds if $|z_1| < |z_2|$. Thus, trajectories of the mean values of x and y in the quasicoherent states with $j = 0$ coincide with the classical trajectories, if equation (3.571) and the inequality $|z_1| < |z_2|$ hold true.

For states with $j = 1$ these conditions are

$$\begin{aligned} &z_2\Delta_\mu(|z_2|, |z_1|) = (x_0 - i\epsilon y_0)\, m\sqrt{\frac{b}{2}} \to R_0 = \frac{|z_2|\Delta_\mu(|z_2|, |z_1|)}{m}\sqrt{\frac{2}{b}}; \\ &|z_1| = \sqrt{\frac{\gamma}{2}}R = Rm\sqrt{\frac{b}{2}} \to R = \frac{|z_1|}{m}\sqrt{\frac{2}{b}}\,. \end{aligned} \tag{3.573}$$

For the states with $j = 0$, the inequality $R > R_0$ holds. In the case under consideration, this inequality has the form

$$|z_2|\Delta_\mu(|z_2|, |z_1|) < |z_1| \,, \tag{3.574}$$

and holds true if $|z_1| > |z_2|$. Thus, trajectories of the mean values of x and y in the quasicoherent states with $j = 1$ coincide with the classical trajectories if equation (3.573) and the inequality $|z_1| > |z_2|$ hold true.

One can easily check that trajectories of the mean values in the mixed semicoherent states do not coincide with the classical trajectories.

Time dependent quasicoherent states of spinning particles in the MSF

Let us search for solutions of the Dirac equation (3.535) with the integral of motion λ given by (3.561). Taking into account equation (3.540), we obtain a set of equations for the spinor $\psi(x)$,

$$\begin{aligned} \frac{2i\lambda}{m}\frac{\partial\psi(x)}{\partial u_0} &= \left\{1 + b[2\hat{N}_1 + (1+f)(1-\epsilon\sigma_3)]\right\}\psi(x) \,, \\ \frac{2i}{m\lambda}\frac{\partial\psi(x)}{\partial u_3} &= \psi(x) \,. \end{aligned} \tag{3.575}$$

To solve this set of equations we use the following Ansatz:

$$\begin{aligned} \psi(x) &= U_\lambda(u_3, u_0)V_\lambda, \\ U_\lambda(u_3, u_0) &= \frac{1}{2}\left[1 + \epsilon\sigma_3 + (1-\epsilon\sigma_3)\exp\left(-\frac{imb}{\lambda}u_0\right)\right]\exp\left[-\frac{im(\lambda^2 u_3 + u_0)}{2\lambda}\right], \end{aligned} \tag{3.576}$$

where V_λ is a spinor that satisfies the following equations:

$$i\frac{\partial V_\lambda}{\partial u_0} = \mathcal{H}^B V_\lambda, \quad i\frac{\partial V_\lambda}{\partial u_3} = 0 \,, \tag{3.577}$$

with $\mathcal{H}^B$ given in (3.563). It follows from (3.577) that the spinor V_λ does not depend on the variable u_3, i.e. $V_\lambda = V_\lambda(u_0, \varphi, \rho)$. Taking into account the above analysis of equation (3.563), we can conclude that this spinor can be taken in the form

$$V_\lambda(u_0, \varphi, \rho) = V_\lambda^{(j)}\Psi^{(j)}_{z_1^{(\lambda)}(u_0), z_2}(\varphi, \rho) \,, \tag{3.578}$$

where $V_\lambda^{(j)}$ is a constant complex spinor and functions $\Psi^{(j)}_{z_1^{(\lambda)}(u_0), z_2}(\varphi, \rho)$ are defined in (3.552) with account of (3.566).

Thus, quasicoherent states of an electron are expressed via quasicoherent states of a spinless particle. If one considers the scalar product on the light-cone plane, then the mean trajectories of the electron in the (x, y)-plane coincide with the corresponding classical trajectories similar to the spinless case.

A complete study of semiclassical and coherent states of nonrelativistic and relativistic particles in the MSF is presented in Refs [63, 64].

3.6.5 Aharonov–Bohm field and additional electromagnetic fields

Exact solutions of relativistic wave equations with an external field that represents a combination of the AB field and additional electromagnetic fields have an undoubted physical interest. Additional fields can emphasize or even reinforce some specific manifestations of the Aharonov–Bohm effect. Below, we demonstrate that besides the known cases, there are broad classes of additional fields, for which exact solutions of the relativistic wave equations exist. Among these new additional fields there are physically interesting electric fields acting during a finite time, or localized in a finite space region. We stress that in [65, 66] all electromagnetic fields that include the AB field as a part and that allow exact solutions of the relativistic wave equations by separating variables are described. In this Section these results are presented. In what follows, we consider only the nontrivial AB field with nonzero mantissa μ. As was already said, in such a case cylindrical and spherical coordinates are physically preferable. Namely, in these coordinates relativistic wave equations with the AB field and some additional fields allow separation of the variables, and exact solutions can be obtained.

Structure of additional electromagnetic fields – cylindrical coordinates

Let us write potentials A_ν of the additional field in the form

$$eA_0 = f_0(r, z, x_0), \quad e\mathbf{A} = \frac{f_2(r)}{r}\mathbf{e}_\varphi - f_1(r, z, x_0)\mathbf{k}\,, \tag{3.579}$$

where $f_k, k = 0, 1, 2$ are some arbitrary functions of the indicated arguments and r, z, φ are cylindrical coordinates. Thus, potentials of the AB field (3.465) can be considered as a special case of (3.579) for $f_0(rz, x_0) = f_1(r, z, x_0) = 0$, and $f_2(r) = \Phi/2\pi = \text{const}$.

Electric $\mathbf{E}$ and magnetic $\mathbf{H}$ fields that correspond to potentials (2.93) are

$$\begin{aligned} e\mathbf{E} &= -\partial_r f_0(r, z, x_0)\mathbf{e}_r + [\partial_0 f_1(r, z, x_0) - \partial_z f_0(r, z, x_0)]\mathbf{k}, \\ e\mathbf{H} &= \partial_r f_1(r, z, x_0)\mathbf{e}_\varphi + \frac{f_2'(r)}{r}\mathbf{k}; \ \mathbf{e}_r = \mathbf{i}\cos\varphi + \mathbf{j}\sin\varphi\,. \end{aligned} \tag{3.580}$$

In such fields, the K–G and the Dirac equations have operators $\hat{L}_z = -i\partial_\varphi$ and $\hat{J}_z = \hat{L}_z + 1/2\,\Sigma_3$, respectively, as integrals of motion.

Let us look for solutions of these equations, which are eigenfunctions of these operators with the respective eigenvalues

$$L_z = l - l_0; \quad J_z = l - l_0 - 1/2, \quad l \in \mathbb{Z}\,.$$

For the K–G equations such solutions have the form

$$\Psi_K(x) = \exp(iQ)\psi_K(r, z, x_0), \quad Q = (l - l_0)\varphi\,, \tag{3.581}$$

where the functions $\psi_K(r, z, x_0)$ satisfy the equation

$$\left\{\pi_0^2 + \partial_r^2 + \frac{1}{r}\partial_r - \frac{[f_2(r) - l - \mu]^2}{r^2} - \pi_3^2 - m^2\right\}\psi_K(r, z, x_0) = 0\,. \tag{3.582}$$

Here

$$\pi_0 = i\partial_0 - f_0(r, z, x_0), \quad \pi_3 = i\partial_z - f_1(r, z, x_0)\,. \tag{3.583}$$

For the Dirac equation, such solutions have the form

$$\Psi_{D;l}(x) = e^{iQ}\begin{pmatrix} e^{-i\varphi}\psi_1(r, z, x_0) \\ i\psi_2(r, z, x_0) \\ e^{-i\varphi}\psi_3(r, z, x_0) \\ i\psi_4(r, z, x_0) \end{pmatrix}, \tag{3.584}$$

where the functions $\psi_s(r, z, x_0)$ satisfy the following set of equations:

$$\overline{D}\,\overline{\Phi} = 0, \overline{\Phi} = \begin{pmatrix} \psi_1(r, z, x_0) \\ \psi_2(r, z, x_0) \\ \psi_3(r, z, x_0) \\ \psi_4(r, z, x_0) \end{pmatrix}, \tag{3.585}$$

with the matrix operator $\overline{D}$ having the form

$$\overline{D} = \rho_3\pi_0 + i\rho_2\Sigma_3\pi_3 + \rho_2\Sigma_2\left(\partial_r + \frac{1}{2r}\right) + i\rho_2\Sigma_1\frac{f_2(r) - l - \mu + 1/2}{r} - m\mathbb{I}\,. \tag{3.586}$$

Here ρ_k and Σ_k are Dirac matrices in the standard representation, see Section A.2.1.

Exact solutions of equations (3.582) and (3.585) are known only for two types of function $f_2(r)$:

$$\begin{aligned} &(a) \quad f_2(r) = \gamma r; \\ &(b) \quad f_2(r) = \gamma r^2, \quad \gamma = \text{const}\,, \end{aligned} \tag{3.587}$$

and for two types of the functions f_0 and f_1 considered below.

Case I

In this case the functions f_0 and f_1 depend on r alone ($f_{0,1} = f_{0,1}(r)$) and are linearly dependent.

For such $f_{0,1}$, let us consider solutions of equations (3.582) and (3.585) that are eigenfunctions for both operators $i\partial_0$ and $i\partial_z$ with eigenvalues k_0, and k_3, respectively. Such solutions have the form (3.581) and (3.584) with

$$Q = (l - l_0)\varphi - k_0 x_0 - k_3 z, \pi_0 = k_0 - f_0(r)\,,$$
$$\pi_3 = k_3 - f_1(r); \quad \psi_K = \psi_K(r), \quad \psi_s = \psi_s(r)\,.$$

Equations for the functions $\psi_K(r)$ and $\psi_s(r)$ hold the form (3.582) and (3.585) with the substitution $\partial_r \to d/dr$.

Let us suppose that the functions $f_0(r)$ and $f_1(r)$ are linearly dependent. Then, with the help of a Lorentz transformation, which does not change the function $f_2(r)$, one can reduce the problem to the following nonequivalent subcases:

$$(1) \quad f_0(r) = f(r) \neq 0, 6 \qquad f_1(r) = 0\,; \tag{3.588}$$
$$(2) \quad f_0(r) = 0, \qquad f_1(r) = f(r) \neq 0\,; \tag{3.589}$$
$$(3) \quad f_0(r) = \epsilon f_1(r) = f(r), \qquad \epsilon = \pm 1\,. \tag{3.590}$$

They are considered separately below.

Subcase 1

This subcase is characterized by the conditions

$$f_0(r) = f(r) \neq 0, \quad f_1(r) = 0\,; \tag{3.591}$$
$$e\mathbf{E} = -f'(r)\mathbf{e}_r, \qquad e\mathbf{H} = r^{-1} f_2'(r)\mathbf{k}\,. \tag{3.592}$$

For such fields, the Dirac equation admits a spin integral of motion $T_1 = m\rho_3\Sigma_3 - k_3\rho_1$. We consider solutions that are its eigenfunctions,

$$T_1 \Psi_{D;l}(x) = \zeta\lambda_1 \Psi_{D;l}(x), \quad \zeta = \pm 1, \quad \lambda_1 = \sqrt{m^2 + k_3^2}\,. \tag{3.593}$$

It follows from (3.593) (with the use of (3.584)) that

$$\psi_1(r) = a\varphi_1(r), \quad \psi_2(r) = -b\varphi_2(r), \quad \psi_3(r) = b\varphi_1(r), \quad \psi_4(r) = -a\varphi_2(r)\,;$$
$$a = \lambda_1 + m + k_3 + \zeta(\lambda_1 + m - k_3), \quad b = \lambda_1 + m - k_3 - \zeta(\lambda_1 + m + k_3)\,.$$

Equations for functions $\varphi_1(r)$ and $\varphi_2(r)$ follow from (3.585),

$$[k_0 - f(r) - \zeta\lambda_1]\varphi_1(r) - \left[\frac{f_2(r) - l - \mu}{r} - \frac{d}{dr}\right]\varphi_2(r) = 0,$$
$$\left[\frac{f_2(r) - l - \mu + 1}{r} + \frac{d}{dr}\right]\varphi_1(r) - [k_0 - f(r) + \zeta\lambda_1]\varphi_2(r) = 0\,. \tag{3.594}$$

Then the K–G equation (3.582) takes the form

$$\left\{\frac{d^2}{dr^2}+\frac{1}{r}\frac{d}{dr}+[k_0-f(r)]^2-\frac{[f_2(r)-l-\mu]^2}{r^2}-k_3^2-m^2\right\}\psi_K(r)=0\,. \qquad (3.595)$$

Solutions of the set (3.594) and equation (3.595) (for nonzero $f(r)$) are known only for

$$f_2(r)=\gamma r,\quad f(r)=\frac{\alpha}{r};\quad \alpha\neq 0,\quad \gamma=\text{const}\,.$$

These solutions have the form

$$\begin{aligned}
\varphi_1(r)&=\sqrt{q_{1(-)}\omega_1(\zeta)}\,v_1(x)+\sqrt{q_{1(+)}\omega_1(-\zeta)}\,v_2(x)\,,\\
\varphi_2(r)&=\sqrt{q_{1(+)}\omega_1(\zeta)}\,v_1(x)+\sqrt{q_{1(-)}\omega_1(-\zeta)}\,v_2(x)\,,\\
\psi_K(r)&=v_0(x)\,,
\end{aligned} \qquad (3.596)$$

where

$$q_{1(\pm)}=l+\mu-1/2\pm\sqrt{(l+\mu-1/2)^2-\alpha^2},E=\sqrt{m^2+k_3^2+\gamma^2-k_0^2}\,,$$

$$\omega_1(\zeta)=\gamma\alpha-k_0\,(l+\mu-1/2)+\zeta\lambda_1\sqrt{(l+\mu-1/2)^2-\alpha^2},\quad x=2rE\,.$$

The functions $v_s(x)$, $s=0,1,2$, have similar structure,

$$v_s(x)=AI_{p_s,n_s}(x)+BI_{n_s,p_s}(x)\,, \qquad (3.597)$$

where A and B are arbitrary constants, and $I_{p,n}(x)$ the Laguerre functions, the subscripts of the Laguerre functions p_s and n_s, $s=0,\ 1,\ 2$, having the form

$$n_s=\frac{1}{E}\left\{\gamma\left[l+\mu-\frac{s(3-s)}{4}\right]-k_0\alpha\right\}+\frac{(3s+1)(s-2)}{4}-\sqrt{\left[l+\mu-\frac{s(3-s)}{4}\right]^2-\alpha^2},$$

and

$$p_s=\frac{1}{E}\left\{\gamma\left[l+\mu-\frac{s(3-s)}{4}\right]-k_0\alpha\right\}+\frac{(2-3s)(s-1)}{4}+\sqrt{\left[l+\mu-\frac{s(3-s)}{4}\right]^2-\alpha^2}\,. \qquad (3.598)$$

For $n_s\in\mathbb{Z}_+$, the functions (3.598) are square-integrable for $A\neq 0$ and $B=0$, and the energy k_0 is quantized. In the case of the Dirac equation, we find

$$\begin{aligned}
k_0&=\frac{1}{N^2+\alpha^2}\left[\alpha\gamma\,(l+\mu-1/2)+N\sqrt{(N^2+\alpha^2)(m^2+k_3^2+\gamma^2)-\gamma^2\,(l+\mu-1/2)^2}\right],\\
N&=n+\sqrt{(l+\mu-1/2)^2-\alpha^2};\quad (l+\mu-1/2)^2\geq\alpha^2;\quad n_1=n-1,\ n_2=n\in\mathbb{Z}_+\,,
\end{aligned} \qquad (3.599)$$

whereas in the case of the K–G equation we have

$$\begin{aligned}
k_0&=\frac{1}{\overline{N}^2+\alpha^2}\left[\alpha\gamma\,(l+\mu)+\overline{N}\sqrt{(\overline{N}^2+\alpha^2)(m^2+k_3^2+\gamma^2)-\gamma^2\,(l+\mu)^2}\right],\\
\overline{N}&=n+1/2+\sqrt{(l+\mu)^2-\alpha^2},\quad (l+\mu)^2\geq\alpha^2,\quad n_0=n\,.
\end{aligned} \qquad (3.600)$$

Subcase 2

This subcase is characterized by the conditions

$$f_0(r) = 0, \quad f_1(r) = f(r) \neq 0; \quad \mathbf{E} = 0, \quad e\mathbf{H} = f'(r)\mathbf{e}_\varphi + r^{-1} f_2'(r)\mathbf{k}\,.$$

Thus, we are dealing with a pure magnetic field. In such a field, the Dirac equation admits a spin integral of motion $T_2 = (\mathbf{\Sigma P})$. Then we can impose an additional condition on the wave function,

$$(\mathbf{\Sigma P})\Psi_{D;l}(x) = \zeta\lambda_2\Psi_{D;l}(x), \quad \zeta = \pm 1, \lambda_2 = \sqrt{k_0^2 - m^2}\,. \tag{3.601}$$

Equations (3.601) and (3.585) are consistent if we set

$$\psi_1(r) = \sqrt{k_0 + m}\,\overline{\varphi}_1(r), \quad \psi_2(r) = \sqrt{k_0 + m}\,\overline{\varphi}_2(r)\,,$$
$$\psi_3(r) = \zeta\sqrt{k_0 - m}\,\overline{\varphi}_1(r), \quad \psi_4(r) = \zeta\sqrt{k_0 - m}\,\overline{\varphi}_2(r)$$

in (3.585), where the functions $\overline{\varphi}_1(r)$, and $\overline{\varphi}_2(r)$ obey equations that are similar to those in (3.594),

$$[\zeta\lambda_2 + k_3 - f(r)]\overline{\varphi}_1(r) + \left[\frac{f_2(r) - l - \mu}{r} - \frac{d}{dr}\right]\overline{\varphi}_2(r) = 0,$$
$$\left[\frac{f_2(r) - l - \mu + 1}{r} + \frac{d}{dr}\right]\overline{\varphi}_1(r) + [\zeta\lambda_2 - k_3 + f(r)]\overline{\varphi}_2(r) = 0\,. \tag{3.602}$$

Now, the K–G equation (3.582) takes the form

$$\left\{\frac{d^2}{dr^2} + \frac{1}{r}\frac{d}{dr} - [k_3 - f(r)]^2 - \frac{[f_2(r) - l - \mu]^2}{r^2} + k_0^2 - m^2\right\}\psi_K(r) = 0\,. \tag{3.603}$$

Equations (3.602) and (3.603) have exact solutions (for nonzero $f(r)$) only for

$$f(r) = \frac{\alpha}{r}, \quad f_2(r) = \gamma r; \quad \alpha \neq 0, \quad \gamma = \text{const}\,.$$

Such solutions have the form (they are similar to those in (3.596)

$$\overline{\varphi}_1(r) = \sqrt{q_{2(-)}\omega_2(\zeta)}\,v_1(x) - \zeta\sqrt{q_{2(+)}\omega_2(-\zeta)}\,v_2(x)\,,$$
$$\overline{\varphi}_2(r) = \sqrt{q_{2(+)}\omega_2(\zeta)}\,v_1(x) + \zeta\sqrt{q_{2(-)}\omega_2(-\zeta)}\,v_2(x), \quad \psi_K(r) = v_0(x);$$
$$q_{2(\pm)} = \sqrt{(l + \mu - 1/2)^2 + \alpha^2} \pm (l + \mu - 1/2)\,, \quad E = \sqrt{m^2 + k_3^2 + \gamma^2 - k_0^2}\,,$$
$$\omega_2(\zeta) = \lambda_2\sqrt{(l + \mu - 1/2)^2 + \alpha^2} + \zeta[k_3\,(l + \mu - 1/2) - \gamma\alpha], \quad x = 2rE\,. \tag{3.604}$$

The functions $v_s(x)$, $(s = 0, 1, 2)$ are given by expressions (3.597), with the following indices of the Laguerre functions:

$$n_s = \frac{1}{E}\left\{\gamma\left[l + \mu - \frac{s(3 - s)}{4}\right] + k_3\alpha\right\} + \frac{(3s + 1)(s - 2)}{4} - \sqrt{\left[l + \mu - \frac{s(3 - s)}{4}\right]^2 + \alpha^2}\,,$$

$$p_s = \frac{1}{E}\left\{\gamma\left[l + \mu - \frac{s(3 - s)}{4}\right] + k_3\alpha\right\} + \frac{(2 - 3s)(s - 1)}{4} + \sqrt{\left[l + \mu - \frac{s(3 - s)}{4}\right]^2 + \alpha^2}\,.$$

For nonnegative integer n_s the functions (3.597) are square-integrable for $A \neq 0$ and $B = 0$, and the energy k_0 is quantized ($n_1 = n - 1, \quad n_2 = n_0 = n \in \mathbb{Z}_+$),

$$k_0^2 = m^2 + k_3^2 + \gamma^2 - \left[\frac{\gamma (l + \mu - \tau/2) + k_3 \alpha}{n + (1 - \tau)/2 + \sqrt{(l + \mu - \tau/2)^2 + \alpha^2}} \right]^2 , \tag{3.605}$$

where $\tau = 1$ for the Dirac equation and $\tau = 0$ for the K–G equation.

Subcase 3

This subcase is characterized by the conditions

$$f_0(r) = f(r) \neq 0, \quad f_1(r) = \epsilon f(r), \quad \epsilon^2 = 1;$$
$$e\mathbf{E} = f'(r)(\epsilon\mathbf{k} - \mathbf{e}_r), \quad e\mathbf{H} = \epsilon f'(r)\mathbf{e}_\varphi + r^{-1} f_2'(r)\mathbf{k}$$

and the fields have the following properties:

$$\mathbf{H} = [\mathbf{n} \times \mathbf{E}] + \mathbf{n}(\mathbf{nH}), \quad \mathbf{E} = -[\mathbf{n} \times \mathbf{H}] + \mathbf{n}(\mathbf{nE}), \quad \mathbf{n} = -\epsilon\mathbf{k}\,. \tag{3.606}$$

If equations (3.606) hold, then the bispinor $\overline{\Phi}(r)$ in equation (3.585) can be represented as

$$\overline{\Phi}(r) = \begin{pmatrix} \left[k_0 - \epsilon k_3 + m - \epsilon \hat{Q}\sigma_3\right] V(r) \\ \left[\epsilon(k_0 - \epsilon k_3 - m)\sigma_3 - \hat{Q}\right] V(r) \end{pmatrix}, \quad V(r) = \begin{pmatrix} v_1(r) \\ v_2(r) \end{pmatrix},$$
$$\hat{Q} = \sigma_1 \frac{f_2(r) - l - \mu + 1/2}{r} - i\sigma_2 \left(\frac{d}{dr} + \frac{1}{2r} \right). \tag{3.607}$$

The functions $v_1(r)$, $v_2(r)$ and $\psi_K(r) = v_0(r)$ obey the equations

$$\left\{ \frac{d^2}{dr^2} + \frac{1}{r}\frac{d}{dr} - 2(k_0 - \epsilon k_3) f(r) - \frac{[f_2(r) - l - \mu + \tau_s]^2}{r^2} + \delta_s \frac{f_2'(r)}{r} \right.$$
$$\left. + k_0^2 - k_3^2 - m^2 \right\} v_s(r) = 0,$$
$$s \in \mathbb{Z}_+; \quad \tau_s = s(2 - s); \quad \delta_s = \frac{s(5 - 3s)}{2}\,. \tag{3.608}$$

Equations (3.608) have exact solutions for the following two types of the functions $f(r)$ and $f_2(r)$:

(a)

$$f_2(r) = \gamma r, \quad f(r) = \frac{\alpha}{r} + \frac{\beta}{r^2};$$
$$x = 2rE, \quad \alpha, \beta, \gamma = \text{const};$$
$$E = \sqrt{m^2 + k_3^2 + \gamma^2 - k_0^2}\,; \tag{3.609}$$

(b)

$$f_2(r) = \gamma r^2, \quad f(r) = \alpha r^2 + \frac{\beta}{r^2};$$
$$x = r^2 E_0, \quad \alpha, \beta, \gamma = \text{const};$$
$$E_0 = \sqrt{\gamma^2 + 2\alpha(k_0 - \epsilon k_3)} \,. \tag{3.610}$$

As before, the functions $v_s(x)$, $(s = 0, 1, 2)$ are given by expressions (3.597), where the indices of the Laguerre functions n_s, p_s have the form:

(a)

$$p_s = \frac{b_s}{E} - \frac{1}{2} + a_s, \quad n_s = \frac{b_s}{E} - \frac{1}{2} - a_s,$$
$$b_s = \gamma(l + \mu - \tau_s + \frac{\delta_s}{2}) - \alpha(k_0 - \epsilon k_3) \,,$$
$$a_s = \sqrt{(l + \mu - \tau_s)^2 + 2\beta(k_0 - \epsilon k_3)} \,; \tag{3.611}$$

(b)

$$p_s = \frac{\tilde{b}_s}{4E_0} - \frac{1}{2} + \frac{a_s}{2}, \quad n_s = \frac{\tilde{b}_s}{4E_0} - \frac{1}{2} - \frac{a_s}{2},$$
$$\tilde{b}_s = 2\gamma(l + \mu - \tau_s + \delta_s) + k_0^2 - k_3^2 - m^2,$$
$$a_s = \sqrt{(l + \mu - \tau_s)^2 + 2\beta(k_0 - \epsilon k_3)} \,. \tag{3.612}$$

For $n_s \in \mathbb{Z}_+$, the functions (3.597) with indices (3.612) are square-integrable for $A \neq 0$ and $B = 0$, and the energy k_0 is quantized ($n_1 = n - 1, \quad n_2 = n_0 = n \in \mathbb{Z}_+$). In the general case explicit expressions for the energy k_0 do not exist. Such expressions can be written for fields of the type (a) as a root of an algebraic equation of degree six, and for fields of the type (b) as a root of an algebraic equation of degree eight.

Case II

In this case, the functions f_0 and f_1 depend on z and x_0, and do not depend on r ($f_{0,1} = f_{0,1}(z, x_0)$.

As follows from (2.94), additional fields have the form

$$e\mathbf{E} = F\mathbf{k}, \quad e\mathbf{H} = G\mathbf{k}; \quad F = \partial_0 f_1(z, x_0) - \partial_z f_0(z, x_0), \quad G = f_2'(r) r^{-1} \,. \tag{3.613}$$

These are longitudinal electromagnetic fields, the electric and magnetic fields are parallel to the axis z, and, besides, $\mathbf{H} = \mathbf{H}(r)$ and $\mathbf{E} = \mathbf{E}(z, x_0)$. For such fields, solutions of the K–G equation (3.582) can be found in the form

$$\psi_K(r, z, x_0) = w_0(z, x_0) v_0(r) \,. \tag{3.614}$$

As to solutions of the Dirac equation (3.585), we represent the bispinor $\overline{\Phi}(r, z, x_0)$ in the form

$$\overline{\Phi}(r, z, x_0) = \begin{pmatrix} \left[m + \pi_0 + \pi_3 + \hat{Q}\sigma_3\right] w_1(z, x_0) V(r) \\ \left[(m - \pi_0 - \pi_3)\sigma_3 - \hat{Q}\right] w_1(z, x_0) V(r) \end{pmatrix}, \tag{3.615}$$

where

$$\pi_0 = i\partial_0 - f_0(z, x_0), \quad \pi_3 = i\partial_z - f_1(z, x_0),$$
$$V(r) = \begin{pmatrix} v_1(r) \\ v_2(r) \end{pmatrix}, \quad \hat{Q} = \sigma_1 \frac{f_2(r) - l - \mu + 1/2}{r} - i\sigma_2 \left(\frac{d}{dr} + \frac{1}{2r} \right).$$

The functions $v_s(r)$, $s = 0, 1, 2$, and functions $w_\nu(z, x_0)$, $\nu = 0, 1$, in equations (3.614) and (3.615) obey the equations

$$\left\{ \frac{d^2}{dr^2} + \frac{1}{r}\frac{d}{dr} - \frac{[f_2(r) - l - \mu + s(2-s)]^2}{r^2} + \frac{s(5-3s)}{2}\frac{f_2'(r)}{r} + k_\perp^2 \right\} v_s(r) = 0, \tag{3.616}$$

$$\left\{ \pi_0^2 - m^2 - \pi_3^2 - k_\perp^2 + i\nu[\partial_z f_0(z, x_0) - \partial_0 f_1(z, x_0)] \right\} w_\nu(z, x_0) = 0, k_\perp^2 = \text{const}. \tag{3.617}$$

Exact solutions of equation (3.616) (for $f_2(r) \neq 0$) are known in the two cases (3.587). For $f_2(r) = \gamma r$, ($\gamma = \text{const}$) they have the form (3.597) if we set their $x = 2r\sqrt{\gamma^2 - k_\perp^2}$ and

$$p_s = \frac{\gamma}{\sqrt{\gamma^2 - k_\perp^2}} \left[l + \mu + \frac{s(s-3)}{4} \right] - \frac{1}{2} + |l + \mu - s(2-s)|,$$
$$n_s = \frac{\gamma}{\sqrt{\gamma^2 - k_\perp^2}} \left[l + \mu + \frac{s(s-3)}{4} \right] - \frac{1}{2} - |l + \mu - s(2-s)|. \tag{3.618}$$

For $\gamma^2 > k_\perp^2$, $\gamma(l+\mu-\tau) \geq 0$ ($\tau = 1/2$ for the Dirac equation and $\tau = 0$ for the K–G equation) the quantity $k_\perp^2$ is quantized and the wave functions (3.597) with subindices are square-integrable for $A \neq 0$, $B = 0$,

$$k_\perp^2 = \gamma^2 \left[1 - \frac{(l + \mu - \tau)^2}{(n + 1/2 + |l + \mu|)^2} \right], \quad n_0 = n_2 = n \in \mathbb{Z}_+, \quad n_1 = n - \operatorname{sgn} l. \tag{3.619}$$

The second case $f_2(r) = \gamma r^2$ corresponds to the MSF studied in detail in Section 3.6.3.

Let us consider solutions of the equation (3.617). A wide class of its solutions can be found if both the functions $f_{0,1}(z, x_0)$ depend on one variable only: either $f_{0,1} = f_{0,1}(z)$ (in such a case, without loss of generality, we can set $f_0(z) = f(z) \neq 0$, $f_1(z) = 0$), or $f_{0,1} = f_{0,1}(x_0)$ (here also without loss of generality, we can set $f_0(x_0) = 0$, $f_1(x_0) = f(x_0) \neq 0$). In the first case, we can choose $w_\nu(x, x_0)$ as an eigenfunction of $i\partial_0$, so that we find

$$i\partial_0 w_\nu(x, x_0) = k_0 w_\nu(x, x_0) \Rightarrow w_\nu(x, x_0) = \exp(-ik_0 x_0) w_\nu(x),$$
$$\left\{ \frac{d^2}{dx^2} + [k_0 - f(x)]^2 - m^2 - k_\perp^2 + i\nu f'(x) \right\} w_\nu(x) = 0. \tag{3.620}$$

In the second case, $w_\nu(z,x)$ can be chosen as an eigenfunction of $i\partial_z$ and we find

$$i\partial_z w_\nu(z,x) = k_3 w_\nu(z,x) \Rightarrow w_\nu(z,x) = \exp(-ik_3 z) w_\nu(x)\,,$$
$$\left\{\frac{d^2}{dx^2} + [k_3 - f(x)]^2 + m^2 + k_\perp^2 + i\nu f'(x)\right\} w_\nu(x) = 0\,. \qquad (3.621)$$

Equations (3.620) and (3.621) are one-dimensional Schrödinger equations. Their exact solutions are known for the following functions $f(x)$:

$$f(x) = \alpha x;\quad f(x) = \alpha/x;\quad f(x) = \alpha\exp(\beta x);$$
$$f(x) = \alpha\tan(\beta x);\quad f(x) = \alpha\tanh(\beta x);\quad f(x) = \alpha\coth(\beta x)\,, \qquad (3.622)$$

see Chapter 9.

Finally, there are two cases when the functions $f_{0,1}(z,x_0)$ depend essentially on both arguments z, x_0, and one can find exact solutions of equation (3.617). Let us consider these cases.

(1) Let $f_0(z,x_0) = f_1(z,x_0) = 1/2\, f(\xi)$, $\xi = x_0 - z$; in this case $F = f'(\xi)$ in equation (3.613). Then $w_\nu(z,x_0)$ can be taken as an eigenfunction of the operator $i(\partial_0 + \partial_z)$ with the eigenvalue λ, which implies

$$w_\nu(z,x_0) = [\lambda - f(\xi)]^{-\frac{1+\nu}{2}}\exp(iS)\,,$$
$$S = -\frac{1}{2}\left[\lambda(x_0+z) + (m^2+k_\perp^2)\int\frac{d\xi}{\lambda - f(\xi)}\right]. \qquad (3.623)$$

(2) Let

$$f_0(z,x_0) = -\frac{z}{\bar\xi}f(\bar\xi),\quad f_1(z,x_0) = \frac{x_0}{\bar\xi}f(\bar\xi);\quad \bar\xi = x_0^2 - z^2\,,$$

in this case $F = 2f'(\bar\xi)$ in equation (3.613). Then $w_\nu(z,x_0)$ can be taken as an eigenfunction of the operator $\hat q = i(z\partial_0 + x_0\partial_z)$ (this operator is an integral of motion for the K–G and Dirac equations) with the eigenvalue λ, which implies

$$\hat q w_\nu(z,x_0) = \lambda w_\nu(z,x_0) \Rightarrow w_\nu(z,x_0) = \left(\frac{x_0 - z}{x_0 + z}\right)^{i\frac{\lambda}{2}} w_\nu(\bar\xi)\,. \qquad (3.624)$$

Substituting (3.624) into (3.617), we find an equation for the function $w_\nu(\bar\xi)$,

$$4\bar\xi^2 w_\nu''(\bar\xi) + 4\bar\xi w_\nu'(\bar\xi) + \left\{[\lambda - f(\bar\xi)]^2 + \bar\xi[m^2 + k_\perp^2 + 2i\nu f'(\bar\xi)]\right\} w_\nu(\bar\xi) = 0. \quad (3.625)$$

Equation (3.625) can be solved exactly in two cases:

(a)

$$f(\bar\xi) = \alpha\bar\xi,\quad \alpha = \text{const}\,,$$

which corresponds to a constant and uniform electric field.

(b)

$$f(\bar\xi) = \alpha\sqrt{|\bar\xi|},\quad \alpha = \text{const}\,.$$

In these cases solutions have the form (3.597),

$$w_\nu(\bar{\xi}) = AI_{p,n}(x) + BI_{n,p}(x), \quad A,\, B = \text{const}\,, \tag{3.626}$$

where one has to set respectively: in the case (a)

$$x = -i\alpha\bar{\xi}, \quad p = \frac{i(m^2 + k_\perp^2) - 2\alpha(1+\nu)}{4\alpha}, \quad n = p - i\lambda\,;$$

in the case (b)

$$x = 2i\sqrt{\alpha^2|\bar{\xi}| + \bar{\xi}(m^2 + k_\perp^2)}\,,$$

$$p = \frac{\alpha(\nu + 2i\lambda)}{\sqrt{\alpha^2 + \varepsilon(m^2 + k_\perp^2)}} - 1/2 + i\lambda, \quad n = p - 2i\lambda, \quad \varepsilon = \operatorname{sgn}\bar{\xi}\,.$$

Thus, with the above consideration, all exactly solvable additional electromagnetic fields in the cylindric coordinates have been exhausted.

Structure of additional electromagnetic fields – spherical coordinates

In the spherical coordinates r, θ, φ ($x = r\sin\theta\cos\varphi$, $y = r\sin\theta\sin\varphi$, $z = r\cos\theta$), potentials of the AB field (2.87) have the form

$$A_\nu^{\text{AB}} = \left(A_0^{\text{AB}}, -\mathbf{A}^{\text{AB}}\right), \quad \mathbf{A}^{\text{AB}} = \frac{\Phi}{2\pi r\sin\theta}\mathbf{e}_\varphi, \quad \mathbf{e}_\varphi = -\mathbf{i}\sin\varphi + \mathbf{j}\cos\varphi\,. \tag{3.627}$$

Let us choose potentials A_ν of additional fields in the following form:

$$eA_0 = f_0(r), \quad e\mathbf{A} = \frac{f_1(\cos\theta)}{r\sin\theta}\mathbf{e}_\varphi\,, \tag{3.628}$$

where $f_k, k = 0, 1$, are arbitrary functions of the indicated arguments. Thus, the potentials of the AB field (3.627) can be considered as a particular case of potentials of the additional field (3.628) for $f_0(r) = 0$ and $f_1(\cos\theta) = \Phi/2\pi = \text{const}$.

Electric and magnetic fields that correspond to (3.628) have the form

$$e\mathbf{E} = -f_0'(r)\mathbf{e}_r, \quad e\mathbf{H} = -\frac{f_1'(\cos\theta)}{r^2}\mathbf{e}_r, \quad \mathbf{e}_r = \sin\theta(\mathbf{i}\cos\varphi + \mathbf{j}\sin\varphi) + \mathbf{k}\cos\theta\,. \tag{3.629}$$

One can find exact solutions of the K–G and Dirac equations for

$$f_1(\cos\theta) = \alpha\cos\theta + \beta, \quad f_0(r) = \frac{\gamma}{r}, \quad \alpha, \beta, \gamma = \text{const}\,. \tag{3.630}$$

In this case the complete external electromagnetic field is a combination of the AB field with the Coulomb field and the field of a magnetic monopole. The corresponding exact solutions were studied in the works [3, 56, 111, 203, 214, 334].

4 Particles in fields of special structure

4.1 Introduction

In this Chapter we present exact solutions of the K–G and Dirac equations (and solutions of the corresponding classical equations of motion) with external fields of various configurations that are not considered in Chapter 3.

For every field under consideration, three (out of the complete set of) mutually commuting integrals of motion are explicitly known. The spin integral of motion is not known in every case. Nevertheless, classification of the Dirac wave functions according to the spin states turns out to always be possible. The three explicitly known integrals of motion always have their direct classical meaning. The Lorentz equations can be explicitly integrated one time and are reduced to a set of first-order differential equations. Henceforth, we always write this set of equations, referring to it also as the Lorentz system or as the set of first integrals. Moreover, this first-order set always allows an explicit integration in quadratures.

All the solutions of the K–G and Dirac equations to be listed below make complete and orthogonal systems either on the plane $x^0 = \text{const}$, or on the light-cone plane $u^0 = \text{const}$. These facts are not verified each time, nor are the corresponding normalizing factors calculated. The discussion of the listed solutions and their properties is but briefly presented because they are very numerous.

Each class of solutions is placed in the corresponding subsection. The notation of different subsections are independent, unless an indication is given to the contrary.

Solutions of every special class can be most simply written in some curvilinear system of coordinates (namely, the one whose application results in the separation of variables in the K–G equation for the given class of fields). In this Chapter, for the sake of convenience, we shall write the index of a curvilinear coordinate as a subscript (in contrast to Chapter 2 where curvilinear coordinates are denoted by u^μ). We indicate each time the relation between the curvilinear coordinates u_μ and Minkowski coordinates x^μ.

Electromagnetic potentials are given by their covariant components relative to the corresponding curvilinear reference frame and are denoted as $\tilde{A}_\mu$ in contrast with potentials A_μ in the Minkowski reference frame. For the field strengths we most often indicate the Minkowski components (as functions of u_μ). Sometimes, when it is more convenient, we show their projections onto the basis unit vectors of the cylindrical reference frame. As a rule, the major part of the problems is, in the end, reduced to a solution of the one-dimensional stationary Schrödinger equation. To avoid repetition, we have gathered all the known potentials allowing exact solutions of the one-dimensional stationary Schrödinger equation, as well as the solutions themselves, in the Chapter 9. After having reduced a problem to the one-dimensional Schrödinger equation, we refer to the items of such a Chapter, where the corresponding solutions and potentials are written.

Note, finally, that all the fields considered here allow the separation of variables in the K–G equation, but do not allow such separation directly in the Dirac equation, see Ref. [27] and Refs. [26, 299] respectively.

We finally note that in this Chapter the algebraic charge e is absorbed into the field strengths and potentials, i.e. $\mathbf{E} \longrightarrow e\mathbf{E}$, $\mathbf{H} \longrightarrow e\mathbf{H}$, $A_\mu \longrightarrow eA_\mu$.

4.2 Crossed electromagnetic fields

4.2.1 General

In this section, we consider the classical and quantum motion of charged particles in *crossed electromagnetic fields*. We recall that crossed fields are electromagnetic fields whose strengths $\mathbf{E}$ and $\mathbf{H}$ are orthogonal to each other, and besides are equal to each other in absolute magnitude and orthogonal to the constant unit vector $\mathbf{n}$, see (2.99) and (2.100). Some general properties of the motion of a charged particle in crossed electromagnetic fields can be investigated without specializing the space and time dependence of the fields. In what follows, we shall stick to a special direction of the vector $\mathbf{n}$, namely to that of the Cartesian axis z, i.e. $\mathbf{n} = (0, 0, 1)$. This does not imply any loss of generality. Then one has for the crossed electromagnetic fields

$$E_x = H_y, \quad E_y = -H_x, \quad E_z = H_z = 0\,, \tag{4.1}$$

i.e. the fields can be determined by fixing E_x and E_y. Using the light-cone variables (2.37),

$$u_0 = x^0 - x^3, \quad u_3 = x^0 - x^3\,, \tag{4.2}$$

$$u_1 = x^1, \qquad u_2 = x^2\,, \tag{4.3}$$

we obtain from the Maxwell equations that, with these variables, the covariant components of electromagnetic potentials of an arbitrary crossed field can always be defined in the form

$$\tilde{A}_0 = \tilde{A}_0(u_0, u_1, u_2)\,, \quad \tilde{A}_1 = \tilde{A}_2 = \tilde{A}_3 = 0\,, \tag{4.4}$$

where $\tilde{A}_0$ is an arbitrary function of the indicated arguments. Hence we have

$$E_x = -\partial_{u_1}\tilde{A}_0\,, \quad E_y = -\partial_{u_2}\tilde{A}_0\,. \tag{4.5}$$

For the crossed fields, the vector of the electric current density is expressed via the charge density ρ as

$$j^\mu = n^\mu \rho, \quad j^0 = \rho, \quad \mathbf{j} = \mathbf{n}\rho, \quad \rho = -(\partial_{u_1 u_1} + \partial_{u_2 u_2})\tilde{A}_0\,. \tag{4.6}$$

As already noted, the quantity $\lambda = (nP)$ is an integral of motion in the crossed fields, see Section 2.3.5. The Lorentz equations of motion for the crossed fields can be written as

$$\lambda \mathbf{r}_\perp'' - \mathbf{E} = 0, \quad u_3' = \mathbf{r}_\perp'^2 + m^2\lambda^{-2},$$
$$m\dot{u}_0 = \lambda, \quad \mathbf{r}_\perp = \left(x^1, x^2, 0\right). \tag{4.7}$$

The primes here denote derivatives with respect to the light-cone variable u_0, and the dot in derivative with respect to the proper time. It follows from (4.7) that

$$mu_0 = \lambda\tau, \tag{4.8}$$

i.e. u_0 is proportional to the proper time τ. In order to solve the classical equations of motion it is sufficient to integrate two second-order and one first-order equations (first line in equations (4.7)).

The classical action can be found in the form

$$S = -\lambda u_3/2 - \bar{S}(u_0, u_1, u_2), \tag{4.9}$$

where $\bar{S}$ obeys the equation

$$\frac{\partial \bar{S}}{\partial u_0} = (2\lambda)^{-1}\left[(\nabla_\perp \bar{S})^2 + m^2\right] + \tilde{A}_0, \quad \nabla_\perp = \left(\partial_{u_1}, \partial_{u_2}, 0\right). \tag{4.10}$$

Equation (4.10) is the nonrelativistic Hamilton–Jacobi equation [230] for two-dimensional motion of a particle with an effective mass λ, in a potential field

$$V = \tilde{A}_0 + (2\lambda)^{-1} m^2. \tag{4.11}$$

Solutions of the K–G equation φ, which are eigenfunctions for the operator $\hat{\lambda} = (n\hat{P})$ (with the eigenvalues λ) can be represented in the form

$$\varphi = \exp(-i\lambda u_3/2)\Phi(u_0, u_1, u_2), \tag{4.12}$$

where the function $\Phi(u_0, u_1, u_2)$ is a solution of the equation

$$i\frac{\partial \Phi}{\partial u_0} = \hat{H}\Phi, \quad \hat{H} = (2\lambda)^{-1}\left(\hat{\mathbf{P}}_\perp^2 + m^2\right) + \tilde{A}_0, \quad \hat{\mathbf{P}}_\perp = -i\nabla_\perp \tag{4.13}$$

that has the form of a two-dimensional Schrödinger equation for a particle with a mass λ and with the potential (4.11).

Solutions of the Dirac equation can be expressed in terms of solutions of the K–G equation using the standard trick (2.110).

The current densities (in Cartesian coordinates) are:

$$j^\mu = \varphi^* \hat{P}^\mu \varphi + (\hat{P}^\mu \varphi)^* \varphi = \lambda^{-1}\Phi^* \bar{j}^\mu \Phi,$$
$$\bar{j}^\mu = \left\{\hat{\mathbf{P}}_\perp^2 + m^2 + \lambda^2,\ 2\lambda\hat{\mathbf{P}}_\perp + \mathbf{n}(\hat{\mathbf{P}}_\perp^2 + m^2 - \lambda^2)\right\} \tag{4.14}$$

for scalar wave functions (4.12), and

$$j^{\mu} = \Psi^{+}\gamma^{0}\gamma^{\mu}\Psi = 2\Phi^{*}\bar{j}^{\mu}\Phi v^{+}v \tag{4.15}$$

for spinor wave functions Ψ.

Thus, in the case under consideration, the relativistic quantum-mechanical problem is reduced to a solution of equation (4.13). Further details of the classical and quantum motion will be specified for each type of crossed fields.

When analyzing the motion in special crossed fields, one may need to use more complicated curvilinear coordinates than the light-cone variables. Sometimes we define electromagnetic potentials in gauges that differ from (4.4).

In all the exact solvable cases listed below, the following facts always take place.

The classical action has the following form:

$$S = -\frac{\lambda}{2}u_3 - \Gamma + \int \sqrt{R_1(u_1)}du_1 + \int \sqrt{R_2(u_2)}du_2\,, \tag{4.16}$$

where $R_1(u_1)$ and $R_2(u_2)$ are some functions of the indicated variables and the function Γ is given explicitly in each case. The classically admissible ranges of motion are determined by the conditions

$$R_1(u_1) \geq 0, \quad R_2(u_2) \geq 0\,. \tag{4.17}$$

The scalar wave functions with definite λ can be represented as

$$\varphi(x) = Q^{-1/4}\exp(-i\lambda u_3/2 - i\Gamma)\psi_1(u_1)\psi_2(u_2)\,, \tag{4.18}$$

where $Q = Q(u_0)$ is a function of u_0 only, whereas the functions $\psi_s(u_s)$, $(s = 1, 2)$ are to be found as solutions of certain ordinary linear differential equations of the order not higher than second.

The spinor wave functions constructed from the corresponding scalar ones using the standard trick (2.110) have the form

$$\begin{aligned} \Psi &= Q^{-1/4}\exp(-iu_3/\lambda - i\Gamma)K\psi_1(u_1)\psi_2(u_2)v\,, \\ K &= \begin{pmatrix} m + \lambda - \sigma_3(\boldsymbol{\sigma}\hat{\mathbf{F}}) \\ (m-\lambda)\sigma_3 - (\boldsymbol{\sigma}\hat{\mathbf{F}}) \end{pmatrix}, \end{aligned} \tag{4.19}$$

where $\hat{\mathbf{F}}$ is an operator[1] orthogonal to $\mathbf{n}$, i.e. $\hat{\mathbf{F}} = (\hat{F}_1, \hat{F}_2, 0)$. The arbitrary two-component spinor v allows us to classify solutions (4.19) according to spin, and can generally be determined by choosing the spin operator (2.151) in the form (2.155). Equations (4.8), and (4.16)–(4.19) remain valid throughout this Section. Equations (4.14) and (4.15) also hold after the substitution $\hat{\mathbf{P}}_{\perp} \to -\hat{\mathbf{F}}$.

1 In the gauge (4.4) $\mathbf{F} = i\nabla_{\perp}$. Unless the opposite is indicated, we shall take the Cartesian components of $\mathbf{F}$.

4.2.2 Stationary crossed fields

Some exact solutions for *stationary crossed fields* were found in the works cited in the Introduction to the book (Chapter 1). A general study of solutions in these fields has been covered in Refs. [29, 30, 33].

Type I

At this point, we use the light-cone variables (4.2), (4.3) and define potentials in the corresponding reference frame as

$$-\tilde{A}_0 = f_1(u_1) + f_2(u_2), \quad \tilde{A}_1 = \tilde{A}_2 = \tilde{A}_3 = 0\,. \tag{4.20}$$

Here $f_1(u_1)$ and $f_2(u_2)$ are arbitrary functions of the indicated arguments. Such a choice of potentials corresponds to the stationary crossed fields

$$E_x = H_y = f_1', \quad E_y = -H_x = f_2'\,. \tag{4.21}$$

The charge density ρ from (4.6) is $\rho = f_1'' + f_2''$.

First integrals of the Lorentz equations are

$$\begin{aligned} m^2\dot{u}_1^2 &= 2\lambda f_1 + (\lambda k_1 - m^2 + k_2)/2\,, \\ m^2\dot{u}_2^2 &= 2\lambda f_2 + (\lambda k_1 - m^2 - k_2)/2\,, \\ m\dot{u}_3 &= k_1 + 2(f_1 + f_2), \quad k_2 = m^2(\dot{u}_1^2 - \dot{u}_2^2) + 2\lambda(f_2 - f_1)\,. \end{aligned} \tag{4.22}$$

Since the fields under consideration are stationary and depend only on u_1 and u_2, the momenta P_0 and P_3 are conserved. Indeed, equations (4.22) and (4.7) imply

$$2P_0 = \lambda + k_1, \quad 2P_3 = \lambda - k_1, \quad \lambda = P_0 + P_3, \quad k_1 = P_0 - P_3\,. \tag{4.23}$$

It follows from (4.22) that u_1 and u_2 obey the same type of equations,

$$m^2\dot{u}_s^2 = R_s(u_s), \quad R_s(u_s) = 2\lambda f_s(u_s) + [\lambda k_1 - m^2 - (-1)^s k_2]/2, \quad s = 1, 2\,, \tag{4.24}$$

which can be integrated

$$\tau = m\int \frac{du_1}{\sqrt{R_1(u_1)}} = m\int \frac{du_2}{\sqrt{R_2(u_2)}}\,. \tag{4.25}$$

Integrating the last equation (4.22), we obtain

$$u_3 = \frac{k_1\tau}{m} + 2\int \frac{f_1(u_1)du_1}{\sqrt{R_1(u_1)}} + 2\int \frac{f_2(u_2)du_2}{\sqrt{R_2(u_2)}}\,. \tag{4.26}$$

The classical action has the form (4.16) with $\Gamma = k_2 u_0/2$.

The wave functions corresponding to the states characterized by definite values of λ, k_1, and k_2 are determined by relations (4.18) and (4.19), where $Q = 1$, and the functions $\psi_s(u_s)$ are solutions of the equations

$$\psi_s''(u_s) + R_s(u_s)\psi_s(u_s) = 0, \quad s = 1, 2\,. \tag{4.27}$$

The operators in (4.19) are $\hat{F}_s = i\partial_s$, $s = 1, 2$.

Thus, the solution of the quantum problem is reduced to integrating the one-dimensional stationary Schrödinger equation (9.1), which we write here as

$$\psi''(x) + R(x)\psi(x) = 0, \quad R(x) = E - V(x)\,, \tag{4.28}$$

where

$$V(x) = -2\lambda f(x), \quad E = (\lambda k_1 - m^2 \pm k_2)/2\,. \tag{4.29}$$

As was demonstrated, exact solutions can be found for some *exactly-solvable potentials (ESP)* $V(x)$. Functions f corresponding to these ESP are listed below:

$$\begin{aligned}
&\text{ESP I:} && f(x) = ax\,,\\
&\text{ESP II:} && f(x) = ax^2 + bx\,,\\
&\text{ESP III:} && f(x) = a/x + b/x^2\,,\\
&\text{ESP IV:} && f(x) = a/x^2 + bx^2\,,\\
&\text{ESP V:} && f(x) = a\exp(-2cx) + b\exp(2cx)\,,\\
&\text{ESP VI:} && f(x) = \frac{a}{\sin^2 cx} + \frac{b}{\cos^2 cx}\,,\\
&\text{ESP VII:} && f(x) = a\tan^2 cx + b\tan cx\,,\\
&\text{ESP VIII:} && f(x) = a\tanh^2 cx + b\tanh cx\,,\\
&\text{ESP IX:} && f(x) = a\coth^2 cx + b\coth cx\,,\\
&\text{ESP X:} && f(x) = \frac{a + b\cosh cx}{\sinh^2 cx}\,,\\
&\text{ESP XI:} && f(x) = \frac{a + b\sinh cx}{\cosh^2 cx}\,.
\end{aligned} \tag{4.30}$$

Here a, b, and $c > 0$ are arbitrary real constants.

We can take any function from (4.30) for $f_1(u_1)$, taking another, also from (4.30), for $f_2(u_2)$.

To find the energy spectrum in some cases can be a rather complicated task. Let us set, for example,

$$f_1(u_1) = -V_1 u_1^2, \quad f_2(u_2) = -V_2\left[\left(\frac{u_2}{c}\right)^2 + \left(\frac{c}{u_2}\right)^2\right], \quad V_{1,2} = \text{const} > 0\,. \tag{4.31}$$

Using solutions that are represented for ESP II and IV, see 9 (Appendix 3), we find

$$\psi_1(u_1) = U_{n_1}(z_1), \quad \psi_2(u_2) = z_2^{\frac{1+2\mu}{4}} e^{-\frac{z_2}{2}} L_{n_2}^{\mu}(z_2),$$
$$z_1 = u_1(2\lambda V_1)^{1/4}, \quad cz_2 = u_2^2\sqrt{2\lambda V_2}, \quad \mu = (1/2)\sqrt{1 + 8c^2\lambda V_2},$$
$$P_0^2 - P_3^2 - m^2 + k_2 = 2(2n_1 + 1)\sqrt{2(P_0 + P_3)V_1},$$
$$P_0^2 - P_3^2 - m^2 - k_2 = (4/c)(1 + \mu + 2n_2)\sqrt{2(P_0 + P_3)V_2}. \tag{4.32}$$

The determination of P_0 as a function of the integers n_1 and n_2 from these expressions is reduced to solution of a fourth-order algebraic equation.

For the first time, solutions in the fields under consideration were found in [29, 30, 33].

Type II

At this point, we use light-cone coordinates (4.2) and the coordinates u_1 and u_2,

$$x^1 = u_1 \cos u_2, \quad x^2 = u_1 \sin u_2. \tag{4.33}$$

In the corresponding reference frame, we define potentials as follows:

$$-\tilde{A}_0 = f_1(u_1) + f_2(u_2)u_1^{-2}, \quad \tilde{A}_1 = \tilde{A}_2 = \tilde{A}_3 = 0, \tag{4.34}$$

where $f_1(u_1)$ and $f_2(u_2)$ are arbitrary functions of the indicated arguments ($f_2(u_2$ has to be periodic with the period 2π). Projections of the electromagnetic fields onto unit vectors of the cylindrical reference frame ($r = u_1, \quad \varphi = u_2$) are

$$E_r = H_\varphi = f_1' - f_2\left(2u_1^3\right)^{-1}, \quad E_\varphi = -H_r = f_2' u_1^{-3}, \quad E_z = H_z = 0. \tag{4.35}$$

The charge density ρ (4.6) reads $\rho = (u_1^4 f_1'' + u_1^3 f_1' + f_2'' + 4f_2)u_1^{-4}$. Free electromagnetic fields correspond to the choice

$$f_1 = a \ln u_1, \quad f_2 = b\cos(2u_2 + c),$$

with a, b, and c being arbitrary constants.

First integrals of the Lorentz equations are

$$m^2\dot{u}_1^2 = R_1(u_1) = 2\lambda f_1 + k_2 u_1^{-2} + \lambda k_1 - m^2,$$
$$m^2 u_1^4 \dot{u}_2^2 = R_2(u_2) = 2\lambda f_2 - k_2, \quad m\dot{u}_3 = k_1 + 2(f_1 + f_2 u_1^{-2}).$$

They can be integrated,

$$\int \frac{du_2}{\sqrt{R_2(u_2)}} = \int \frac{du_1}{u_1^2\sqrt{R_1(u_1)}}, \quad \tau = m\int \frac{du_1}{\sqrt{R_1(u_1)}},$$
$$u_3 = \frac{k_1\tau}{m} + 2\int \frac{f_1 du_1}{\sqrt{R_1(u_1)}} + 2\int \frac{f_2 du_2}{\sqrt{R_2(u_2)}}.$$

It is obvious that the momentum components P_0 and P_3 are conserved for the fields under consideration, relations (4.23) hold true again.

The classical action has the form (4.16) with $\Gamma = k_1 u_0/2$.

Solutions of the K–G and Dirac equations are determined by equations (4.18) and (4.19) where one should set

$$Q = 1, \quad \hat{F}_1 = i(\cos u_2 \partial_{u_1} - u_1^{-1} \sin u_2 \partial_{u_2}), \ \hat{F}_2 = i(\sin u_2 \partial_{u_1} + u_1^{-1} \cos u_2 \partial_{u_2}).$$

The functions $\psi_s(u_s)$, $s = 1, 2$, are solutions to the equations

$$\psi_1'' + u_1^{-1}\psi_1' + R_1\psi_1 = 0, \quad \psi_2'' + R_2\psi_2 = 0. \tag{4.36}$$

The first of these equations is reduced to the form

$$\chi'' + \left[R_1 + \left(4u_1^2\right)^{-1}\right]\chi = 0,$$

by the substitution $\psi_1 = \chi/\sqrt{u_1}$. Thus, we again arrive at the one-dimensional stationary Schrödinger equation (9.1). Exact solutions can be found either with the choice $f_1 = \alpha u_1^2 + \beta u_1^{-2}$ (ESP IV) or with $f_1 = \alpha u_1^{-1} + \beta u_1^{-2}$ (ESP III). The function f_2 can be chosen from among those of (4.30).

Exact solutions in the fields (4.34) were first found in [37, 42].

Type III

At this point, we use the light-cone coordinates (4.2) and coordinates u_1 and u_2,

$$u_1 = r - x^1, \quad u_2 = r + x^1, \quad r = \sqrt{x_1^2 + x_2^2}. \tag{4.37}$$

In the corresponding reference frame, we define potentials as

$$\tilde{A}_0 = -\frac{f_1(u_1) + f_2(u_2)}{u_1 + u_2}, \quad \tilde{A}_1 = \tilde{A}_2 = \tilde{A}_3 = 0, \tag{4.38}$$

where $f_s(u_s)$, $s = 1, 2$, are arbitrary functions of their arguments. The Cartesian components of the electromagnetic fields are

$$\begin{aligned} &E_x = H_y = -\left[(f_1 + f_2)x^1 + (u_1 f_1' + u_2 f_2')r\right]/2r^3, \\ &E_y = -H_x = -x^2\left[f_1 + f_2 - r(f_1' + f_2')\right]/2r^3, \quad E_z = H_z = 0. \end{aligned} \tag{4.39}$$

The charge density ρ (4.6) is

$$\rho = (2u_1 f_1'' + f_1' + 2u_2 f_2'' + f_2)/2r^2 - (2u_1 f_1' - f_1 + 2u_2 f_2' - f_2)/2r^3.$$

The fields under considerations are free provided that

$$f_s(u_s) = \alpha_s \sqrt{u_s}, \quad s = 1, 2. \tag{4.40}$$

Note that the free fields defined by (4.40) are particular cases of the fields defined by (4.43).

First integrals of the Lorentz equations are

$$m^2\dot{u}_s^2 = 16u_s^2 R_s(u_s)(u_1+u_2)^{-2}, \quad m\dot{u}_3 = k_1 + 2\left(f_1+f_2\right)\left(u_1+u_2\right)^{-1} ;$$
$$4u_s R_s(u_s) = (\lambda k_1 - m^2)u_s + 2\lambda f_s(u_s) + (-1)^s k_2, \quad s = 1, 2 . \tag{4.41}$$

The classical action has the form (4.16) with $\Gamma = k_1 u_0/2$.

Solutions of the K–G equations have the form of (4.18), where one should set $Q = 1$ and make the functions $\psi_s(u_s)$ obey the equations

$$2u_s\psi_s''(u_s) + \psi_s'(u_s) + 2u_s R_s(u_s)\psi_s(u_s) = 0, \quad s = 1, 2 . \tag{4.42}$$

These equations are exactly solved only for two kinds of the functions f_s,

$$\text{(a)} \qquad f_s = \alpha_s + \beta_s u_s + \gamma\sqrt{u_s},$$
$$\text{(b)} \qquad f_s = \alpha_s + \beta_s/u_s + \gamma_s u_s . \tag{4.43}$$

The substitution $u = x^2$ reduces each equation (4.42) with the functions (4.43) in them to the one-dimensional Schrödinger equation (9.1) with ESP II and IV.

Solutions of the Dirac equation have the form (4.19) with

$$\hat{F}_1 = (i/r)\,(u_2\partial_{u_2} - u_1\partial_{u_1}), \quad \hat{F}_2 = (i/r)\,x^2(\partial_{u_1} + \partial_{u_2}) .$$

Exact solutions in the fields under consideration were first found in [48].

The cases considered above do not exhaust all the types of stationary crossed fields allowing exact solutions. For example, some stationary crossed fields may be particular cases of more general fields considered below.

4.2.3 Nonstationary crossed fields

Some exact solutions for *nonstationary crossed fields* were found in the works cited in the Sections 4.1 and 4.2.2. A general study of solutions in these fields has been covered in Refs. [30, 31, 33, 37, 42, 43, 48].

Type I

At this point, we use the light-cone coordinates (4.2) and (4.3). In the corresponding reference frame, the electromagnetic potentials are defined as follows:

$$\tilde{A}_0 = -a_1(u_1), \quad \tilde{A}_2 = -a_2(u_0), \quad \tilde{A}_1 = \tilde{A}_3 = 0 . \tag{4.44}$$

This is a superposition of a constant nonuniform crossed field (with $E_x \neq 0$) and a plane-wave field running along z and polarized along y, i.e. with $E_y \neq 0$,

$$E_x = H_y = a_1'(u_1), \quad E_y = -H_x = -a_2'(u_0), \quad E_z = H_z = 0 . \tag{4.45}$$

For the fields under consideration, the charge density ρ (4.6) is $\rho = a_1''$.

First integrals of the Lorentz equations are

$$m^2\dot{u}_1^2 = R_1(u_1) = 2\lambda a_1(u_1) + mk_1\,,$$
$$m\dot{u}_2 + k_2 + a_2(u_0) = 0\,,$$
$$m\lambda\dot{u}_3 = R_1 + (k_2 + a_2)^2 + m^2\,.$$

They can be integrated

$$\tau = m\int \frac{du_1}{\sqrt{R_1(u_1)}}, \quad u_2 = -\frac{1}{\lambda}\int (k_2 + a_2)du_0\,,$$
$$u_3 = \frac{m}{\lambda}\tau + \frac{1}{\lambda}\int \frac{du_1}{\sqrt{R_1(u_1)}} + \frac{1}{\lambda^2}\int (k_2 + a_2)^2 du_0\,.$$

The classical action has the form (4.16) with

$$\Gamma = k_2 u_2 + \frac{1}{2\lambda}\left[(m^2 + mk_1)u_0 + \int (k_2 + a_2)^2 du_0\right], \quad R_2 = 0\,.$$

Solutions of the K–G and Dirac equations are determined by equations (4.18) and (4.19) where one should set

$$Q = 1, \quad \psi_2 = 1, \quad \hat{F}_1 = i\partial_1, \quad \hat{F}_2 = k_2 + a_2\,,$$

and choose the function $\psi_1(u_1)$ to be a solution of the equation

$$\psi_1'' + R_1\psi_1 = 0\,.$$

This equation is of the type (4.28). Therefore, with the function $a_2(u_0)$ arbitrary, the solution can be found for such functions $a_1(u_1)$, which allow explicit solutions of equation (4.28). Note, finally, that the Volkov solutions for a charge in a linearly-polarized plane-wave field (see Section 3.2) follow from the above solutions at $a_1 = 0$.

Exact solutions in the fields under consideration have been studied in Refs. [30, 37].

Type II

At this point, we use the coordinates u_μ,

$$u_1 = x^1, \quad u_2 = (x^2)^2[(x^0 - x^3)^2 + a]^{-1}, \quad u_0 = x^0 - x^3\,,$$
$$u_3 = x^0 + x^3 - (x^0 - x^3)(x^2)^2[(x^0 - x^3)^2 + a]^{-1}\,, \tag{4.46}$$

where a is an arbitrary constant. In the corresponding reference frame, we define potentials as follows:

$$\tilde{A}_0 = f(u_2)\left(a + u_0^2\right)^{-1}, \quad \tilde{A}_1 = g(u_0), \quad \tilde{A}_2 = \tilde{A}_3 = 0\,, \tag{4.47}$$

where f and g are arbitrary functions. The Cartesian components of the electromagnetic fields are

$$E_x = H_y = -g', \quad E_y = -H_x = 2x^2 f'\left(a + u_0^2\right)^{-1}, \quad E_z = H_z = 0\,. \tag{4.48}$$

The charge density ρ (4.6) is $\rho = 2(2u_2 f'' + f')(a + u_0^2)^{-2}$, for the fields under consideration. The free fields (for $f = \alpha\sqrt{u_2}$) correspond to the specific case of a plane wave.

First integrals of the Lorentz equations are

$$\begin{aligned}
&m\dot{u}_1 + P_1 = 0, \quad m^2(a + u_0)^2 \dot{u}_2^2 = 16u_2^2 R_2(u_2)\,,\\
&m\lambda\dot{u}_3 = m^2 + P_1^2 + \left(2\lambda f - 2a\lambda^2 u_2 + k_2\right)\left(a + u_0^2\right)^{-1};\\
&4u_2 R_2(u_2) = 2\lambda f - a\lambda^2 u_2 + k_2, \quad P_1 = k_1 + g\,.
\end{aligned}$$

They can be integrated

$$\begin{aligned}
&\lambda u_1 = -\int P_1 du_0, \quad \lambda \int \frac{du_2}{u_2\sqrt{R_2(u_2)}} = 4\int \frac{du_0}{a + u_0^2}\,,\\
&u_3 = \frac{1}{\lambda^2}\int \left(m^2 + P^2 + \frac{k_2}{a + u_0^2}\right) du_0 + \frac{1}{2}\int \frac{f - a\lambda u_2}{u_2\sqrt{R_2(u_2)}} du_2\,.
\end{aligned}$$

The classical action has the form (4.16) with

$$\Gamma = k_1 u_1 + \frac{1}{2\lambda}\int \left(m^2 + P^2 + \frac{k_2}{a + u_0^2}\right) du_0, \quad R_1 = 0\,.$$

Solutions of the K–G and Dirac equations are given by equations (4.18) and (4.19), where one should set

$$Q = a + u_0^2, \quad \psi_1 = 1, \quad \hat{F}_1 = P_1, \quad \hat{F}_2 = u_2(2i\partial_{u_2} - \lambda u_0)/x^2\,,$$

and make the function $\psi_2(u_2)$ obey the equation

$$\psi_2'' + (2u_2)^{-1}\psi_2' + R_2\psi_2 = 0\,, \tag{4.49}$$

which can be reduced to the form

$$\psi_2'' + 4z^2 R_2(\pm z^2)\psi_2 = 0\,,$$

by the change of the variable $u_2 = \pm z^2$. Thus, we have again an equation of the type of (4.28). If $a \neq 0$, exact solutions can be found for the two choices of the function f,

$$f = \alpha\sqrt{u_2} + \beta u_2, \quad \text{or} \quad f = \alpha u_2 + \beta/u_2\,,$$

see ESP II and IV for Equation (9.1).

For $a = 0$, the function f can be chosen from (4.30). The function g remains arbitrary.

Solutions in the fields under consideration were first found in the works [31, 37].

Type III

At this point, we use the coordinates u_μ,

$$u_1 = x^1, \quad u_2 = (a + 2x^2 u_0)u_0^{-2}, \quad u_0 = x^0 - x^3,$$
$$u_3 = x^0 + x^3 - (2ax^0 + u_0 x_2^2)u_0^{-2} - (2/3)\, a^2 u_0^{-3}, \tag{4.50}$$

where a is an arbitrary constant. In the corresponding reference frame, we define potentials to be

$$\tilde{A}_0 = -u_0^{-2} f(u_2), \quad \tilde{A}_1 = -g(u_0), \quad \tilde{A}_2 = \tilde{A}_3 = 0. \tag{4.51}$$

The Cartesian components of the electromagnetic fields are

$$E_x = H_y = -g', \quad E_y = -H_x = 2u_0^{-3} f', \quad E_z = H_z = 0. \tag{4.52}$$

The charge density ρ (4.6) is $\rho = 4u_0^{-4} f''$, for the fields under consideration. Here the free fields are plane waves of a particular form.

First integrals of the Lorentz equations are

$$m\dot{u}_1 + P_1 = 0, \quad m^2 u_0^4 \dot{u}_2^2 = 16R_2(u_2),$$
$$m\lambda\dot{u}_3 = m^2 + P_1^2 + (4R_2 + a\lambda^2 u_2)u_0^{-2},$$
$$P_1 = k_1 + g, \quad 4R_2(u_2) = a\lambda^2 u_2 + 4k_2 + 2\lambda f.$$

They can be integrated,

$$u_1 = -\frac{1}{\lambda}\int P_1 du_0, \quad \int \frac{du_2}{\sqrt{R_2(u_2)}} = \frac{4}{\lambda u_0},$$
$$u_3 = \frac{1}{\lambda^2}\int (m^2 + P_1^2) du_0 - \int \frac{a\lambda^2 u_2 + 4R_2}{4\lambda\sqrt{R_2(u_2)}} du_2.$$

The classical action has the form (4.16) with

$$\Gamma = k_1 u_1 + \frac{m^2}{2\lambda} u_0 - \frac{2k_2}{\lambda u_0} + \frac{1}{2\lambda}\int P_1^2 du_0, \quad R_1 = 0.$$

Solutions of the K–G and Dirac equations are given by the equations (4.18) and (4.19) where one should set

$$Q = u_0^2, \quad \psi_1 = 1, \quad \psi_2'' + R_2\psi_2 = 0,$$
$$\hat{F}_1 = P_1, \quad \hat{F}_2 = 2iu_0^{-1}\partial_{u_2} - \lambda(au_0^{-2} + u_2)/2. \tag{4.53}$$

For $a \neq 0$, solutions can only be found provided that $f = \alpha u_2 + \beta u_2^2$, see ESR II for Equation (9.1). If, however, $a = 0$, then f can be chosen from (4.30).

Solutions in the fields under consideration were first found in the works [31, 37].

Type IV

Define the curvilinear coordinates u_μ,

$$u_1 = x^1, \quad u_2 = u_0^{-1} x_2^2, \quad u_3 = x^0 + x^3 - (2u_0)^{-1} x_2^2, \quad u_0 = x^0 - x^3 . \tag{4.54}$$

In the corresponding reference frame, we choose the potentials as

$$\tilde{A}_0 = -u_0^{-1} f(u_2), \quad \tilde{A}_1 = -g(u_0), \quad \tilde{A}_2 = \tilde{A}_3 = 0 . \tag{4.55}$$

The Cartesian components of the electromagnetic fields are

$$E_x = H_y = -g', \quad E_y = -H_x = 2x^2 f' u_0^{-2}, \quad E_z = H_z = 0 . \tag{4.56}$$

The charge density ρ (4.6) is $\rho = 2(f' + 2u_2 f'')u_0^{-2}$ for the fields under consideration. Free fields correspond to $f = \alpha\sqrt{u_2}$ and are plane-wave fields of a particular form.

First integrals of the Lorentz equations are

$$\begin{aligned} &m\dot{u}_1 + P_1 = 0, \quad m^2 u_0^2 \dot{u}_2^2 = 16 u_2^2 R_2(u_2) , \\ &m\lambda\dot{u}_3 = m^2 + P_1^2 + (4\lambda f + \lambda^2 u_2 + 2k_2)(2u_0)^{-1} , \\ &P_1 = k_1 + g, \quad 16 u_2 R_2(u_2) = 8\lambda f + \lambda^2 u_2 + 4k_2 . \end{aligned}$$

They can be integrated,

$$\begin{aligned} &u_1 = -\frac{1}{\lambda}\int P_1 du_0, \quad \int \frac{du_2}{u_2\sqrt{R_2(u_2)}} = \frac{4}{\lambda}\ln|u_0| , \\ &u_3 = \frac{1}{\lambda^2}\int (m^2 + P_1^2) du_0 + \int \frac{16R_2 + \lambda^2}{16\lambda\sqrt{R_2}} du_2 . \end{aligned} \tag{4.57}$$

The classical action has the form (4.16) with

$$\Gamma = k_1 u_1 + \frac{k_2}{2\lambda}\ln|u_0| + \frac{1}{2\lambda}\int (m^2 + P_1^2) du_0, \quad R_1 = 0 .$$

Solutions of the K–G and Dirac equations are given by equations (4.18) and (4.19), where one should set

$$\begin{aligned} &Q = u_0, \quad \psi_1 = 1, \quad \psi_2'' + (2u_2)^{-1}\psi_2' + R_2\psi_2 = 0 , \\ &\hat{F}_1 = P_1, \quad \hat{F}_2 = x^2(4i\partial_{u_2} - \lambda)(2u_0)^{-1} . \end{aligned} \tag{4.58}$$

Therefore, the quantum mechanical problems have been reduced to solving an equation of the type (4.49). An explicit solution is possible (see ESP II and IV) with the function g taken arbitrarily and the function f chosen as one of the two types:

$$f = \alpha\sqrt{|u_2|}, \quad f = \alpha u_2 + \beta/u_2 .$$

Solutions in the fields under consideration were first found in the works [31, 37].

Type V

At this point, we use the coordinates u_μ,

$$u_1 = x^1, \quad u_2 = x^2 - au_0^2/4, \quad u_3 = x^0 + x^3 - ax^2u_0 + a^2u_0^3/6, \quad u_0 = x^0 - x^3\,,$$

where a is a constant. In the corresponding reference frame, we define potentials as

$$\tilde{A}_0 = -f(u_2), \quad \tilde{A}_1 = -g(u_0), \quad \tilde{A}_2 = \tilde{A}_3 = 0\,. \tag{4.59}$$

The Cartesian components of the electromagnetic fields are

$$E_x = H_y = -g', \quad E_y = -H_x = f', \quad E_z = H_z = 0\,.$$

The charge density ρ from (4.6) is $\rho = f''$ for the fields under consideration.

At $a = 0$ we are left with the fields studied in the Type I.

First integrals of the Lorentz equations are

$$m\dot{u}_1 + P_1 = 0, \quad m^2\dot{u}_2^2 = R_2(u_2) = 2\lambda f - a\lambda^2 u_2 + 4k_2\,,$$
$$m\lambda\dot{u}_3 = m^2 + 4k_2 + P_1^2 + 2\lambda f - 2a\lambda^2 u_2, \quad P_1 = k_1 + g\,.$$

They can be integrated,

$$\tau = m\int\frac{du_2}{\sqrt{R_2(u_2)}}, \quad u_1 = -\frac{1}{\lambda}\int P_1 du_0\,,$$
$$u_3 = \frac{1}{\lambda^2}\int(m^2 + 4k_2 + P_1^2)du_0 + 2\int\frac{f - a\lambda u_2}{\sqrt{R_2(u_2)}}du_2\,. \tag{4.60}$$

The classical action has the form (4.16) with

$$\Gamma = k_1u_1 + \frac{1}{2\lambda}\int(m^2 + 4k_2 + P_1^2)du_0, \quad R_1 = 0\,.$$

Solutions of the K–G and Dirac equations are given by equations (4.18) and (4.19), where one should set

$$Q = 1, \quad \psi_1 = 1, \quad \psi_2'' + R_2\psi_2 = 0; \quad \hat{F}_1 = P_1, \quad \hat{F}_2 = i\partial_{u_2}\,. \tag{4.61}$$

At $a \neq 0$ and with the function g arbitrary, the solutions for ψ_2 can be found for $f = \alpha u_2 + \beta u_2^2$, see ESP I and II from Chapter 9.

Solutions in the fields under consideration were first found in Refs. [31, 37].

Type VI

Here we use the coordinates u_μ,

$$u_s = x^s + \alpha_s(x^0 - x^3)^2, \quad s = 1,2\,, u_0 = x^0 - x^3\,,$$
$$u_3 = x^0 + x^3 + 4(x^0 - x^3)(\alpha_1x^1 + \alpha_2x^2) + (8/3)\,(\alpha_1^2 + \alpha_2^2)(x^0 - x^3)^3\,, \tag{4.62}$$

where α_s are some constants. In the corresponding reference frame, we define potentials as

$$-\tilde{A}_0 = f_1(u_1) + f_2(u_2), \quad \tilde{A}_1 = \tilde{A}_2 = \tilde{A}_3 = 0\,. \tag{4.63}$$

The Cartesian components of the electromagnetic fields are

$$E_x = H_y = f_1', \quad E_y = -H_x = f_2', \quad E_z = H_z = 0\,. \tag{4.64}$$

The charge density ρ (4.6) is $\rho = f_1'' + f_2''$ for the fields under consideration. The fields are free providing $f_s' = \gamma_s + (-1)^s \beta u_s$, $s = 1, 2$.

First integrals of the Lorentz equations are

$$\begin{aligned} &m^2\dot{u}_s^2 = R_s(u_s), \quad m\dot{u}_3 = 2(f_1 + f_2 + 4\lambda\alpha_1 u_1 + 4\lambda\alpha_2 u_2 + k_2)\,, \\ &R_s(u_s) = 2\lambda f_s + 4\lambda\alpha_s^2 u_s - \lambda k_2 - m^2/2 + (-1)^s k_1, \quad s = 1, 2\,. \end{aligned} \tag{4.65}$$

They can be integrated,

$$\begin{aligned} \tau &= m\int \frac{du_s}{\sqrt{R_s(u_s)}}, \quad s = 1, 2\,, \\ u_3 &= 2k_2\tau + 2\sum_{s=1,2}\int \frac{f_s + 4\lambda\alpha_s u_s}{\sqrt{R_s(u_s)}} du_s\,. \end{aligned} \tag{4.66}$$

The classical action has the form (4.16) with $\Gamma = k_2 u_0$.

Solutions of the K–G and Dirac equations have the form (4.18) and (4.19) provided that one sets

$$Q = 1, \quad \hat{F}_1 = i\partial_{u_1} + 2\lambda\alpha_1 u_1, \quad \hat{F}_2 = i\partial_{u_2} + 2\lambda\alpha_2 u_2\,, \tag{4.67}$$

and makes the functions $\psi_s(u_s)$ obey the equations

$$\psi_s'' + R_s(u_s)\psi_s = 0, \quad s = 1, 2\,, \tag{4.68}$$

each being a one-dimensional Schrödinger equation (9.1) with

$$E_s = \lambda k_2 - m^2/2 + (-1)^s k_1, \quad V_s = -2\lambda f_s + 4\lambda^2\alpha_s u_s\,. \tag{4.69}$$

If $\alpha_s \neq 0$, solutions of (4.68) can be found for the functions f_s no more than quadratic in u_s and, in particular, for free fields, see ESP I and II. If, however, $\alpha_s = 0$, we are left with the fields considered in Section 4.2.2, item Type I.

Solutions in the fields under consideration were first found in [48].

Type VII

There are several types of crossed fields where one uses very similar formal expressions describing the classical and quantum motion. It is therefore convenient to consider them in a unique way.

At this point, we use the coordinates u_0, u_1 and u_2,

$$u_0 = x^0 - x^3, \quad u_1 = \left(x^1\right)^2 \varphi_1, \quad u_2 = x^2\varphi_2 + b\varphi_3^2/2\,, \tag{4.70}$$

where b is an arbitrary constant. The coordinate u_3, the functions φ_s, $s = 1, 2, 3$, and the parameters μ_1, μ_2, the constant a (that may be either fixed or arbitrary) are defined for each type of field in the following way:

(a)

$$\begin{aligned}
&u_3 = x^0 + x^3 - u_0u_1 - [(u_0x^2 + cx^2 + b)^2 - b/2]\varphi_2^3\,;\\
&\varphi_1 = \left(u_0^2 - a\right)^{-1}, \quad \varphi_2 = \varphi_3 = \left(u_0 + c\right)^{-1}, \quad a, c \text{ are arbitrary;}\\
&\mu_1 = u_0, \quad \mu_2 = u_0x^2 + cx^2 + b\,;
\end{aligned} \tag{4.71}$$

(b)

$$\begin{aligned}
&u_3 = x^0 + x^3 + 2bu_0x^2 + \frac{2}{3}b^2u_0^3 - u_0u_1\,;\\
&\varphi_1 = \left(u_0^2 - a\right)^{-1}, \quad \varphi_2 = 1, \quad \varphi_3 = u_0\,;\\
&\mu_1 = u_0, \quad \mu_2 = -bu_0\,;
\end{aligned} \tag{4.72}$$

(c)

$$\begin{aligned}
&u_3 = x^0 + x^3 - u_1/2 - u_0^{-3}[(u_0x^2 + b)^2 - b^2/3]\,;\\
&\varphi_1 = \left(u_0 + c\right)^{-1}, \quad \varphi_2 = \varphi_3 = u_0^{-1}, \quad a = 1/4\,;\\
&\mu_1 = 1/2, \quad \mu_2 = u_0x^2 + b\,;
\end{aligned} \tag{4.73}$$

(d)

$$\begin{aligned}
&u_3 = x^0 + x^3 - u_1/2 + 2bu_0u_2 - \left(b^2u_0^3\right)/3\,;\\
&\varphi_1 = u_0^{-1}, \quad \varphi_2 = 1, \quad \varphi_3 = u_0, \quad a = 1/4\,;\\
&\mu_1 = 1/2, \quad \mu_2 = -bu_0\,.
\end{aligned} \tag{4.74}$$

In the corresponding reference frames, potentials for all the cases can be taken in the same form

$$-\tilde{A}_0 = \varphi_1 f_1(u_1) + \varphi_2^2 f_2(u_2), \quad \tilde{A}_1 = \tilde{A}_2 = \tilde{A}_3 = 0\,. \tag{4.75}$$

The Cartesian components of the electromagnetic fields are

$$E_x = H_y = 2x^1\varphi_1^2 f_1', \quad E_y = -H_x = \varphi_2^3 f_2', \quad E_z = H_z = 0\,.$$

The charge density ρ (4.6) is $\rho = 2\varphi_1^2(f_1' + 2u_1 f_1'') + \varphi_2^4 f_2''$, for the fields under consideration. The free fields are particular cases of the plane waves.

First integrals of the Lorentz equations are

$$\begin{aligned}
&m^2\dot{u}_1^2 = 16u_1^2\varphi_1^2 R_1(u_1), \quad m^2\dot{u}_2^2 = \varphi_2^4 R_2(u_2)\,,\\
&m\lambda\dot{u}_3 = m^2 + u_1\varphi_1(4R_1 + a\lambda^2) + \varphi_2^2(R_2 + 2b\lambda^2u_2)\,;\\
&4u_1R_1(u_1) = 4k_1 + 2\lambda f_1 + a\lambda^2u_1, \quad R_2(u_2) = 2(2k_2 + \lambda f_2 + b\lambda^2u_2)\,.
\end{aligned}$$

They can be integrated,

$$\int \frac{du_1}{u_1\sqrt{R_1}} = \frac{4}{\lambda}\int \varphi_1 du_0, \quad \int \frac{du_2}{\sqrt{R_2}} = \frac{1}{\lambda}\int \varphi_2^2 du_0\,,$$
$$u_3 = \frac{m}{\lambda}\tau + \frac{1}{4\lambda}\int \frac{4R_1 + a\lambda^2}{\sqrt{R_1}}du_1 + \frac{1}{\lambda}\int \frac{R_2 + 2b\lambda^2 u_2}{\sqrt{R_2}}du_2\,. \tag{4.76}$$

The classical action has the form (4.16) with

$$\Gamma = \frac{m^2}{2\lambda}u_0 + \frac{2}{\lambda}\left(k_1\int \varphi_1 du_0 + k_2\int \varphi_2^2 du_0\right).$$

Solutions of the K–G and Dirac equations are given by equations (4.18) and (4.19) where one should set

$$Q = \varphi_1\varphi_2^2, \quad \hat{F}_1 = x^1\varphi_1(2i\partial_{u_1} - \lambda\mu_1), \quad \hat{F}_2 = \varphi_2(i\partial_{u_2} - \lambda\varphi_2\mu_2)\,,$$
$$\psi_1'' + (2u_1)^{-1}\psi_1' + R_1\psi_1 = 0, \quad \psi_2'' + R_2\psi_2 = 0\,. \tag{4.77}$$

The equation for ψ_2 is already a one-dimensional Schrödinger equation, whereas that for ψ_1 is reduced to it by the substitution $u_1 = \pm(u_1')^2$. Solutions for these equations can be found provided that the functions f_1 and f_2 are quadratic in their arguments (see ESP I and II from Chapter 9).

It is worth emphasizing that in spite of the close mathematical resemblance in the description of the motion in the fields (4.75) with different functions φ_1 and φ_2, these fields are essentially different and do not reduce to one another even when particular cases are considered.

Solutions in the fields under consideration were first found in [37, 43].

Type VIII

There are a few other types of external fields for which the motion can be described on a mathematically common basis. We consider them together below.

At this point, we use the coordinates u_0, u_1 and u_2,

$$u_0 = x^0 - x^3, \quad u_s = x_s^2\varphi_s(u_0), \quad s = 1, 2\,. \tag{4.78}$$

The functions $\varphi_s(u_0)$, μ_1 and μ_2, the coordinate u_3, and constants c, a_1 and a_2 have the following form in three different cases:

(a)

$$u_3 = x^0 + x^3 - u_0u_1 - u_2(u_0 + c)\,;$$
$$\varphi_1 = \left(u_0^2 - a_1\right)^{-1}, \quad \varphi_2 = \left[(u_0 + c)^2 - a_2\right]^{-1};$$
$$\mu_1 = u_0, \quad \mu_2 = u_0 + c\,; \tag{4.79}$$

(b)

$$u_3 = x^0 + x^3 - u_0 u_1 - u_2/2\,;$$
$$\varphi_1 = \left(u_0^2 - a_1\right)^{-1}, \quad \varphi_2 = \left(u_0 + c\right)^{-1}, \quad a_2 = 1/4\,;$$
$$\mu_1 = u_0, \quad \mu_2 = 1/2\,; \tag{4.80}$$

(c)

$$u_3 = x^0 + x^3 - (u_1 + u_2)/2\,;$$
$$\varphi_1 = u_0^{-1}, \quad \varphi_2 = \left(u_0 + c\right)^{-1}, \quad a_1 = a_2 = 1/4\,;$$
$$\mu_1 = \mu_2 = 1/2\,. \tag{4.81}$$

In the corresponding reference frames, potentials for all the cases can be chosen in the same form,

$$-\tilde{A}_0 = \varphi_1 f_1(u_1) + \varphi_2 f_2(u_2), \quad \tilde{A}_1 = \tilde{A}_2 = \tilde{A}_3 = 0\,. \tag{4.82}$$

The Cartesian components of the electromagnetic fields are

$$E_x = H_y = 2x^1\varphi_1^2 f_1', \quad E_y = -H_x = 2x^2\varphi_2^2 f_2', \quad E_z = H_z = 0\,. \tag{4.83}$$

The charge density ρ (4.6) is $\rho = 2\sum_{s=1,2}\varphi_s^2(f_s' + 2u_s f_s'')$ for the fields under consideration. The free fields are particular cases of the plane waves.

First integrals of the Lorentz equations are

$$m^2\dot{u}_s^2 = 16u_s^2 R_s(u_s), \quad m\lambda\dot{u}_3 = m^2 + \sum_{s=1,2} u_s(4R_s + \lambda^2 a_s)\varphi_s\,;$$
$$4u_s R_s(u_s) = 4k_s + 2\lambda f_s + a_s\lambda^2 u_s, \quad s = 1, 2\,.$$

They can be integrated,

$$\int \frac{du_s}{u_s\sqrt{R_s(u_s)}} = \frac{4}{\lambda}\int \varphi_s du_0, \quad u_3 = \frac{m}{\lambda}\tau + \frac{1}{4\lambda}\sum_{s=1,2}\int \frac{4R_s + \lambda^2 a_s}{\sqrt{R_s}}du_s\,.$$

The classical action has the form (4.16) with

$$\Gamma = \frac{m^2}{2\lambda}u_0 + \frac{2}{\lambda}\sum_{s=1,2} k_s \int \varphi_s du_0\,.$$

Solutions of the K–G and Dirac equations are determined by the expressions (4.18) and (4.19) where one should set

$$Q = (\varphi_1\varphi_2)^{-1}, \quad \hat{F}_1 = x^1\varphi_1(2i\partial_{u_1} - \lambda\mu_1), \quad \hat{F}_2 = x^2\varphi_2(2i\partial_{u_2} - \lambda\mu_2)\,,$$
$$\psi_s'' + (2u_s)^{-1}\psi_s' + R_s\psi_s = 0\,. \tag{4.84}$$

The equations for $\psi_s(u_s)$, $s = 1, 2$ are reduced to the one-dimensional Schrödinger equations (4.77) by the substitution $u_s = \pm(u_s')^2$. Exact solutions can be found for the functions $f_s(u_s)$ quadratic in the variables u_s, see ESP I and II for Equation (9.1).

Solutions in these fields were first obtained in the works [37, 43].

Type IX

At this point, we use the coordinates u_0 and u_1,

$$u_0 = x^0 - x^3, \quad u_1 = \left(2u_0x^1 + a_1\right)/2u_0^2 . \tag{4.85}$$

The coordinates u_2 and u_3 have different forms in two cases under consideration:

(a)

$$\begin{aligned} &u_2 = x^2\varphi + a_2\varphi^2/2, \quad \varphi = (u_0 + b)^{-1} , \\ &u_3 = x^0 + x^3 + [a_1^2 - 3(x^1u_0 + a_1)^2]/3u_0^3 + a_2^2/3 - \varphi(x^2 + a_2\varphi)^2 ; \end{aligned} \tag{4.86}$$

(b)

$$\begin{aligned} &u_2 = x^2 + a_2u_0^2/2, \quad \varphi = 1 , \\ &u_3 = x^0 + x^3 + [a_1^2 - 3(x^1u_0 + a_1)^2]/2u_0^3 + 2a_2x^2u_0 + (3/2)\, a_2^2u_0^3 , \end{aligned} \tag{4.87}$$

where a_1, a_2 and b are some constants. In the corresponding reference frame, we define potentials to be

$$-\tilde{A}_0 = u_0^{-2} f_1(u_1) + \varphi^2 f_2(u_2), \quad \tilde{A}_1 = \tilde{A}_2 = \tilde{A}_3 = 0 . \tag{4.88}$$

The Cartesian components of the electromagnetic fields are

$$E_x = H_y = u_0^{-3} f_1', \quad E_y = -H_x = \varphi^3 f_2', \quad E_z = H_z = 0 .$$

The charge density ρ (4.6) is $\rho = u_0^{-4} f_1'' + \varphi^3 f_2''$ for the fields under consideration. The free fields are the plane waves.

First integrals of the Lorentz equations are

$$\begin{aligned} &m^2u_0^4\dot{u}_1^2 = R_1(u_1), \quad m^2\dot{u}_2^2 = \varphi^4 R_2(u_2) , \\ &m\lambda\dot{u}_3 = m^2 + u_0^{-2}(R_1 + 2a_1\lambda^2u_1) + (R_2 + 2a_2\lambda^2u_2)\varphi^2 , \\ &R_s(u_s) = 2\lambda f_s(u_s) + 2a_s\lambda^2u_s + 4k_s, \quad s = 1, 2 . \end{aligned}$$

They can be integrated,

$$\begin{aligned} &\lambda \int \frac{du_1}{\sqrt{R_1}} = -\frac{1}{u_0}, \quad \lambda \int \frac{du_2}{\sqrt{R_2}} = \int \varphi^2 du_0 , \\ &u_3 = \frac{m}{\lambda}\tau + \frac{1}{\lambda} \sum_{s=1,2} \int \frac{R_s + 2a_s\lambda^2u_s}{\sqrt{R_s}} du_s . \end{aligned} \tag{4.89}$$

The classical action has the form (4.16) with

$$\Gamma = \frac{m^2}{2\lambda}u_0 - \frac{2k_1}{\lambda u_0} + \frac{2k_2}{\lambda} \int \varphi^2 du_0 .$$

Solutions of the K–G and Dirac equations are given by equations (4.18) and (4.19), in which one should set

$$\begin{aligned} &Q = u_0^2\varphi^{-2}, \quad \psi_s'' + R_s\psi_s = 0 , \\ &\hat{F}_1 = u_0^{-2}(iu_0\partial_{u_1} - \lambda u_1u_0^2 - \lambda a_1/2), \quad \hat{F}_2 = i\varphi\partial_{u_2} - G , \end{aligned} \tag{4.90}$$

where for the case (a) $G = \lambda u_2 + \lambda a_2\varphi^2/2$, and for the case (b) $G = -\lambda a_2 u_0$. The problem is again reduced to solving the one-dimensional Schrödinger equation. If $a_s \neq 0$, solutions can be found for the functions $f_s(u_s)$ being no higher than a second-power polynomial in u_s. If, however, $a_s = 0$, one can select for $f_s(u_s)$ any function from (4.30).

The corresponding solutions were considered in [37, 42].

Type X

At this point, we use the light-cone coordinates (4.2) and the coordinates u_1 and u_2,

$$u_1 = q^{-1}x^1, \quad u_2 = x^2,$$
$$q = q(u_0) = |au_0^2 + bu_0 + c|, \tag{4.91}$$

where a , b, and c are constants. In the corresponding reference frame, we define potentials as

$$\tilde{A}_0 = q^{-2}f(u_1), \quad -\tilde{A}_2 = u_2 g_1(u_0) + g_2(u_0), \quad \tilde{A}_1 = \tilde{A}_3 = 0\,. \tag{4.92}$$

The Cartesian components of the electromagnetic fields are

$$E_x = H_y = -q^{-3}f', \quad -E_y = H_x = u_2 g_1' + g_2', \quad E_z = H_z = 0\,. \tag{4.93}$$

The charge density ρ (4.6) is $-\rho = q^{-4}f'' + g_1'$ for the fields under consideration. The free fields correspond to $f'' = -\alpha$, $g_1' = \alpha q^{-4}$, where α is an arbitrary constant.

First integrals of the Lorentz equations are

$$m^2q^4\dot{u}_1^2 = R_1(u_1) = k_1 - 2\lambda f - \kappa\lambda^2 u_1, \quad m\dot{u}_2 = \lambda\varphi^{-1}\varphi' u_2 - \chi - k_2\varphi^{-1},$$
$$m\lambda\dot{u}_3 = m^2 + m^2q^2\dot{u}_1^2 + m\lambda\frac{d}{d\tau}(qq'u_1^2) - \kappa\lambda^2q^{-2}u_1^2 + m^2\dot{u}_2^2, \quad 4\kappa = 4ac - b^2\,.$$

They can be integrated,

$$\lambda\int\frac{du_1}{\sqrt{R_1}} = \int q^{-2}du_0, \quad u_2 = -\frac{\varphi}{\lambda}\int\varphi^{-1}(\chi + k_2\varphi^{-1})du_0\,,$$
$$u_3 = \frac{m^2}{\lambda^2}u_0 + qq'u_1^2 + \int\frac{R_1 - \kappa\lambda^2u_1^2}{\lambda\sqrt{R_1}}du_1 + u_2u_2' + \frac{1}{\lambda}\int(u_2^2g_1' + u_2g_2')du_0\,.$$

The classical action has the form (4.16) with

$$\Gamma = (2\lambda)^{-1}m^2u_0 + k_2u_2\varphi^{-1} - (1/2)(\lambda\varphi^{-1}\varphi' + g_1)u_2^2 + (\chi - g_2)u_2$$
$$+ \frac{1}{2\lambda}\int[k_1q^{-2} + (\chi + k_2\varphi^{-1})^2]du_0 - \frac{\lambda}{2}qq'u_1^2, \quad R_2 = 0\,.$$

The prime marks the derivative with respect to u_0. The functions χ and φ are related to one another as

$$\chi = \chi(u_0) = \varphi^{-1}\int g_2'\varphi du_0\,, \tag{4.94}$$

and $\varphi = \varphi(u_0)$ is a nonzero particular solution of the equation

$$\lambda\varphi'' + g_1'\varphi = 0\,. \tag{4.95}$$

Solutions of the K–G and Dirac equations are given by equations (4.18) and (4.19) where one should set

$$\begin{aligned} &Q = q^2\varphi^2, \quad \psi_2 = 1, \quad \psi_1'' + R_1\psi_1 = 0\,,\\ &\hat{F}_1 = iq^{-1}\partial_{u_1} - \lambda q' u_1, \quad \hat{F}_2 = k_2\varphi^{-1} + \chi - g_2 - (\lambda\varphi^{-1}\varphi' + g_1)u_2\,. \end{aligned} \tag{4.96}$$

The differential equations (4.95) and (4.96) are specific cases of the one-dimensional Schrödinger equation (9.1). It should be borne in mind that only a particular solution of (4.95) is needed.

Solutions in the fields under consideration were first found in [37, 40].

Type XI

At this point, we use the light-cone coordinates (4.2), and coordinates u_1 and u_2 related to the Minkowski coordinates x^μ by equation (4.91) but the function $q = q(u_0)$ differs from (4.91) and is specified below. In the corresponding reference frame, we define potentials as

$$\tilde{A}_0 = x_1^2 f(u_0)/2 - a x_1^{-2}/2, \quad -\tilde{A}_2 = u_2 g_1(u_0) + g_2(u_0), \quad \tilde{A}_1 = \tilde{A}_3 = 0\,. \tag{4.97}$$

The Cartesian components of the electromagnetic fields are

$$-E_x = -H_y = x^1 f + a(x^1)^{-3}, \quad -E_y = H_x = x^2 g_1' + g_2', \quad E_z = H_z = 0\,, \tag{4.98}$$

where a is an arbitrary constant. The charge density ρ (4.6) is $\rho = 3ax_1^{-4} - f - g_1'$ for the fields under consideration. If $a \neq 0$, there are no free fields. If $a = 0$, the fields (4.97) make up a special case of those considered below in Section 4.4.3 (fields of Type VII).

First integrals of the Lorentz equations are

$$\begin{aligned} &m^2q^4\dot{u}_1^2 = k_1 + a\lambda u_1^{-2}, \quad m\varphi\dot{u}_2 = \lambda\varphi' u_2 - \varphi\chi - k_2\,,\\ &m\lambda\dot{u}_3 = m^2 + (k_1 + a\lambda u_1^{-2})q^{-2} + \lambda^2 q'^2 u_1^2 + 2m\lambda u_1 q q' \dot{u}_1\,. \end{aligned}$$

They can be integrated,

$$\begin{aligned} &k_1 u_1^2 = k_1^2 t^2 - a\lambda, \quad t = t(u_0) = \frac{1}{\lambda}\int\frac{du_0}{q^2}, \quad u_2 = -\frac{\varphi}{\lambda}\int\varphi^{-1}(\chi + k_2\varphi^{-1})du_0\,,\\ &u_3 = \frac{m^2}{\lambda^2}u_0 + \frac{k_1}{\lambda}t + k_1^{-1}qq'(k_1^2t^2 - a\lambda) + \frac{ak_1}{\lambda}\int(k_1^2t^2 - a\lambda)^{-1}dt\\ &\quad + \frac{1}{\lambda k_1}\int(a\lambda - k_1^2t^2)q^2 f du_0\,. \end{aligned} \tag{4.99}$$

The classical action has the form (4.16) with

$$\Gamma = k_2 u_2 \varphi^{-1} - \frac{1}{2}(\lambda\varphi^{-1}\varphi' + g_1)u_2^2 + (\chi - g_2)u_2 + \frac{1}{2\lambda}\int [k_1 q^{-2} + (\chi + k_2\varphi^{-1})^2]du_0 - \frac{\lambda}{2}qq'u_1^2 . \quad (4.100)$$

Here the function χ is defined by the relation (4.94) in terms of the function φ, which is a particular solution of equation (4.95), the function q is a nonzero particular solution of the equation

$$\lambda q'' + fq = 0 . \quad (4.101)$$

We see that the choice of curvilinear coordinates is in this case itself determined by the fields and even by the values of the integrals of motion of the problem.

Solutions of the K–G and Dirac equations are given by equations (4.18) and (4.19), where one should set

$$Q = q^2\varphi^2, \quad \psi_2(u_1) = 1 ,$$
$$\psi_1(u_1) = \sqrt{u_1}\left[r_1 J_\nu(\sqrt{k_1}u_1) + r_2 J_{-\nu}(\sqrt{k_1}u_1)\right], \quad 4\nu^2 = 1 - 4a\lambda . \quad (4.102)$$

The operators $\hat{F}_1$ and $\hat{F}_2$ are given by equations (4.96). In (4.102), $J_\nu(x)$ are the Bessel functions, r_1 and r_2 are constants, whose ratio fixes the charge flux as $u_1 \to \infty$ and in the possible condensation point $u_1 = 0$.

These solutions were first found in [37, 40].

Type XII

At this point, we use the light-cone coordinates (4.2) and coordinates u_1 and u_2 given by

$$x^1 = qu_1 \cos u_2, \quad x^2 = qu_1 \sin u_2 . \quad (4.103)$$

In the corresponding reference frame, we define potentials as

$$\tilde{A}_0 = \left[r^4 f(u_0) - a\right]/2r^2, \quad \tilde{A}_1 = \tilde{A}_2 = \tilde{A}_3 = 0, \quad r^2 = x_1^2 + x_2^2 = q^2u_1^2 , \quad (4.104)$$

where a is an arbitrary constant. The function $q = q(u_0)$ will be specified below. The electromagnetic fields are given by their projections onto the basis vectors of the cylindrical reference frame ($u_2 = \varphi$)

$$-E_r = -H_\varphi = rf + ar^{-3}, \quad E_\varphi = H_r = E_z = H_z = 0 . \quad (4.105)$$

The charge density ρ (4.6) is $\rho = 2(ar^{-4} - f)$ for the fields under consideration. There are no free fields in this case.

First integrals of the Lorentz equations are

$$m^2q^4\dot{u}_1^2 = R_1(u_1) = k_1 + (a\lambda - k_2^2)u_1^{-2}, \quad mq^2u_1^2\dot{u}_2 + k_2 = 0 ,$$
$$m\lambda\dot{u}_3 = m^2 + \lambda^2q'^2u_1^2 + 2m\lambda qq'u_1\dot{u}_1 + k_1q^{-2} + a\lambda q^{-2}u_1^{-2} .$$

They can be integrated,

$$k_1 u_1 = k_1^2 t^2 + k_2^2 - a\lambda, \quad u_2 = \int \frac{k_1 k_2}{k_1^2 t^2 + k_2^2 - a\lambda} dt\,,$$
$$u_3 = \frac{m^2}{\lambda^2} u_0 + \int \frac{k_1^2 k_2^2 + [\lambda q q'(k_1^2 t^2 + k_2^2 - a\lambda) + k_1^2 t]^2}{k_1 q^2 (k_1^2 t^2 + k_2^2 - a\lambda)} du_0\,. \tag{4.106}$$

The classical action has the form (4.16) with

$$\Gamma = k_2 u_2 - \frac{\lambda}{2} q q' u_1^2 + \frac{1}{2} k_1 t + \frac{m^2}{2\lambda} u_0, \quad R_2 = 0\,.$$

The function $t = t(u_0)$ is given by equation (4.99). The function $q = q(u_0)$ is a particular solution of equation (4.101).

Solutions of the K–G and Dirac equations are given by equations (4.18) and (4.19) in which one should set

$$Q = q^4, \quad \psi_2 = 1, \quad \psi_1 = aJ_\nu(z) + bJ_{-\nu}(z), \quad z = \sqrt{k_1 u_1}, \quad \nu^2 = k_2^2 - a\lambda\,,$$
$$\hat{F}_1 = q^{-1}\left[\left(i\partial_{u_1} - \lambda q q' u_1\right)\cos u_2 - k_2 u_1^{-1} \sin u_2\right]\,,$$
$$\hat{F}_2 = q^{-1}\left[\left(i\partial_{u_1} - \lambda q q' u_1\right)\sin u_2 + k_2 u_1^{-1} \cos u_2\right]\,, \tag{4.107}$$

where a and b are arbitrary constants. In quantum theory, k_2 is an integer.

These solutions were first obtained in the works [37, 40]. The cases considered in this section do not exhaust all the crossed fields which admit exact solutions of the relativistic wave equations. Namely, some fields of more complicated structure to be studied below coincide, in a number of particular cases, with crossed fields. Nonetheless, the hitherto known general classes of crossed fields admitting exact solutions have been presented here completely.

4.3 Longitudinal electromagnetic fields

4.3.1 General

In this section, we consider the classical and quantum motion of charged particles in *longitudinal electromagnetic fields*. We recall that electric and magnetic vectors for such fields obey the relations (2.101) ($\mathbf{E} = \mathbf{n}E$, $\mathbf{H} = \mathbf{n}H$) in an inertial reference frame. Here $\mathbf{n}$ is a constant unit vector. In what follows, the vector $\mathbf{n}$ is, without any loss of generality, directed along the z-axis, i.e. $\mathbf{n} = (0, 0, 1)$. Then, the Maxwell equations imply that

$$E = E(x^0, x^3), \quad H = H(x^1, x^2)\,, \tag{4.108}$$

i.e. that the magnetic field has to be stationary, while the electric field may not be. For the electric current density we find

$$\rho = \partial_3 E, \quad \mathbf{j} = -[\mathbf{n} \times \nabla H] - \mathbf{n}\partial_0 \mathrm{E}\,, \tag{4.109}$$

which indicates that free longitudinal fields must be stationary and homogeneous. Evidently, the longitudinal fields can, in the general case, be represented by potentials of the form

$$\begin{aligned} A_0 &= A_0(x^0, x^3), \quad A_3 = A_3(x^0, x^3), \quad E = \partial_0 A_3 - \partial_3 A_0\,, \\ A_1 &= A_1(x^1, x^2), \quad A_2 = A_2(x^1, x^2), \quad H = \partial_2 A_1 - \partial_1 A_2\,, \end{aligned} \tag{4.110}$$

with A_μ being arbitrary functions of the indicated arguments.

The Lorentz equations with the longitudinal fields are separated into two independent sets of equations in the planes x^0, x^3 and x^1, x^2,

$$m\ddot{x}^0 - E\dot{x}^3 = 0, \quad m\ddot{x}^3 - E\dot{x}^0 = 0\,, \tag{4.111}$$

$$m\ddot{x}^1 - H\dot{x}^2 = 0, \quad m\ddot{x}^2 + H\dot{x}^1 = 0\,. \tag{4.112}$$

Each set has a first integral,

$$m^2(\dot{x}_0^2 - \dot{x}_3^2) = m^2 + k_1^2, \quad m^2(\dot{x}_1^2 + \dot{x}_2^2) = k_1^2\,. \tag{4.113}$$

Thus, the electric field determines the motion in the plane x^0, x^3. Further, we call this the longitudinal motion. The magnetic field governs the motion in the plane x^1, x^2. We call this the transversal motion. The longitudinal and transversal motions are independent.

The classical action can be represented as a sum of two terms

$$S(x) = S_{03}(x^0, x^3) + S_{12}(x^1, x^2)\,, \tag{4.114}$$

which separately describe the longitudinal motion and the transversal motion. These terms satisfy two independent equations

$$\begin{aligned} &\left[\partial_0 S_{03} + A_0(x^0, x^3)\right]^2 - \left[\partial_3 S_{03} + A_3(x^0, x^3)\right]^2 = m^2 + k_1^2\,, \\ &\left[\partial_1 S_{12} + A_1(x^1, x^2)\right]^2 + \left[\partial_2 S_{12} + A_2(x^1, x^2)\right]^2 = k_1^2\,. \end{aligned} \tag{4.115}$$

Solutions $\varphi(x)$ of the K–G equation corresponding to definite integrals of motion k_1 can be represented in the following form:

$$\varphi(x) = \phi(x^0, x^3)\Phi(x^1, x^2)\,, \tag{4.116}$$

where the function $\phi(x^0, x^3)$ is a solution of the following equation for the longitudinal motion:

$$\left(\hat{P}_0^2 - \hat{P}_3^2 - m^2 - k_1^2\right)\phi(x^0, x^3) = 0\,, \tag{4.117}$$

and the function $\Phi(x^1, x^2)$ is a solution of the equation

$$\left(\hat{P}_1^2 + \hat{P}_2^2 - k_1^2\right)\Phi(x^1, x^2) = 0\,, \tag{4.118}$$

which is a two-dimensional stationary Schrödinger equation for the the transversal motion.

Solutions of the Dirac equation can be written in the form

$$\Psi(x) = \begin{pmatrix} Q(x^1, x^2)\left[m + \hat{P}_0 + \hat{P}_3 - ik_1\sigma_2\right] \\ Q(x^1, x^2)\left[(m - \hat{P}_0 - \hat{P}_3)\sigma_3 - k_1\sigma_1\right] \end{pmatrix} \psi(x^0, x^3)v\,, \tag{4.119}$$

where

$$Q = \begin{pmatrix} f_1(x^1, x^2) & 0 \\ 0 & f_{-1}(x^1, x^2) \end{pmatrix}, \tag{4.120}$$

v is an arbitrary two-component spinor, the function $\psi(x^0, x^3)$ obeys the equation analogous to (4.117):

$$\left(\hat{P}_0^2 - \hat{P}_3^2 - m^2 - k_1^2 - iE\right)\psi(x^0, x^3) = 0\,, \tag{4.121}$$

while the functions $f_s(x^1, x^2)$, $s = \pm 1$, are solutions of the set of equations

$$(\boldsymbol{\sigma}\hat{\mathbf{P}}_\perp)f = k_1 f, \quad \hat{\mathbf{P}}_\perp = \left(\hat{P}_1, \hat{P}_2, 0\right), \quad f = \begin{pmatrix} f_1 \\ f_{-1} \end{pmatrix}. \tag{4.122}$$

The latter set can also be written in another equivalent form

$$(\boldsymbol{\sigma}\hat{\mathbf{P}}_\perp)Q = k_1 Q\sigma_1\,.$$

Making the substitution

$$f = \left(\boldsymbol{\sigma}\hat{\mathbf{P}}_\perp + k_1\right)u, \quad u(x^1, x^2) = \begin{pmatrix} u_1(x^1, x^2) \\ u_{-1}(x^1, x^2) \end{pmatrix} \tag{4.123}$$

in equation (4.122), we find that the functions $u_s(x^1, x^2)$ satisfy independent second-order differential equations,

$$\left(\hat{P}_1^2 + \hat{P}_2^2 - k_1^2 + sH\right)u_s(x^1, x^2) = 0, \quad s = \pm 1\,, \tag{4.124}$$

which are similar to equation (4.118).

Thus, the quantum problem, the same as the classical one, has been reduced to solving independent equations for the longitudinal and the transversal motions.

To specify spin states, we chose the conserved spin integral for the fields under consideration in the following form:

$$\begin{aligned} \hat{L} &= \left\{m\,(\boldsymbol{\sigma}\mathbf{n}) + \rho_2(\mathbf{n}[\boldsymbol{\sigma}\hat{\mathbf{P}}_\perp])\right\}\cos\gamma - \rho_3(\mathbf{n}[\boldsymbol{\sigma}\hat{\mathbf{P}}_\perp])\sin\gamma \\ &= (\boldsymbol{\sigma}\mathbf{n})\left[m\cos\gamma + i(\boldsymbol{\sigma}\hat{\mathbf{P}}_\perp)\left(\rho_3\sin\gamma - \rho_2\cos\gamma\right)\right], \quad \gamma = \text{const}\,, \end{aligned}$$

see Section 2.4. We subject solutions of the Dirac equation (4.119) to the equation

$$\hat{L}\Psi = \zeta L\Psi, \quad \zeta = \pm 1\,, \tag{4.125}$$

which implies $L = \sqrt{m^2\cos^2\gamma + k_1^2}$, and

$$(\boldsymbol{\sigma}\mathbf{l})v = \zeta v, \quad \mathbf{l} = -L^{-1}(k_1\cos\gamma, k_1\sin\gamma, -m\cos\gamma), \quad \mathbf{l}^2 = 1\,. \tag{4.126}$$

Solutions of this equation are known, see (2.155).

Solutions of equations (4.117), (4.121) and (4.118), (4.124) have only been found for the fields that admit separation of variables in these equations. Four classes of electric, and two classes of magnetic fields of this sort were pointed out in [26, 27]. Some particular solutions in these fields were studied in [10, 28, 32, 278].

4.3.2 Longitudinal motion in the electric field

We consider possible solutions of equations (4.117) and (4.121) that determine the longitudinal motion in the electric field. As shown in [27], there are four classes of electric fields admitting the separation of variables in these equations. For each class, we list potentials, the electric field itself, and solutions of classical and quantum equations of motion.

Type I

At this point, we use the Minkowski coordinates and define potentials as

$$A_0 = -A(x^3), \quad A_3 = 0\,. \tag{4.127}$$

The corresponding electric field reads $E(x^3) = A'(x^3)$. By primes, here and in what follows, derivatives with respect to the indicated arguments are denoted.

First integrals of the Lorentz equations (4.111) are

$$m\dot{x}^0 - k_0 - A(x^3) = 0, \quad m^2\dot{x}_3^2 = R(x^3)\,,$$
$$R(x^3) = \left[k_0 + A(x^3)\right]^2 - m^2 - k_1^2\,.$$

They can be integrated,

$$x^0 = \int \frac{k_0 + A}{\sqrt{R}} dx^3, \quad \tau = m\int \frac{dx^3}{\sqrt{R}}\,.$$

The longitudinal part of the classical action is

$$S_{03}(x^0, x^3) = -k_0 x^0 + \int \sqrt{R} dx^3\,.$$

Solutions of the quantum longitudinal equations (4.117) and (4.121) can be represented as

$$\phi(x^0, x^3) = \exp(-ik_0x^0)\tilde{\phi}(x^3), \quad \psi(x^0, x^3) = \exp(-ik_0x^0)\tilde{\psi}(x^3)\,,$$

where each of the functions $\tilde{\phi}(x^3)$ and $\tilde{\psi}(x^3)$ satisfies the one-dimensional Schrödinger equation,

$$\tilde{\phi}''(x^3) + R\tilde{\phi}(x^3) = 0, \quad \tilde{\psi}''(x^3) + (R - iA')\tilde{\psi}(x^3) = 0\,.$$

Both equations admit exact solutions for the following choices of the function $A(x^3)$:

$$\begin{aligned}
A(x) &= \alpha x && \text{(ESP II)}\,,\\
A(x) &= \alpha / x && \text{(ESP III)}\,,\\
A(x) &= \alpha \exp \beta x && \text{(ESP V)}\,,\\
A(x) &= \alpha \tan \beta x && \text{(ESP VII)}\,,\\
A(x) &= \alpha \tanh \beta x && \text{(ESP VIII)}\,,\\
A(x) &= \alpha \coth \beta x && \text{(ESP IX)}\,,
\end{aligned} \tag{4.128}$$

where α and β are arbitrary constants, and corresponding ESP for equation (9.1) are indicated in the brackets.

Type II

At this point, we use the Minkowski coordinates and define potentials as

$$A_3 = -A(x^0), \quad A_0 = 0 \,. \tag{4.129}$$

The corresponding electric field reads $E(x^0) = A'(x^0)$.

First integrals of the Lorentz equations (4.111) have the form

$$m^2\dot{x}_0^2 = R(x^0), \quad m\dot{x}^3 + k_3 + A(x^0) = 0\,,$$
$$R(x^0) = m^2 + k_1^2 + \left[k_3 + A(x^0)\right]^2 .$$

They can be integrated,

$$x^3 = -\int \frac{k_3 + A}{\sqrt{R}} dx^0, \quad \tau = m\int \frac{dx^0}{\sqrt{R}} \,.$$

The longitudinal part of the classical action is

$$S_{03}(x^0, x^3) = -k_3 x^3 - \int \sqrt{R} dx^0 \,.$$

Solutions of the quantum longitudinal equations (4.117) and (4.121) can be represented as

$$\phi(x^0, x^3) = \exp(-ik_3 x^3)\tilde{\phi}(x^0), \quad \psi(x^0, x^3) = \exp(-ik_3 x^3)\tilde{\psi}(x^0)\,,$$

where each the functions $\tilde{\phi}(x^0)$ and $\tilde{\psi}(x^0)$ satisfies the one-dimensional Schrödinger equation,

$$\tilde{\phi}''(x^0) + R\tilde{\phi}(x^0) = 0, \quad \tilde{\psi}''(x^0) + (R - iA')\tilde{\psi}(x^0) = 0\,.$$

Both equations admit exact solutions for the function $A(x^0)$ of the form (4.128).

Type III

At this point, we use the light-cone coordinates (4.2) and define potentials in the corresponding reference frame as

$$\tilde{A}_0 = 0, \; \tilde{A}_3 = -A(u_0)/2\,. \tag{4.130}$$

The corresponding electric field is $E(u_0) = -A'(u_0)$.

First integrals of the Lorentz equations (4.111) are

$$m\dot{u}_0 = \lambda + A(u_0), \quad m\left[\lambda + A(u_0)\right]\dot{u}_3 = m^2 + k_1^2, \quad \lambda = P_0 + P_3\,.$$

They can be integrated,

$$\tau = m \int \frac{du_0}{\lambda + A}, \quad u_3 = (m^2 + k_1^2) \int \frac{du_0}{(\lambda + A)^2}.$$

The longitudinal part of the classical action is

$$S_{03}(x^0, x^3) = -\frac{\lambda}{2} u_3 - \frac{m^2 + k_1^2}{2} \int \frac{du_0}{\lambda + A}.$$

Solutions of the quantum longitudinal equations (4.117) and (4.121) have the form

$$\phi(x^0, x^3) = \frac{\exp iS_{03}(x^0, x^3)}{\sqrt{\lambda + A(x^0, x^3)}},$$
$$\psi(x^0, x^3) = \frac{\exp iS_{03}(x^0, x^3)}{\lambda + A(x^0, x^3)},$$

for arbitrary function $A(u_0)$.

Type IV

At this point, we use the coordinates u_0 and u_3,

$$u_0 = x_0^2 - x_3^2, \quad u_3 = \frac{1}{2} \ln \left| \frac{x^0 + x^3}{x^0 - x^3} \right|. \tag{4.131}$$

In the corresponding reference frame, potentials are defined as

$$\tilde{A}_0 = 0, \quad \tilde{A}_3 = -A(u_0)/2. \tag{4.132}$$

The electric field reads $E(u_0) = -A'(u_0)$.

First integrals of the Lorentz equations (4.111) are

$$m^2 \dot{u}_0^2 = R(u_0), \quad 2mu_0 \dot{u}_3 + k_3 + A(u_0) = 0,$$
$$R(u_0) = [k_3 + A(u_0)]^2 + 4u_0(m^2 + k_1^2).$$

They can be integrated,

$$\tau = m \int \frac{du_0}{\sqrt{R}}, \quad u_3 = -\frac{1}{2} \int \frac{k_3 + A}{u_0 \sqrt{R}} du_0.$$

The longitudinal part of the classical action is

$$-S_{03} = \frac{k_3}{2} u_3 + \frac{1}{4} \int \frac{\sqrt{R}}{u_0} du_0.$$

Solutions of the quantum longitudinal equations (4.117) and (4.121) can be represented as

$$\phi(x^0, x^3) = \exp(-ik_3u_3/2)\tilde{\phi}(u_0)\,,$$
$$\psi(x^0, x^3) = \exp(-ik_3u_3/2)\tilde{\psi}(u_0)\,,$$

where the functions $\tilde{\phi}(u_0)$ and $\tilde{\psi}(u_0)$ satisfy the equations:

$$16u_0^2\tilde{\phi}'' + 16u_0\tilde{\phi}' + R(u_0)\tilde{\phi} = 0\,,$$
$$16u_0^2\tilde{\psi}'' + 16u_0\tilde{\psi}' + \left[R(u_0) - iA'(u_0)\right]\tilde{\psi} = 0\,. \tag{4.133}$$

Solutions of equations (4.133) can be found for $A = \alpha u_0$, which corresponds to a constant and homogeneous electric field, and also for $A = \alpha\sqrt{|u_0|}$. In the latter case, the substitution $u_0 = \pm x^2$ reduces equations (4.133) to the one-dimensional Schrödinger equation (9.1) with the ESP II.

4.3.3 Transversal motion in the magnetic field

Two types of fields allow one to separate variables in equations (4.118) and (4.124). We discuss these two cases below.

Type I

At this point, we use the Minkowski coordinates and define potentials as follows:

$$A_1 = 0, \quad A_2 = -A(x^1)\,. \tag{4.134}$$

The corresponding magnetic field reads $H = A'(x^1)$.

First integrals of the Lorentz equations (4.112) are

$$m^2\dot{x}_1^2 = R(x^1), \quad m\dot{x}^2 + k_2 + A = 0\,,$$
$$R(x^1) = k_1^2 - (k_2 + A)^2\,.$$

They can be integrated,

$$\tau = m\int\frac{dx^1}{\sqrt{R}}, \quad x^2 = -\int(k_2 + A)\sqrt{R}dx^1\,.$$

The transversal part of the classical action is

$$S_{12} = -k_2x^2 + \int\sqrt{R}dx^1\,.$$

Solutions of the quantum transversal equations (4.118) and (4.124) can be represented as

$$\Phi(x^1, x^2) = \exp(-ik_2x^2)\tilde{\Phi}(x^1), \quad u_s(x^1, x^2) = \exp(-ik_2x^2)\tilde{u}_s(x^1), \quad s = \pm 1\,,$$

where each the function $\tilde{\Phi}(x^1)$ and $\tilde{u}_s(x^1)$ obey one-dimensional Schrödinger equations (9.1),

$$\tilde{\Phi}''(x^1) + R\tilde{\Phi}(x^1) = 0, \quad \tilde{u}_s''(x^1) + (R + sA')\tilde{u}_s(x^1) = 0, \quad s = \pm 1 .$$

Both equations admit exact solutions for the functions $A(x^3)$ of the form (4.128).

Type II

At this point, we use polar coordinates $u_1 = r$ and $u_2 = \varphi$,

$$x^1 = u_1 \cos u_2, \quad x^2 = u_1 \sin u_2 .$$

In the corresponding reference frame we define the potentials as

$$\tilde{A}_1 = 0, \quad \tilde{A}_2 = -A(u_1) . \tag{4.135}$$

The magnetic field reads $H = u_1^{-1}A'(u_1)$.

First integrals of the Lorentz equations (4.112) are

$$m^2\dot{u}_1^2 = R(u_1), \quad mu_1^2\dot{u}_2 + k_2 + A = 0 ,$$
$$R(u_1) = k_1^2 - (k_2 + A)^2 u_1^{-2} .$$

They can be integrated,

$$\tau = m\int \frac{du_1}{\sqrt{R}}, \quad u_2 = -\int \frac{k_2 + A}{u_1^2\sqrt{R}} du_1 .$$

The transversal part of the classical action is

$$S_{12} = -k_2 u_2 + \int \sqrt{R} du_1 .$$

Solutions of the quantum transversal equations (4.118) and (4.124) can be represented as

$$\Phi(x^1, x^2) = \exp(-ik_2u_2)\tilde{\Phi}(u_1), \quad k_2 \text{ are integers}$$
$$u_s(x^1, x^2) = \exp(-ik_2u_2)\tilde{u}_s(u_1), \quad s = \pm 1, k_2 \text{ are half integers} ,$$

where the functions $\tilde{\Phi}(u_1)$ and $\tilde{u}_s(u_1)$ obey the equations

$$\tilde{\Phi}''(u_1) + \tilde{\Phi}(u_1)/u_1 + R\tilde{\Phi}(u_1) = 0 ,$$
$$\tilde{u}_s''(u_1) + \tilde{u}_s(u_1)/u_1 + (R + sA'/u_1)\tilde{u}_s(u_1) = 0, \quad s = \pm 1 . \tag{4.136}$$

All the equations (4.136) can be reduced to one-dimensional Schrödinger equations (9.1) for the functions $\sqrt{u_1}\tilde{\Phi}(u_1)$ and $\tilde{u}_s\sqrt{u_1}$. Solutions to such equations can be found for the functions $A(u_1)$ of the form

$$A = \alpha u_1^2 + \beta \quad \text{(ESP IV)} ,$$
$$A = \alpha u_1 + \beta \quad \text{(ESP III)} .$$

It might seem that the addition of the constant β to the function A could not lead to appearance of a new field. Nevertheless, in the case $\beta \neq 0$ an Aharonov–Bohm field additionally appears. This and other cases will be considered elsewhere.

Solutions for $\beta \neq 0$ were first found in [235]. The general study of particle motion in the longitudinal fields was first presented in [47].

4.4 Superposition of crossed and longitudinal fields

4.4.1 General

In this section, we consider the classical and quantum motion of charged particles in a *superposition of crossed and longitudinal electromagnetic fields*. For such a superposition, the magnetic and electric fields are related as

$$\mathbf{H} = [\mathbf{n} \times \mathbf{E}] + (\mathbf{nH})\mathbf{n}, \quad \mathbf{E} = -[\mathbf{n} \times \mathbf{H}] + (\mathbf{nE})\mathbf{n}\,, \tag{4.137}$$

see (2.103) and (2.104). In what follows, the vector $\mathbf{n}$ is, without any loss of generality, directed along the z-axis, i.e. $\mathbf{n} = (0, 0, 1)$. We call the superposition (4.137) the *combined electromagnetic field.*

There are many common features of classical and quantum motion in different types of combined fields. We point out these common features below.

It is convenient in every case to use the light-cone coordinate $u_0 = x^0 - x^3$. The rest of the curvilinear coordinates u_1, u_2, and u_3 are chosen individually for each case. In all the reference frames, which we use in what follows, the potential component $\tilde{A}_3$ can be taken universally as $\tilde{A}_3 = -g(u_0)/2$, where $g(u_0)$ is given in each particular case. The Cartesian component E_z of the electric field in all the cases has the form

$$E_z = -g'(u_0)\,. \tag{4.138}$$

The charge and current densities ρ and $\mathbf{j} = (j_x, j_y, j_z)$ are always such that $\rho = j_z$. The Cartesian components E_x, E_y, H_z and the quantities ρ, j_x, j_y are specified in each case separately.

For every type of combined field considered below, the quantity $\lambda = (nP) = P_0 + P_3$ is an integral of motion[2]. Due to this fact, one of the Lorentz equations always has the form

$$m\dot{u}_0 = \Lambda, \quad \Lambda = \lambda + g(u_0)\,, \tag{4.139}$$

and hence there exists the universal relation

$$\tau = m \int \frac{du_0}{\Lambda} \tag{4.140}$$

2 This fact does not mean that all the combined fields necessarily admit λ as an integral of motion. It is, perhaps, a mere coincidence that λ is conserved just for those fields of type (2.103), for which exact solutions are known.

between u_0 and the proper time τ ($\Lambda \geq 0$ in classical theory).

The classical action for a particle in all the combined fields under consideration can be written in the form

$$S(u) = S_0(u) + \int \sqrt{R(u_1)}du_1 \,,$$
$$S_0(u) = -\frac{\lambda}{2}u_3 - \Gamma(u), \quad R(u_1) \geq 0 \,, \tag{4.141}$$

where the functions $R(u_1)$ and $\Gamma(u)$ are listed below for each type of the combined field.

Solutions of the K–G equation with a combined field can be written in the form

$$\varphi(u) = \sqrt{\frac{q(u_0)}{\Lambda}} \exp(iS_0)\Phi(u_1) \,, \tag{4.142}$$

where the functions $q(u_0)$ and $\Phi(u_1)$ are listed below for each type of the combined field.

Solution of the Dirac equation with a combined field can be written as

$$\Psi(u) = \frac{\sqrt{q(u_0)}}{\Lambda} \exp(iS_0)\hat{K} \left[\sum_{\zeta=\pm 1} (1 + \zeta\sigma_3)\, \psi_\zeta(u_1) \exp i\zeta t(u_0) \right] v \,,$$
$$\hat{K} = \begin{pmatrix} m + \Lambda - \sigma_3(\boldsymbol{\sigma}\hat{\mathbf{F}}) \\ (m - \Lambda)\sigma_3 - (\boldsymbol{\sigma}\hat{\mathbf{F}}) \end{pmatrix} , \tag{4.143}$$

where v is a constant two-component spinor, $\hat{\mathbf{F}} = (\hat{F}_1, \hat{F}_2, 0)$, the functions $t(u_0)$ and $\psi_\zeta(u_1)$, $\zeta = \pm 1$, and the operators $\hat{F}_1$ and $\hat{F}_2$ are listed below for each type of the combined field.

4.4.2 Crossed and longitudinal electric field

We consider here the combined field for which $(\mathbf{nH}) = 0 \Longrightarrow H_z = 0$. This means that the longitudinal field in the superposition represents only an electric field. For this type of the combined field, solutions of the Dirac equation take the form

$$\Psi = \frac{\sqrt{q(u_0)}}{\Lambda} \exp(iS_0)\hat{K}\Phi(u_1)v \,, \tag{4.144}$$

where the functions $\Phi(u_1)$ are listed in the corresponding items below. Solutions in a constant electric field and collinear plane-wave field were studied in detail in Ref. [248].

Type I

At this point, we use the light-cone coordinates (4.2), (4.3) and define potentials in the corresponding reference frame as

$$\tilde{A}_0 = 0, \quad \tilde{A}_s = -a_s(u_0), \quad s = 1, 2, \quad \tilde{A}_3 = -g(u_0)/2\,, \tag{4.145}$$

where $a_s(u_0)$ and $g(u_0)$ are arbitrary functions. In this case

$$E_x = H_y = -a_1'(u_0), \quad E_y = -H_x = -a_2'(u_0), \quad H_z = 0; \quad \rho = g''(u_0)\,. \tag{4.146}$$

The field under consideration is a superposition of a plane-wave field and an electric field (4.138) parallel to it.

The first integrals of the Lorentz equations are

$$m\dot{u}_1 + P_1 = 0, \quad m\dot{u}_2 + P_2 = 0, \quad P_1 = k_1 + a_1\,,$$
$$m\Lambda\dot{u}_3 = m^2 + P_1^2 + P_2^2, \quad P_2 = k_2 + a_2\,.$$

They can be integrated,

$$u_1 = \int \frac{P_1}{\Lambda} du_0, \quad u_2 = -\int \frac{P_2}{\Lambda} du_0, \quad u_3 = \int \frac{m^2 + P_1^2 + P_2^2}{\Lambda^2} du_0\,.$$

The classical action has the form (4.141) with

$$\Gamma = k_1 u_1 + k_2 u_2 + \int \frac{m^2 + P_1^2 + P_2^2}{2\Lambda} du_0, \quad R = 0\,.$$

Solutions of the K–G and Dirac equations are given by equations (4.142) and (4.144), where one should set

$$\Phi = q = 1, \quad \hat{F}_1 = \hat{P}_1, \quad \hat{F}_2 = \hat{P}_2\,.$$

These solutions were first found in [28]. Detailed investigation of these solutions can also be found in [103, 248].

Type II

At this point, we use the coordinates u_μ,

$$u_0 = x^0 - x^3, \quad u_1 = u_0^{-1} x^1, \quad u_2 = x^2, \quad u_3 = u_0^{-1}(x_0^2 - x_1^2 - x_3^2)\,. \tag{4.147}$$

We define potentials in the corresponding reference frame as

$$\tilde{A}_0 = 0, \quad \tilde{A}_s = -a_s(u_0), \quad s = 1, 2, \quad \tilde{A}_3 = -g(u_0)/2\,, \tag{4.148}$$

where $a_s(u_0)$ and $g(u_0)$ are arbitrary functions. In this case

$$\begin{aligned} &E_x = H_y = u_1 g'(u_0) - u_0^{-1} a_1'(u_0), \quad E_y = -H_x = -a_2'(u_0)\,, \\ &H_z = 0; \quad \rho = g''(u_0) + u_0^{-1} g'(u_0)\,. \end{aligned} \tag{4.149}$$

For $g' = \alpha/u_0$, where α is an arbitrary constant, the fields are free, but are not plane waves.

The first integrals of the Lorentz equations are

$$m u_0^2 \dot{u}_1 + P_1 = 0, \quad m\dot{u}_2 + P_2 = 0, \quad P_1 = k_1 + a_1\,,$$
$$m\Lambda \dot{u}_3 = m^2 + u_0^{-2}P_1^2 + P_2^2, \quad P_2 = k_2 + a_2\,.$$

They can be integrated,

$$u_1 = -\int \frac{P_1}{u_0^2\Lambda}du_0, \quad u_2 = -\int \frac{P_2}{\Lambda}du_0, \quad u_3 = \int \frac{m^2 + u_0^{-2}P_1^2 + P_2^2}{\Lambda^2}du_0\,.$$

The classical action has the form (4.141) with

$$\Gamma = k_1u_1 + k_2u_2 + \int \frac{m^2 + u_0^{-2}P_1^2 + P_2^2}{2\Lambda}du_0, \quad R = 0\,.$$

Solutions of the K–G and Dirac equations are given by equations (4.142) and (4.144) where one should set

$$q = u_0^{-1}, \quad \Phi = 1, \quad \hat{F}_1 = u_0^{-1}\hat{P}_1 - u_1\Lambda, \quad \hat{F}_2 = \hat{P}_2\,.$$

These solutions were first found in [28].

Type III

At this point, we use the coordinates u_μ,

$$u_0 = x^0 - x^3, \quad u_1 = q(u_0)[(u_0 + a)x^1 - bx^2], \quad u_2 = q(u_0)(u_0x^2 - bx^1)\,,$$
$$u_3 = x^0 + x^3 + q(u_0)[2bx^1x^2 - u_0(x_1^2 + x_2^2) - ax_1^2]\,, \tag{4.150}$$

where

$$q(u_0) = \left(u_0^2 + au_0 - b^2\right)^{-1}\,,$$

and a and b are arbitrary constants. We define potentials in the corresponding reference frame as

$$\tilde{A}_0 = 0, \quad \tilde{A}_s = -a_s(u_0), \quad s = 1, 2, \quad \tilde{A}_3 = -g(u_0)/2\,, \tag{4.151}$$

where $a_s(u_0)$ and $g(u_0)$ are arbitrary functions. In this case

$$E_x = H_y = bq(u_0)a_2'(u_0) + u_1g'(u_0) - q(u_0)(u_0 + a)a_1'(u_0)\,,$$
$$E_y = -H_x = bq(u_0)a_1'(u_0) + 2g'(u_0) - q(u_0)u_0a_2'(u_0), \quad H_z = 0\,;$$
$$\rho = g''(u_0) + q(u_0)(2u_0 + a)g'(u_0)\,. \tag{4.152}$$

For $g'(u_0) = \alpha q(u_0)$, where α is an arbitrary constant, the fields are free, but are not plane waves.

The first integrals of the Lorentz equations are

$$mq^{-2}\dot{u}_1 - b(2u_0 + a)P_2 + [b^2 + (u_0 + a)^2]P_1 = 0\,,$$
$$mq^{-2}\dot{u}_2 + (u_0^2 + b^2)P_2 - b(2u_0 + a)P_1 = 0, \quad P_1 = k_1 + a_1, \quad P_2 = k_2 + a_2\,,$$
$$m\Lambda\dot{u}_3 = m^2 + q^2[(u_0 + a)P_1 - bP_2]^2 + q^2(bP_1 - u_0P_2)^2\,.$$

They can be integrated,

$$u_1 = \int \frac{q^2}{\Lambda}\{b(2u_0 + a)P_2 - [b^2 + (u_0 + a)^2]P_1\}du_0\,,$$
$$u_2 = \int \frac{q^2}{\Lambda}[b(2u_0 + a)P_1 - (u_0^2 + b^2)P_2]du_0\,,$$
$$u_3 = \int \{m^2 + q^2[(u_0 + a)P_1 - bP_2]^2 + q^2(bP_1 - u_0P_2)^2\}\frac{du_0}{2\Lambda}\,.$$

The classical action has the form (4.141) with

$$\Gamma = \int \left\{m^2 + q^2[(u_0 + a)P_1 - bP_2]^2 + q^2(bP_1 - u_0P_2)^2\right\}\frac{du_0}{2\Lambda}$$
$$+ k_1u_1 + k_2u_2, \quad R = 0\,.$$

Solutions of the K–G and Dirac equations are given by equations (4.142) and (4.144), where one should set

$$\hat{F}_1 = q(u_0)(u_0 + a)\hat{P}_1 - bq(u_0)\hat{P}_2 - u_1\Lambda\,,$$
$$\hat{F}_2 = q(u_0)u_0\hat{P}_2 - bq(u_0)\hat{P}_1 - u_2\Lambda, \quad \Phi = 1\,.$$

These solutions were first found in [28].

Type IV

At this point, we use the light-cone coordinates (4.2) and (4.3) and define potentials in the corresponding reference frame as

$$\tilde{A}_0 = -a\exp(bu_1), \quad \tilde{A}_1 = -b^{-1}g'(u_0), \quad \tilde{A}_2 = -f(u_0), \quad \tilde{A}_3 = -g(u_0)/2\,, \quad (4.153)$$

where a and $b \neq 0$ are arbitrary constants, $g(u_0)$ and $f(u_0)$ are arbitrary functions. In this case

$$E_x = H_y = ab\exp(bu_1) - b^{-1}g''(u_0), \quad E_y = -H_x = -f'(u_0)\,,$$
$$H_z = 0; \quad \rho = g''(u_0) + ab\exp(bu_1)\,. \quad (4.154)$$

No free fields exist.

The first integrals of the Lorentz equations are

$$\left[m\dot{u}_1 + b^{-1}g'(u_0)\right]^2 = k_1 + 2a\Lambda\exp(bu_1), \quad m\dot{u}_2 + Q(u_0) = 0,$$
$$m\Lambda\dot{u}_3 = m^2 + Q^2(u_0) + b^{-2}\left[2m\chi'(\tau)\chi^{-1}(\tau) + g'(u_0)\right]^2,$$

where $Q(u_0) = k_2 + f(u_0)$, and

$$\chi(\tau) = \begin{cases} \frac{1}{\sqrt{|k_1|}} \begin{cases} \sinh\omega\tau, & k_1 > 0, \quad a > 0 \\ \cosh\omega\tau, & k_1 > 0, \quad a < 0 \\ \cos\omega\tau, & k_1 < 0, \quad a > 0 \end{cases} \\ \frac{b\tau}{2m}, \quad k_1 = 0, \quad a > 0 \end{cases},$$

with $\omega = b\sqrt{|k_1|}/2m$. They can be integrated,

$$bu_1 = -\ln(2|a|\Lambda\chi^2), \quad u_2 = -\int \frac{Q}{\Lambda}du_0,$$
$$u_3 = \int \frac{(m^2 + Q^2)b^2 + (2m\chi'\chi^{-1} + g')^2}{b^2\Lambda^2}du_0.$$

The classical action has the form (4.141) with

$$\Gamma = k_2u_2 + \int \frac{m^2 + Q^2 + b^{-2}g'^2 + k_1}{2\Lambda}du_0,$$
$$\int \sqrt{R}du_1 = \int \sqrt{4ma\exp b\mu + k_1}d\mu, \quad \mu = u_1 + b^{-1}(\ln\Lambda - \ln 2m).$$

Solutions of the K–G and Dirac equations are given by equations (4.142) and (4.144) with

$$q = 1, \quad \hat{F}_1 = \frac{iz}{2}\partial_z + b^{-1}g', \quad \hat{F}_2 = Q, \quad bz = \sqrt{8a|\Lambda|}\exp(bu_1/2),$$
$$\Phi(u_1) = r_1J_\nu(z) + r_2J_{-\nu}(z), \quad b\nu = 2\sqrt{-k_1},$$

where r_1 and r_2 are arbitrary constants.

Note that every combined field of the type I–IV contains arbitrary functions (three functions of type I–III and two functions of type IV). Nevertheless, solutions of both classical and quantum equations of motion are reduced to an integration without specifying these functions.

4.4.3 Crossed and longitudinal electric and magnetic fields

Type I

At this point, we use the coordinates u_μ,

$$u_0 = x^0 - x^3, \qquad u_1 = qr^2,$$
$$u_2 = \arctan\left(x^2/x^1\right), \quad u_3 = x^0 + x^3 - u_0u_1, \tag{4.155}$$

where

$$q = (u_0^2 + a)^{-1}, \quad r^2 = x_1^2 + x_2^2\,,$$

and a is an arbitrary constant. We define potentials in the corresponding reference frame as

$$\begin{aligned} \tilde{A}_0 &= -q[f_1(u_1) + ag(u_0)u_1], \quad \tilde{A}_1 = 0\,, \\ \tilde{A}_2 &= -f_2(u_1) - u_1 g_1(u_0), \qquad \tilde{A}_3 = -g(u_0)/2\,. \end{aligned} \tag{4.156}$$

Components of the electromagnetic fields and currents are given as projections onto the basis vectors of the cylindrical reference frame

$$\begin{aligned} E_r &= H_\varphi = qr[2q(f_1' + ag) + g'u_0]\,, \\ E_\varphi &= -H_r = qr[2qu_0(g_1 + f_2') - g_1'], \quad H_z = 2q(g_1 + f_2')\,, \\ \rho &= g'' + 2qu_0 g' + 4q^2(u_1 f_1'' + f_1'), \quad j_r = 0, \quad j_\varphi = 4q^2 r f_2''\,. \end{aligned} \tag{4.157}$$

The fields are source-free provided that

$$\begin{aligned} f_1' &= \alpha/u_1 + \beta, \quad f_2'' = 0\,, \\ g &= \begin{cases} \gamma(1 - 2aq) + \delta q u_0 - \beta/2a, & a \neq 0 \\ \gamma/u_0 - \beta/u_0^2 + \delta, & a = 0 \end{cases}, \end{aligned}$$

where α, β, γ and δ are constants.

The functions f_1, f_2, g_1 and g must be subjected to one of the following conditions:

$$\begin{aligned} &\text{(i)} && g = g_1 = 0\,, \\ &\text{(ii)} && f_1 = f_2 = 0, \quad g_1 = \sqrt{ag^2 + b}\,, \\ &\text{(iii)} && f_1 = cu_1/2, \quad f_2 = 0, \quad g_1 = \sqrt{cg}, \quad a = 0\,, \\ &\text{(iv)} && f_1 = \sqrt{a} f_2\, g_1 = \sqrt{a} g\,, \quad a \geq 0\,, \end{aligned} \tag{4.158}$$

where b and c are arbitrary constants.

The first integrals of the Lorentz equations are

$$\begin{aligned} m^2\dot{u}_1^2 &= 16q^2u_1^2 R(u_1), \quad mu_1\dot{u}_2 + q(k_2 + f_2 + u_1 g_1) = 0\,, \\ m\dot{u}_3 &= 2q(f_1 - a\lambda u_1) + \left[m^2 + 2q(2k_1 + k_2 g_1)\right]\Lambda^{-1}\,, \\ 4u_1 R(u_1) &= 4k_1 - 2\lambda f_1 - a\lambda^2 u_1 - 4bu_1 - (k_2 + f_2)^2 u_1^{-1}\,. \end{aligned}$$

They can be integrated,

$$u_2 = -t(u_0) - \int \frac{k_2 + f_2}{4u_1^2\sqrt{R}} du_1, \quad t(u_0) = \int \frac{qg_1}{\Lambda} du_0 ,$$
$$u_3 = \int \frac{f_1 - a\lambda u_1}{2u_1\sqrt{R}} du_1 + \int \frac{m^2 + 2q(2k_1 + k_2 g_1)}{\Lambda^2} du_0 ,$$
$$\int \frac{du_1}{u_1\sqrt{R}} = \int \frac{4q}{\Lambda} du_0 .$$

The classical action has the form (4.141) with

$$\Gamma = k_2 u_2 + \int \frac{m^2 + 2q(2k_1 + k_2 g_1)}{2\Lambda} .$$

One must set $b = 0$ in every case, except the case (ii) in (4.158).

Solutions of the K–G and Dirac equations are given by equations (4.142) and (4.143) where the functions $\Phi(u_1)$ and $\psi_\zeta(u_1)$ satisfy the equations

$$u_1\Phi'' + \Phi' + u_1 R(u_1)\Phi = 0, \quad u_1\psi_\zeta'' + \psi_\zeta' + [u_1 R(u_1) + \zeta f_2'/2]\psi_\zeta = 0 , \tag{4.159}$$

with k_2 being an integer. These equations can be reduced to the one-dimensional Schrödinger equations (4.36). In this case

$$\hat{F}_1 = \mu_1 \cos u_2 - \mu_2 \sin u_2, \quad \hat{F}_2 = \mu_1 \sin u_2 + \mu_2 \cos u_2 ,$$
$$\mu_1 = r^{-1}u_1\left(2i\partial_{u_1} - \Lambda u_0\right), \quad \mu_2 = r^{-1}(k_2 + f_2 + u_1 g_1) .$$

Solutions in the fields under consideration were first found in [32].

Type II

At this point, we use the coordinates u_μ,

$$u_0 = x^0 - x^3, \quad u_1 = u_0^{-1}r^2, \quad u_2 = \arctan\left(x^2/x^1\right) ,$$
$$u_3 = x^0 + x^3 - u_1/2, \quad r^2 = x_1^2 + x_2^2, \quad u_2 = \varphi . \tag{4.160}$$

We define potentials in the corresponding reference frame as

$$\tilde{A}_0 = -u_0^{-1} f_1(u_1), \quad \tilde{A}_2 = -f_2(u_1), \quad \tilde{A}_1 = \tilde{A}_3 = 0 \quad (g = 0) , \tag{4.161}$$

where $f_1(u_1)$ and $f_2(u_1)$ are arbitrary functions.

In this case

$$E_r = H_\varphi = 2ru_0^{-2} f_1', \quad E_\varphi = -H_r = 2ru_0^{-2} f_2', \quad H_z = 2u_0^{-1} f_2', \quad E_z = 0 ;$$
$$\rho = 4u_0^{-2}(u_1 f_1'' + f_2'), \quad j_\varphi = 4ru_0^{-2} f_2'', \quad j_r = 0 . \tag{4.162}$$

The fields are free provided that $f_1' = \alpha/u_1$, and $f_2'' = 0$, where α is an arbitrary constant.

The first integrals of the Lorentz equations are

$$m^2 u_0^2 \dot{u}_1^2 = 16u_1^2 R(u_1), \quad m u_0 u_1 \dot{u}_2 + k_2 + f_2 = 0\,,$$
$$m\lambda u_0 \dot{u}_3 = m^2 u_0 + k_1 + 2\lambda f_1 + \lambda u_1/2\,,$$
$$16 u_1 R(u_1) = \lambda^2 u_1 + 4k_1 + 8\lambda f_1 - 4(k_2 + f_2)^2 u_1^{-1}\,.$$

They can be integrated,

$$m u_0 = \lambda \tau = \exp\left(\frac{\lambda}{4}\int \frac{du_1}{u_1\sqrt{R}}\right), \quad u_2 = -\int \frac{k_2 + f_2}{4u_1^2\sqrt{R}} du_1\,,$$
$$u_3 = \frac{m^2}{\lambda^2} u_0 + \int \frac{2k_1 + 4\lambda f_1 + \lambda^2 u_1}{8\lambda u_1 \sqrt{R}} du_1\,.$$

The classical action has the form (4.141) with

$$\Gamma = k_2 u_2 + (m^2 u_0 + k_1 \ln|u_0|)/2\lambda\,.$$

Solutions of the K–G and Dirac equations are given by equations (4.142) and (4.143) where the functions Φ and ψ_ζ obey equations (4.159), k_2 is an integer, and one should set

$$\hat{F}_1 = \mu_1 \cos u_2 - \mu_2 \sin u_2, \quad \hat{F}_2 = \mu_1 \sin u_2 + \mu_2 \cos u_2\,,$$
$$\mu_1 = r\,(2u_0)^{-1}\,(4i\partial_{u_1} - \lambda), \quad \mu_2 = r^{-1}(k_2 + f_2), \quad q = u_0^{-1}, \quad t = 0\,.$$

These solutions were first found in [42].

Type III

At this point, we use the coordinates (4.160) and define potentials in the corresponding reference frame as

$$\tilde{A}_0 = g u_1/4u_0, \quad \tilde{A}_2 = -u_1 f(u_0)/2, \quad \tilde{A}_1 = 0, \quad \tilde{A}_3 = -g(u_0)/2\,, \tag{4.163}$$

and functions $g(u_0)$ and $f(u_0)$ are related as $f^2 + g^2 = a^2$, where a is an arbitrary constant. In this case

$$E_r = H_\varphi = r(u_0 g' - g)\left(2u_0^2\right)^{-1}, \quad E_\varphi = -H_r = r(f - u_0 f')\left(2u_0^2\right)^{-1}\,,$$
$$H_z = -u_0^{-1} f; \quad \rho = u_0^{-2}(u_0 g' + u_0^2 g'' - g), \quad j_{x,y} = 0\,. \tag{4.164}$$

The field is free provided that $g = \alpha u_0 + \beta u_0^{-1}$, where α and β are arbitrary constants.

The first integrals of the Lorentz equations are

$$
\begin{aligned}
&m^2u_0^2\dot{u}_1^2 = 16u_1^2R(u_1) = (\lambda^2 - a^2)u_1^2 + 4k_1u_1 - 4k_2^2\,,\\
&2mu_0u_1\dot{u}_2 + 2k_2 + fu_1 = 0, \quad k_1^2 \geq (a^2 - \lambda^2)k_2^2\,,\\
&2m\Lambda u_0\dot{u}_3 = \lambda\Lambda u_1 + 2fk_2 + 2k_1 + 2m^2u_0\,.
\end{aligned} \tag{4.165}
$$

They can be integrated,

$$
(\lambda^2 - a^2)u_1 = 2\sqrt{k_1^2 + k_2^2(\lambda^2 - a^2)}\varphi(\chi) - 2k_1, \quad u_2 = -k_2\int\frac{d\chi}{u_1} - t(u_0)\,,
$$

$$
u_3 = \frac{\lambda}{2}\int u_1 d\chi + \int\frac{m^2u_0 + k_2 f + k_1}{u_0\Lambda^2}du_0, \quad t(u_0) = \int\frac{f du_0}{2u_0\Lambda}, \quad \chi = \int\frac{du_0}{u_0\Lambda}\,,
$$

$$
\varphi(\chi) = \begin{cases} (\operatorname{sgn} u_0)\cosh\sqrt{\lambda^2 - a^2}\chi\,, & \lambda^2 > a^2 \\ \cos\sqrt{a^2 - \lambda^2}\chi\,, & \lambda^2 < a^2 \\ (k_1^2\chi^2 + k_2^2)k_1^{-1}\,, & \lambda^2 = a^2 \end{cases}.
$$

The classical action has the form (4.141) with

$$
\Gamma = k_2u_2 + \int\frac{m^2u_0 + k_2 f + k_1}{2u_0\Lambda}du_0\,.
$$

Solutions of the K–G and Dirac equations are determined by the expressions (4.142) and (4.143) where the function Φ is a solution of the first equation (4.159), k_2 are integer, and one should set

$$
\begin{aligned}
&\hat{F}_1 = \mu_1\cos u_2 - \mu_2\sin u_2, \quad \hat{F}_2 = \mu_1\sin u_2 + \mu_2\cos u_2\,,\\
&\mu_1 = ru_0^{-1}(4i\partial_{u_1} - \Lambda), \quad \mu_2 = (2r)^{-1}(k_2 + fu_1)\,, \quad \psi_\zeta = \Phi, \quad q = u_0^{-1}\,.
\end{aligned} \tag{4.166}
$$

Once $R(u_1)$ is specified in (4.165), such a solution can be written explicitly. For doing this, we introduce a variable z,

$$
z = (\epsilon/2)\sqrt{|\lambda^2 - a^2|}u_1, \quad \epsilon = \operatorname{sgn} u_0, \quad z \geq 0\,. \tag{4.167}
$$

Then we find for $a^2 > \lambda^2$

$$
\Phi(u_1) = z^\mu \exp(-z/2)\,L_n^{2\mu}(z), \quad \mu = |k_2|/2\,, \tag{4.168}
$$

where k_1 is quantized as $k_1 = \epsilon\sqrt{a^2 - \lambda^2}(2n + 1 + 2\mu)$, $n \in \mathbb{N}_+$.

For $\lambda^2 > a^2$ we have

$$
\Phi(u_1) = z^\mu\Phi(\alpha, 1 + 2\mu; iz)\exp(-iz/2)\,, \quad 2\alpha = 1 + 2\mu + i\epsilon k_1(\lambda^2 - a^2)^{-1/2}\,, \tag{4.169}
$$

where $\Phi(\alpha, \gamma; x)$ is the confluent hypergeometric function.

Solutions in the fields under consideration were first found in the works [37, 42].

Type IV

At this point, we use the coordinates (4.103) with $q = 1$ and define the potentials in the corresponding reference frame as

$$\tilde{A}_0 = -2cu_1^2, \quad \tilde{A}_1 = 0, \quad \tilde{A}_2 = -2u_1^2 f(u_0), \quad \tilde{A}_3 = -g(u_0)/2, \quad g(u_0) = c^{-1} f^2(u_0), \tag{4.170}$$

where c is an arbitrary constant. In this case

$$E_r = H_\varphi = 4cu_1, \quad E_\varphi = -H_r = -2u_1 f', \quad E_z = -2c^{-1} ff', \quad H_z = 4f;$$
$$j_{x,y} = 0, \quad \rho = 2c^{-1}(4c^2 + f'^2 + ff''). \tag{4.171}$$

The fields are free providing $f = \sqrt{\alpha - (\beta + 2cu_0)^2}$, where α and β are arbitrary constants.

The first integrals of the Lorentz equations are

$$m^2 \dot{u}_1^2 = R(u_1) = k_1 + 4c\lambda u_1^2 - k_2^2 u_1^{-2}, \quad mu_1^2 \dot{u}_2 + k_2 + 2fu_1^2 = 0,$$
$$m\Lambda \dot{u}_3 = m^2 + k_1 + 4fk_2 + 4cu_1^2 \Lambda.$$

They can be integrated,

$$8c\lambda u_1^2 = \sqrt{k_1^2 + 16c\lambda k_2^2 \varphi(\tau)} - k_1, \quad u_2 = -\frac{t(u_0)}{2} - \frac{k_2}{m} \int \frac{d\tau}{u_1^2},$$
$$u_3 = \int \frac{m^2 + k_1 + 4fk_2}{\Lambda^2} du_0 + \frac{4c}{m} \int u_1^2 d\tau, \quad t(u_0) = 2 \int f\Lambda^{-1} du_0,$$
$$\varphi(\tau) = \begin{cases} \cosh \omega\tau, c\lambda > 0, \\ \cos \omega\tau, c\lambda < 0, \quad k_1 > m\omega|k_2|, \end{cases}, \quad \omega = 4m^{-1}\sqrt{|c\lambda|}.$$

The classical action has the form (4.141) with

$$\Gamma = k_2 u_2 + \int \frac{m^2 + k_1 + 4fk_2}{2\Lambda} du_0.$$

Solutions of the K–G and Dirac equations are given by equations (4.142) and (4.143) where one should set

$$\hat{F}_1 = \mu_1 \cos u_2 - \mu_2 \sin u_2, \quad \hat{F}_2 = \mu_1 \sin u_2 + \mu_2 \cos u_2,$$
$$\mu_1 = i\partial_{u_1}, \quad \mu_2 = k_2 u_1^{-1} + 2fu_1, \quad q = 1, \quad \Phi = \psi_\zeta.$$

In quantum theory, k_2 are integers. The function Φ is a solution of equation (4.159). Such a solution can be explicitly written as: For $c\lambda < 0$,

$$\Phi = z^\mu \exp(-z/2) L_n^{2\mu}(z), \quad z = m\omega u_1^2/2, \; k_1 = m\omega(2n + 2\mu + 1), \quad \mu = |k_2|/2;$$

for $c\lambda > 0$

$$\Phi = z^\mu \exp(-iz/2)\Phi(\alpha, 1 + 2\mu; iz), \quad 2\alpha = 1 + 2\mu + ik_1(m\omega)^{-1}, \tag{4.172}$$

where $\Phi(\alpha, \gamma; x)$ is the confluent hypergeometric function and $n \in \mathbb{N}_+$.

Solutions in the fields under consideration were first found in the work [32].

Type V

At this point, we use the coordinates u_μ,

$$u_1 = q(u_0 x^2 - ax^1),\quad u_2 = q(u_0 x^1 + ax^2)\,,$$
$$u_3 = x^0 + x^3 - x_1^2 u_0^{-1} + u_2^2(a^2 - u_0^2)u_0^{-1},\quad q^{-1} = u_0^2 + a^2\,, \tag{4.173}$$

where a is an arbitrary constant. We define the potentials in the corresponding reference frame as

$$\tilde{A}_0 = -q[f_2(u_1) + 2au_1(f_1(u_1) + g_1(u_0))],\quad \tilde{A}_1 = 0\,,$$
$$\tilde{A}_2 = 2ag(u_0)u_1 - f_1(u_1) - g_1(u_0),\quad \tilde{A}_3 = -g(u_0)/2\,. \tag{4.174}$$

In this case

$$E_x = H_y = u_2 g' - qu_1(2ag - f_1') - qu_0 g_1' - aq^2[f_2' + 2au_1 f_1' + 2a(f_1 + g_1)]\,,$$
$$E_y = -H_x = u_1 g' + qu_2(2ag - f_1') + aqg_1' + u_0 q^2[f_2' + 2au_1 f_1' + 2a(f_1 + g_1)]\,,$$
$$H_z = q(2ag - f_1')\,;$$
$$\rho = g'' + 2u_0 qg' + 4a^2q^2 g + q^2[f_2'' + 2af_1' + (2au - u_0u_2)f_1'']\,,$$
$$j_x = -u_0 q^2 f_1'',\quad j_y = -aq^2 f_1''\,. \tag{4.175}$$

The fields are free provided that

$$\text{(i)}\quad f_1'' = 0,\quad 2ag + \gamma = (1 - 2qa^2)\alpha + \beta qu_0,\quad 2af_1' + f_2'' = 2a\gamma,\quad a \neq 0\,,$$
$$\text{(ii)}\quad f_2'' = 4\delta,\quad g = -\delta u_0^{-2} + \alpha u_0^{-1} + \beta,\quad a = 0\,,$$

where α, β, γ and δ are arbitrary constants.

One of the five conditions

$$\begin{aligned}
&\text{(i)} && f_1 = g = 0\,,\\
&\text{(ii)} && g_1 = 0,\quad f_2 = -2au_1 f_1\,,\\
&\text{(iii)} && g = g_1 = 0\,,\\
&\text{(iv)} && f_1 = f_2 = 0,\quad gg_1 = b\,,\\
&\text{(v)} && g_1 = cg,\quad f_2 = cf_1,\quad a = 0\,,
\end{aligned} \tag{4.176}$$

(here b and c are constants) must be fulfilled for obtaining solutions in what follows.

The first integrals of the Lorentz equations are

$$m^2\dot{u}_1^2 = q^2 R(u_1),\quad m\dot{u}_2 + q(k_1 + f_1 + g_1 + 2a\lambda u_1) = 0\,,$$
$$m\Lambda\dot{u}_3 = m^2 + q[(k_1 + g_1)^2 + k_2] + 2q\Lambda[f_2 - 2au_1(k_1 + 2a\lambda u_1)]\,,$$
$$R(u_1) = 2k_1^2 + k_2 + 2\lambda f_2 + 4abu_2 - (k_1 + f_1)^2 - (2a\lambda u_1 + k_1)^2\,.$$

Here and in what follows, we set $b = 0$ in R in all the cases except for case (iv) in (4.176). The first integrals can be integrated,

$$\int \frac{du_1}{\sqrt{R}} = \int \frac{q du_0}{\Lambda}, \quad u_2 = -\int \frac{k_1 + 2a\lambda u_1 + f_1}{\sqrt{R}} du_1 - \int \frac{q g_1 du_0}{\Lambda},$$
$$u_3 = \int \frac{m^2 + qk_2 + q(k_1 + g_1)^2}{\Lambda^2} du_0 + 2\int \frac{f_2 - 2au_1(k_1 + 2a\lambda u_1)}{\sqrt{R}} du_1 .$$

The classical action has the form (4.141) with

$$\Gamma = k_1 u_2 + \int \frac{m^2 + q[(k_1 + g_1)^2 + k_2]}{2\Lambda} du_0 .$$

Solutions of the K–G and Dirac equations are given by equations (4.142) and (4.143) where the functions $\Phi(u_1)$ and $\psi_\zeta(u_1)$ satisfy the equations

$$\Phi'' + R\Phi = 0, \quad \psi_\zeta'' + [R - \zeta(2a\lambda + f_1')]\psi_\zeta = 0, \quad \zeta = \pm 1 , \tag{4.177}$$

with k_2 being an integer, and one should set $\tan t = -au_0^{-1}$ and

$$\hat{F}_1 = qu_0(k_1 + 2a\lambda u_1 + f_1 + g_1) - \Lambda u_2 - iaq\partial_{u_1} ,$$
$$\hat{F}_2 = aq(k_1 + 2a\lambda u_1 + f_1 + g_1) - \Lambda u_1 - iqu_0\partial_{u_2} .$$

If $a \neq 0$, solutions of equations (4.177) can be found for a linear function f_1 and a quadratic function f_2 (see ESP I and II for equation (9.1)), provided, of course, that one of the conditions (4.176) is fulfilled. If $a = 0$, one can choose f_2 from (4.30) and simultaneously select f_1 so that (4.176) be fulfilled.

Solutions in the fields under consideration were first found in Ref. [32].

Type VI

Now we are going to consider the motion of a charge in a combined field that is a superposition of a plane-wave propagating along the z-axis, a longitudinal constant and uniform magnetic field and a arbitrary running electric field. When the electric field is zero, this is the so-called *Redmond configuration*. The classical motion in the latter case was studied in [278], and the quantum problem was first solved in [273]. With a nonzero electric field the classical and quantum problems were first solved in [25] (see also [37]). If the electric field is constant and uniform, such a superposition is called a BGY superposition of electromagnetic fields. This field is a rather particular case of fields considered in what follows. Solutions in the BGY field were studied in Section 3.3.

At this point, we use the light-cone coordinates (4.2), (4.3) and define potentials in the corresponding reference frame as

$$\tilde{A}_0 = 0, \quad \tilde{A}_1 = -a_1(u_0) - u_2 H, \quad \tilde{A}_2 = -a_2(u_0), \quad \tilde{A}_3 = -g(u_0)/2 , \tag{4.178}$$

where $H > 0$ is a constant, and $a_1(u_0), a_2(u_0)$ and $g(u_0)$ are arbitrary functions. In this case

$$E_x = H_y = -a_1', \quad E_y = -H_x = -a_2', \quad H_z = -H; \quad \rho = g'' . \tag{4.179}$$

The first integrals of the Lorentz equations are

$$\begin{aligned} &m\dot{u}_1 + k_1 + a_1 + Hu_2 = 0, \quad m^2\dot{r}^2 + H^2r^2 = H^2Q^2 , \\ &m\Lambda\dot{u}_3 = m^2 + |i\sqrt{H}Q\exp[2i(t - t_0)] + \sqrt{2H}\chi - a_1 - ia_2|^2 , \\ &k_1, \quad Q, \quad t_0 = \text{const} , \end{aligned} \tag{4.180}$$

where

$$\begin{aligned} r &= \sqrt{H}u_2 + \sqrt{2}\chi_1(u_0) + k_1/\sqrt{H} , \\ t(u_0) &= \frac{H}{2}\int\frac{du_0}{\Lambda} = \frac{H}{2m}\tau , \end{aligned}$$

and the complex function $\chi(u_0) = \chi_1(u_0) + i\chi_2(u_0)$ is defined as

$$\chi(u_0) = -i\sqrt{2H}\exp(2it)\int\frac{(a_1 + ia_2)\exp(-2it)}{2\Lambda}du_0 . \tag{4.181}$$

It satisfies the following equation:

$$2\Lambda\chi' - 2iH\chi + i\sqrt{2H}(a_1 + ia_2) = 0 . \tag{4.182}$$

Equations (4.180) are readily integrated to give

$$\begin{aligned} &\sqrt{H}(u_1 - \bar{u}_1) = Q\cos 2(t - t_0) + \sqrt{2}\chi_2, \quad \bar{u}_1 = \text{const} , \\ &\sqrt{H}u_2 + k_1/\sqrt{H} = Q\sin 2(t - t_0) - \sqrt{2}\chi_1 , \\ &u_3 = \int\frac{m^2 + |i\sqrt{H}Q\exp[2i(t - t_0)] + \sqrt{2H}\chi - a_1 - ia_2|^2}{\Lambda^2}du_0 . \end{aligned} \tag{4.183}$$

Note the relation following from (4.183)

$$H(u_1 + iu_2) = H\bar{u}_1 - ik_1 + \sqrt{H}Q\exp[2i(t - t_0)] - i\sqrt{2H}\chi . \tag{4.184}$$

The classical action is given by equation (4.141) where one should set

$$\begin{aligned} &\Gamma = -\sqrt{2}r\chi_2 + \int(2\Lambda^{-1}[m^2 + HQ^2 + (a_1 - \sqrt{2H}\chi_1)^2 + a_2^2 - 2H\chi_2^2]du_0 , \\ &\int\frac{du_1}{\sqrt{R}} = -\frac{Q^2}{2}\arcsin\frac{r}{Q} + \frac{r}{2}\sqrt{Q^2 - r^2}, \quad r^2 < Q^2 . \end{aligned} \tag{4.185}$$

Solutions of the K–G and Dirac equations are given by equations (4.142) and (4.143) with

$$\begin{aligned} &\hat{F}_1 = \sqrt{H}r + a_1 - \sqrt{2H}\chi_1, \quad \hat{F}_2 = i\sqrt{H}\partial_r + a_2 - \sqrt{2H}\chi_2 , \\ &q = 1, \quad \Phi = \psi_\zeta = U_n(r) , \end{aligned} \tag{4.186}$$

where $U_n(x)$ are Hermite functions. In quantum theory, Q is quantized $Q^2 = 2n + 1$, $n \in \mathbb{Z}_+$.

If the plane-wave and electric fields are switched off ($a_1 = a_2 = g = 0$) these solutions are reduced to solutions in the uniform magnetic field considered in Section 3.4.

Type VII

At this point, we use the light-cone coordinates (4.2–4.3) and define potentials in the corresponding reference frame as

$$2\tilde{A}_0 = \psi_{11}(u_0)u_1^2 + 2\psi_{12}(u_0)u_1u_2 + \psi_{22}(u_0)u_2^2, \quad \tilde{A}_3 = -g(u_0)/2\,,$$
$$\tilde{A}_1 = -G_1(u_0) - H(u_0)u_2, \quad \tilde{A}_2 = H(u_0)u_1 - G_2(u_0)\,, \tag{4.187}$$

where $\psi_{ij} = \psi_{ji}$, G_i, $i, j = 1, 2$, g, and H are arbitrary functions of u_0. In this case

$$-E_x = -H_y = G_1' + (\psi_{12} + H')u_2 + \psi_{11}u_1, \quad H_z = -2H\,,$$
$$-E_y = H_x = G_2' + (\psi_{12} - H')u_1 + \psi_{22}u_2; \quad \rho = g'' - \psi_{11} - \psi_{22}\,. \tag{4.188}$$

Note that the fields under consideration contain seven arbitrary functions of u_0, out of which only three are involved in the expressions for the current. The free fields are determined by the conditions $g'' = \psi_{11} + \psi_{22}$ and hence contain six arbitrary functions of u_0. Evidently, the fields of the type VI are particular cases of the fields under consideration.

The classical action has the form (4.141) with

$$2\Gamma = u^+ fu + \chi^+ u + u^+\chi - G^+ u - u^+ G + \int \frac{m^2 + \chi^+\chi}{\Lambda} du_0\,. \tag{4.189}$$

Here

$$\psi = \begin{pmatrix} \psi_{11} & \psi_{12} \\ \psi_{21} & \psi_{22} \end{pmatrix}, \quad f = \begin{pmatrix} f_{11} & f_{12} \\ f_{21} & f_{22} \end{pmatrix},$$
$$G = \begin{pmatrix} G_1 \\ G_2 \end{pmatrix}, \quad \chi = \begin{pmatrix} \chi_1 \\ \chi_2 \end{pmatrix}, \quad u = \begin{pmatrix} u_1 \\ u_2 \end{pmatrix}, \tag{4.190}$$

where $f^T = f \Longrightarrow f_{ij} = f_{ji}$ and χ_i are certain real functions of u_0 that obey the following equations:

$$\Lambda(f' - \psi) - (f - iH\sigma_2)(f + iH\sigma_2) = 0\,, \tag{4.191}$$
$$\Lambda(\chi' - G') - (f - iH\sigma_2)\chi = 0\,, \tag{4.192}$$

where the prime denotes the derivative with respect to u_0. The functions f_{ij} satisfy a set of three nonlinear first-order differential equations. Equations (4.192) make a linear inhomogeneous system of first-order differential equations for χ_i with the function f

considered as already known. We need a particular solution of the set (4.191) and the general solution of (4.192). The latter has the following structure:

$$\chi = \sum_{j=1}^{2} k_j \chi_{(j)} + \bar{\chi}\,, \tag{4.193}$$

where $\chi_{(j)}$ is a fundamental set of solutions of the homogeneous equations (4.192), and $\bar{\chi}$ is a particular solution of the inhomogeneous equation (4.192), constants k_j being integrals of motion. Thus, the classical action depends on three integrals of motion: λ, k_1 and k_2.

First integrals to the Lorentz equations are easy to find provided that the solutions of equations (4.191) and (4.192) are known. Such a set of first integrals is

$$\Lambda u' + (f + iH\sigma_2)u + \chi = 0, \quad u_3' = m^2\Lambda^{-2} + u'^T u'\,. \tag{4.194}$$

Solutions of the K–G and Dirac equations are given by equations (4.142) and (4.143) where one should set

$$q = \exp\left(\left(\int \frac{f_{11} + f_{22}}{\Lambda} du_0\right)\right), \quad t = -\int \frac{H}{\Lambda} du_0, \quad \Phi(u_1) = \psi_\zeta(u_1) = 1\,,$$
$$\hat{F}_1 = f_{11}u_1 + (f_{12} + H)u_2 + \chi_1, \quad \hat{F}_2 = (f_{12} - H)u_1 + f_{22}u_2 + \chi_2\,. \tag{4.195}$$

We see that relativistic wave functions are expressed in terms of solutions of equations (4.191) and (4.192). We discuss below possible solutions of these equations.

If a solution of (4.191) is known, a solution of (4.192) and (4.194) can be obtained by an integration. To see this, we linearize the set (4.191) using the substitution

$$f = -\Lambda \exp(i\sigma_2 t) Z' Z^{-1} \exp(-i\sigma_2 t)\,, \tag{4.196}$$

where $Z(u_0)$ is a real 2×2 matrix. This matrix obeys the equation

$$\Lambda^2 Z'' + \Lambda\Lambda' Z' + (H^2 + \Lambda\bar{\psi})Z = 0, \quad \bar{\psi} = \exp(-i\sigma_2 t)\psi \exp(i\sigma_2 t)\,. \tag{4.197}$$

We seek for such Z that satisfy the condition

$$J = Z'Z^{-1} = \left(Z^T\right)^{-1}\left(Z^T\right)' = J^T\,, \tag{4.198}$$

which is needed to provide $f^T = f$. For such solutions, the matrix J obeys the equation

$$J' + J^2 + \Lambda'\Lambda^{-1}J + (H^2 + \Lambda\bar{\psi})\Lambda^{-2} = 0\,. \tag{4.199}$$

The matrix J^T obeys the same equation, since $(\bar{\psi})^T = \bar{\psi}$. Therefore, there exist solutions of (4.199) with the property $J = J^T$. Once (4.197) is satisfied, equation (4.191) is fulfilled identically and $f = f^T$. Using (4.198), we find from (4.192) and (4.191) that

$$\chi = \exp(i\sigma_2 t)\left(Z^T\right)^{-1}\left[k + \int \left(Z^T\right)\exp(-i\sigma_2 t)G' du_0\right],$$
$$u = \exp(i\sigma_2 t)Z\left[r - \int Z^{-1}\exp(-i\sigma_2 t)\chi\Lambda^{-1} du_0\right],$$
$$k = \begin{pmatrix} k_1 \\ k_2 \end{pmatrix}, \quad r = \begin{pmatrix} r_1 \\ r_2 \end{pmatrix}, \tag{4.200}$$

where r_i are arbitrary constants. The obtained solution for χ really has the structure (4.193).

Thus, both the classical and quantum problems have been reduced to finding a particular real solution of set of second-order differential equations (4.197). At present four types of fields are known for which this set can be transformed (by a special transformation) into a single second-order equation. Then, specifying the fields, one can obtain exact solutions of this equation. We list these cases below, specifying the fields and the corresponding transformations.

(i) The matrices ψ and Z are multiples of the 2×2 unit matrix $\mathbf{I}$,

$$\psi = \varphi(u_0)\mathbf{I}, \quad Z = z(u_0)\mathbf{I}.$$

In such a case, the function $z(u_0)$ obeys the equation

$$\Lambda^2 z'' + \Lambda\Lambda' z' + (H^2 + \Lambda\varphi)z = 0, \tag{4.201}$$

and $\varphi(u_0)$ is arbitrary. Using the substitution $y = z\sqrt{\Lambda}$, we reduce (4.201) to the one-dimensional Schrödinger equation for the function $y(u_0)$,

$$4y'' + (4H^2 + 4\Lambda\varphi + \Lambda'^2 - 2\Lambda\Lambda')\Lambda^{-2} y = 0. \tag{4.202}$$

In this case the matrix f is also a multiple of unity, and it is expressed in terms of z or y as follows:

$$f = -\left(\Lambda z' z^{-1}\right)\mathbf{I} = \left(\Lambda'/2 - \Lambda y'/y\right)\mathbf{I}.$$

(ii) The matrices ψ and Z are diagonal,

$$\psi = \operatorname{diag}\left(\psi_{11}, \psi_{22}\right), \quad Z = \operatorname{diag}\left(z_{11}, z_{22}\right),$$

and $H = 0$. In such a case, the functions z_{ss}, $s = 1, 2$, obey the equations analogous to (4.201),

$$\Lambda^2 z''_{ss} + \Lambda\Lambda' z'_{ss} + \Lambda\psi_{ss} z_{ss} = 0, \quad s = 1, 2.$$

The matrix f is also diagonal in this case,

$$f = \operatorname{diag}\left(f_{11}, f_{22}\right), \quad f_{ss} = -\Lambda z'_{ss} z_{ss}^{-1}.$$

(iii) The matrices ψ and Z have the form

$$\psi = \varphi(1+\sigma_3)/2 - H'\sigma_1\,,$$
$$Z = \exp(-i\sigma_2 t)\left[z\frac{(1+\sigma_3)}{2} + \frac{1-\sigma_3}{2} - 2(\sigma_1 - i\sigma_2)\int \frac{H}{\Lambda}du_0\right].$$

In such a case, the function $z = z(u_0)$ satisfies equation (4.201) where one has to make the replacement $H \to 2H$, and $\varphi(u_0)$ is arbitrary. In this case, the matrix f takes the form

$$f = -\Lambda z'(1+\sigma_3)/2z - H\sigma_1\,.$$

(iv) The matrices ψ and Z have the form

$$\psi = \varphi\frac{(1+\sigma_3)}{2} - \frac{g'}{(u_0+a)}\frac{(1-\sigma_3)}{2} - \left[H' + \frac{2H}{u_0+a}\right]\sigma_1\,,$$
$$Z = z\exp(-i\sigma_2 t)\frac{(1+\sigma_3)}{2} + (u_0+a)\left[\frac{1-\sigma_3}{2}\right.$$
$$\left. + (\sigma_1 - i\sigma_2)\int \frac{zH du_0}{\Lambda(u_0+a)}\right],$$

where a is an arbitrary constant. In such a case, the function $z = z(u_0)$ satisfies equation (4.201) where one has to make the replacement $H \to 2H$ and $\varphi(u_0)$ is arbitrary. In this case, the matrix f takes the form

$$f = -\frac{z'\Lambda}{2z}(1+\sigma_3) - \frac{\Lambda(1-\sigma_3)}{2(u_0+a)} - H\sigma_1\,.$$

Thus, in all the above cases the problem has been reduced to solving equation (4.201) for $z(u_0)$, or equation (4.202) for $y(u_0)$. We consider below a possible exact solution of these equations, pointing in each case to the corresponding function H, φ and g. No other solutions of Equation (4.201) different from those listed in (1) and (2) (that contain arbitrary functions) are known. Some solutions listed below are complex. Once equations (4.201) and (4.202) are linear, the real and imaginary parts are each a solution. Note that in the case $g = 0$, equations (4.201) and (4.202) coincide. In all the cases listed below, we denote $u_0 = \xi$.

(1) For $\varphi = 0$, $H = \text{const}$, and g arbitrary a solution of (4.201) has the form

$$z = \cos t\,. \tag{4.203}$$

In particular, this case gives rise to solutions for the combined field that are different from the ones considered in type VI of this section. At $g = 0$, new solutions for the uniform magnetic field are produced that are expressed in terms of elementary functions.

(2) For

$$\varphi = -g'(\xi + b)^{-1}, \quad b = \text{const}, \quad H = 0\,,$$

and g; arbitrary a solution of (4.201) has the form

$$z = \xi + b\,. \tag{4.204}$$

(3) In this item we take $g = 0$ and

$$H = \frac{\sqrt{|G|}}{2\xi(a + b\xi^r)}, \quad \varphi = -\frac{a_2 - b_2\xi^r + c_2\xi^{2r}}{4\xi^2(a + b\xi^r)^2}, \quad G = a_1 - b_1\xi^r + c_1\xi^{2r}\,, \tag{4.205}$$

where $a, b, r \neq 0, a_s, b_s$, and c_s, $s = 1, 2$, are arbitrary constants.

For $a \neq 0$, a solution of equation (4.202) has the form

$$y = \sqrt{\xi}(A\varphi_\mu + B\varphi_{-\mu}), \quad \mu = (2ar\lambda)^{-1}\sqrt{a^2\lambda^2 + \epsilon a_1 + \lambda a_2}, \quad \epsilon = \operatorname{sgn} G\,, \tag{4.206}$$

where A and B are constants chosen so that y is everywhere continuous together with its first derivatives, the points where ϵ changes its sign included. For $b \neq 0$,

$$\varphi_\mu = (1 - x)^\nu x^\mu F(\alpha, \beta; 1 + 2\mu; x), \quad x = -b\xi^r/a\,, \tag{4.207}$$

where

$$\nu = \frac{1}{2} + \frac{\sqrt{r^2a^2b^2\lambda^2 + (\epsilon a_1 + \lambda a_2)b^2 + (\epsilon b_1 + \lambda b_2)ab + (\epsilon c_1 + \lambda c_2)a^2}}{2rab\lambda}\,,$$

$$\delta = (2rb\lambda)^{-1}\sqrt{b^2\lambda^2 + \epsilon c_1 + \lambda c_2}, \quad \alpha = \mu + \nu + \delta, \quad \beta = \mu + \nu - \delta\,,$$

and $F(\alpha, \beta; \gamma; x)$ is the Gauss hypergeometric function. For $b = 0$, but $\epsilon c_1 + \lambda c_2 \neq 0$,

$$\varphi_\mu = \exp(-x/2)x^\mu\Phi(\alpha_1, 1 + 2\mu; \; x), \quad x = (ar\lambda)^{-1}\sqrt{\epsilon c_1 + \lambda c_2}\xi^r\,,$$

$$\alpha_1 = 1/2 + \mu - (\epsilon b_1 + \lambda b_2)\left(4ar\lambda\sqrt{\epsilon c_1 + \lambda c_2}\right)^{-1}\,, \tag{4.208}$$

where $\Phi(\alpha, \gamma; x)$ is the confluent hypergeometric function. For $b = \epsilon c_1 + \lambda c_2 = 0$, $\epsilon b_1 + \lambda b_2 \neq 0$, we have

$$\varphi_\mu = J_{2\mu}(\alpha_2\xi^{\frac{r}{2}}), \quad \alpha_2 = (ar\lambda)^{-1}\sqrt{\epsilon b_1 + \lambda b_2}\,, \tag{4.209}$$

where $J_\mu(x)$ are Bessel functions. For $b = \epsilon c_1 + \lambda c_2 = \epsilon b_1 + \lambda b_2 = 0$,

$$\varphi_\mu = \xi^{\mu r}\,.$$

The case $a = 0, b \neq 0$ is reduced to the case $a \neq 0, \quad b = 0$ by the transformation $r \to -r$.

(4) In this item we take $g = 0$ and

$$H = \frac{\sqrt{a_1\xi^2 + b_1\xi + c_1}}{a\xi^2 + b\xi + c}, \quad \delta = b^2 - 4ac, \quad \varphi = \frac{a_2\xi^2 + b_2\xi + c_2}{(a\xi^2 + b\xi + c)^2}. \tag{4.210}$$

For $a \neq 0$, a solution of equation (4.202) has the form

$$\begin{aligned} &y = A\varphi_\mu + B\varphi_{-\mu}, \quad \varphi_\mu = (1-x)^{1/2+\mu} F(\alpha, \beta; 1 + 2\mu; x), \\ &x = (\delta + 2a\xi + b)(2\delta)^{-1}, \end{aligned} \tag{4.211}$$

where

$$\begin{aligned} &\mu = (2a\delta\lambda)^{-1}\Big[a^2\lambda^2\delta^2 + 2a(b+\delta)(\epsilon b_1 + \lambda b_2) - 4a^2(\epsilon c_1 + \lambda c_2) \\ &\quad - (b+\delta)^2(\epsilon a_1 + \lambda a_2)\Big]^{1/2}, \quad \gamma = (2a\lambda)^{-1}\sqrt{a^2\lambda^2 - 4(\epsilon a_1 + \lambda a_2)}, \\ &\alpha = 1/2 + \mu + \nu + \gamma, \quad \beta = 1/2 + \mu + \nu - \gamma, \quad \nu = -\mu(-\delta), \end{aligned}$$

for $\delta \neq 0$. If $\delta = 0$, the problem is reduced to that considered in item 3 for $r = 1$. For $a = 0$, the problem is also reduced to that considered in item 3.

For $a = b = 0$ and $\epsilon a_1 + \lambda a_2 \neq 0, c \neq 0$, a solution of equation (4.202) has the form

$$\begin{aligned} &y = AD_p(x) + BD_p(-x), \quad x = \left(2\xi + \frac{\epsilon b_1 + \lambda b_2}{\epsilon a_1 + \lambda a_2}\right)\frac{(-\epsilon a_1 - \lambda a_2)^{1/4}}{\sqrt{2c\lambda}}, \\ &p = -1/2 + i\Big[4(\epsilon c_1 + \lambda c_2)(\epsilon a_1 + \lambda a_2) \\ &\quad - (\epsilon b_1 + \lambda b_2)^2\Big]\left(8c\lambda\sqrt{(\epsilon a_1 + \lambda a_2)^3}\right)^{-1}, \end{aligned} \tag{4.212}$$

where $D_p(x)$ are the Weber parabolic cylinder functions.

For $a = b = \epsilon a_1 + \lambda a_2 = 0$ and $\epsilon b_1 + \lambda b_2 \neq 0, c \neq 0$, a solution of equation (4.202) has the form

$$y = x^{1/3}[AJ_{1/3}(x) + BJ_{-1/3}(x)], \quad x = \frac{2\sqrt{[(\epsilon b_1 + \lambda b_1)\xi + \epsilon c_1 + \lambda c_2]^3}}{3c\lambda(\epsilon b_1 + \lambda b_2)}. \tag{4.213}$$

Note that a solution for constant H and φ can be obtained from the above consideration by choosing

$$a = a_s = b = b_s = 0, \quad s = 1, 2, \quad c \neq 0.$$

Such a solution has the form

$$y = A\exp(i\omega\xi) + B\exp(-i\omega\xi), \quad \lambda\omega = \sqrt{H^2 + \lambda\varphi}. \tag{4.214}$$

The case $\epsilon a_1 + \lambda a_2 = \epsilon b_1 + \lambda b_2$ is of interest. The corresponding solution can be written via elementary functions as

$$y = \sqrt{a\xi^2 + b\xi + c}[A\exp(\eta x) + B\exp(-\eta x)] \,,$$
$$x = \int \frac{d\xi}{a\xi^2 + b\xi + c}, \quad \eta = \frac{1}{2\lambda}\sqrt{\delta^2\lambda^2 - |c_1| - \lambda c_2} \,. \tag{4.215}$$

(5) In this item we take $g = 0$ and

$$H = \sqrt{|G|}, \quad G = a_1 \exp(2c\xi) - b_1 \exp(c\xi) + c_1 \,,$$
$$\varphi = -a_2 \exp(2c\xi) + b_2 \exp(c\xi) - c_2, \quad c \neq 0 \,. \tag{4.216}$$

For $\epsilon a_1 + \lambda a_2 \neq 0$, $\epsilon = \operatorname{sgn} G$, a solution of equation (4.202) has the form

$$y = \exp(-x/2)(A\varphi_\mu + B\varphi_{-\mu}), \quad \varphi_\mu = x^\mu \Phi(\alpha, \quad 1 + 2\mu; \quad x) \,,$$
$$x = \frac{2}{c\lambda}\sqrt{\epsilon a_1 + \lambda a_2}\exp(c\xi), \quad \mu = \frac{\sqrt{\epsilon c_1 + \lambda c_2}}{c\lambda} \,,$$
$$\alpha = \frac{1}{2} + \mu - \frac{\epsilon b_1 + \lambda b_2}{2c\lambda\sqrt{\epsilon a_1 + \lambda a_2}} \,, \tag{4.217}$$

where $\Phi(\alpha, \gamma; x)$ is the confluent hypergeometric function.

The case $\epsilon a_1 + \lambda a_2 = 0$ and $\epsilon b_1 + \lambda b_2 \neq 0$ is reduced to the previous one by the replacement $c \to 2c$.

(6) In this item we take $g = 0$ and

$$H = \sqrt{|G|}, \quad G = a_1 \tanh^2 c\xi + b_1 \tanh c\xi + c_1 \,,$$
$$\varphi = a_2 \tanh^2 c\xi + b_2 \tanh c\xi + c_2 \,, c \neq 0 \,. \tag{4.218}$$

A solution of equation (4.202) reads

$$y = (1 - x)^\nu (A\varphi_\mu + B\varphi_{-\mu}) \,,$$
$$\varphi_\mu = x^\mu F(\alpha, \beta; 1 + 2\mu; x), \quad x = (1 + \tanh\xi)/2 \,, \tag{4.219}$$

where $F(\alpha, \beta; \gamma; x)$ is the Gauss hypergeometric function, and

$$\mu = (2c\lambda)^{-1}\sqrt{\epsilon(b_1 - a_1 - c_1) + \lambda(b_2 - a_2 - c_2)}, \quad \alpha = 1/2 + \mu + \nu + \delta \,,$$
$$\nu = (2c\lambda)^{-1}\sqrt{-\epsilon(a_1 + b_1 + c_1) - \lambda(a_2 + b_2 + c_2)}, \quad \beta = 1/2 + \mu + \nu - \delta \,,$$
$$\delta = (2c\lambda)^{-1}\sqrt{c^2\lambda^2 - 4\epsilon a_1 - 4\lambda a_2}, \quad \epsilon = \operatorname{sign} G \,.$$

(7) In this item we take $g = 0$ and

$$H = \sqrt{|G|}, \quad G = a_1 \coth^2 c\xi + b_1 \coth c\xi + c_1 \,,$$
$$\varphi = a_2 \coth^2 c\xi + b_2 \coth c\xi + c_2, \quad c \neq 0 \,. \tag{4.220}$$

After the change of the variable $2x = 1 + \coth c\xi$, we obtain solutions that coincide, up to all the other notations, with the ones considered in type VI of this section.

(8) In this item we take $g = 0$ and

$$H = \sqrt{|G|}, \quad G = a_1 \sin^{-2} c\xi + b_1 \cos^{-2} c\xi + c_1,$$
$$\varphi = a_2 \sin^{-2} c\xi + b_2 \cos^{-2} c\xi + c_2, \quad c \neq 0 . \tag{4.221}$$

A solution of (4.202) has the form

$$y = \sqrt{\sin 2c\xi} \cos^\nu c\xi (A\varphi_\mu + B\varphi_{-\mu}) ,$$
$$\varphi_\mu = \sin^\mu c\xi F(\alpha, \beta; 1 + \mu; \sin^2 c\xi) , \tag{4.222}$$

where

$$\mu = (2c\lambda)^{-1} \sqrt{c^2\lambda^2 - 4\epsilon a_1 - 4\lambda a_2}, \quad 2\alpha = 1 + \mu + \nu + \delta , \tag{4.223}$$
$$\nu = (2c\lambda)^{-1} \sqrt{c^2\lambda^2 - 4\epsilon b_1 - 4\lambda b_2}, \quad 2\beta = 1 + \mu + \nu - \delta ,$$
$$\delta = (c\lambda)^{-1} \sqrt{\epsilon c_1 + \lambda c_2}, \quad \epsilon = \operatorname{sign} G ,$$

and $F(\alpha, \beta; \gamma; x)$ is the Gauss hypergeometric function. No other cases of exact solutions for $g = 0$ have been found as yet.

(9) For

$$g = a\xi^r, \quad H = \xi^{-1}\sqrt{|G|}, \quad G = a_1 + b_1\xi^r + c_1\xi^{2r} + ac_2\xi^{3r} ,$$
$$\varphi = \xi^{-2}(a_2 + b_2\xi^r - \epsilon c_2\xi^{2r}), \quad r \neq 0 , \tag{4.224}$$

the problem of solving Equation (4.202) is reduced to the one studied in item 3 above.

For special values of r, a solution is possible for some additional functions H and φ. For instance, for $g = a\xi^2$ and

$$H = H_1 = \sqrt{|G|}, \quad G = a_1 + b_1\xi + c_1\xi^2 + a\xi^2(b_2\xi + c_2\xi^2) ,$$
$$\varphi = \varphi_1 = a_2 - \epsilon b_2\xi - \epsilon c_2\xi^2, \quad \epsilon = \operatorname{sgn} G , \tag{4.225}$$

the problem of solving Equation (4.202) is reduced to the one studied in item 4 above. For $g = a\xi^2$, one can also choose

$$H = \xi^{-1} H_1, \quad \varphi = \xi^{-2}\varphi_1 . \tag{4.226}$$

By using the change of the variable $a\xi \to \lambda\xi^{-1}$ it is an easy matter to establish that equation (4.201) with φ and H given as in (4.225) is reduced to equation (4.201) with $\varphi = \varphi_1$ and $H = H_1$ from (4.225). A solution is again obtained thereof.

Finally, if $g = a\xi^{-2}$, the choice

$$H = \xi^{-3}\sqrt{|G|}, \quad G = a_1\xi^4 + b_1\xi^3 + c_1\xi^2 - ab_2\xi - ac_2\,,$$
$$\varphi = \xi^{-4}(a_2\xi^2 + \epsilon b_2\xi + \epsilon c_2 + 2a), \quad \epsilon = \operatorname{sgn} G\,, \tag{4.227}$$

leads us to the case represented in item 4.

(10) For

$$g = a\exp(c\xi), \quad H = \sqrt{|G|}, \quad G = a_1 + b_1\exp(c\xi) + c_1\exp(2c\xi) + ac_2\exp(3c\xi)\,,$$
$$\varphi = a_2 + b_2\exp(c\xi) - \epsilon c_2\exp(2c\xi), \quad c \neq 0\,, \tag{4.228}$$

the problem of solving equation (4.202) can be reduced (by an appropriate shift of ξ) either to the one represented in item 6 if $a\lambda > 0$, or to the one represented in item 7 if $a\lambda < 0$.

Exact solutions in the fields under consideration were first found in Ref. [39], see also [37].

Only one type of field for which solutions of the set (4.197) were obtained is known, however the transformation that reduces this set to one second-order differential equation was not found.

Let us specify functions that define the field (4.188) as follows:

$$g(u_0) = 0, \quad H(u_0) = H = \text{const}, \quad r_{11}(u_0) = c_1 + c_2\cos\omega u_0,$$
$$r_{22}(u_0) = c_1 - c_2\cos\omega u_0, \quad r_{12}(u_0) = c_2\sin\omega u_0; \quad \omega, c_{1,2} = \text{const}\,. \tag{4.229}$$

Solutions of equation (4.197) with such a field, for the special case $H = c_1 = F_i(u_0) = 0$, were found in [97], and in the general case in [62].

One can easily see that the set (4.197) with the field (4.229) can be written as

$$\lambda^2 Z'' + \left[H^2 - \lambda c_1 - \lambda c_2(\boldsymbol{\sigma}\mathbf{l})\right] Z = 0, \quad \mathbf{l} = (\sin\Omega u_0, 0, \cos\Omega u_0)\,,$$
$$t(u_0) = -\lambda^{-1}Hu_0, \quad \Omega = \omega - 2\lambda^{-1}H\,. \tag{4.230}$$

For $c_2 = 0$ this equation is reduced to a second-order differential equation considered above, that is why we study now only the case $c_2 \neq 0$. By straightforward calculations, one can check that the matrix elements

$$Z_{11} = A\left[\alpha\cos\alpha u_0\cos\frac{\Omega u_0}{2} + \left(\frac{\Omega}{2} + \frac{\gamma + c_2}{\lambda\Omega}\right)\sin\alpha u_0\sin\frac{\Omega u_0}{2}\right],$$
$$Z_{21} = A\left[\alpha\cos\alpha u_0\sin\frac{\Omega u_0}{2} - \left(\frac{\Omega}{2} + \frac{\gamma + c_2}{\lambda\Omega}\right)\sin\alpha u_0\cos\frac{\Omega u_0}{2}\right],$$
$$Z_{12} = B\left[\beta\sin\beta u_0\sin\frac{\Omega u_0}{2} + \left(\frac{\Omega}{2} - \frac{\gamma + c_2}{\lambda\Omega}\right)\cos\beta u_0\cos\frac{\Omega u_0}{2}\right],$$
$$Z_{22} = -B\left[\beta\sin\beta u_0\cos\frac{\Omega u_0}{2} - \left(\frac{\Omega}{2} - \frac{\gamma + c_2}{\lambda\Omega}\right)\cos\beta u_0\sin\frac{\Omega u_0}{2}\right],$$
$$\alpha^2 = \frac{H^2 - \lambda c_1}{\lambda^2} + \frac{\Omega^2}{4} + \frac{\gamma}{\lambda}, \quad \beta^2 = \frac{H^2 - \lambda c_1}{\lambda^2} + \frac{\Omega^2}{4} - \frac{\gamma}{\lambda},$$
$$\gamma^2 = c_2^2 + (H^2 - \lambda c_1)\Omega^2, \quad A,\ B = \text{const}\,, \tag{4.231}$$

satisfy equations (4.230) and condition (4.198). One can choose arbitrary signs for the quantities α, β, and γ. The quantity γ can be either real or purely imaginary, the quantities α and β can be complex. In the latter case, real and imaginary parts of matrix elements (4.231) are also solutions of equations (4.230).

Solutions (4.231) have a limit as $\Omega \longrightarrow 0$ if the sign of γ (γ is real as $\Omega \longrightarrow 0$) is chosen so as to satisfy the condition $c_2\gamma < 0$. As a result of such a limiting process, we obtain (after redefining constants A and B)

$$Z(u_0) = \begin{pmatrix} A\cos\alpha u_0 & 0 \\ 0 & B\sin\beta u_0 \end{pmatrix},$$
$$\alpha^2 = (H^2 - \lambda c_1 - \lambda c_2)\lambda^{-2}, \quad \beta^2 = (H^2 - \lambda c_1 + \lambda c_2)\lambda^{-2}. \tag{4.232}$$

A calculation of the quantities $f(u_0)$, $\chi(u_0)$ and $u(u_0)$ using equations (4.196) and (4.200) is simple, so we do not present it here.

Type VIII

At this point, we use the coordinates u_μ,

$$u_0 = x^0 - x^3, \quad u_1 = [a + 2x^1(x^0 - x^3)](x^0 - x^3)^{-2},$$
$$u_2 = x^2(x^0 - x^3 + b)^{-1}, \quad u_3 = x^0 + x^3 + b - x_2^2(x^0 - x^3 + b)^{-1}$$
$$- [2ax^1 + x_1^2(x^0 - x^3)](x^0 - x^3)^{-2} - (2/3)\,a^2(x^0 - x^3)^3, \tag{4.233}$$

where a and b are arbitrary constants. We define potentials in the corresponding reference frame as

$$\tilde{A}_0 = [au_1 g(u_0) - \varphi(u_1)]\,u_0^{-2}, \quad \tilde{A}_1 = 0,$$
$$\tilde{A}_2 = -(u_0 + b)^2 u_0^{-2} f(u_1) - g_1(u_0), \quad \tilde{A}_3 = -g(u_0)/2, \tag{4.234}$$

where $g(u_0)$, $g_1(u_0)$, $\varphi(u_1)$, and $f(u_1)$ are some functions of the indicated arguments. In this case

$$E_x = H_y = 2(\varphi' - ag)u_0^{-3} - (u_1 u_0^2 + a)g'/2u_0^2 - 2u_2 u_0^{-3}(u_0 + b)f',$$
$$E_y = -H_x = 2bu_0^{-3} f - g_1'(u_0 + b)^{-1} + u_2 g' + (u_1 u_0^2 + a)(u_0 + b)u_0^{-5} f',$$
$$H_z = 2(u_0 + b)u_0^{-3} f', \quad j_x = 0, \quad j_y = -4(u_0 + b)u_0^{-4} f'',$$
$$\rho = 4u_0^{-4}\varphi'' - 4u_2(u_0 + b)u_0^{-4} f'' + (2u_0 + b)u_0^{-1}(u_0 + b)^{-1} g' + g''. \tag{4.235}$$

The fields are free if

$$f'' = \varphi'' = 0, \quad g' = \nu u_0^{-1}(u_0 + b)^{-1},$$

where ν is a constant.

Exact solutions can be obtained for the functions $f\varphi$, g, and g_1 subjected to one of the following conditions:

$$\begin{aligned}
&\text{(i)} \quad 4\mu\varphi = f^2, \quad g = \mu b(2u_0 + b)u_0^{-2}, \quad g_1 = 0, \quad a = 0, \quad b \neq 0\,,\\
&\text{(ii)} \quad g = g_1 = 0, \quad b = 0\,,\\
&\text{(iii)} \quad f = \mu u_1, \quad \varphi = 0, \quad 2\mu g_1 + ag^2 = 0, \quad b = 0\,,
\end{aligned} \tag{4.236}$$

where μ is a constant.

The first integrals of the Lorentz equations are

$$\begin{aligned}
&m^2u_0^4\dot{u}_1^2 = 16R(u_1)\,, 4R(u_1) = k_2 + 2\lambda\varphi - f^2 - 2k_1 f + a\lambda^2 u_1\,,\\
&m\dot{u}_2 + (k_1 + g_1)(u_0 + b)^{-2} + u_0^{-2} f = 0\,,\\
&m\Lambda\dot{u}_3 = k_2 u_0^{-2} + (k_1 + g_1)^2(u_0 + b)^{-2} + 2u_0^{-2}(\varphi + a\lambda u_2)\Lambda + m^2\,.
\end{aligned} \tag{4.237}$$

They can be integrated,

$$\begin{aligned}
&\int \frac{du_1}{\sqrt{R}} = \int \frac{4du_0}{u_0^2\Lambda}, \quad u_2 = -\int \frac{(k_1 + g_1)du_0}{\Lambda(u_0 + b)^2} - \int \frac{fdu_1}{4\sqrt{R}}\,,\\
&u_3 = \int \left[\frac{(k_1 + g_1)^2}{(u_0 + b)^2} + \frac{k_2}{u_0^2} + m^2\right]\frac{du_0}{\Lambda^2} + \int \frac{\varphi + a\lambda u_1}{2\sqrt{R}}du_1\,.
\end{aligned} \tag{4.238}$$

The classical action has the form (4.141) with

$$\Gamma = k_1 u_2 + \int \left[\frac{(k_1 + g_1)^2}{(u_0 + b)^2} + \frac{k_2}{u_0^2} + m^2\right]\frac{du_0}{2\Lambda}\,. \tag{4.239}$$

Solutions of the K–G equation have the form (4.142) where the function $\Phi(u_1)$ satisfies the one-dimensional Schrödinger equation

$$\Phi'' + R\Phi = 0 \tag{4.240}$$

and $q = u_0^{-1}(u_0 + b)^{-1}$.

If $a \neq 0$, solutions of equation (4.240) can be found provided that f is a linear, and φ is a quadratic function of u_1 (conditions (4.236) must be fulfilled). In such a case, the problem is reduced to ESP I and II for Equation (9.1).

If $a = 0$, then exact solutions of (4.240) are possible for the following f (besides the case of linear f and quadratic φ):

$$\begin{aligned}
&\text{(i)} \quad f(x) = \alpha + \beta \exp cx, \quad \text{(ESP V)}\,,\\
&\text{(ii)} \quad f(x) = \alpha + \beta/x, \quad \text{(ESP III)}\,,\\
&\text{(iii)} \quad f(x) = \alpha + \beta \tanh cx, \quad \text{(ESP VIII)}\,,\\
&\text{(iv)} \quad f(x) = \alpha + \beta \tan cx, \quad \text{(ESP VII)}\,,\\
&\text{(v)} \quad f(x) = \alpha + \beta \coth cx, \quad \text{(ESP IX)}\,,
\end{aligned} \tag{4.241}$$

with α, β, and c being constants. In the brackets, we point to the corresponding ESP for equation (9.1). The functions $\varphi(u_1)$ have respectively to be taken in the form (conditions (4.236) must be fulfilled)

$$\begin{aligned}
&\text{(i)} && \varphi(x) = \alpha_1 + \beta_1 \exp cx + \gamma \exp 2cx\,, \\
&\text{(ii)} && \varphi(x) = \alpha_1 + \beta_1 x^{-1} + \gamma x^{-2}\,, \\
&\text{(iii)} && \varphi(x) = \alpha_1 + \beta_1 \tanh cx + \gamma \tanh^2 cx\,, \\
&\text{(iv)} && \varphi(x) = \alpha_1 + \beta_1 \tan cx + \gamma \tan^2 cx\,, \\
&\text{(v)} && \varphi(x) = \alpha_1 + \beta_1 \coth cx + \gamma \coth^2 cx\,.
\end{aligned}$$

Solutions of the Dirac equation have the form (4.143) where $\psi_\zeta = \psi_\zeta(u_0, u_1)$, $\zeta = \pm 1$, are solutions of the equation

$$\left[2\partial_{u_1u_1} + 2R + \zeta f' + i\Lambda u_0^2 \partial_{u_0} + \zeta u_0^{-1}(bf' - \Lambda u_0^3 t')\right] \psi_\zeta = 0\,, \tag{4.242}$$

and one has to set

$$\begin{aligned}
\hat{F}_1 &= 2iu_0^{-1}\partial_{u_1} - (u_1u_0^2 + a)\Lambda(2u_0^2)^{-1}\,, \\
\hat{F}_2 &= (k_1 + g_1)(u_0 + b)^{-1} + (u_0 + b)u_0^{-2} f - \Lambda u_2\,.
\end{aligned} \tag{4.243}$$

If the function f is linear, we choose

$$t = t(u_0) = bf' \int \frac{du_0}{\Lambda u_0^3} + f' \int \frac{du_0}{\Lambda u_0^2}\,, \tag{4.244}$$

and $\psi_\zeta = \psi_\zeta(u_1) = \Phi(u_1)$, where $\Phi(u_1)$ is a solution of equation (4.240). If $b = 0$, after setting $t = 0$, we find that ψ_ζ depend on u_1 alone and satisfy the equations

$$\psi_\zeta'' + (R + \zeta f'/2)\psi_\zeta(u_1) = 0\,. \tag{4.245}$$

Equations (4.245) allow, at $a \neq 0$, exact solutions for linear f and quadratic φ, whereas at $a = 0$ the function f can be taken from (4.241) and φ chosen respectively.

If, however, $b \neq 0$, (this is case (i) in (4.236), with $a = 0$), and f is not linear, it is not possible to separate the variables u_0 and u_1 in (4.242). In this case, even if an explicit solution of equation (4.240) can be found for certain f and φ, this does not yet guarantee the possibility of finding some solutions of equation (4.242). Moreover, at $a = 0$ and $b \neq 0$, we failed to find solutions of equation (4.242) for any of the cases (4.241), although solutions of equation (4.240) and, hence, of the K–G equation are known in such cases.

Solutions in the fields under consideration were first studied in [53].

Type IX

At this point, we use coordinates u_μ,

$$u_0 = x^0 - x^3, \quad u_1 = x^1, \quad u_2 = x^2\left(x^0 - x^3\right)^{-1}, \quad u_3 = x^0 + x^3 - x_2^2\left(x^0 - x^3\right)^{-1}, \tag{4.246}$$

and define potentials in the corresponding reference frame as

$$\tilde{A}_0 = -\epsilon f^2(u_1), \quad \epsilon = \pm 1, \quad \tilde{A}_1 = 0, \quad \tilde{A}_2 = -2au_0^2 f(u_1)\,,$$
$$\tilde{A}_3 = -g(u_0)/2, \quad g(u_0) = 2\epsilon a^2 u_0^2\,, \tag{4.247}$$

where $f(u_1)$ is an arbitrary function. In this case

$$E_x = H_y = 2(\epsilon f - au_0u_2)f', \quad E_z = -4\epsilon a^2 u_0\,,$$
$$E_y = -H_x = 4a(\epsilon au_0u_2 - f), \quad H_z = 2au_0 f'\,,$$
$$j_x = 0, \quad j_y = -2au_0 f'', \quad \rho = 2(\epsilon f - au_0u_2)f'' + 2\epsilon f'^2 + 8\epsilon a^2\,. \tag{4.248}$$

There are no free fields at $a \neq 0$. If, however, $a = 0$, we come to the crossed stationary fields studied above.

The first integrals of the Lorentz equations are

$$mu_0^2\dot{u}_2 + k_2 + 2au_0^2 f = 0, \quad m^2\dot{u}_1^2 = R(u_1)\,,$$
$$m\dot{u}_3 = 2\epsilon f^2 + (k_2u_0^{-2} + k_1 + m^2)\Lambda^{-1}\,,$$

where

$$R(u_1) = 2\epsilon\lambda f^2 - 4ak_2 f + k_1\,. \tag{4.249}$$

They can be integrated,

$$\tau = \int \frac{m du_1}{\sqrt{R}}, \quad u_2 = \frac{k_2}{\lambda u_0} + \frac{2\epsilon a^2 k_2}{m\lambda}\tau - 2a\int \frac{f du_1}{\sqrt{R}}\,,$$
$$u_3 = \frac{(k_1 + m^2)\lambda - 2\epsilon a^2 k_2^2}{2\lambda^2\Lambda} - \frac{k_2^2}{\lambda^2 u_0} + \frac{(k_1 + m^2)\lambda - 6\epsilon a^2 k_2^2}{2m\lambda^2}\tau + 2\epsilon\int \frac{f^2 du_1}{\sqrt{R}}\,,$$
$$2\sqrt{2}au_0 = \sqrt{\lambda}[(1+\epsilon)\tan\omega\tau + (1-\epsilon)\tanh\omega\tau], \quad m\omega = a\sqrt{2\lambda}\,.$$

The classical action has the form (4.141) with R given by (4.249) and

$$\Gamma = k_2u_2 + \int \frac{k_2^2 + (k_2 + m^2)u_0^2}{2u_0^2\Lambda}du_0\,.$$

Solutions of the K–G equation have the form (4.142), where $q = u_0^{-1}$ and the function Φ satisfies equation (4.240) with R and Λ given by (4.249).

Solutions of the Dirac equation have the form (4.143), with

$$\hat{F}_1 = i\partial_{u_1}, \quad \hat{F}_2 = k_2u_0^{-1} + 2au_0 f - \Lambda u_2\,,$$

and with the functions ψ_ζ satisfying the equation

$$[\partial_{u_1u_1} + R + 2i\Lambda\partial_{u_0} + 2\zeta(au_0 f' - \Lambda t')]\psi_\zeta = 0\,. \tag{4.250}$$

In the case $a \neq 0$, solutions of equation (4.240) can be explicitly found providing f is either a linear function of u_1 or one of the functions from (4.241). As for equation (4.250) and, consequently, the Dirac equation, one only can find solutions for linear f. In this case, we put

$$t(u_0) = (4a)^{-1} f' \ln|\Lambda| , \tag{4.251}$$

and then take both functions ψ_ζ as depending on u_1 alone and satisfying equation (4.240). If, on the contrary, f is selected from among the functions (4.241), we cannot find any solution of (4.250).

Solutions in the fields under consideration were first found in the work [53].

Type X

At this point, we use coordinates u_μ,

$$u_0 = x^0 - x^3, \quad u_1 = \left(x^0 - x^3\right)^{-1} x^1, \quad u_2 = x^2, \quad u_3 = x^0 + x^3 - x_1^2 \left(x^0 - x^3\right)^{-1} \tag{4.252}$$

and define potentials in the corresponding reference frame as

$$\begin{aligned} &\tilde{A}_0 = -\epsilon u_0^{-2} f^2(u_1), \quad \epsilon = \pm 1, \quad \tilde{A}_1 = 0 , \\ &\tilde{A}_2 = -2a u_0^{-2} f(u_1), \quad \tilde{A}_3 = -g(u_0)/2, \quad g(u_0) = 2\epsilon a^2 u_0^{-2} , \end{aligned} \tag{4.253}$$

where $f(u_1)$ is an arbitrary function. In this case

$$\begin{aligned} &E_x = H_y = 2\epsilon u_0^{-3}\left(ff' - 2a^2 u_1\right), \quad E_z = 4\epsilon a^2 u_0^{-3} , \\ &E_y = -H_x = 2a u_0^{-3}(u_1 f' + 2f), \quad H_z = 2a u_0^{-3} f' ; \\ &j_x = 0, \quad j_y = -2a u_0^{-4} f'', \quad \rho = 2\epsilon u_0^{-4}(4a^2 + f'^2 + ff'') . \end{aligned} \tag{4.254}$$

There are no free fields for $a \neq 0$. If $a = 0$ we face a particular case of fields already considered in type III of the Section 4.2.3. Below we shall only deal with the case $a \neq 0$.

The first integrals of the Lorentz equations are

$$m^2 u_0^4 \dot{u}_1^2 = R(u_1), \quad m\dot{u}_2 + k_2 + 2a u_0^{-2} f = 0, \quad m\Lambda u_0^2 \dot{u}_3 = 2\epsilon\Lambda f^2 + k_1 + (m^2 + k_1^2)u_0^2 ,$$

where

$$\Lambda = \lambda + 2\epsilon a^2 u_0^{-2}, \quad R(u_1) = 2\epsilon\lambda f^2 - 4a k_2 f + k_1 . \tag{4.255}$$

They can be integrated,

$$\begin{aligned} &\int \frac{du_1}{\sqrt{R}} = \int \frac{du_0}{\Lambda u_0^2}, \quad u_2 = 2a \int \frac{f du_0}{\sqrt{R}} - k_2 \int \frac{du_0}{\Lambda} , \\ &u_3 = 2\epsilon \int \frac{f^2 du_1}{\sqrt{R}} + \int \frac{(m^2 + k_2^2)u_0^2 + k_1}{u_0^2 \Lambda^2} du_0 . \end{aligned}$$

The classical action has the form (4.141) with R given by (4.255) and

$$\Gamma = k_2 u_2 + \int \frac{(m^2 + k_2^2)u_0^2 + k_1}{2u_0^2 \Lambda} du_0 \,.$$

Solutions of the K–G equation have the form (4.142) where $q = u_0^{-1}$, and the function Φ satisfies equation (4.240) with R and Λ given by (4.255). Thus, one can see that solutions of the K–G equation can be found provided that f is taken as a linear function of u_1 or it is selected as one of the functions (4.241).

Solutions of the Dirac equation have the form (4.143) with

$$\hat{F}_1 = iu_0^{-1}\partial_{u_1} - \Lambda u_1, \quad \hat{F}_2 = k_2 + 2au_0^{-2} f \,,$$

and ψ_ζ satisfying the equation

$$[\partial_{u_1 u_1} + R - 2iu_0^2 \Lambda \partial_{u_0} + 2\zeta(u_0^2 \Lambda t' - au_0^{-1} f')]\psi_\zeta = 0 \,. \tag{4.256}$$

If f is a linear function of u_1, we take

$$t(u_0) = af' \int \frac{du_0}{u_0^3 \Lambda}$$

and every $\psi_\zeta = \Phi(u_1)$ as satisfying equation (4.240). When, however, we selected f from (4.241) the expressions (4.241), we failed to find any solution of equation (4.256).

Solutions in the fields under consideration were first found in [53].

4.5 Fields of nonstandard structure

A few exact solutions of the relativistic wave equations are known for fields that are not particular cases of the general field (4.137) considered above. These are spherically-symmetric fields already treated in Section 3.5, and four types of fields (we call them *fields of nonstandard structure*), for which solutions are presented below in this section. The latter solutions were obtained in Refs [46, 48].

Type I

At this point, we use coordinates u_μ,

$$\begin{aligned} u_0 &= x^0 - x^3, \quad u_1 = x^0 + x^3 + 2a(x^0 - x^3)\left[(a/3)(x^0 - x^3)^2 - x^1\right] , \\ u_2 &= x^2, \qquad\quad u_3 = (a/2)(x^0 - x^3)^2 - x^1 \,, \end{aligned} \tag{4.257}$$

where a is a constant. We define potentials in the corresponding reference frame as

$$\tilde{A}_s = -a_s(u_3), \quad s = 0, 1, 2, \quad \tilde{A}_3 = 0 \,. \tag{4.258}$$

In this case

$$
\begin{aligned}
E_x &= (2ax^1 - 1)a_1' - a_0', \quad E_y = -H_x = -au_0a_2', \\
H_y &= (1 + 2ax^1)a_1' - a_0', \quad E_z = -2au_0a_1', \quad H_z = -a_2', \\
\rho &= 4aa_1' + (a^2u_0^2 + 2au_3 + 1)a_1'' + a_0'', \quad j_x = 2au_0a_1'', \\
j_y &= -a_2'', \quad j_z = 4aa_1' + (a^2u_0^2 + 2au_3 - 1)a_1'' + a_0''.
\end{aligned} \tag{4.259}
$$

The fields are free if

$$a_0 = \beta_1 u_3 - 2\alpha a u_3^2, \quad a_1 = \alpha u_3, \quad a_2 = \gamma u_3, \tag{4.260}$$

where α, β_1, and γ are arbitrary constants. These free fields are not plane waves.

The first integrals of the Lorentz equations are

$$
\begin{aligned}
&m\dot{u}_0 = 2P_1, \quad m\dot{u}_1 = 2P_0 + 8au_3P_1, \\
&m\dot{u}_2 + P_2 = 0, \quad m^2\dot{u}_3^2 = R(u_3), \quad P_s = k_s + a_s, \quad s = 0, 1, 2, \\
&R(u_3) = 8au_3P_1^2 + 4P_0P_1 - P_2^2 - m^2.
\end{aligned}
$$

They can be integrated,

$$
\begin{aligned}
u_0 &= 2\int \frac{P_1 du_3}{\sqrt{R}}, \quad u_1 = 2\int \frac{P_0 + 4au_3P_1}{\sqrt{R}} du_3, \\
u_2 &= -\int \frac{P_2 du_3}{\sqrt{R}}, \quad \tau = m\int \frac{du_3}{\sqrt{R}}.
\end{aligned}
$$

The classical action reads

$$S = -\Gamma + \int \frac{du_3}{\sqrt{R}}, \quad \Gamma = k_0u_0 + k_1u_1 + k_2u_2.$$

Solutions φ of the K–G equation have the form

$$\varphi = \exp(-i\Gamma)\phi(u_3), \tag{4.261}$$

where the function $\phi(u_3)$ obeys the equation $\phi'' + R\phi = 0$. Solutions of this equation can only be found for the free fields, i.e. for the functions $a_s(u_3)$, $s = 0, 1, 2$, defined in (4.260). They are

$$
\begin{aligned}
&\phi(u_3) = \begin{cases} U_n(x), \quad \mu = (2n+1)\omega^2, \quad n = 0, 1, 2, \ldots; & \omega^4 > 0 \\ D_p(x\sqrt{2}), \quad 2p = \mu\omega^{-2} - 1; & \omega^4 < 0 \end{cases}, \\
&x = \omega(u_3 - \nu), \quad \omega^4 = \gamma^2 - 4\alpha\beta, \quad \nu = (2\alpha k_0 + 2\beta k_1 - \gamma k_2)\omega^{-4}, \\
&\mu = 4k_0k_1 - k_2^2 - m^2 + \omega^4\nu^2, \quad \beta = 2ak_1 + \beta_1,
\end{aligned}
$$

where $U_n(x)$ are the Hermite functions, see Section B.2, and $D_p(x)$ are the Weber parabolic cylinder functions.

Solutions Ψ of the Dirac equation have the form

$$\Psi = \exp(-i\Gamma)K\chi(u_3), \quad K = [2 + ia(1 + \rho_1\Sigma_3)\Sigma_2 u_0](1 - i\Sigma_3), \tag{4.262}$$

where K is a nonsingular matrix,

$$K^{-1} = \frac{(1 + i\Sigma_3)}{8}[2 - ia(1 + \rho_1\Sigma_3)\Sigma_2 u_0],$$

and bispinor $\chi(u_3)$ obeys the equation

$$\hat{D}\chi = 0,$$
$$\hat{D} = i\rho_1\Sigma_2\partial_{u_3} + (1 - \rho_1\Sigma_3)P_0 + [1 + \rho_1\Sigma_3 + 2au_3(1 - \rho_1\Sigma_3)]P_1 + \rho_1\Sigma_1 P_2 - m\rho_3 .$$

The bispinor χ can be found solely for the free fields, i.e. for the functions $a_s(u_3)$, $s = 0, 1, 2$, defined in (4.260). Such solutions have the form

$$\begin{aligned}\chi = \{&f(x)\sqrt{\mu} - ig(x)\rho_1\Sigma_2[(1 - \rho_1\Sigma_3)(k_0 + \nu\beta)\\ &+ (1 + \rho_1\Sigma_3)(k_1 + \nu\alpha) + \rho_1\Sigma_1(k_2 + \nu\gamma) - m\rho_3]\}\\ &\times [\alpha + \beta - (\alpha - \beta)\rho_1\Sigma_3 - \gamma\rho_1\Sigma_1 + i\omega^2\rho_1\Sigma_2]V, \quad \omega \neq 0, ,\end{aligned} \tag{4.263}$$

where, without loss of generality, the constant bispinor V can be chosen as

$$V = \begin{pmatrix} v \\ 0 \end{pmatrix} .$$

The functions $f(x)$ and $g(x)$ satisfy the equations

$$f' = \epsilon g - xf, \quad g' = xg - \omega^{-1}\sqrt{\mu}f .$$

For $\omega^4 > 0$:

$$f = U_n(x), \quad g = U_{n-1}(x), \quad \mu = 2n\omega^2, \quad n \in \mathbb{N}, \quad v \text{ is arbitrary},$$
$$f = U_0(x), \quad g = 0, \quad \mu = 0, \quad v = \left(\frac{(k_2 + \nu\omega^2)(\alpha - \beta) - (\omega^2 - \gamma)(k_1 - k_0)}{m(\alpha + \beta) + (k_2 + \nu\gamma)\omega^2}\right),$$

and for $\omega^4 < 0$:

$$f = D_r(x\sqrt{2}), \quad g = \sqrt{r}D_{r-1}(x\sqrt{2}), \quad 2r = \mu\omega^{-2} .$$

Solutions of the K–G and Dirac equations have to be considered separately for $\omega = 0$. Here $\gamma^2 = 4\alpha\beta$ and, therefore, α and β have the same sign. In this case, $\phi(u_3)$ from (4.261) obeys the equation

$$\phi''(u_3) - (\nu_0 u_3 - \mu_0)\phi(u_3) = 0 ,$$
$$\nu_0 = 2\gamma k_2 - 4\alpha k_0 - 4\beta k_1, \quad \mu_0 = 4k_0 k_1 - k_2^2 - m^2 .$$

It is easy to see that $\phi(u_3)$ is expressed through the Airy function $\Phi(z)$,

$$\phi(u_3) = \Phi(z), \quad z = \nu_0^{1/3}(u_3 - \mu_0/\nu_0) .$$

The function $\chi(u_3)$ from (4.262) is

$$\chi(u_3) = \phi(u_3) A + \phi'(u_3) B ,$$

where A and B are constant bispinors having the structure

$$A = \begin{pmatrix} A_1 \\ A_2 \end{pmatrix}, \quad B = \begin{pmatrix} B_1 \\ B_2 \end{pmatrix} .$$

Here, the spinors $A_s, B_s, s = 1, 2$, are

$$\begin{aligned} A_1 &= [2\alpha k_0 - 2\beta k_1 + m(\alpha - \beta)]\sigma_1 v + [(\alpha + \beta)k_2 - \gamma(k_0 + k_1 + m)]\sigma_3 v , \\ A_2 &= [\gamma(k_1 - k_0) + (\beta - \alpha)k_2] v - i[\nu_0/2 + m(\alpha + \beta)]\sigma_2 v , \\ B_1 &= (\alpha + \beta) v, \quad B_2 = (\beta - \alpha)\sigma_3 v - \gamma\sigma_1 v , \end{aligned}$$

where v is an arbitrary constant spinor.

Type II

We consider below axial-symmetric fields. At this point we use the following coordinates

$$x^0 = u_0, \quad x^1 = u_1 \cos u_2, \quad x^2 = u_1 \sin u_2, \quad x^3 = u_3, \quad (u_1 = r, \ u_2 = \varphi) . \tag{4.264}$$

Electromagnetic potentials in the corresponding reference frame are

$$\tilde{A}_s = -a_s(u_1), \quad s = 0, 2, 3, \quad \tilde{A}_1 = 0 . \tag{4.265}$$

In this case

$$\begin{aligned} &E_r = a_0', \quad E_\varphi = E_z = 0, \quad H_\varphi = -a_3', \quad H_z = u_1^{-1} a_2' , \\ &j_r = 0, \quad j_\varphi = u_1^{-2}(u_1 a_2'' - a_2'), \quad j_z = -u_1^{-1} a_3' - a_3'', \quad \rho = u_1^{-1} a_0' + a_0'' . \end{aligned} \tag{4.266}$$

The fields are free if

$$a_0' = \alpha u_1^{-1}, \quad a_2' = \beta u_1, \quad a_3' = \gamma u_1^{-1} , \tag{4.267}$$

where α, β, γ are constants.

The first integrals of the Lorentz equations read

$$\begin{aligned} &m^2 \dot{u}_1^2 = R(u_1), \quad m u_1^2 \dot{u}_2 + P_2 = 0, \quad m\dot{u}_0 = P_0, \quad m\dot{u}_3 + P_3 = 0 , \\ &R(u_1) = P_0^2 - P_3^2 - u_1^{-2} P_2^2 - m^2, \quad P_s = k_s + a_s(u_1), \quad s = 0, 2, 3 . \end{aligned}$$

They can be integrated,

$$\tau = m\int \frac{du_1}{\sqrt{R}}, \quad u_0 = \int \frac{P_0 du_1}{\sqrt{R}},$$
$$u_2 = -\int \frac{P_2 du_1}{u_1^2\sqrt{R}}, \quad u_3 = -\int \frac{P_3 du_1}{\sqrt{R}}.$$

The classical action S and solutions φ of the K–G equation have the form

$$S = -\Gamma + \int \sqrt{R} du_1, \quad \Gamma = k_0 u_0 + k_2 u_2 + k_3 u_3,$$
$$\Phi = \exp(-i\Gamma)\chi(u_1), \quad \chi'' + u_1^{-1}\chi' + R\chi = 0, \tag{4.268}$$

where k_2 are integers.

Solutions Ψ of the Dirac equation can be represented as

$$\Psi = \exp(-i\Gamma)(1 + i\Sigma_3)\exp\left(-\frac{i}{2}u_2\Sigma_3\right)\psi(u_1), \tag{4.269}$$

where k_2 are half-integers and the bispinor $\psi(u_1)$ satisfies the equation

$$\hat{D}\psi = 0,$$
$$\hat{D} = P_0 + i\rho_1\Sigma_2\hat{P}_1 - \rho_1\Sigma_1 u_1^{-1}P_2 + \rho_1\Sigma_3 P_3 - m\rho_3, \quad \hat{P}_1 = \partial_{u_1} + (2u_1)^{-1}. \tag{4.270}$$

Solutions of equations (4.268) and (4.270) are hitherto only known for linearly-dependent potentials a_0 and a_3. In this case, it is evident that the fields can be reduced to the following three nonequivalent types by appropriate Lorentz transformations:

(i) $a_0 = 0, a_3 = A(u_1)$. This is a purely magnetic field. For this field, the spinor wave function can be subjected to the additional condition (in accordance with the results stated in Section 2.4)

$$(\mathbf{\Sigma}\hat{\mathbf{P}})\Psi = \zeta\lambda_1\Psi, \quad \zeta = \pm 1, \quad \lambda_1 = \sqrt{k_0^2 - m^2}. \tag{4.271}$$

Using this condition, we obtain from (4.270) the box form for ψ

$$\psi = \begin{pmatrix} \sqrt{k_0 + m} \\ \zeta\sqrt{k_0 - m} \end{pmatrix} f(u_1), \quad f(u_1) = \begin{pmatrix} f_1 \\ f_2 \end{pmatrix}, \tag{4.272}$$

where the bispinor f satisfies the equation

$$\left(\hat{P}_1 + u_1^{-1}P_2\sigma_3 + P_3\sigma_1 - i\zeta\lambda_1\sigma_2\right) f = 0. \tag{4.273}$$

(ii) $a_3 = 0, a_0 = A(u_1)$. Using the results stated in Section 2.4, we find that this field admits a spin operator T,

$$T\Psi = \zeta\lambda_2\Psi, \quad T = (m\rho_3\Sigma_3 - \rho_1 P_3), \quad \lambda_2 = \sqrt{m^2 + k_3^2}, \tag{4.274}$$

whence, with the use of equation (4.270), we obtain

$$\psi = \begin{pmatrix} m + \zeta(\lambda_2 - k_3)\sigma_3 \\ -\zeta m + (\lambda_2 - k_3)\sigma_3 \end{pmatrix} \varphi(u_1), \quad \varphi(u_1) = \begin{pmatrix} \varphi_1 \\ \varphi_2 \end{pmatrix}, \tag{4.275}$$

where the spinor φ is a solution of the equation

$$\left(\hat{P}_1 + u_1^{-1} P_2 \sigma_3 + \lambda_2 \sigma_1 + i\zeta P_0 \sigma_2\right) \varphi = 0 . \tag{4.276}$$

(iii) $a_0 = -a_3 = A(u_1)$. This is a combined field of the form (2.103). For the function ψ we obtain

$$\psi = \hat{K}[(1 + \sigma_3)\chi_1 + (1 - \sigma_3)\chi_{-1}]v , \tag{4.277}$$

where the operator $\hat{K}$ is given by (4.143) with

$$\Lambda = \lambda = k_0 + k_3, \quad \hat{F}_1 = -u_1^{-1} P_2, \quad \hat{F}_2 = i\hat{P}_1 , \tag{4.278}$$

the bispinor v being constant and arbitrary. The functions χ_ζ obey the equations

$$\chi_\zeta'' + u_1^{-1}\chi_\zeta' + [k_0^2 - k_3^2 - m^2 + 2\lambda A - u_1^{-2}(P_2 + \zeta/2)^2 + \zeta u_1^{-1} a_2']\chi_\zeta = 0 . \tag{4.279}$$

Solutions of equations (4.268) and of the sets (4.273) and (4.276) are known for

$$A(u_1) = \alpha u_1^{-1}, \quad a_2 = \gamma u_1, \quad \alpha, \gamma = \text{const} . \tag{4.280}$$

In this case

$$\chi = x^\mu \exp(-x/2)\Phi(\mu + \nu + 1/2, 1 + 2\mu; x) ,$$
$$\nu = b/a, \quad x = 2au_1, \quad a = \sqrt{\gamma^2 + m^2 + k_3^2 - k_0^2} .$$

For the fields of the type (i):

$$f_1 = \psi_1 \sqrt{\mu + k_2} + \psi_{-1} \sqrt{\mu - k_2} ,$$
$$f_2 = \psi_1 \sqrt{\mu - k_2} - \psi_{-1} \sqrt{\mu + k_2}, \quad \mu = \sqrt{k_2^2 + \alpha^2} . \tag{4.281}$$

For the fields of the type (ii):

$$\varphi_1 = \psi_\epsilon \sqrt{|k_2| + \mu} + \psi_{-\epsilon} \sqrt{|k_2| - \mu}, \qquad \mu = \sqrt{k_2^2 - \alpha^2} ,$$
$$\varphi_2 = \zeta\psi_\epsilon \sqrt{|k_2| - \mu} + \zeta\psi_{-\epsilon} \sqrt{|k_2| + \mu}, \quad \epsilon = \operatorname{sgn} k_2 . \tag{4.282}$$

The functions ψ_ϵ, $\epsilon = \pm 1$, have the form

$$\psi_\epsilon = B_\epsilon x^{\mu+\epsilon/2} \exp(-x/2)\Phi(\mu + \nu + 1/2 + \epsilon/2, 1 + 2\mu + \epsilon; x) ,$$
$$B_\epsilon = (2\mu a)^{-1} (1 + \epsilon)c + (1 - \epsilon)(1 + 2\mu) ,$$
$$\nu = b/a, \quad x = 2au_1, \quad a = \sqrt{\gamma^2 + m^2 + k_3^2 - k_0^2}, \quad \epsilon = \pm 1 , \tag{4.283}$$

where $\Phi(\alpha, \gamma; x)$ is the confluent hypergeometric function and the constants b and c should be taken as

$$\begin{aligned} &\text{(i)} \quad c = k_2 k_3 + \xi\mu\lambda_1 - \alpha\gamma, \quad b = \gamma k_2 + \alpha k_3 ; \\ &\text{(ii)} \quad c = \alpha\gamma - k_0 k_3 - \xi\mu\lambda_2, \quad b = \gamma k_2 - \alpha k_0 . \end{aligned} \tag{4.284}$$

For the fields of the type (iii), solutions can be obtained in two cases:
(a)

$$A = \alpha u_1^{-1} + \beta u_1^{-2}, \quad a_2 = \gamma u_1, \quad \alpha, \beta, \gamma = \text{const} . \tag{4.285}$$

In this case solutions are given by

$$\begin{aligned} &\chi = x^{\mu} \exp(-x/2)\Phi(\mu + \nu + 1/2, 1 + 2\mu; x), \quad \mu = \sqrt{k_2^2 - 2\lambda\beta} , \\ &\chi_\zeta = x^{\mu} \exp(-x/2)\Phi(\mu + \nu + 1/2, 1 + 2\mu; x), \quad \mu = \sqrt{(k_2 + \zeta/2)^2 - 2\lambda\beta} , \\ &x = 2au_1, \quad a = \sqrt{\gamma^2 + m^2 + k_3^2 - k_0^2}, \quad \nu = \left(\gamma k_2 - \alpha\lambda\right)/a . \end{aligned}$$

If $\mu^2 > 0$, $\nu < 0$ in (4.283), ν is quantized, $-\nu = \mu + n + (1 + \epsilon)/2$, $\quad n \in \mathbb{Z}_+$, and the confluent hypergeometric function Φ is reduced to the Laguerre polynomial [191].
(b)

$$A = \alpha u_1^{-2} + \beta u_1^2, \quad a_2 = \gamma u_1^2, \quad \alpha, \beta, \gamma = \text{const} . \tag{4.286}$$

Equations which appear here coincide with those dealt with in Section 4.2.2 (fields of Type II), apart from some redefinition of the constants.

Particular fields of this sort are the constant and homogeneous magnetic field and some fields of special configurations studied in [10, 197–199]. The general study of relativistic wave equations with axial-symmetric fields was done in the work [46].

Type III
At this point, we use coordinates u_μ,

$$u_0 = x_0^2 - x_3^2, \quad u_1 = x^1, \quad u_2 = x^2, \quad u_3 = \frac{1}{2} \ln \left| \frac{x^0 + x^3}{x^0 - x^3} \right| , \tag{4.287}$$

and define potentials in the corresponding reference frame as

$$\tilde{A}_0 = 0, \quad \tilde{A}_s = -a_s(u_0), \quad s = 1, 2, 3 .$$

In this case

$$\begin{aligned} &E_x = -2x^0 a_1', \quad E_y = -2x^0 a_2', \quad E_z = -2a_3' , \\ &H_x = 2x^3 a_2', \quad H_y = -2x^3 a_1', \quad H_z = 0 ; \\ &j_{x,y} = 4(u_0 a_{1,2}'' + a_{1,2}'), \quad j_z = 4x^0 a_3'', \quad \rho = 4x^3 a_3'' . \end{aligned} \tag{4.288}$$

The fields are free provided that $a_3'' = 0$, $a_r' = u_0^{-1}\alpha_r$, $r = 1, 2$, where α_r are arbitrary constants.

The first integrals of the Lorentz equations read

$$m^2\dot{u}_0^2 = 16u_0^2 R(u_0), \quad m\dot{u}_{1,2} + P_{1,2} = 0, \quad mu_0\dot{u}_3 + P_3 = 0\,,$$
$$P_s = k_s + a_s(u_0), \quad 4u_0 R(u_0) = P_1^2 + P_2^2 + u_0^{-1}P_3^2 + m^2\,.$$

They can be integrated,

$$\tau = m\int \frac{du_0}{4u_0\sqrt{R}}, \quad u_{1,2} = -\int \frac{P_{1,2}du_0}{4u_0\sqrt{R}}, \quad u_3 = -\int \frac{P_3 du_0}{4u_0^2\sqrt{R}}\,.$$

The classical action S and solutions φ of the K–G equation have the form

$$S = -\Gamma - \int \sqrt{R}du_0, \quad \Gamma = k_1u_1 + k_2u_2 + k_3u_3\,,$$
$$\varphi = \exp(-i\Gamma)\Phi(u_0)\,, \tag{4.289}$$

where the function $\Phi(u_0)$ satisfies the ordinary differential equation

$$u_0\Phi'' + \Phi' + 4u_0 R\Phi = 0\,. \tag{4.290}$$

When the potentials a_1 and a_2 are linearly dependent, solutions of equation (4.290) can be found. In this case one can set $a_1 = 0$ without any loss of generality. The remaining functions are chosen to be

$$a_2 = \alpha/\sqrt{|u_0|}, \quad a_3 = \beta\sqrt{|u_0|}, \quad \alpha\,,\beta = \text{const}\,. \tag{4.291}$$

The substitution $|u_0| = x^2$ reduces equation (4.290) to the one-dimensional Schrödinger equation (9.1) with ESP III.

Solutions Ψ of the Dirac equation have the form

$$\Psi = \exp(-i\Gamma)UU_1\phi(u_0)\,, \tag{4.292}$$

where the matrices U and U_1 are

$$U = u_0^{-1/4}\left(\cosh\frac{u_3}{2} + \rho_1\Sigma_3\sinh\frac{u_3}{2}\right), \quad U_1 = 2 + i(1-\epsilon\delta)\rho_1\Sigma_3\,,$$
$$U^{-1} = u_0^{1/4}\left(\cosh\frac{u_3}{2} - \rho_1\Sigma_3\sinh\frac{u_3}{2}\right), \quad 8U_1^{-1} = 3 + \epsilon\delta - i(1-\epsilon\delta)\rho_1\Sigma_3\,,$$
$$\epsilon = \operatorname{sgn}(x^0 - x^3), \quad \delta = \operatorname{sgn}(x^0 + x^3) \tag{4.293}$$

and the bispinor $\phi(u_0)$ satisfies the following equations:

$$\left[\mu\sqrt{|u_0|}(i\partial_{u_0} + \rho_1\Sigma_3P_3/2u_0) + \rho_1\Sigma_1P_1 + \rho_1\Sigma_2P_2 - m\rho_3\right]\phi = 0\,,$$
$$\mu = \delta + \epsilon - i(\delta - \epsilon)\,. \tag{4.294}$$

The spin integral of motion can be chosen to be

$$\hat{L} = m\rho_3\Sigma_1 - \rho_1 P_1\,, \tag{4.295}$$

see Section 3.4. We require Ψ to be eigenvectors for this operator. Then

$$\hat{L}\Psi = \zeta L\Psi, \quad L = \sqrt{m^2 + k_1^2}, \quad \zeta = \pm 1\,, \tag{4.296}$$

and the spinor $\phi(u_0)$ has the form

$$\phi(u_0) = \begin{pmatrix} (k_1 - \zeta L)\psi_1 + m\psi_2 \\ -m\psi_1 - (k_1 - \zeta L)\psi_2 \\ (k_1 - \zeta L)\psi_1 - m\psi_2 \\ m\psi_1 - (k_1 - \zeta L)\psi_2 \end{pmatrix}, \tag{4.297}$$

where functions $\psi_1(u_0)$ and $\psi_2(u_0)$ satisfy the set of equations

$$\left[\mu\sqrt{|u_0|}\left(i\partial_{u_0} + \frac{P_3}{2u_0}\sigma_3\right) + \zeta L\sigma_1 - P_2\sigma_2\right]\psi = 0, \quad \psi = \begin{pmatrix}\psi_1 \\ \psi_2\end{pmatrix}. \tag{4.298}$$

Solutions of this set are only found for the fields (4.291) (the squaring of the set (4.298) results in equations such as (4.290)).

Type IV

At this point, we use coordinates u_μ,

$$u_0 = x_0^2 - x_1^2 - x_2^2 - x_3^2, \quad u_3 = u_0(x^0 - x^3)^{-2}, \quad u_s = x^s\left(x^0 - x^3\right)^{-1}, \quad s = 1, 2\,. \tag{4.299}$$

We define potentials in the corresponding reference frame as

$$\tilde{A}_0 = \tilde{A}_3 = 0, \quad \tilde{A}_s = -a_s(u_s), \quad s = 1, 2\,. \tag{4.300}$$

In this case

$$\begin{aligned}
E_x &= [2u_1u_2a_2' - (1 - u_3 - u_1^2 + u_2^2)a_1']u_3u_0^{-1}\,,\\
E_y &= [2u_1u_2a_1' - (1 - u_3 + u_1^2 - u_2^2)a_2']u_3u_0^{-1}\,,\\
E_z &= -2(u_1a_1' + u_2a_2')u_3u_0^{-1}\,,\\
H_x &= -[(1 + u_3 - u_1^2 + u_2^2)a_2' + 2u_1u_2a_1']u_3u_0^{-1}\,,\\
H_y &= [(1 + u_3 + u_1^2 - u_2^2)a_1' + 2u_1u_2a_2']u_3u_0^{-1}\,,\\
H_z &= 2(u_2a_1' - u_1a_2')u_3u_0^{-1}\,;\\
j_{x,y} &= -4(u_3a_{1,2}'' + a_{1,2}')u_3\left[u_0(x^0 - x^3)\right]^{-1}, \quad \rho = j_z = u_1j_x + u_2j_y\,.
\end{aligned} \tag{4.301}$$

The fields are source-free for $a_s' = \alpha_s u_3^{-1}$, where α_s are arbitrary constants.

The first integrals of the Lorentz equations are

$$m^2\dot{u}_0^2 = 4(k_0 + m^2 u_0), \quad m u_0 \dot{u}_s + u_s P_s = 0, \quad m^2 u_0^2 \dot{u}_3^2 = 4u_3^2 R(u_3)\,,$$
$$P_s = k_s + a_s(u_3), \quad R(u_3) = k_0 - u_3(P_1^2 + P_2^2)\,.$$

They can be integrated,

$$u_0 = (\tau + \tau_0) - \frac{k_0}{m^2}, \quad u_s = -\int \frac{P_s du_3}{2\sqrt{R}}\,,$$
$$\int \frac{du_3}{u_3\sqrt{R}} = \int \frac{m d\tau}{m^2(\tau+\tau_0)^2 - k_0}\,.$$

The classical action S and solutions φ of the K–G equation have the form

$$S = -\Gamma - \int \frac{\sqrt{k_0 + m^2 u_0}}{2u_0} du_0 + \int \frac{\sqrt{R}}{2u_3} du_3, \quad \Gamma = k_1 u_1 + k_2 u_2\,,$$
$$\varphi = \exp(-i\Gamma) g(u_0) f(u_3)\,, \tag{4.302}$$

where the functions $g(u_0)$ and $f(u_3)$ are solutions to the ordinary differential equations

$$4u_0^2 g'' + 8u_0 g' + (m^2 u_0 + k_0) g = 0\,, \tag{4.303}$$
$$4u_3^2 f'' + R(u_3) f = 0\,. \tag{4.304}$$

It is straightforward to find the general solution of equation (4.303). This is

$$g(u_0) = [r_1 J_\mu(x) + r_2 J_{-\mu}(x)] x^{-1}, \quad x = m\sqrt{u_0}, \quad \mu = \sqrt{1 - k_0}\,, \tag{4.305}$$

where $J_\mu(x)$ are the Bessel functions, and r_s, $s = 1, 2$ are arbitrary constants.

Solutions Ψ of the Dirac equation have the form

$$\Psi = \exp(-i\Gamma) U \phi\,(u_0, u_3)\,, \tag{4.306}$$

where the nonsingular matrix U is defined as

$$U = [(1 + \rho_1\Sigma_3)(\sqrt{u_3} - ip) + 1 - \rho_1\Sigma_3] u_0^{-3/4} u_3^{1/4}, \quad p = u_2\Sigma_1 - u_1\Sigma_2\,,$$
$$U^{-1} = (1/4)\,[(1 + \rho_1\Sigma_3)(1 + ip) + \sqrt{u_3}(1 - \rho_1\Sigma_3)]\,(u_0/u_3)^{3/4}\,. \tag{4.307}$$

The bispinor ϕ should be chosen as

$$\phi = \begin{pmatrix} g_1(u_0) f_1(u_3) \\ g_1(u_0) f_2(u_3) \\ g_2(u_0) f_1(u_3) \\ g_2(u_0) f_2(u_3) \end{pmatrix}, \tag{4.308}$$

where the functions $g_s(u_0)$ and $f_s(u_3)$, $s = 1, 2$, satisfy the following set of equations:

$$(2iu_0\partial_{u_0} - \epsilon m\sqrt{u_0}\sigma_3 - q_1\sigma_1)g = 0\,, \tag{4.309}$$

$$[2iu_3\sigma_3\partial_{u_3} + \sqrt{u_3}(\sigma_1 P_1 + \sigma_2 P_2) + q_2]f = 0\,, \tag{4.310}$$

$$g = \begin{pmatrix} g_1(u_0) \\ g_2(u_0) \end{pmatrix}, \quad f = \begin{pmatrix} f_1(u_3) \\ f_2(u_3) \end{pmatrix},$$

where q_1 and q_2 are arbitrary constants. The set (4.309) has the following general solution:

$$g_1(u_0) = g(\epsilon, x), \quad g_2(u_0) = g(-\epsilon, x), \quad x = m\sqrt{u_0}\,,$$
$$g(\epsilon, x) = \epsilon \exp(-i\epsilon x)[r_1\varphi(q_1, x) + r_2\varphi(-q_1, x)]\,,$$
$$\varphi(q_1, x) = x^{iq_1}\Phi(iq_1, 1 + 2iq_1; 2i\epsilon x), \quad \epsilon = \operatorname{sgn}(x^0 - x^3)\,, \tag{4.311}$$

where r_1 and r_2 are arbitrary constants, and $\Phi(\alpha, \beta; x)$ is the confluent hypergeometric function.

Equation (4.304) and the set (4.310) can be solved for $a_s(u_3) = \alpha_s/\sqrt{|u_3|}$, where α_s are arbitrary constants. In this case, equation (4.304) (after the change of variable $|u_3| = x^2$) is reduced to the one-dimensional Schrödinger equation (9.1) with ESP III. The same reduction is available for the set (4.310) after squaring.

Solutions for the fields under consideration were first found in the work [46, 48].

To conclude the study of known exact solutions of the relativistic wave equations, we must stress once again that the principal criterion for our selection was the possibility of solving exactly the Dirac equation. Evidently, for many of the fields considered here, several nonequivalent complete systems of exact solutions can be found explicitly. For instance, in every system of curvilinear coordinates considered in this book, one can find complete systems of solutions of the free Dirac (or K–G) equation corresponding to different (nonequivalent) complete sets of operator-valued integrals of motion (operators belonging to different sets, generally, do not commute among themselves). We have not, as a rule, displayed all such solutions, but have confined ourselves to presenting a few complete systems.

Undoubtedly, further investigations will result in finding new exact solutions in the fields not indicated here. In this connection, it would be of special interest to find at least one system of solutions in a field that would not allow the separation of variables in the K–G equation (i.e. those that do not belong to the classes listed in Refs. [26, 27]). Such an example might be useful for finding an extension to the separation of variables concept.

It would be equally interesting to find an example of a field for which exact solutions of the Dirac equation are essentially different from those of the K–G equation. We mean, for instance, a situation where solutions corresponding to the same integrals of motion could be expressed in terms of transcendental functions of a different nature, or else where the spectrum of admissible values of integrals of motion would be essentially different for the K–G and Dirac equation. In this respect, the fields considered in

Section 4.4 are of great interest. They may be thought of as the most likely candidates. It has been noted above, however, that in the cases where the variables in the Dirac equation cannot be separated, exact solutions to this equation could not be found either, although exact solutions of the K–G equation have been obtained for a few fields of this sort. Consequently, we must state that examples of such fields allowing explicit solutions of the Dirac equation are not yet found.

5 Dirac–Pauli equation and its solutions

5.1 Introduction

In 1941 Pauli [262] suggested an extension of the Dirac equation. The corresponding extended equation was later understood as approximately describing a charged (or neutral) spinor particle whose magnetic moment differs from the Bohr magneton by a term that has been given the name of "anomalous magnetic moment". This equation can readily be further extended to include the terms responsible for the possible anomalous electric moment of the particle. (The normal "electric" moment of a particle within standard QED is equal to zero, which is an inevitable consequence of the requirement that parity be conserved.) Such an equation has the form

$$\left(\gamma^\mu \hat{P}_\mu - m + \frac{1}{2}\mu_1 \sigma^{\mu\nu} F_{\mu\nu} - \frac{1}{2}\mu_2 \sigma^{\mu\nu} \tilde{F}_{\mu\nu}\right) \Psi(x) = 0 , \tag{5.1}$$

and is conventionally called the *Dirac–Pauli equation*, see for example [244]. Here $F_{\mu\nu}$ is the electromagnetic field strength tensor, $\tilde{F}_{\mu\nu}$ is the tensor dual to $F_{\mu\nu}$ as defined in (2.2), and $\sigma^{\mu\nu}$ is the spin tensor (A.50). The quantity μ_1 characterizes the anomalous (additional to the Bohr magneton) magnetic moment, and μ_2 designates the anomalous electric moment. In what follows we shall set

$$\mu_1 = \mu \cos\alpha, \quad \mu_2 = \mu \sin\alpha, \quad \mu = \sqrt{\mu_1^2 + \mu_2^2}, \quad 0 \le \alpha < 2\pi . \tag{5.2}$$

At $\alpha = 0$ ($\mu_2 = 0$) the Dirac–Pauli equation is reduced to the above mentioned Pauli equation [262]. For $\mu_2 \neq 0$, the Dirac–Pauli equation leads to nonconserving parity, in disagreement with quantum electrodynamics. This equation is quite reasonable however within, for instance, electroweak theory.

The Dirac–Pauli equation (5.1) can be written in the form of the Schrödinger equation,

$$i\frac{\partial \Psi(x)}{\partial t} = \mathcal{H}\Psi(x) , \tag{5.3}$$

with the Hamiltonian

$$\begin{aligned} \mathcal{H} = &\left(\boldsymbol{\alpha}\hat{\mathbf{P}}\right) + m\rho_3 + eA_0 \\ &+ \mu_1 \left[\rho_3 \left(\boldsymbol{\Sigma}\mathbf{H}\right) + \rho_2 \left(\boldsymbol{\Sigma}\mathbf{E}\right)\right] + \mu_2 \left[\rho_3 \left(\boldsymbol{\Sigma}\mathbf{E}\right) - \rho_2 \left(\boldsymbol{\Sigma}\mathbf{H}\right)\right] . \end{aligned}$$

The Dirac–Pauli equation can be generalized to include particles with arbitrary spin (see, e.g. [210, 271]). We confine ourselves, however, to considering only the case of spin $1/2$ particles.

There are but a few exact solutions of the Dirac–Pauli equation known to date. The majority of works [11, 13, 112, 210, 221, 222, 237, 238, 313, 314, 326, 328, 329, 335] are devoted to investigations of solutions of equation (5.1) with a constant and uniform magnetic field, and specifically to the study of how the anomalous moments affect the electron spin precession. Exact solutions of equation (5.1) with such a field were first obtained for $\mu_2 = 0$ in the work [326], and for $\mu_2 \neq 0$ in Refs [237, 238].

Exact solutions of the Dirac–Pauli equations with an electromagnetic plane wave field for $\mu_2 = 0$ were first obtained in [13, 237, 313, 314, 326], and then studied in detail in Refs [73, 108, 250, 323, 324], and then studied in detail in the works [18, 19, 83, 271]. The first solution for $\mu_2 \neq 0$ was found in the work [19]. More sophisticated fields and corresponding exact solutions were considered in [45, 227]. Solutions of the Dirac–Pauli equation with a spherically-symmetric electrostatic field have been studied in [80, 325]. The progress in this case was, however, limited to the separation of variables in spherical coordinates, whereas exact solutions of the radial equation have not been found in a closed form.

In obtaining exact solutions of the Dirac–Pauli equation, an important role was assigned to the spin operators that are integrals of motion. Such operators have been found in a few external fields, see [44, 221]. For instance, for constant and uniform longitudinal fields defined in Section 4.3.1 the spin integral of motion $\hat{M}$ has the following form:

$$\hat{M} = (\Pi \mathbf{H} - \Phi \mathbf{E}) \cos\alpha + (\Pi \mathbf{E} + \Phi \mathbf{H}) \sin\alpha \,,$$
$$\Pi = m\boldsymbol{\Sigma} + \rho_2 \, [\boldsymbol{\Sigma}\times\mathbf{P}] \,, \quad \Phi = \rho_3 \, [\boldsymbol{\Sigma}\times\mathbf{P}] \,. \tag{5.4}$$

5.2 Constant and uniform magnetic field

Consider the Dirac–Pauli equation with a constant and uniform magnetic field for a particle with the negative electric charge $e = -|e|$. Without loss of generality, we chose the magnetic field $\mathbf{H}$ to be directed along the z-axis, i.e. $\mathbf{H} = (0, 0, H)$, $H > 0$. The electromagnetic potentials are chosen as

$$A^0 = A^2 = A^3 = 0, \quad A^1 = -Hx^2 \,. \tag{5.5}$$

As for the normal particle (a particle without anomalous momenta) in such a field (see Section 3.4), we can, in this case, refer to the energy E, and the momenta along the axes x^1 and x^3, as integrals of motion. Therefore, we can subject the wave function $\Psi(x)$ to the equations

$$i\partial_0 \Psi = E\Psi, \quad i\partial_1 \Psi = k_1 \Psi, \quad i\partial_3 \Psi = k_3 \Psi \,, \tag{5.6}$$

to be added to the Dirac-Pauli equation. Then, equations (5.6) imply

$$\Psi = \exp(-i\Gamma)\,\psi, \quad \Gamma = Ex^0 + k_1 x^1 + k_3 x^3 \,, \tag{5.7}$$

where the bispinor ψ only depends on x^2 and has to be determined from the Dirac–Pauli equation. As for the normal particle, we represent this bispinor as follows:

$$\psi(\xi) = \mathbb{T}_n(\xi)\, C \,, \tag{5.8}$$

where

$$\mathbb{T}_n(\xi) = \begin{pmatrix} T_n(\xi) & 0 \\ 0 & T_n(\xi) \end{pmatrix}, \quad T_n(\xi) = \begin{pmatrix} U_{n-1}(\xi) & 0 \\ 0 & U_n(\xi) \end{pmatrix}, \tag{5.9}$$

$U_n(\xi)$, $n \in \mathbb{Z}_+$, are the Hermite functions (see Section B.2),

$$\xi = -\left(\frac{k_1}{\sqrt{\gamma}} + \sqrt{\gamma}x^2\right), \quad \gamma = |e|\,H\,,$$

and C is a constant bispinor. The substitution of (5.7) and (5.8) into the Dirac–Pauli equation leads to the following linear homogeneous set of algebraic equations for the bispinor C:

$$\begin{aligned}&\left[E - m\rho_3 + \sqrt{2\gamma n}\rho_1\Sigma_1 + k_3\rho_1\Sigma_3\right.\\ &\left.-\mu H\Sigma_3\left(\rho_3\cos\alpha - \rho_2\sin\alpha\right)\right]C = 0\,.\end{aligned} \tag{5.10}$$

As mentioned in the previous section, in the case under consideration the spin-operator integral of motion has the form (5.4). We demand that the bispinor (5.7) should be an eigenvector for this operator,

$$M\Psi = \zeta\lambda\Psi, \quad \zeta = \pm 1, \quad \lambda > 0\,.$$

That implies the following equation for C:

$$\left[m\Sigma_3\cos\alpha + \sqrt{2\gamma n}\Sigma_1\left(\rho_2\cos\alpha + \rho_3\sin\alpha\right) - \zeta\lambda\right]C = 0\,. \tag{5.11}$$

The sets of equations (5.10) and (5.11) can be solved together to yield the solution

$$C = \frac{1}{2}\begin{pmatrix} \zeta\sqrt{\frac{1}{2}\left(1+\zeta\frac{m}{\lambda}\cos\alpha\right)}\left(\sqrt{1-\frac{k_3}{E}}e^{-i\epsilon_1} + \zeta\sqrt{1+\frac{k_3}{E}}e^{i\epsilon_1}\right) \\ \sqrt{\frac{1}{2}\left(1-\zeta\frac{m}{\lambda}\cos\alpha\right)}\left(\sqrt{1-\frac{k_3}{E}}e^{i\epsilon_2} - \zeta\sqrt{1+\frac{k_3}{E}}e^{-i\epsilon_2}\right) \\ \zeta\sqrt{\frac{1}{2}\left(1+\zeta\frac{m}{\lambda}\cos\alpha\right)}\left(\sqrt{1-\frac{k_3}{E}}e^{-i\epsilon_1} - \zeta\sqrt{1+\frac{k_3}{E}}e^{i\epsilon_1}\right) \\ -\sqrt{\frac{1}{2}\left(1-\zeta\frac{m}{\lambda}\cos\alpha\right)}\left(\sqrt{1-\frac{k_3}{E}}e^{i\epsilon_2} + \zeta\sqrt{1+\frac{k_3}{E}}e^{-i\epsilon_2}\right) \end{pmatrix}, \tag{5.12}$$

where

$$\epsilon_1 = \frac{\alpha - \zeta\varphi}{2}, \quad \epsilon_2 = \frac{\alpha + \zeta\varphi}{2}, \quad \sin\varphi = \frac{m\sin\alpha}{\sqrt{E^2 - k_3^2}}, \quad -\pi/2 < \varphi < \pi/2\,. \tag{5.13}$$

For the absolute value of the quantum number λ, we obtain

$$\lambda = \sqrt{m^2\cos^2\alpha + 2\gamma n}, \quad n \in \mathbb{Z}_+\,, \tag{5.14}$$

whereas the energy E is determined by the expression

$$E = \sqrt{k_3^2 + m^2\sin^2\alpha + \left(\lambda + \zeta\mu H\right)^2}\,. \tag{5.15}$$

The bispinor (5.12) is normalized as $C^{+}C = 1$. Then the spinor $\psi(\xi)$ is normalized by the condition

$$\int_{-\infty}^{\infty} \psi^{+}(\xi)\,\psi(\xi)\,d\xi = 1\,. \tag{5.16}$$

The most interesting physical result here is the removal of the spin degeneracy, viewed from the fact that now the energy eigenvalues depend on the spin quantum number ζ. We see that the energy of the spinning particle depends on the spin orientation ζ. Thus, when the anomalous moments are taken into account, the splitting of the two-fold-degenerate energy levels of the spinning particle occurs in a magnetic field.

It is worth noting that equations (5.12) and (5.13) retain their validity for $\mu = 0$ and α arbitrarily, which determines in this case the solution of the Dirac equation (with no anomalous moments) to be the eigenfunction for the spin operator (5.4) at $\mu_1 = \mu_2 = 0$.

5.3 Plane-wave field

In this section we find exact solutions of the Dirac–Pauli equation for a particle which moves in the field of a plane electromagnetic wave running along a direction $\mathbf{n}$, where the constant unit vector $\mathbf{n}$ can, without loss of generality, be taken to be parallel to the z-axis, i.e. we set $\mathbf{n} = (0, 0, 1)$. The electromagnetic field potentials are chosen to be

$$\begin{aligned} &A^{\mu} = (0, \mathbf{A}(u_0))\,, \quad u_0 = (nx) = x^0 - (\mathbf{nr})\,,\\ &(nA) = (\mathbf{nA}) = 0, \quad n^{\mu} = (1, \mathbf{n})\,. \end{aligned} \tag{5.17}$$

As it was mentioned, for example, in Section 3.2, in case the anomalous moments are absent one can take the generalized momenta of the particle for integrals of motion. Then the solution of the Dirac equation can be written as

$$\Psi = \exp(iS)\begin{pmatrix} m + \lambda + (\sigma\mathbf{n})(\sigma F) \\ (m-\lambda)(\sigma\mathbf{n}) + (\sigma F) \end{pmatrix} V\,, \tag{5.18}$$

where S is the classical action

$$S = -(px) - \frac{1}{2\lambda}\int^{u_0} \left[2\,|e|\,(\mathbf{A}(u_0)\,\mathbf{p}) + e^2\mathbf{A}^2(u_0)\right] du_0\,, \tag{5.19}$$

$\lambda = (np)$ is the integral of motion defined in Section 3.2, p^{μ} is the generalized momentum, subject to the relation $p^2 = m^2$, and

$$F = \mathbf{p}_{\perp} + |e|\,\mathbf{A}(u_0)\,, \quad \mathbf{p}_{\perp} = \mathbf{p} - \mathbf{n}(\mathbf{np})\,, \quad (\mathbf{n}F) = 0\,. \tag{5.20}$$

Choosing λ and $\mathbf{p}_\perp$ as independent integrals of motion and using $p^2 = m^2$, we obtain

$$p^0 = \frac{m^2 + \mathbf{p}_\perp^2 + \lambda^2}{2\lambda}, \quad (\mathbf{np}) = \frac{m^2 + \mathbf{p}_\perp^2 - \lambda^2}{2\lambda}. \tag{5.21}$$

If (5.21) is substituted into (5.19), equation (5.18) will correspond exactly to the Volkov solution given in Section 3.2. The spinor V must be constant. It is only under this condition that equation (5.18) is a solution of the Dirac equation (without anomalous moments).

In agreement with the suggestion of the works [262, 328], one should seek solutions of the Dirac–Pauli equation corresponding to definite values of the conserving quantities p^μ in the form (5.18) and (5.19), with the spinor V now depending on u_0. By substituting (5.18) into (5.1) one concludes that (5.18) is indeed a solution of the Dirac–Pauli equation provided that $V(u_0)$ satisfies the equation

$$iV' = \mu(\boldsymbol{\sigma}\mathbf{F})\, V, \tag{5.22}$$

where

$$\mathbf{F} = \mathbf{H}\cos\alpha + \mathbf{E}\sin\alpha = \left[\mathbf{A}'(u_0)\,\mathbf{n}\right]\cos\alpha - \mathbf{A}'(u_0)\sin\alpha,$$
$$\mathbf{E} = -\mathbf{A}'(u_0), \quad \mathbf{H} = [\mathbf{n}\times\mathbf{E}] = \left[\mathbf{A}'(u_0)\,\mathbf{n}\right].$$

The prime denotes the derivative with respect to u_0. Equation (5.22) is the Schrödinger equation with the expression $\mu(\boldsymbol{\sigma}\mathbf{F})$ for the Hamiltonian[1]. If $\alpha = 0$ (i.e. if the anomalous magnetic moment is present alone), the Hamiltonian has the form of the pure-spin Pauli term responsible for the magnetic moment interaction with the external field. If the rotation by the angle α in the plane $\mathbf{H}, \mathbf{E}$ (i.e. around the axis $\mathbf{n}$) is performed, the vector $\mathbf{F}$ will coincide with the vector $\mathbf{H}$. This provides a visualization of the geometric sense of the angle α. It follows naturally from (5.22) that

$$\mathbf{F}^2 = \mathbf{H}^2 = \mathbf{E}^2. \tag{5.23}$$

Since $\mathbf{F}$ is a real vector, the operator $\mu(\boldsymbol{\sigma}\mathbf{F})$ is Hermitian and the quantity V^+V is constant (i.e. independent of the "time" u_0).

Equation (5.22) may be referred to as a system of two ordinary first-order differential equations for the spinor components $V_1(u_0)$ and $V_2(u_0)$. Expressing $\mathbf{F}$ in terms of the components of the vector $\mathbf{E}$, one can represent this system in the form

$$iV_1' = -\mu f V_2, \quad iV_2' = -\mu f^* V_1,$$
$$f = \left(E_y + iE_x\right)e^{i\alpha}. \tag{5.24}$$

1 Equation (5.22) is also called the spin equation. Its complete analysis and corresponding exact solutions are given in Chapter A.

One can also seek solutions of equation (5.22) in the matrix form

$$V = [ia + (\boldsymbol{\sigma}\mathbf{b})]\, V_0, \quad a = a(u_0), \quad \mathbf{b} = \mathbf{b}(u_0)\,, \tag{5.25}$$

where the function $a(u_0)$ and the vector $\mathbf{b}(u_0)$ make a special solution of the set of equations

$$a' = -\mu\,(\mathbf{F}\mathbf{b})\,, \quad \mathbf{b}' = \mu a\mathbf{F} + \mu\,[\mathbf{F}\times\mathbf{b}]\ . \tag{5.26}$$

In (5.25), V_0 is a constant spinor. The quantity V^+V is constant if and only if the conditions

$$a\mathbf{b}^* - a^*\mathbf{b} + [\mathbf{b}^*\times\mathbf{b}] = 0, \quad aa^* + (\mathbf{b}\mathbf{b}^*) = \text{const} \tag{5.27}$$

are fulfilled. The second of these conditions is a consequence of the system (5.26); the first condition is satisfied provided that it is satisfied at any fixed moment of time u_0, referred to as an initial moment. From (5.25) one finds

$$V^+V = [aa^* + (\mathbf{b}\mathbf{b}^*)]\, V_0^+V_0\,. \tag{5.28}$$

In cases of practical interest we can consider μ as small. Then, solutions of equation (5.22) can be sought in the form of a formal series in powers of μ. Within linear accuracy of μ, one has

$$V = [1 + i\mu\,(\boldsymbol{\sigma}\mathbf{R})]\, V_0, \quad V^+V = V_0^+V_0\,,$$
$$\mathbf{R} = [\mathbf{n}\times\mathbf{A}]\cos\alpha + \mathbf{A}\sin\alpha\,. \tag{5.29}$$

Two cases are known, for which equations (5.22) are solved exactly.

Consider a linearly polarized plane wave. In this case, the direction of the vector $\mathbf{A}(u_0)$ does not depend on u_0 and one can set

$$\mathbf{A}\,(u_0) = A\,(u_0)\,\mathbf{e}, \quad (\mathbf{n}\mathbf{e}) = 0, \quad \mathbf{e}^2 = 1\,, \tag{5.30}$$

where $\mathbf{e}$ is a constant unit vector orthogonal to the vector $\mathbf{n}$ and $A(u_0)$ is an arbitrary function of u_0. For the vector $\mathbf{F}$, one obtains from (5.22)

$$\mathbf{F} = -A'\,(u_0)\,\mathbf{l}, \quad \mathbf{l}^2 = 1, \quad (\mathbf{n}\mathbf{l}) = 0\,,$$
$$\mathbf{l} = [\mathbf{n}\times\mathbf{e}]\cos\alpha + \mathbf{e}\sin\alpha\,, \tag{5.31}$$

where $\mathbf{l}$ is also a constant unit vector orthogonal to $\mathbf{n}$ (to be obtained from the vector $\mathbf{e}$ by a rotation through the angle α around the axis $\mathbf{n}$). In this case, equations (5.22) can be easily solved for any function $A(u_0)$ to give

$$V = \frac{1}{2}\left\{\sum_{\xi=\pm1}[1 + \xi\,(\boldsymbol{\sigma}\mathbf{l})]\exp\left[i\xi\mu A\,(u_0)\right]\right\} V_0\,,$$
$$V^+V = V_0^+V_0\,. \tag{5.32}$$

It should be stressed that if $A(u_0)$ is a periodic function of u_0, the spinor $V(u_0)$ is also periodic with the same period. Under the corresponding quantization conditions imposed on $\mathbf{p}_\perp$ and λ, the wave function (5.18) is also periodic in this case.

Consider now a plane wave having a generalized circular polarization. In other words, define $\mathbf{A}(u_0)$ as

$$\mathbf{A}(u_0) = A_0\left(-\sin\varphi, \cos\varphi, 0\right), \quad \varphi = \varphi\left(u_0\right), \tag{5.33}$$

where A_0 is a constant, and $\varphi(u_0)$ is an arbitrary function of u_0. From (5.33) it follows directly that

$$\begin{aligned} &\mathbf{E}\left(u_0\right) = A_0\varphi'\left(\cos\varphi, \sin\varphi, 0\right), \quad \mathbf{A}^2\left(u_0\right) = A_0^2, \\ &\mathbf{H}\left(u_0\right) = A_0\varphi'\left(-\sin\varphi, \cos\varphi, 0\right), \quad \mathbf{E}^2 = \mathbf{H}^2 = A_0^2\varphi'^2. \end{aligned} \tag{5.34}$$

The wave is circularly polarized in the usual sense, providing φ is a linear function of u_0. Otherwise, the lengths of the vectors $\mathbf{E}$ and $\mathbf{H}$ are not constant (although the length of the potential vector is), and the wave does not possess thereby a definite circular polarization in the usual sense. For the vector $\mathbf{F}$ it may be seen from (5.22) that

$$\mathbf{F} = A_0\psi'\left(-\sin\psi, \cos\psi, 0\right), \quad \psi = \psi\left(u_0\right) = \varphi\left(u_0\right) - \alpha. \tag{5.35}$$

Equation (5.25) gives, in this case, a solution to equation (5.22) providing one sets

$$\begin{aligned} &\mathbf{b}\left(u_0\right) = \left(\sqrt{\omega_1}\sin\omega_2\psi, \quad \sqrt{\omega_1}\cos\omega_2\psi, \quad \epsilon\sqrt{\omega_2}\cos\omega_1\psi\right), \\ &a\left(u_0\right) = -\epsilon\sqrt{\omega_2}\sin\omega_1\psi, \quad \epsilon = \operatorname{sgn}\mu A_0, \\ &\omega_1 = \frac{q+1}{2}, \quad \omega_2 = \frac{q-1}{2}, \quad q = \sqrt{1+4\mu^2A_0^2}, \quad V^+V = qV_0^+V_0. \end{aligned} \tag{5.36}$$

As for the function $\varphi(u_0)$, it remains arbitrary. A physically important peculiarity of this solution is that when the potential $\mathbf{A}(u_0)$ (5.33) is periodic, the function $a(u_0)$ and the vector $\mathbf{b}(u_0)$ (as well as the spinor V thereof) are generally no longer periodic functions of u_0 (they are, instead, the so-called almost-periodic functions). Therefore, the presence of anomalous moments results, in this case, in solutions that do not possess the periodicity properties even though the wave is periodic. No other types of external fields of the form (5.33) have yet been found for which exact solutions of the Dirac–Pauli equation for a charged particle are known.

5.4 Superposition of a plane-wave field and a parallel electric field

Retaining the notation from the preceding section, we choose the electromagnetic potential to be

$$A^\mu = \left(\frac{g\left(u_0\right)}{2}, \mathbf{A}\left(u_0\right) - \mathbf{n}\frac{g\left(u_0\right)}{2}\right), \quad (\mathbf{n}\mathbf{A}) = 0, \tag{5.37}$$

where the vector $\mathbf{A}(u_0)$ is arbitrary, and $g(u_0)$ is an arbitrary function of u_0. The field strengths corresponding to the potentials (5.37) are

$$\mathbf{E} = -\mathbf{A}'(u_0) + \mathbf{n}g'(u_0), \quad \mathbf{H} = -\left[\mathbf{n} \times \mathbf{A}'\right] . \tag{5.38}$$

Thus, we have here a superposition of a plane-wave field and parallel to it a running longitudinal electric field.

Solutions of the Dirac–Pauli equation with the fields (5.38) were first found in the work [45]. For these fields, the quantities $\mathbf{p}_\perp$ and λ (see the preceding section for their definition) remain integrals of motion. The wave function satisfying equation (5.1) and corresponding to definite values of the conserved quantities $\mathbf{p}_\perp$ and λ will be sought using the Ansatz

$$\Psi = \exp\left[-\lambda x^0 + i(\mathbf{p}_\perp \mathbf{r}) - i\tilde{S}(u_0)\right] \mathcal{R} V(u_0) , \tag{5.39}$$

where the operator $\mathcal{R}$ can be written in the following box form:

$$\mathcal{R} = \frac{1}{B} \begin{pmatrix} m + B + (\boldsymbol{\sigma}\mathbf{n})(\boldsymbol{\sigma}\mathbf{F}) - \mu\left[i\cos\alpha + (\boldsymbol{\sigma}\mathbf{n})\sin\alpha\right] g'(u_0) \\ (m - B)(\boldsymbol{\sigma}\mathbf{n}) + (\boldsymbol{\sigma}\mathbf{F}) - \mu\left[i(\boldsymbol{\sigma}\mathbf{n})\cos\alpha + \sin\alpha\right] g'(u_0) \end{pmatrix} . \tag{5.40}$$

In (5.40) the two-dimensional vector $\mathbf{F}$ is defined by equation (5.20) and $B = \lambda + |e|g(u_0)$. The function $\tilde{S}(u_0)$ in (5.39) has the form

$$\tilde{S}(u_0) = \int \frac{m^2 + \mathbf{F}^2 + \mu^2 g'^2}{2B} du_0 . \tag{5.41}$$

To make the functions (5.39)–(5.41) satisfy the Dirac–Pauli equation (5.1), one should subject the spinor $V(u_0)$ to the equation

$$\begin{aligned} iV'(u_0) &= \mu(\boldsymbol{\sigma}\tilde{\mathbf{F}}) V(u_0) , \\ \tilde{\mathbf{F}} &= -A'(u_0)\sin\alpha - \left[\mathbf{n} \times \mathbf{A}'\right]\cos\alpha \\ &\quad + \{[\mathbf{n} \times \mathbf{F}]\cos\alpha + (\mathbf{F} + m\mathbf{n})\sin\alpha\} g'(u_0) B^{-1} . \end{aligned} \tag{5.42}$$

If we set $g(u_0) = 0$, we are left with the problem considered in the previous section with the wave function (5.39) turning into (5.18). Therefore, our task now is to study only the case $g(u_0) \neq 0$.

If $\mathbf{A}(u_0)$ is set equal to zero, only the running longitudinal electric field remains. Then we get from (5.42), with $g(u_0)$ arbitrary, the following solution for $V(u_0)$:

$$\begin{aligned} V(u_0) &= \left\{[\omega - (\boldsymbol{\sigma}\boldsymbol{\omega})] B^{i\nu} + [\omega + (\boldsymbol{\sigma}\boldsymbol{\omega})] B^{-i\nu}\right\} V_0 , \\ \boldsymbol{\omega} &= \left[\mathbf{n} \times \mathbf{p}_\perp\right]\cos\alpha + (\mathbf{p}_\perp + m\mathbf{n})\sin\alpha, \quad \omega = |\boldsymbol{\omega}| , \quad \nu = \mu\omega/|e| , \end{aligned} \tag{5.43}$$

where V_0 is a constant spinor. It also turns out to be possible to find a solution for the linearly polarized plane wave with the potential

$$\mathbf{A}(u_0) = \mathbf{l}g(u_0), \quad (\mathbf{nl}) = 0 ,$$

where $\mathbf{l}$ is a unit constant vector. The solution in this case has the form of (5.43) with the substitution

$$\mathbf{p}_{\perp} \to \mathbf{p}_{\perp} - \lambda \mathbf{l} \,.$$

No other types of external fields have yet been found for which exact solutions of the Dirac–Pauli equation for a charged particle are known.

6 Propagators of relativistic particles

6.1 Introduction

Matrix functions $S(x, y)$ that satisfy the *inhomogeneous Dirac equation*

$$\hat{\mathfrak{D}} S(x, y) = -\delta(x - y) \tag{6.1}$$

are called *Green's functions* of this equation. Here $\hat{\mathfrak{D}} = \gamma^{\mu}\hat{P}_{\mu} - m$ is the Dirac operator (2.55).

Analogously, functions $D(x, y)$ that satisfy the inhomogeneous K–G equation

$$\hat{\mathfrak{K}} D(x, y) = \delta(x - y) \tag{6.2}$$

are called Green's functions of this equation. Here the operator $\hat{\mathfrak{K}} = \hat{P}^2 - m^2$ is the K–G operator (2.28).

It is worth noticing that equations (6.1) and (6.2) have each an infinite set of solutions. Solutions of equation (6.1) differ from each other by adding matrix functions $S(x, y)$ that satisfy the homogeneous Dirac equation

$$\hat{\mathfrak{D}} S(x, y) = 0\,, \tag{6.3}$$

whereas solutions of equation (6.2) differ from each other by adding functions $D(x, y)$ that satisfy the homogeneous K–G equation

$$\hat{\mathfrak{K}} D(x, y) = 0\,. \tag{6.4}$$

Solutions of equations (6.3) and (6.4) are often called homogeneous solutions. The evolution functions $G(x, y)$ of the Dirac and K–G equations, defined in Sections 2.3.2 and 2.2.2 by equations (2.78) and (2.43), respectively, are examples of the homogeneous solutions.

Sometimes to find a special Green's function from equations (6.1) or (6.2), it is enough to fix boundary conditions and initial data, or to specify the space to which the solutions should belong, or else to modify, in a certain manner, the Dirac operator $\hat{\mathfrak{D}}$ in equation (6.1) or the K–G operator $\hat{\mathfrak{K}}$ in equation (6.2) so that the corresponding modified equation would already have unique solutions, which is equivalent to the requirement that the corresponding homogeneous equations (6.3) and (6.4) should have no solutions.

Green's functions of relativistic wave equations are important in QED and relativistic quantum mechanics. The point is that in QED with an external field the particle propagators serving the perturbation expansion are just Green's functions of the Dirac (the spinor electrodynamics) or the K–G (the scalar electrodynamics) equations with the external field, when the latter is treated exactly. Such a perturbation theory was formulated by Furry [149] for external electromagnetic fields that do not violate the vacuum stability, i.e. do not create real particles from the vacuum. It is known as the Furry picture. For external fields of a general form, especially those which are able to violate the stability of the vacuum, the corresponding theory was formulated in Refs. [171] (see [146, 158–161] for the further development). In this theory, various propagators are involved which are either the Green's functions of equations (6.1) and (6.2), or satisfy the homogeneous equations (6.3) or (6.4). These propagators are certain matrix elements of quantum field operators or are expressed in terms of commutators or anticommutators of the latter. These expressions are single-valued and serve to calculate the propagators in external fields. For instance, these representations most naturally give rise to expressions for the propagators or, what is the same, for definite Green's functions in the form of sums over exact solutions of the Dirac and K–G equations in the external fields, see [146]. On the other hand, once the propagators are Green's functions of the Dirac or K–G equations, or solutions of equations (6.3) and (6.4), it is tempting to try to find them directly by solving these equations, since this method often happens to be simpler than the method of determination of propagators by summing over exact solutions. This problem, unfortunately, has not been solved as yet in a general form. To be more precise, the ways to fix solutions for Green's functions that would correspond to definite propagators involved in the perturbation theory of QED with an external field have not been found in the general case. For this reason, summation over exact solutions of the relativistic wave equations remains to date the only method that enables us to obtain a definite propagator in an unambiguous way. At present, methods based on the direct solution of the equations should be confronted with this method. It should be noted that the commutation function of the quantized spinor field interacting with an arbitrary external electromagnetic field was studied in Ref. [154]. It was shown that the proper time representation is suitable in any dimension. With its use all the light cone singularities of the function were found explicitly to generalize the Fock formula in four dimensions, and a path integral representation was constructed.

Now we shall review the works published to date concerning determination of Green's functions of relativistic wave equations in external electromagnetic fields.

In Ref. [143], the evolution function of the K–G equation, i.e. the commutator function of the scalar field, was found in the presence of a constant and homogeneous electromagnetic field by directly solving the equation with the help of the proper-time method. The commutator functions of the scalar and spinor fields were found in Refs. [166, 167] for a constant magnetic field and in Ref. [248] for a constant electric field. The same results were obtained in Refs. [84, 85] by the method of canonical operator [243]. In Ref. [294] the operator proper-time method was used for finding the causal Green's functions of the Dirac and K–G equations in a constant and homogeneous field, as well as in the plane-wave field. The same Green's functions were determined in Refs. [117, 166, 167] with the help of the commutator functions, and in Refs. [248, 253, 254] using summation over exact solutions. In the plane-wave field, the causal Green's functions were found by a continual integration method in Ref. [144].

Analogous results were obtained in Refs. [76, 279] by somewhat different means. For a constant and homogeneous field and for its combination with the plane-wave field, the causal Green's functions were found in Ref. [182] by the path integration method and then in Ref. [101] by the proper-time method. In a constant and homogeneous magnetic field combined with the plane wave propagating along the magnetic field, the causal Green's function of the K–G equation was found in Ref. [259] by directly solving the equation, and in Ref. [182] by path integration. In a constant and homogeneous electric field combined with a plane wave propagating along it, the causal Green's functions of the Dirac and K–G equations were found in Refs. [248, 249]. The method of coherent states was also used for calculating these functions in external fields of several configurations [122–124, 241]. The finding of the Green's function of the Dirac equation in the Coulomb field of a nucleus was considered in Refs. [70, 209, 345]. The method of eigenfunctions was exploited for calculating the causal Green's functions in the constant and homogeneous field, as well as in the plane-wave field in Refs. [280–286]. In Ref. [114] the external field in the form of a running electric wave was studied, for which case the causal Green's function was found using a continual integration method.

The authors of Refs. [150–152, 340] calculated all the propagators of QED with external field creating pairs, introduced within the general approach of Refs. [146, 171]. For an external field, the constant field combined with the plane-wave field was taken. The calculations were performed using the method of summation of exact solutions of the relativistic wave equations. On this basis, indications for finding these propagators from the solutions of equations for the Green's functions were given wherever possible.

In the following sections we consider some methods of finding particle propagators and calculate them using the combination of a constant field with that of a plane wave as an example of the external field.

6.2 Proper-time representations for particle propagators

6.2.1 General

To begin with, we consider the general scheme for solving the Dirac equation (6.1) for Green's function using *Schwinger's proper-time method* [294]. Within this method Green's function $S(x, y)$ is treated as a matrix element of a certain operator denoted here as $\hat{S}$:

$$S(x,y) = \langle x|\,\hat{S}\,|y\rangle \ . \tag{6.5}$$

The function $S(x, y)$ also contains spinor indices (that we do not display); correspondingly, the states $\langle x|$ and $|y\rangle$ in (6.5) are labelled not only by the space-time coordinates, but also by the spinor indices. To display this dependence we shall sometimes use the more detailed way of writing

$$|x\rangle = |x, \alpha\rangle\,, \quad \alpha = 1, 2, 3, 4\,.$$

These states are also considered to be eigenstates for some self-adjoint operators X^μ, and form a complete orthonormalized set

$$\begin{aligned} &X^\mu\,|x\rangle = x^\mu\,|x\rangle\,, \quad \langle x, \alpha|y, \beta\rangle = \delta_{\alpha,\beta}\delta\,(x-y)\,,\\ &[X^\mu, X^\nu]_- = 0, \quad \sum_\alpha \int |x, \alpha\rangle\,\langle x, \alpha|\,dx = I\,. \end{aligned} \tag{6.6}$$

The quantities $\hat{\mathcal{P}}_\mu\delta_{\alpha,\beta}\delta(x-y)$ and $\gamma^\mu_{\alpha\beta}\delta(x-y)$ can be represented as matrix elements of some operators Π_μ and Γ^μ with respect to the states introduced

$$\begin{aligned} \hat{\mathcal{P}}_\mu\delta_{\alpha,\beta}\delta\,(x-y) &= \langle x, \alpha|\,\Pi_\mu\,|y, \beta\rangle\,,\\ \gamma^\mu_{\alpha\beta}\delta\,(x-y) &= \langle x, \alpha|\,\Gamma^\mu\,|y, \beta\rangle\,. \end{aligned} \tag{6.7}$$

It can be easily established that the operators X^μ, Π_μ and Γ^μ satisfy the following commutation relations:

$$\begin{aligned} &\left[\Pi_\mu, X^\nu\right]_- = i\delta^\nu_\mu, \quad \left[\Pi_\mu, \Pi_\nu\right]_- = -ieF_{\mu\nu}(X)\,,\\ &\left[\Gamma^\mu, \Gamma^\nu\right]_+ = 2\eta^{\mu\nu}, \quad \left[\Pi_\mu, \Gamma^\nu\right]_- = \left[\Gamma^\mu, X^\nu\right]_- = 0\,. \end{aligned} \tag{6.8}$$

Equation (6.1) for Green's function $S(x, y)$ is equivalent to the equation $(\Pi - m)S = -1$ for the operator S, where $\Pi = \Pi_\mu\Gamma^\mu$. After squaring this equation by the substitution $S = (\Pi + m)\Delta$, we are left with the equation $(\Pi^2 - m^2)\Delta = -1$ for the operator Δ, whose formal solution has the form

$$\Delta = -\left(\Pi^2 - m^2\right)^{-1}\,. \tag{6.9}$$

It is worth noting that the inverse operator on the right hand side of (6.9) is defined up to an additive term Δ_0 obeying the homogeneous equation $(\Pi^2 - m^2)\Delta_0 = 0$. If the operator $\Pi^2 - m^2$ is modified, for example by adding an infinitesimal negative imaginary part to the mass squared ($m^2 \to m^2 - i\epsilon$), the corresponding homogeneous equation no longer has solutions. Then the operator (denoted in what follows as Δ^c) inverse to $\Pi^2 - m^2 + i\epsilon$,

$$\Delta^c = \left(\Pi^2 - m^2 + i\epsilon\right)^{-1}\,, \tag{6.10}$$

is already single-valued.

Calculations with special fields when confronted with the results obtained using the method of summation over exact solutions, show that this form of the operator Δ^c leads to Green's function of the Dirac equation corresponding to the causal propagator of the perturbative QED. This is one of the reasons why Green's function corresponding to the operator Δ^c is called the causal Green's function and is designated as $S^c(x, y)$.

In what follows, we shall be dealing with just this function. It should be emphasized that the proof of the fact that $m^2 \to m^2 - i\epsilon$ picks out just the causal Green's function of quantum electrodynamics in an arbitrary external field is as yet lacking.

Let us write the operator Δ^c in an exponential form, or, as is conventionally stated, in the form of an integral over proper time s

$$\Delta^c = i\int_0^\infty \exp(-\epsilon s)\, U(s) ds,$$
$$U(s) = \exp\{i\mathcal{H}s\}, \quad \mathcal{H} = \left(\Pi^2 - m^2\right). \tag{6.11}$$

Now it can be seen that the problem of finding the function $S^c(x, y)$ is reduced to calculating the matrix element

$$\Delta^c(x, y, s) = \langle x| U(s) |y\rangle, \tag{6.12}$$

which is the evolution function in the proper time s. Once the latter is known, the causal function is determined as

$$S^c(x, y) = i\left(\gamma^\mu \hat{P}_\mu + m\right) \int_0^\infty \exp(-\epsilon s)\, \Delta^c(x, y, s)\, ds. \tag{6.13}$$

The operator $U(s)$ can be thought of as an operator of evolution in the proper time s for a system described by the Hamiltonian function $\mathcal{H}$. If now the operators X^μ, Π_μ and Γ^μ are referred to as Schrödinger operators, then their Heisenberg representation is

$$X^\mu(s) = U^{-1}(s) X^\mu U(s), \quad \Pi_\mu(s) = U^{-1}(s)\Pi_\mu U(s), \quad \Gamma^\mu(s) = U^{-1}(s)\Gamma^\mu U(s).$$

The Heisenberg operators satisfy the equations

$$\frac{dX^\mu(s)}{ds} = 2\Pi^\mu(s), \quad \frac{d\Pi_\mu(s)}{ds} = 2eF_{\mu\nu}(X(s))\,\Pi^\nu(s) + ie\frac{\partial F_{\mu\nu}(X(s))}{\partial X_\nu(s)} + \frac{e}{2}\Sigma^{\alpha\beta}(s)\frac{\partial F_{\alpha\beta}(X(s))}{\partial X^\mu(s)},$$
$$\frac{d\Gamma^\mu(s)}{ds} = 2eF^\mu_\nu(X(s))\,\Gamma^\nu(s), \quad \Sigma^{\alpha\beta}(s) = \frac{i}{2}\left[\Gamma^\alpha(s), \Gamma^\beta(s)\right]_-, \tag{6.14}$$

with the initial data $X^\mu(0) = X^\mu$, $\Pi_\mu(0) = \Pi_\mu$, $\Gamma^\mu(0) = \Gamma^\mu$.

Let us also introduce the bra and ket states depending on the proper time s,

$$\langle x(s)| = \langle x| U(s), \quad |x(0)\rangle = |x\rangle, \langle x(s)| X^\mu(s) = x^\mu \langle x(s)|.$$

Then

$$\Delta^c(x, y, s) = \langle x(s)| y(0). \tag{6.15}$$

The explicit form (6.11) of the evolution operator $U(s)$ implies

$$i\frac{d}{ds}\langle x(s)|\, y(0)\rangle = -\langle x(s)|\,\mathcal{H}\,|x(0)\rangle\ . \tag{6.16}$$

If now one succeeds, by using solutions of equations (6.14), to express the Hamiltonian $\mathcal{H}$ in terms of the operators $X^\mu(s)$ and $X^\mu(0) = X^\mu$ alone, with every operator $X^\mu(s)$ placed to the left of all the operators X^μ, then one obtains the ordinary differential equation for the evolution function $\Delta^c(x, y, s)$

$$i\frac{d}{ds}\Delta^c(x, y, s) = Q(x, y, s)\,\Delta^c(x, y, s)\ , \tag{6.17}$$

where $Q(x, y, s)$ is a function of the arguments indicated. This equation, when combined with the relations

$$\langle x(s)|\,\Pi_\mu(s)\,|y(0)\rangle = \left(i\frac{\partial}{\partial x^\mu} - e\mathcal{A}_\mu(x)\right)\Delta^c(x, y, s)\ , \tag{6.18}$$

$$\langle x(s)|\,\Pi_\mu(0)\,|y(0)\rangle = \left(-i\frac{\partial}{\partial y^\mu} - e\mathcal{A}_\mu(y)\right)\Delta^c(x, y, s) \tag{6.19}$$

and the initial condition $\Delta^c(x, y, 0) = \delta(x - y)$ completely determines the evolution function.

6.2.2 Proper-time representations in a constant uniform field and a plane wave field

Here we find the proper-time representation for spinor causal Green's function $S^c(x, y)$ in the external field given as a combination of a constant field and a plane-wave field. In doing this, we follow Ref. [152].

The potentials for this field are chosen as

$$\mathcal{A}_\mu(x) = -\frac{1}{2}\mathcal{F}_{\mu\nu}x^\nu + f_\mu(nx)\ , \tag{6.20}$$

where $\mathcal{F}_{\mu\nu}$ is the field strength tensor of the constant electromagnetic field with its two nonzero invariants I_1 and I_2,

$$I_1 = \frac{1}{2}\mathcal{F}_{\mu\nu}\mathcal{F}^{\mu\nu}, \quad I_2 = \frac{1}{4}\mathcal{F}^*_{\mu\nu}\mathcal{F}^{\mu\nu}\ . \tag{6.21}$$

The eigenvalues E and H of the field strength tensor

$$\begin{aligned}
\mathcal{F}_{\mu\nu}n^\nu &= -En_\mu, & \mathcal{F}_{\mu\nu}\bar{n}^\nu &= E\bar{n}_\mu,\\
\mathcal{F}_{\mu\nu}m^\nu &= iHm_\mu, & \mathcal{F}_{\mu\nu}\bar{m}^\nu &= -iH\bar{m}_\mu,\\
\mathcal{F}^*_{\mu\nu}n^\nu &= -Hn_\mu, & \mathcal{F}^*_{\mu\nu}\bar{n}^\nu &= H\bar{n}_\mu,\\
\mathcal{F}^*_{\mu\nu}m^\nu &= iEm_\mu, & \mathcal{F}_{\mu\nu}\bar{m}^\nu &= -iE\bar{m}_\mu\ ,
\end{aligned} \tag{6.22}$$

are expressed in terms of I_1 and I_2 as[1]

$$E = \sqrt{\sqrt{\left(\frac{I_1}{2}\right)^2 + I_2^2} - \frac{I_1}{2}}\,, H = \sqrt{\sqrt{\left(\frac{I_1}{2}\right)^2 + I_2^2} + \frac{I_1}{2}}\,. \tag{6.23}$$

The eigenvectors $n, \bar{n}, m$, and $\bar{m}$ are isotropic:

$$n^2 = \bar{n}^2 = m^2 = \bar{m}^2 = 0, \quad (n\bar{n}) = 2, (m\bar{m}) = -2\,,$$
$$(nm) = (\bar{n}m) = (n\bar{m}) = (\bar{n}\bar{m}) = 0\,.$$

The functions $f_\mu(nx)$ give the potential of the plane wave propagating along the spatial part $\boldsymbol{n}$ of the vector n and obey the conditions $nf(nx) = \bar{n}f(nx) = 0$. The total electromagnetic field tensor corresponding to the potentials (6.20) has the form

$$F_{\mu\nu} = \mathcal{F}_{\mu\nu} + \Psi_{\mu\nu}(nx)\,, \quad \Psi_{\mu\nu}(nx) = n_\mu f'_\nu(nx) - n_\nu f'_\mu(nx)\,. \tag{6.24}$$

Since the invariants I_1 and I_2 of the tensor $\mathcal{F}_{\mu\nu}$ are both nonzero, there exists a special Lorentz frame where the electric and magnetic fields of $\mathcal{F}_{\mu\nu}$ are parallel to one another and to the spatial part $\boldsymbol{n}$ of the four-vector n. In this frame, the total field (6.20) corresponds to constant and parallel electric and magnetic fields and a plane wave propagating along their common direction, while E and H, equation (6.23), are the strengths of the constant electric and magnetic fields, respectively. It is useful to consider the following combinations of the eigenvectors $n, \bar{n}, m$, and $\bar{m}$:

$$n^{(\epsilon)} = \frac{1}{2}(n + \epsilon\bar{n})\,, \quad \epsilon = \pm$$
$$m^{(+)} = \frac{1}{2}(m + \bar{m})\,, \quad m^{(-)} = \frac{i}{2}(\bar{m} - m)\,. \tag{6.25}$$

The following relations hold true:

$$\left(n^{(\epsilon)}\right)^2 = \epsilon, \quad \left(m^{(\epsilon)}\right)^2 = -1,$$
$$\left(nm^{(\epsilon)}\right) = \left(\bar{n}m^{(\epsilon)}\right) = \left(m^{(+)}m^{(-)}\right) = 0,$$
$$\mathcal{F}_{\mu\nu}m^{(\epsilon)\nu} = -\epsilon H m_\mu^{(-\epsilon)}\,. \tag{6.26}$$

In terms of the vectors introduced, the potentials (6.20) and the tensor (6.24) can be written in the following way:

$$\begin{aligned} A_\mu(x) &= \frac{1}{4}\left[n_\mu(\bar{n}x) - \bar{n}_\mu(nx)\right]E \\ &\quad - \frac{1}{2}\left[m_\mu^{(-)}\left(m^{(+)}x\right) - m_\mu^{(+)}\left(m^{(-)}x\right)\right]H + f_\mu(nx)\,, \\ F_{\mu\nu}(x) &= \frac{1}{2}\left(\bar{n}_\mu n_\nu - n_\mu\bar{n}_\nu\right)E \\ &\quad + \left(m_\mu^{(-)}m_\nu^{(+)} - m_\mu^{(+)}m_\nu^{(-)}\right)H + \Psi_{\mu\nu}(x)\,. \end{aligned} \tag{6.27}$$

1 The square roots in (6.23) are to be understood in an algebraic way: $\sqrt{x^2} = x$ and not $|x|$.

The vector $n_\mu^{(+)}$ is time-like, and the scalar product $n^{(+)}x$ is just the time variable in the special Lorentz frame introduced.

Now, let us write equations (6.14) for the field, given by the potentials (6.20), taking into account that the field is sourceless $\partial_\mu F^{\mu\nu} = 0$,

$$\frac{dX^\mu(s)}{ds} = 2\Pi^\mu(s),$$
$$\frac{d\Pi_\mu(s)}{ds} = 2e\left[\mathfrak{F}_{\mu\nu} + \Psi_{\mu\nu}(nX(s))\right]\Pi^\nu(s) + \frac{e}{2}\Sigma^{\alpha\beta}(s)\frac{\partial\Psi_{\alpha\beta}(nX(s))}{\partial X^\mu(s)},$$
$$\frac{d\Gamma^\mu(s)}{ds} = 2e\eta^{\mu\nu}\left[\mathfrak{F}_{\nu\lambda} + \Psi_{\nu\lambda}(nX(s))\right]\Gamma^\lambda(s)\,.$$

Their solution is readily found to be

$$n\Pi(s) = \exp\{2eEs\}\, n\Pi(0)\,,$$
$$nX(s) - nX(0) = -(eE)^{-1}\left[1 - \exp\{2eEs\}\right] n\Pi(0)\,,$$
$$n\Gamma(s) = \exp\{2eEs\}\, n\Gamma(0)\,,$$
$$\Pi^\perp(s) = \exp\{2e\mathfrak{F}s\}\,\Pi^\perp(0) - e\int_0^s \exp\{2e\mathfrak{F}(s-u)\}\,\frac{df(nX(u))}{du}du,$$
$$X^\perp(s) - X^\perp(0) = (e\mathfrak{F})^{-1}\Bigg[\left(\exp\{2e\mathfrak{F}s\} - 1\right)\Pi^\perp(0) - e\int_0^s \left(\exp\{2e\mathfrak{F}(s-u)\} - 1\right)\frac{df(nX(u))}{du}du\Bigg],$$
$$\Gamma^\perp(s) = \exp\{2e\mathfrak{F}s\}\,\Gamma^\perp(0) - (n\Pi(0))^{-1}(n\Gamma(0)) \times e\int_0^s \exp\{2e\mathfrak{F}(s-u)\}\,\frac{df(nX(u))}{du}du,$$
$$\tilde{n}\Pi(s) = \exp\{-2eEs\}\Bigg[\tilde{n}\Pi(0) + (n\Pi(0))^{-1} e\int_0^s \Bigg\{2\Pi^\perp(u)\frac{df(nX(u))}{du} + i(n\Gamma(0))\,\Gamma^\perp(0)\exp\{2e(E-\mathfrak{F})u\}\,\frac{df'(nX(u))}{du}\Bigg\}du\Bigg],$$

$$\bar{n}X(s) - \bar{n}X(0) = -(eE)^{-1}\Bigg[(\exp\{-2eEs\} - 1)\bar{n}\Pi(0) + (n\Pi(0))^{-1} e\int_0^s [\exp\{-2eEs\} - \exp\{-2eEu\}] \times \left(2\Pi^{\perp}(u)\frac{df(nX(u))}{du} + in\Gamma(0)\Gamma^{\perp}(0) \times \exp\{2e(E-\mathfrak{F})u\}\frac{df'(nX(u))}{du}\right)du\Bigg],$$

$$\bar{n}\Gamma(s) = \exp\{-2eEs\}\bar{n}\Gamma(0) + (n\Pi(0))^{-1}\exp\{-2eEs\} \times 2e\int_0^s \Gamma^{\perp}(u)\frac{df(nX(u))}{du}du, \tag{6.28}$$

where the four-vectors marked by the superscript $\perp$ are defined in the following way:

$$K^{\perp} = -\sum_{\epsilon=\pm}\left(Km^{(\epsilon)}\right)m^{(\epsilon)}, \quad K = \left(X^{\mu}, \Pi_{\mu}, \Gamma^{\mu}\right),$$

i.e. they are projections of the vectors X^{μ}, Π_{μ} and Γ^{μ} onto the two-dimensional space spanned by the vectors $m^{(\epsilon)}$, equation (6.25). Using (6.28) one can find the commutation relations

$$\begin{aligned}
[nX(s), nX(0)]_- &= [nX(s), n\Pi(0)]_- = [nX(0), n\Pi(0)]_- = 0,\\
[\bar{n}X(s), nX(0)]_- &= 2i(eE)^{-1}\left(1 - \exp\{-2eEs\}\right),\\
\left[X^{\perp}_{\mu}(s), X^{\perp}_{\nu}(0)\right]_- &= i\left[\left(e\mathfrak{F}^{\perp}\right)^{-1}\left(\exp\{2e\mathfrak{F}^{\perp}s\} - 1\right)\right]_{\mu\nu},\\
\left[nX(s), X^{\perp}_{\mu}(0)\right]_- &= 0, \quad \left[\Gamma^{\perp}_{\mu}(s), \Gamma^{\perp}_{\nu}(0)\right]_+ = 2\left[\exp\{2e\mathfrak{F}^{\perp}s\}\right]_{\mu\nu},
\end{aligned} \tag{6.29}$$

where

$$\mathfrak{F}^{\perp}_{\mu\nu} = \mathfrak{F}_{\mu\nu} + \frac{1}{2}E\left(n_{\mu}\bar{n}_{\nu} - \bar{n}_{\mu}n_{\nu}\right) = H\left(m^{(-)}_{\mu}m^{(+)}_{\nu} - m^{(+)}_{\mu}m^{(-)}_{\nu}\right).$$

After expressing the operators $\Pi(s)$, $\Pi(0)$, $\Gamma(s)$, $\Gamma(0)$ from (6.28) in terms of $X_{\mu}(s)$ and $X_{\mu}(0)$ and using the commutation relations (6.29) we obtain

$$\begin{aligned}
\langle x(s)|\,\Pi^2\,|y(0)\rangle = \langle x(s)|\,y(0)\rangle \Big\{ & Z_1(s) + Z_2(s) + [x - y + \mathfrak{Z}(s)]\\
& \times \frac{(e\mathfrak{F})^2}{4\sinh^2 e\mathfrak{F}s}[x - y + \mathfrak{Z}(s)] + \frac{ie^2}{2}\mathrm{tr}\mathfrak{F}\coth\mathfrak{F}s\\
& - \frac{e}{2}\sigma^{\mu\nu}\left[\mathfrak{F}_{\mu\nu} + \Psi_{\mu\nu}\left(nx'\right)\right]\Big\},
\end{aligned} \tag{6.30}$$

where

$$\mathfrak{J}(s) = j(s) + \mathfrak{F}^{-1}\left[1 - \exp\{2e\mathfrak{F}s\}\right] f(ny),$$

$$j(s) = 2e\int_0^s \exp\{2e(s-u)\mathfrak{F}\} f(nx(u))\,du,$$

$$nx(u) = ny + n(x-y)\frac{1-\exp\{2eEu\}}{1-\exp\{2eEs\}},$$

$$Z_1(s) = 2e^2\int_0^s \frac{\exp\{-2eEu\}-\exp\{-2eEs\}}{\exp\{-2eEs\}-1}$$

$$\times\frac{df(nx(u))}{du}\left[\mathfrak{F}\exp\{2e\mathfrak{F}(u-s)\}\left(1-\exp\{-2e\mathfrak{F}(s)\}\right)^{-1}(x-y+\mathfrak{J}(s))\right.$$

$$\left.-\int_0^u \exp\{2e\mathfrak{F}(u-\kappa)\}\frac{df(nx(\kappa))}{d\kappa}d\kappa\right]du,$$

$$Z_2(s) = ie(\gamma n)\int_0^s \frac{\exp\{-2eEu\}-\exp\{-2eEs\}}{\exp\{-2eEs\}-1}\gamma\exp\{-2e(E-\mathfrak{F})u\}\frac{df'(nx(u))}{du}du.$$

By substituting (6.30) into the relation (6.16) we obtain an ordinary differential equation for the function $\Delta^c(x,y,s)$ defined by equations (6.12) and (6.15). Its solution has the form

$$\Delta^c(x,y,s) = C(x,y)\,\omega(s)\chi(x,y,s), \tag{6.31}$$

where $C(x,y)$ is an arbitrary function of x and y, and

$$\omega(s) = \exp\left\{-\frac{ie}{2}\sigma_{\mu\nu}\mathfrak{F}^{\mu\nu}s\right\} + e(\gamma n)\gamma\exp\{e\mathfrak{F}s\}$$

$$\times\int_0^s \exp\{-2e\mathfrak{F}u\}\frac{df(nx(u))}{du}du\frac{\sinh eEs}{eEn(x-y)}$$

$$\chi(x,y,s) = (4\pi)^{-2}\left(\det\frac{\sinh e\mathfrak{F}s}{e\mathfrak{F}}\right)^{-1/2}\exp\left\{i\phi(s) + \frac{ie}{2}(x-y)\mathfrak{F}j(s)\right.$$

$$\left.-\frac{ie}{4}(x-y+j(s))\mathfrak{F}\coth e\mathfrak{F}s\,(x-y+j(s)) - im^2s\right\},$$

$$\phi(s) = e^2\int_0^s f(nx(u))\left[f(nx(u)) + \mathfrak{F}j(u)\right]du.$$

The function $C(x,y)$ is determined using the conditions (6.18) and (6.19) to be

$$C(x,y) = -i\exp\left\{\frac{ie}{2}x\mathfrak{F}y\right\}. \tag{6.32}$$

By summarizing the above results in agreement with equation (6.13) we find the expression for the causal Green's function of the Dirac equation in a constant homogeneous field combined with the plane-wave field as

$$S^{c}(x,y) = \left(\hat{P}_{\mu}\gamma^{\mu} + m\right)\exp\left\{\frac{ie}{2}x\mathfrak{F}y\right\}\int_{0}^{\infty}\exp(-\epsilon s)\,\omega(s)\chi(x,y,s)\,ds\,. \tag{6.33}$$

In a similar manner, one can find the causal Green's function $D^{c}(x,y)$ of the K–G equation. For the external field under consideration it has the form

$$D^{c}(x,y) = \exp\left\{\frac{ie}{2}x\mathfrak{F}x\right\}\int_{0}^{\infty}\exp(-\epsilon s)\,\chi(x,y,s)\,ds\,. \tag{6.34}$$

This result can be obtained from (6.33) by setting $\hat{P}_{\mu}\gamma^{\mu} + m \to 1, \quad \omega(s) \to 1.$

6.3 Path-integrals for particle propagators

Propagators of relativistic particles in external fields (electromagnetic, non-Abelian or gravitational ones) contain important information about the quantum behavior of these particles. Moreover, if such propagators are known in an arbitrary external field one can find exact one-particle Green functions in the corresponding quantum field theory by calculating functional integrals over all external fields, see [100]. The Dirac propagator in an external electromagnetic field differs from that of a scalar particle in that it has a complicated spinor structure. The problem of its path integral representation has been attracting researchers' attention for a long time. Thus, Feynman who had first written his path integral for the probability amplitude in nonrelativistic quantum mechanics [134, 136] and then wrote a path integral for the causal Green function of K–G equation (scalar particle propagator) [137], had also made an attempt to derive a representation for the Dirac propagator via a Bosonic path integral [138]. After the introduction of the integral over Grassmann variables by Berezin, it became possible to present this propagator via both Bosonic and Grassmann variables, the latter describing the spinning degrees of freedom. Representations of this kind have been discussed in the literature for a long time in different contexts [6, 9, 76, 102, 129, 147, 202, 204, 207, 257]. Nevertheless, attempts to write the Dirac propagator via only a Bosonic path integral continued. For instance, Polyakov [267] assumed that the propagator of a free Dirac electron in $D = 3$ Euclidean space-time can be presented by means of a Bosonic path integral similar to the scalar particle case, modified by the so-called *spin factor* (*SF*). This idea was further developed in [224] by writing a SF for Dirac fermions that interacts with a non-Abelian gauge field in D-dimensional Euclidean space-time. In those representations the SF itself was presented via some additional Bosonic path integrals and its γ-matrix structure was not defined explicitly. Surprisingly, it was shown in [181, 182] that all Grassmann integrations in the representation of the Dirac propagator in an arbitrary external field in 3 + 1 dimensions can be fulfilled, so that an expression for the SF was derived as a given functional of the Bosonic trajectory. Having such representation with the SF, one can use it to calculate the propagator in some specific cases of external fields. This way of calculation provides automatically an explicit spinor structure of the propagators which can be used for concrete calculations in QED in the Furry picture.

In the work [172] propagators of spinning particles in arbitrary dimensions were presented by path integrals. It turns out that the problem has different solutions in even and odd dimensions. Below, we construct path integral representations for spinless and spinning particle propagators and use them for calculating the propagators in a constant uniform electromagnetic field and in its combination with a plane wave.

6.3.1 Path integral for K–G propagator

First, we discuss briefly a path integral representation for scalar particle propagator [172], we call it *K–G propagator* in what follows. We consider a spinless charged particle placed into an arbitrary external electromagnetic field with potentials A_μ. As it is known, the K–G propagator is the causal Green function $D^c(x, y)$ of the K–G equation,

$$\left(\hat{\mathcal{P}}^2 - m^2\right) D^c(x, y) = -\delta^D(x - y)\,, \tag{6.35}$$

where $\hat{\mathcal{P}}_\mu = i\partial_\mu - gA_\mu(x)$, $\mu = 0, \ldots, D-1$, and the Minkowski metric tensor is $\eta_{\mu\nu} = \mathrm{diag}(\underbrace{1, -1, \ldots, -1}_{D})$, see for example [100, 292]. Following Schwinger [294], one can represent $D^c(x, y)$ as a matrix element of the operator $\hat{D}^c$,

$$D^c(x, y) = \langle x|\hat{D}^c|y\rangle\,, \tag{6.36}$$

where $|x\rangle$ are eigenvectors for some self-adjoint operators of coordinates $\hat{x}$; the corresponding canonical operators of momenta are $\hat{p}_\mu$, so that:

$$\begin{aligned}
&\hat{x}^\mu|x\rangle = x^\mu|x\rangle, \quad \langle x|y\rangle = \delta^D(x - y), \quad \int |x\rangle\langle x|dx = I, \\
&\left[\hat{p}_\mu, \hat{x}^\nu\right]_- = -i\delta^\nu_\mu, \quad \hat{p}_\mu|p\rangle = p_\mu|p\rangle, \quad \langle p|p'\rangle = \delta^D(p - p')\,, \\
&\int |p\rangle\langle p|dp = I, \quad \langle x|\hat{p}_\mu|y\rangle = -i\partial_\mu\delta^D(x - y), \quad \langle x|p\rangle = \frac{1}{(2\pi)^{D/2}}e^{ipx}\,, \\
&\left[\hat{\Pi}_\mu, \hat{\Pi}_\nu\right]_- = -igF_{\mu\nu}(\hat{x}), \quad \hat{\Pi}_\mu = -\hat{p}_\mu - gA_\mu(\hat{x})\,.
\end{aligned} \tag{6.37}$$

Equation (6.35) implies $\hat{D}^c = \hat{F}^{-1}$, $\hat{F} = m^2 - \hat{\Pi}^2$. Now we use the *Schwinger proper-time representation* for the inverse operator,

$$\hat{F}^{-1} = i \int_0^\infty e^{-i\lambda(\hat{F}-i\epsilon)}\, d\lambda\,, \tag{6.38}$$

where the parameter λ is called the proper-time and the infinitesimal quantity ϵ has to be set equal to zero at the end of calculations. Thus, we get the following integral representation[2] for the propagator (6.36):

$$D^c = D^c(x_{\text{out}}, x_{\text{in}}) = i \int_0^\infty \langle x_{\text{out}}| e^{-i\hat{\mathcal{H}}(\lambda)} |x_{\text{in}}\rangle\, d\lambda\,,$$
$$\hat{\mathcal{H}}(\lambda) = \lambda\left(m^2 - \hat{\Pi}^2\right)\,. \tag{6.39}$$

Then we represent the matrix element, entering in expression (6.39), by means of a path integral, using a discretization procedure, see for example [134]. In doing this, we write the operator exponent $\exp(-i\hat{\mathcal{H}})$ as a product, $\exp(-i\hat{\mathcal{H}}) = [\exp(-i\hat{\mathcal{H}}/N)]^N$, and then we insert $(N-1)$ resolutions of identity $\int |x\rangle\langle x| dx = I$ between all the multipliers $\exp(-i\hat{\mathcal{H}}/N)$. Besides, we introduce N additional integrations over λ_k, $k = 1, \ldots, N$, renominating at the same time λ into λ_0. Thus, we obtain

$$D^c = i \lim_{N\to\infty} \int_0^\infty d\lambda_0 \int_{-\infty}^{+\infty} dx_1 \ldots dx_{N-1} d\lambda_1 \ldots d\lambda_N$$
$$\times \prod_{k=1}^N \langle x_k| e^{-i\hat{\mathcal{H}}(\lambda_k)\Delta\tau} |x_{k-1}\rangle \delta(\lambda_k - \lambda_{k-1})\,, \tag{6.40}$$

where $\Delta\tau = 1/N$, $x_0 = x_{\text{in}}$, $x_N = x_{\text{out}}$. Bearing in mind the further limiting process $N \to \infty$, one can treat the matrix elements from (6.40) approximately as follows:

$$\langle x_k| e^{-i\hat{\mathcal{H}}(\lambda_k)\Delta\tau} |x_{k-1}\rangle \approx \langle x_k| 1 - i\hat{\mathcal{H}}(\lambda_k)\Delta\tau |x_{k-1}\rangle \tag{6.41}$$

and then calculate them using the resolution of identity $\int |p\rangle\langle p| dp = I$. In this connection we notice that originally the operator $\hat{\mathcal{H}}(\lambda)$ has the form symmetric in the operators $\hat{x}$ and $\hat{p}$. Indeed, the only term in $\hat{\mathcal{H}}(\lambda)$ that contains products of these operators is $[\hat{p}_\alpha, A^\alpha(\hat{x})]_+$. One can verify that this is the maximally symmetrized expression that can be constructed from the involved operators (see e.g. a remark in Ref. [341]). Therefore, one can write $\hat{\mathcal{H}}(\lambda) = \text{Sym}_{(\hat{x},\hat{p})}\, \mathcal{H}(\lambda, \hat{x}, \hat{p})$, where $\mathcal{H}(\lambda, x, p)$ is the Weyl symbol[3] of the operator $\hat{\mathcal{H}}(\lambda)$,

$$\mathcal{H}(\lambda, x, p) = \lambda\left(m^2 - \mathcal{P}^2\right), \quad \mathcal{P}_\mu = -p_\mu - gA_\mu(x)\,.$$

2 Here and in what follows we include the factor $-i\epsilon$ in m^2.

3 For the definition of the Weyl symbol of an operator see for example [90].

It is a general statement [90] – easy to check in this special case by direct calculations – that matrix elements (6.41) are expressed in terms of the Weyl symbols in the middle points $\overline{x}_k = (x_k + x_{k-1})/2$. Taking all said into account, one can see that in the limiting process $N \to \infty$ matrix elements (6.41) can be replaced by the expressions

$$\int \frac{dp_k}{(2\pi)^D} \exp\left\{i\left[p_k \frac{x_k - x_{k-1}}{\Delta\tau} - \mathcal{H}(\lambda_k, \overline{x}_k, p_k)\right]\Delta\tau\right\} . \tag{6.42}$$

Using integral representations for the δ-functions in the right hand side of (6.40), we obtain

$$D^c = i \lim_{N\to\infty} \int_0^\infty d\lambda_0 \int_{-\infty}^{+\infty} dx_1 \dots dx_{N-1} \frac{dp_1}{(2\pi)^D} \cdots \frac{dp_N}{(2\pi)^D} d\lambda_1 \dots d\lambda_N$$
$$\times \frac{d\pi_1}{(2\pi)} \cdots \frac{d\pi_N}{(2\pi)} \exp\left\{i \sum_{k=1}^N \left[p_k \frac{x_k - x_{k-1}}{\Delta\tau} - \mathcal{H}(\lambda_k, \overline{x}_k, p_k) + \pi_k \frac{\lambda_k - \lambda_{k-1}}{\Delta\tau}\right]\Delta\tau\right\} . \tag{6.43}$$

Expression (6.43) is to be considered as a definition of the Hamiltonian path integral for the scalar particle propagator $D^c(x, y)$,

$$D^c = i \int_0^\infty d\lambda_0 \int_{x_{\rm in}}^{x_{\rm out}} Dx \int_{\lambda_0} D\lambda \int Dp D\pi \exp\left\{i \int_0^1 [\lambda(\mathcal{P}^2 - m^2) + p\dot{x} + \pi\dot{\lambda}] d\tau\right\} , \tag{6.44}$$

where derivatives with respect to τ are denoted by dots, for example $\dot{x} = dx/d\tau$. The functional integration goes over the trajectories $x^\mu(\tau), p_\mu(\tau), \lambda(\tau), \pi(\tau)$, parameterized by an invariant parameter $\tau \in [0, 1]$ and obeying the boundary conditions $x(0) = x_{\rm in}$, $x(1) = x_{\rm out}$, $\lambda(0) = \lambda_0$.

To go over to the Lagrangian form of the path integral, one has to perform the integration over the momenta p. In fact, the result can be achieved by means of the shift and the replacement, $p_\mu \to -p_\mu - (\dot{x}_\mu/2\lambda) - gA_\mu$, $e = 2\lambda$. Thus, we obtain

$$D^c = \frac{i}{2} \int_0^\infty de_0 \int_{x_{\rm in}}^{x_{\rm out}} Dx \int_{e_0} M(e) De \int D\pi$$
$$\times \exp\left\{i \int_0^1 \left[-\frac{\dot{x}^2}{2e} - e\frac{m^2}{2} - gx_\mu A^\mu + \pi\dot{e}\right] d\tau\right\} , \tag{6.45}$$

where the boundary conditions $x(0) = x_{\rm in}$, $x(1) = x_{\rm out}$, $e(0) = e_0$ are supposed and the measure $M(e)$ has the form

$$M(e) = \int Dp \exp\left\{\frac{i}{2} \int_0^1 ep^2 d\tau\right\} . \tag{6.46}$$

A discussion of the role of the measure (6.46) can be found in Ref. [147].

6.3.2 Path integral for the Dirac propagator in even dimensions

Here we consider path-integral representation for the Dirac propagator in even dimensions D, see [147, 172]. The propagator satisfies the following equation:

$$\left(\hat{\mathcal{P}}_\mu \gamma^\mu - m\right) S^c(x, y) = -\delta^D(x - y)\,. \tag{6.47}$$

In even dimensions a matrix representation of the Clifford algebra with dimensionality $\dim \gamma^\mu = 2^{D/2} = 2^d$ always exists [104]. In other words, γ^μ are $2^d \times 2^d$ matrices. In such dimensions one can introduce another matrix, γ^{D+1}, which anticommutes with all γ^μ (an analog of $\gamma^5 = \gamma^0\gamma^1\gamma^2\gamma^3$ in four dimensions[4]),

$$\gamma^{D+1} = r\gamma^0\gamma^1 \dots \gamma^{D-1}, \quad r = \begin{cases} 1, & \text{if } d \text{ is even} \\ i, & \text{if } d \text{ is odd} \end{cases}, \tag{6.48}$$
$$[\gamma^{D+1}, \gamma^\mu]_+ = 0, \quad \left(\gamma^{D+1}\right)^2 = -1\,.$$

The existence of the matrix γ^{D+1} in even dimensions allows one to pass to a Dirac operator which is homogeneous in γ-matrices. Indeed, let us rewrite equation (6.47) in terms of the propagator $\tilde{S}^c(x, y)$ transformed by γ^{D+1},

$$\tilde{S}^c(x, y) = S^c(x, y)\gamma^{D+1}, \quad \left(\hat{\mathcal{P}}_\mu \tilde{\gamma}^\mu - m\gamma^{D+1}\right) \tilde{S}^c(x, y) = \delta^D(x - y)\,, \tag{6.49}$$

where $\tilde{\gamma}^\mu = \gamma^{D+1}\gamma^\mu$. The matrices $\tilde{\gamma}^\mu$ have the same commutation relations $[\tilde{\gamma}^\mu, \tilde{\gamma}^\nu]_+ = 2\eta^{\mu\nu}$ as the initial ones γ^μ, and anticommute with the matrix γ^{D+1}. The set of $D + 1$ gamma-matrices $\tilde{\gamma}^\nu$ and γ^{D+1} form a representation of the Clifford algebra in odd $2d+1$ dimensions. Let us denote such matrices as Γ^n,

$$\Gamma^n = \begin{cases} \tilde{\gamma}^\mu, & n = \mu = 0, \dots, D-1 \\ \gamma^{D+1}, & n = D \end{cases}, \tag{6.50}$$
$$[\Gamma^k, \Gamma^n]_+ = 2\eta^{kn}, \quad \eta_{kn} = \operatorname{diag}(\underbrace{1, -1, \dots, -1}_{D+1}), \quad k, n = 0, \dots, D\,.$$

In terms of these matrices equation (6.49) takes the form

$$\hat{\mathcal{P}}_n \Gamma^n \tilde{S}^c(x, y) = \delta^D(x - y), \quad \hat{\mathcal{P}}_\mu = i\partial_\mu - gA_\mu(x), \quad \hat{\mathcal{P}}_D = -m\,. \tag{6.51}$$

Similar to (6.36), we present $\tilde{S}^c(x, y)$ as a matrix element of an operator $\hat{S}^c$ (in the coordinate representation (6.37)),

$$\tilde{S}^c_{ab}(x, y) = \langle x|\hat{S}^c_{ab}|y\rangle, \quad a, b = 1, 2, \dots, 2^d\,, \tag{6.52}$$

where the spinor indices a, b are written explicitly for clarity, but will be omitted hereafter. Equation (6.51) implies $\hat{S}^c = \hat{F}^{-1}, \hat{F} = \Pi_n\Gamma^n$, where Π_μ are defined in (6.37),

---

4 In this chapter, we use $\gamma^5 = \gamma^0\gamma^1\gamma^2\gamma^3$ in contrast to $\gamma^5 = -i\gamma^0\gamma^1\gamma^2\gamma^3$ used in all other chapters.

and $\Pi_D = -m$. The operator $\hat{F}$ is homogeneous in γ-matrices. One can consider it as a pure Fermi operator, if one treats γ-matrices as Fermi-operators. Now, instead of the Schwinger proper-time representation (6.38), which is convenient for Bose-type operators, one can use a different representation by means of an integral over the *super-proper time* (λ, χ) of an exponential with an even exponent. Namely, one can write

$$\hat{S}^c = \frac{\hat{F}}{\hat{F}^2} = \int_0^\infty d\lambda \int e^{i[\lambda(\hat{F}^2+i\epsilon)+\chi\hat{F}]} d\chi \,,$$

$$\hat{F}^2 = \Pi^2 - m^2 - \frac{ig}{2} F_{\mu\nu}\Gamma^\mu\Gamma^\nu \,, \tag{6.53}$$

where λ is an even variable and χ is a Grassmann odd variable[5], the latter anticommuting with $\hat{F}$ (with γ-matrices) by the definition. Here and in what follows integrals over Grassmann-odd variables are understood as Berezin's integrals [89]. Representation (6.53) is an analog of the Schwinger proper-time representation for the inverse operator convenient in the Fermi case. Such a representation was introduced for the first time in Ref. [147]. Thus, the Green function (6.52) takes the form

$$\tilde{S}^c = \tilde{S}^c(x_{\text{out}}, x_{\text{in}}) = \int_0^\infty d\lambda \int \langle x_{\text{out}} | e^{-i\hat{\mathcal{H}}(\lambda,\chi)} | x_{\text{in}} \rangle d\chi \,,$$

$$\hat{\mathcal{H}}(\lambda, \chi) = \lambda\left(m^2 - \Pi^2 + \frac{ig}{2} F_{\mu\nu}\Gamma^\mu\Gamma^\nu\right) + \Pi_n\Gamma^n \chi \,. \tag{6.54}$$

Now we present the matrix element entering in the right hand side of (6.54) by means of a path integral. In spite of the fact that the operator $\hat{\mathcal{H}}(\lambda, \chi)$ has γ-matrix structure, it is possible to proceed as in the scalar case. An analog of the formula (6.40) has the form

$$\tilde{S}^c = \lim_{N\to\infty} \int_0^\infty d\lambda_0 \int d\chi_0 \int_{-\infty}^{+\infty} dx_1 \ldots dx_{N-1} d\lambda_1 \ldots d\lambda_N \int d\chi_1 \ldots d\chi_N$$

$$\times \prod_{k=1}^N \langle x_k | e^{-i\hat{\mathcal{H}}(\lambda_k,\chi_k)\Delta\tau} | x_{k-1} \rangle \delta(\lambda_k - \lambda_{k-1}) \delta(\chi_k - \chi_{k-1}) \,. \tag{6.55}$$

The matrix elements in the right hand side of (6.55) can be replaced by the expressions

$$\int \frac{dp_k}{(2\pi)^D} \exp\left\{ i \left[p_k \frac{x_k - x_{k-1}}{\Delta\tau} - \mathcal{H}(\lambda_k, \chi_k, \overline{x}_k, p_k) \right] \Delta\tau \right\} \,, \tag{6.56}$$

5 Notion of Grasmanian variables can be found in [89, 174].

where $\mathcal{H}(\lambda, \chi, x, p)$ is the *Weyl symbol* of the operator $\hat{\mathcal{H}}(\lambda, \chi)$ in the sector of the coordinates and momenta,

$$\mathcal{H}(\lambda, \chi, x, p) = \lambda\left(m^2 - \mathcal{P}^2 + \frac{ig}{2}F_{\mu\nu}\Gamma^\mu\Gamma^\nu\right) + \mathcal{P}_n\Gamma^n\chi$$
$$\mathcal{P}_\mu = -p_\mu - gA_\mu(x), \quad \mathcal{P}_D = -m\,.$$

The multipliers (6.56) are noncommutative due to the γ-matrix structure and are situated in the right hand side of (6.55) so that the numbers k increase from the right to the left. For the two δ-functions that accompany each matrix element (6.56) in (6.55), we use the following integral representations:

$$\delta(\lambda_k - \lambda_{k-1})\delta(\chi_k - \chi_{k-1}) = \frac{i}{2\pi}\int e^{i[\pi_k(\lambda_k-\lambda_{k-1})+\nu_k(\chi_k-\chi_{k-1})]}d\pi_k d\nu_k\,,$$

where ν_k are Grassmann-odd variables. Then we formally attribute the index k, to the γ-matrices, involved in (6.56), and at the same time we attribute the "time" τ_k to all quantities, according to the index k they have, $\tau_k = k\Delta\tau$, so that $\tau \in [0, 1]$. Introducing the T-product, which acts on γ-matrices, it is possible to gather all the expressions in the right hand side of (6.55) in one exponent and deal then with the γ-matrices like with Grassmann odd variables. Thus, we get for the right hand side of equation (6.55)

$$\tilde{S}^c = \mathrm{T}\int_0^\infty d\lambda_0 \int d\chi_0 \int_{x_{\rm in}}^{x_{\rm out}} Dx \int Dp \int_{\lambda_0} D\lambda \int_{\chi_0} D\chi \int D\pi \int D\nu$$
$$\times \exp\left\{i\int_0^1 \left[\lambda\left(\mathcal{P}^2 - m^2 - \frac{ig}{2}F_{\mu\nu}\Gamma^\mu\Gamma^\nu\right) + \chi\mathcal{P}_n\Gamma^n + p\dot{x} + \pi\dot{\lambda} + \nu\dot{\chi}\right]d\tau\right\}, \tag{6.57}$$

where $x(\tau), p(\tau), \lambda(\tau), \pi(\tau)$ are even, and $\chi(\tau), \nu(\tau)$ are Grassmann odd trajectories, obeying the boundary conditions

$$x(0) = x_{\rm in}, \quad x(1) = x_{\rm out}, \quad \lambda(0) = \lambda_0, \quad \chi(0) = \chi_0\,. \tag{6.58}$$

The T-ordering operation acts on γ-matrices which are supposed formally to depend on the time τ. Then equation (6.57) can be transformed as follows:

$$\tilde{S}^c = \int_0^\infty d\lambda_0 \int d\chi_0 \int_{x_{\rm in}}^{x_{\rm out}} Dx \int Dp \int_{\lambda_0} D\lambda \int_{\chi_0} D\chi \int D\pi \int D\nu \exp\left\{i\int_0^1\left[\lambda\left(\mathcal{P}^2 - m^2\right.\right.\right.$$
$$\left.\left.\left. - \frac{ig}{2}F_{\mu\nu}\frac{\delta_l}{\delta\rho_\mu}\frac{\delta_l}{\delta\rho_\nu}\right) + \chi\mathcal{P}_n\frac{\delta_l}{\delta\rho_n} + p\dot{x} + \pi\dot{\lambda} + \nu\dot{\chi}\right]d\tau\right\} \mathrm{T}\exp\int_0^1 \rho_n(\tau)\Gamma^n d\tau\Bigg|_{\rho=0},$$

where Grassmann odd sources $\rho_n(\tau)$ are introduced, and $\partial_\ell/\partial\rho_n$ denotes left derivatives. They anticommute with the γ-matrices by definition. One can present the quantity $\mathrm{T}\exp\int_0^1 \rho_n(\tau)\Gamma^n d\tau$ as a path integral over Grassmann odd trajectories [147, 333],

$$\mathrm{T}\exp\int_0^1 \rho_n(\tau)\Gamma^n d\tau = \exp\left(i\Gamma^n \frac{\partial_l}{\partial\theta^n}\right)$$

$$\times \int_{\psi(0)+\psi(1)=\theta} \exp\left[\int_0^1 \left(\psi_n\dot{\psi}^n - 2i\rho_n\psi^n\right) d\tau + \psi_n(1)\psi^n(0)\right] \mathcal{D}\psi \Bigg|_{\theta=0},$$

$$\mathcal{D}\psi = D\psi \left[\int_{\psi(0)+\psi(1)=0} \exp\left\{\int_0^1 \psi_n\dot{\psi}^n d\tau\right\} D\psi\right]^{-1}, \tag{6.59}$$

where θ^n are Grassmann odd variables, anticommuting with γ-matrices, and $\psi^n(\tau)$ are Grassmann odd trajectories of integration, obeying the boundary conditions, which are indicated below the integration signs. Using representation (6.59), we get the Hamiltonian path integral representation for the Green function in question:

$$\tilde{S}^c = \exp\left(i\Gamma^n \frac{\partial_l}{\partial\theta^n}\right) \int_0^\infty d\lambda_0 \int d\chi_0 \int_{\lambda_0} D\lambda \int_{\chi_0} D\chi \int_{x_{\text{in}}}^{x_{\text{out}}} Dx \int Dp \int D\pi \int D\nu$$

$$\times \int_{\psi(0)+\psi(1)=\theta} \mathcal{D}\psi \exp\left\{i\int_0^1 \left[\lambda\left(\mathcal{P}^2 - m^2 + 2igF_{\mu\nu}\psi^\mu\psi^\nu\right) + 2i\mathcal{P}_n\psi^n\chi\right.\right.$$

$$\left.\left. - i\psi_n\dot{\psi}^n + p\dot{x} + \pi\dot{\lambda} + \nu\dot{\chi}\right] d\tau + \psi_n(1)\psi^n(0)\right\}\Big|_{\theta=0}. \tag{6.60}$$

Integrating over momenta, we derive the path integral in the Lagrangian form,

$$\tilde{S}^c = \exp\left(i\Gamma^n \frac{\partial_\ell}{\partial\theta^n}\right) \int_0^\infty de_0 \int d\chi_0 \int_{e_0} M(e)De \int_{\chi_0} D\chi \int_{x_{\text{in}}}^{x_{\text{out}}} Dx \int D\pi \int D\nu$$

$$\times \int_{\psi(0)+\psi(1)=\theta} \mathcal{D}\psi \exp\left\{i\int_0^1 \left[-\frac{\dot{x}^2}{2e} - \frac{e}{2}m^2 - g\dot{x}A + iegF_{\mu\nu}\psi^\mu\psi^\nu\right.\right.$$

$$\left.\left. + i\left(\frac{\dot{x}_\mu\psi^\mu}{e} - m\psi^D\right)\chi - i\psi_n\dot{\psi}^n + \pi\dot{e} + \nu\dot{\chi}\right] d\tau + \psi_n(1)\psi^n(0)\right\}\Bigg|_{\theta=0}, \tag{6.61}$$

where the measures $M(e)$ and $\mathcal{D}\psi$ are defined by equations (6.46) and (6.59), respectively.

6.3.3 Path integral for the Dirac propagator in odd dimensions

In odd dimensions D a possibility to construct the matrix γ^{D+1} (6.48) does not exist. Hence, the trick, which was used to make the Dirac operator homogeneous in γ-matrices does not work here. Nevertheless, the problem of the path integral construction can be solved, but in a different way [172].

As it is known, in odd dimensions $D = 2d + 1$ there exist two nonequivalent irreducible representations of the Clifford algebra with the dimensionality $2^{[D/2]} = 2^d$. Let us mark these representations by the index $s = \pm$. Thus, we have two nonequivalent sets of γ-matrices which we are going to denote as $\Gamma^n_{(s)}, \quad n = 0, 1, \dots, 2d$ (remark that now we use Latin indices n, k, etc. as Lorentz ones). Such matrices can be constructed for example out of the corresponding matrices in $2d$ dimensions as follows:

$$\Gamma^n_{(s)} = \begin{cases} \gamma^\mu, & n = \mu = 0, \dots, 2d-1 \\ s\gamma^{2d+1}, & n = 2d \end{cases},$$
$$[\Gamma^k_{(s)}, \Gamma^n_{(s)}]_+ = 2\eta^{kn}, \quad \eta_{kn} = \operatorname{diag}\underbrace{(1, -1, \dots, -1)}_{D}, \quad k, n = 0, \dots, D-1\,. \tag{6.62}$$

In odd dimensions there exists also a duality relation which is important for our purposes:

$$\Gamma^n_{(s)} = \frac{sr}{(2d)!}\epsilon^{nk_1\dots k_{2d}}\Gamma_{(s)k_1}\cdots\Gamma_{(s)k_{2d}}, \quad r = \begin{cases} 1, & \text{if } d \text{ is even} \\ i, & \text{if } d \text{ is odd} \end{cases}. \tag{6.63}$$

Here $\epsilon^{nk\dots l}$ is the Levi-Civita tensor density in D dimensions.

The propagator $S^c(x, y)$ obeys the Dirac equation in the dimensions under consideration:

$$\left(\hat{\mathcal{P}}_n\Gamma^n_{(s)} - m\right)S^c(x, y) = -\delta^D(x - y)\,, \tag{6.64}$$

where $\hat{\mathcal{P}}_n = i\partial_n - gA_n(x)$. Thus, we get for the operator $\hat{S}^c$ entering in (6.52), $\hat{S}^c = -\hat{F}^{-1}$, $\hat{F} = \Pi_n\Gamma^n_{(s)} - m$, where all the Π_n are defined by equations (6.37). In the case under consideration it is convenient to present the inverse operator in the following form:

$$\hat{S}^c = \frac{\hat{F}_{(+)}}{-\hat{F}_{(+)}\hat{F}} = s\frac{\hat{A}}{\hat{B}}, \quad \hat{F}_{(+)} = \Pi_n\Gamma^n_{(s)} + m,$$
$$\hat{A} = \frac{r}{(2d)!}\epsilon^{nk_1\dots k_{2d}}\Pi_n\Gamma_{(s)k_1}\cdots\Gamma_{(s)k_{2d}} + sm\,,$$
$$\hat{B} = m^2 - \Pi^2 + \frac{ig}{2}F_{kn}\Gamma^k_{(s)}\Gamma^n_{(s)}\,. \tag{6.65}$$

The form of the operator $\hat{A}$ in (6.65) was obtained from the operator $\hat{F}_{(+)}$ by means of the duality relation (6.63). Now both operators $\hat{A}$ and $\hat{B}$ are even in γ-matrices, so we can treat them as Bose-type operators. For their ratio we are going to use a new kind of integral representation which is a combination of the Schwinger type representation (6.38) for $\hat{B}^{-1}$ and of an additional representation of the operator $\hat{A}$ by means of a

Gaussian integral over two Grassmann variables χ_1 and χ_2 with the involution property $(\chi_1)^+ = \chi_2$. Namely, one can write

$$\hat{S}^c = s\int_0^\infty d\lambda \int e^{-i[\lambda\hat{B}+\chi\hat{A}]}d\chi, \quad \chi = \chi_1\chi_2, \quad d\chi = d\chi_1\, d\chi_2\,. \tag{6.66}$$

Thus, we get the following expression for the propagator:

$$S^c = s\int_0^\infty d\lambda \int \langle x_{\text{out}}|e^{-i\hat{\mathcal{H}}(\lambda,\chi)}|x_{\text{in}}\rangle d\chi, \hat{\mathcal{H}}(\lambda,\chi) = \lambda\Big(m^2 - \Pi^2 + \frac{ig}{2}F_{kn}\Gamma^k_{(s)}\Gamma^n_{(s)}\Big) + \chi\Big(\frac{r}{(2d)!}\epsilon^{nk_1\ldots k_{2d}}\Pi_n\Gamma_{(s)k_1}\ldots\Gamma_{(s)k_{2d}} + sm\Big)\,. \tag{6.67}$$

Starting with this point one can proceed similarly to the even-dimensional case to construct a path integral for the right hand side of equation (6.67). The Hamiltonian form of such path integral reads:

$$S^c = s\,\exp\Big(i\Gamma^n_{(s)}\frac{\partial_l}{\partial\theta^n}\Big)\int_0^\infty d\lambda_0\int d\chi_0\int_{\lambda_0} D\lambda\int_{\chi_0} D\chi\int_{x_{\text{in}}}^{x_{\text{out}}} Dx\int Dp\int D\pi\int D\nu$$
$$\times\int_{\psi(0)+\psi(1)=\theta}\mathcal{D}\psi\,\exp\Bigg\{i\int_0^1\Big[\lambda\big(\mathcal{P}^2 - m^2 + 2igF_{kn}\psi^k\psi^n\big) - \chi\Big(r\frac{(2i)^{2d}}{(2d)!}\epsilon^{nk_1\ldots k_{2d}}\mathcal{P}_n\psi_{k_1}\ldots\psi_{k_{2d}} + sm\Big) - i\psi_n\dot{\psi}^n + p\dot{x} + \pi\dot{\lambda} + \nu\dot{\chi}\Big]d\tau + S\psi_n(1)\psi^n(0)\Bigg\}\Bigg|_{\theta=0}\,, \tag{6.68}$$

where $x(\tau), p(\tau), \lambda(\tau), \pi(\tau)$ are even, and $\psi(\tau), \chi_1(\tau), \chi_2(\tau), \nu_1(\tau), \nu_2(\tau)$ are Grassmann odd trajectories, obeying the boundary conditions

$$x(0) = x_{\text{in}}, \quad x(1) = x_{\text{out}}, \quad \lambda(0) = \lambda_0, \quad \chi(0) = \chi_0, \quad \psi(0) + \psi(1) = \theta\,, \tag{6.69}$$

and the following notations are used:

$$\chi = \chi_1\chi_2, \quad \nu\dot{\chi} = \nu_1\dot{\chi}_1 + \nu_2\dot{\chi}_2, \quad d\chi = d\chi_1\, d\chi_2, \quad D\chi = D\chi_1\, D\chi_2, \quad D\nu = D\nu_1\, D\nu_2\,. \tag{6.70}$$

Integrating over momenta, we get a path integral in the Lagrangian form,

$$
\begin{aligned}
S^c = \frac{s}{2}\exp\left(i\Gamma^n_{(s)}\frac{\partial_\ell}{\partial\theta^n}\right)\int_0^\infty de_0 \int d\chi_0 \int_{e_0} M(e)De \int_{\chi_0} D\chi \int_{x_{\rm in}}^{x_{\rm out}} Dx \int D\pi \int D\nu \\
\times \int_{\psi(0)+\psi(1)=\theta} \mathcal{D}\psi \exp\left\{ i\int_0^1 \left[-\frac{\dot{x}^2}{2e} - \frac{e}{2}m^2 - g\dot{x}_nA^n + iegF_{kn}\psi^k\psi^n \right.\right. \\
\left. - \chi\left(sm + \frac{r}{e}\frac{(2i)^{2d}}{(2d)!}\epsilon^{nk_1\ldots k_{2d}}\dot{x}_n\psi_{k_1}\ldots\psi_{k_{2d}}\right) - i\psi_n\dot{\psi}^n + \pi\dot{e} + \nu\dot{\chi}\right] d\tau \\
\left.\left. + \psi_n(1)\psi^n(0)\right\}\right|_{\theta=0}, \qquad (6.71)
\end{aligned}
$$

where the measure $M(e)$ is defined by equation (6.46) and $e(0) = e_0$.

One can also get a different form of the path integral for the Dirac propagator in odd dimensions. To this end, instead of (6.65), one has to write

$$
\begin{aligned}
\hat{S}^c &= \frac{\hat{F}}{-\hat{F}\hat{F}} = s\frac{\hat{A}}{\hat{B}}, \quad \hat{A} = \frac{r}{(2d)!}\epsilon^{nk_1\ldots k_{2d}}\Pi_n\Gamma_{(s)k_1}\ldots\Gamma_{(s)k_{2d}} - sm, \\
\hat{B} &= m^2 - \Pi^2 + \frac{ig}{2}F_{kn}\Gamma^k_{(s)}\Gamma^n_{(s)} - \frac{2srm}{(2d)!}\epsilon^{nk_1\ldots k_{2d}}\Pi_n\Gamma_{(s)k_1}\ldots\Gamma_{(s)k_{2d}}, \qquad (6.72)
\end{aligned}
$$

and then proceed as before. Thus, we get another form of the Lagrangian path integral

$$
\begin{aligned}
S^c = \frac{s}{2}\exp\left(i\Gamma^n_{(s)}\frac{\partial_\ell}{\partial\theta^n}\right)\int_0^\infty de_0 \int d\chi_0 \int_{e_0} M(e)De \int_{\chi_0} D\chi \int_{x_{\rm in}}^{x_{\rm out}} Dx \int D\pi \int D\nu \\
\times \int_{\psi(0)+\psi(1)=\theta} \mathcal{D}\psi \exp\left\{ i\int_0^1 \left[-\frac{\dot{x}^2}{2e} - \frac{e}{2}m^2 - g\dot{x}_nA^n(x) + iegF_{kn}(x)\psi^k\psi^n + sm\chi \right.\right. \\
\left.\left.\left. -\left(\frac{\chi}{e} - sm\right)\frac{r(2i)^{2d}}{(2d)!}\epsilon^{nk_1\ldots k_{2d}}\dot{x}_n\psi_{k_1}\ldots\psi_{k_{2d}} - i\psi_n\dot{\psi}^n + \pi\dot{e} + \nu\dot{\chi}\right] d\tau + \psi_n(1)\psi^n(0)\right\}\right|_{\theta=0}. \qquad (6.73)
\end{aligned}
$$

6.3.4 Classical and pseudoclassical description of relativistic particles

The path integral representations for particle propagators have also an important heuristic value. They give a possibility to guess the form of actions to describe the particles classically or pseudoclassically if we believe that such representations should have the form $\int \exp(iS)\, D\varphi$. Here φ is a set of variables and S a classical action. Indeed, let us take the simplest example of the scalar particle. Here the path integral representation of the propagator has the form (6.45). The exponent in the integrand of the path integral can be treated as a Lagrangian action of the relativistic spinless particle. This exponent consists of two parts. The first one

$$S = -\int_0^1 \left[-\frac{\dot{x}^2}{2e} - e\frac{m^2}{2} - g\hat{x}_\mu A^\mu \right] d\tau \tag{6.74}$$

is a well-known gauge-invariant (reparametrization-invariant) action of the relativistic spinless particle. The second term in the exponent can be treated as a gauge-fixing term which corresponds to the gauge condition $\dot{e} = 0$. Quantization of the action (6.74) leads to the corresponding quantum theory of a scalar particle [162] which is equivalent to one-particle sector of the scalar quantum field theory. Thus, we have a closed circle: propagator (which is a representative of the one-particle sector of the scalar quantum field theory) – path integral for it – classical action of a point-like particle – quantization – one-particle sector of the scalar quantum field theory.

Let us now turn to the case of spinning particles. Namely, looking on the path integrals (6.61) or (6.73) we can guess the form of the actions for spinning particles in even and odd dimensions. Hence, the exponent in the integrand of the right side of (6.61) can be treated as a pseudoclassical action of the spinning particle in even dimensions. Separating the gauge-fixing terms and boundary terms, we get a gauge-invariant pseudoclassical action

$$\begin{aligned} S &= \int_0^1 \left[-\frac{z^2}{2e} - \frac{e}{2}m^2 - g\dot{x}_\mu A^\mu + iegF_{\mu\nu}\psi^\mu\psi^\nu - im\psi^D\chi - i\psi_n\dot{\psi}^n \right] d\tau\,, \\ z^\mu &= \dot{x}^\mu - i\psi^\mu\chi\,. \end{aligned} \tag{6.75}$$

The action (6.75) is a trivial generalization of the *Berezin–Marinov pseudoclassical action* (see [88]) to D dimensions. The quantization of the action (6.75) reproduces the quantum theory of a spinning particle (in particular, the Dirac equation) in D dimensions.

In odd dimensions, the path integral (6.73) prompts a pseudoclassical action to describe spinning particles in such dimensions, see [172].

6.4 Calculations of Dirac propagators using path integrals

6.4.1 Spin factor in 3 + 1 dimensions

Let us consider the path integral representation (6.61) for the Dirac propagator S^c in $3+1$ dimensions. Here we have $S^c = S^c(x_{\text{out}}, x_{\text{in}}) = -\tilde{S}^c\gamma^5$, where

$$\tilde{S}^c = \exp\left\{i\tilde{\gamma}^n \frac{\partial_\ell}{\partial\theta^n}\right\} \int_0^\infty de_0 \int d\chi_0 \int_{e_0} M(e)De \int_{\chi_0} D\chi \int_{x_{\text{in}}}^{x_{\text{out}}} Dx \int D\pi \int D\nu$$

$$\times \int_{\psi(0)+\psi(1)=\theta} \mathcal{D}\psi \exp\left\{ i \int_0^1 \left[-\frac{\dot{x}^2}{2e} - \frac{e}{2}m^2 - g\dot{x}^\mu A_\mu + iegF_{\mu\nu}\psi^\mu\psi^\nu \right.\right.$$

$$\left.\left. + i\left(\frac{\dot{x}_\mu \psi^\mu}{e} - m\psi^5\right)\chi - i\psi_n\dot{\psi}^n + \pi\dot{e} + \nu\dot{\chi}\right] d\tau + \psi_n(1)\psi^n(0)\right\}\Bigg|_{\theta=0}, \quad (6.76)$$

and

$$\tilde{\gamma}^\mu = \gamma^5\gamma^\mu, \quad \tilde{\gamma}^5 = \tilde{\gamma}^0\tilde{\gamma}^1\tilde{\gamma}^2\tilde{\gamma}^3 = \gamma^0\gamma^1\gamma^2\gamma^3 = \gamma^5,$$

$$[\gamma^m, \gamma^n]_+ = 2\eta^{mn}, \quad \eta^{mn} = \text{diag}(1, -1, -1, -1, -1), \quad m, n = 0, 1, 2, 3, 5.$$

We recall that θ^n are auxiliary Grassmann odd variables, anticommuting by definition with the γ-matrices; $x^\mu(\tau), e(\tau)$ and $\pi(\tau)$ are even trajectories of integration; $\psi^n(\tau)$, $\chi(\tau)$, and $\nu(\tau)$ are Grassmann odd trajectories of integration, satisfying boundary conditions (6.58); the measures $M(e)$ and $\mathcal{D}\psi$ are defined by equations (6.46) and (6.59), respectively.

Let us demonstrate that the propagator (6.76) can be expressed through a Bosonic path integral over the coordinates x alone, see Ref. [181]. For this purpose one needs to perform several functional integrations, in particular, to fulfil all the Grassmann integrations. First, one can integrate over π and ν, and then, using the arising δ-functions, remove the functional integration over e and χ. Thus, we obtain

$$\tilde{S}^c = -\exp\left\{i\tilde{\gamma}^n \frac{\partial_\ell}{\partial\theta^n}\right\} \int_0^\infty de_0 M(e_0) \int_{x_{\text{in}}}^{x_{\text{out}}} Dx \int_{\psi(0)+\psi(1)=\theta} \mathcal{D}\psi \int_0^1 \left(\frac{\dot{x}_\mu\psi^\mu}{e_0} - m\psi^5\right) d\tau$$

$$\times \exp\left\{ i\int_0^1 \left[-\frac{\dot{x}^2}{2e_0} - \frac{e_0}{2}m^2 - g\dot{x}^\mu A_\mu + ige_0F_{\mu\nu}\psi^\mu\psi^\nu - i\psi_n\dot{\psi}^n \right] d\tau + \psi_n(1)\psi^n(0) \right\}\Bigg|_{\theta=0}.$$

Then, changing the integration variables ψ^n to ξ^n as

$$\psi^n = \frac{1}{2}(\xi^n + \theta^n), \quad (6.77)$$

and introducing Grassmann odd sources $\rho_n(\tau)$ for the new variables ξ^n, we get

$$\tilde{S}^c = -\frac{1}{2}\exp\left\{i\tilde{\gamma}^n \frac{\partial_\ell}{\partial\theta^n}\right\} \int_0^\infty de_0 M(e_0) \int_{x_{\text{in}}}^{x_{\text{out}}} Dx \exp\left\{-i\left[\frac{\dot{x}_\mu \star \dot{x}^\mu}{2e_0} + \frac{e_0}{2}m^2 + g\dot{x}^\mu \star A_\mu\right]\right.$$

$$\left. - \frac{ge_0}{4}\theta^\mu \star \mathcal{F}_{\mu\nu} \star \theta^\nu\right\} \left[\frac{\dot{x}_\mu}{e_0} \star \left(\frac{\delta_\ell}{\delta\rho_\mu} + \theta^\mu\right) - m \star \left(\frac{\delta_\ell}{\delta\rho_5} + \theta^5\right)\right] R[x, \rho, \theta]\Bigg|_{\rho=0,\theta=0}, \quad (6.78)$$

where

$$R[x,\rho,\theta] = \int\limits_{\xi(0)+\xi(1)=0} \mathcal{D}\xi \, \exp\left\{\frac{1}{4}\xi_n \star \dot{\xi}^n \right.$$
$$\left. - \frac{ge_0}{4}\xi^\mu \star \mathcal{F}_{\mu\nu} \star \xi^\nu - \frac{ge_0}{2}\theta^\mu \star \mathcal{F}_{\mu\nu} \star \xi^\nu + \rho_n \star \xi^n \right\} ,$$
$$\mathcal{D}\xi = D\xi \left[\int\limits_{\xi(0)+\xi(1)=0} D\xi \, \exp\left\{\frac{1}{4}\xi_n \star \dot{\xi}^n\right\}\right]^{-1} . \tag{6.79}$$

Here condensed notations are used in which $\mathcal{F}_{\mu\nu}$ is understood as a matrix with continuous indices,

$$\mathcal{F}_{\mu\nu}(\tau,\tau') = F_{\mu\nu}(x(\tau))\delta(\tau-\tau') , \tag{6.80}$$

and integration over τ is denoted by star, for example

$$\xi_n \star \dot{\xi}^n = \int_0^1 \xi_n(\tau)\dot{\xi}^n(\tau)d\tau .$$

Sometimes discrete indices will also be omitted. In this case all tensors of second rank have to be understood as matrices with lines marked by the first contravariant indices of the tensors, and with columns marked by the second covariant indices of the tensors.

The Grassmann Gaussian path integral for the quantity $R[x,\rho,\theta]$ in (6.79) can be evaluated straightforwardly [89] to be

$$R[x,\rho,\theta] = \left\{\mathrm{Det}\left[\mathcal{U}^{-1}(0)\mathcal{U}(g)\right]\right\}^{1/2} \exp\left\{J^m \star \mathcal{W}_{mn} \star J^n\right\} , \tag{6.81}$$

where the matrices $\mathcal{W}(g)$ and $\mathcal{U}(g)$ have the form

$$\mathcal{W}_{mn}(g) = \begin{pmatrix} \mathcal{U}_{\mu\nu}(g) & 0 \\ 0 - \delta'(\tau-\tau') & \end{pmatrix} ,$$
$$\mathcal{U}_{\mu\nu}(g) = \eta_{\mu\nu}\delta'(\tau-\tau') - ge_0\mathcal{F}_{\mu\nu}(\tau,\tau') , \tag{6.82}$$

and

$$J_\mu = \rho_\mu + \frac{ge_0}{2}\mathcal{F}_{\mu\nu} \star \theta^\nu , J_5 = \rho_5 .$$

The determinant in (6.81) should be understood as

$$\mathrm{Det}\left[\mathcal{U}^{-1}(0)\mathcal{U}(g)\right] = \exp \mathrm{Tr}\left[\log \mathcal{U}(g) - \log \mathcal{U}(0)\right]$$
$$= \exp\left\{-e_0 \mathrm{Tr} \int_0^g dg' \mathcal{R}(g') \star \mathcal{F}\right\} , \tag{6.83}$$

where the operator matrix $\mathcal{R}(g)$ is inverse to $\mathcal{U}(g)$, and, at the same time, is to be considered as an operator acting in the space of antiperiodic functions,

$$\frac{d}{d\tau}\mathcal{R}_{\mu\nu}(g|\tau,\tau') - ge_0F_\mu^\lambda(x(\tau))\mathcal{R}_{\lambda\nu}(g|\tau,\tau') = \eta_{\mu\nu}\delta(\tau-\tau'),$$
$$\mathcal{R}_{\mu\nu}(g|1,\tau) = -\mathcal{R}_{\mu\nu}(g|0,\tau), \quad \forall\tau \in (0,1). \tag{6.84}$$

Substituting (6.81) and (6.83) into (6.78) and performing then the functional differentiations with respect to ρ_n, we obtain

$$\tilde{S}^c = -\frac{1}{2}\exp\left\{i\tilde{\gamma}^n\frac{\partial_\ell}{\partial\theta^n}\right\}\int_0^\infty de_0M(e_0)\int_{x_{in}}^{x_{out}}Dx\exp\left\{-\frac{i}{2}\left[\frac{\dot{x}^\mu \star \dot{x}^\mu}{e_0} + e_0m^2 + g\dot{x}^\mu A_\mu\right]\right\}$$
$$\times\left[\frac{\dot{x}^\mu}{e_0} \star K_{\mu\nu}\theta^\nu - m\theta^5\right]\left[1 - \frac{ge_0}{4}B_{\alpha\beta}\theta^\alpha\theta^\beta + \frac{g^2e_0^2}{16}B_{\alpha\beta}B^{*\alpha\beta}\theta^0\theta^1\theta^2\theta^3\right]$$
$$\times\exp\left\{-\frac{e_0}{2}\int_0^\infty dg'\mathrm{Tr}\mathcal{R}(g') \star \mathcal{F}\right\}\Bigg|_{\theta=0}, \tag{6.85}$$

where the following notations are used,

$$B_{\mu\nu} = F_{\mu\lambda} \star K^\lambda_{\ \nu}, B^{*\mu\nu} = \frac{1}{2}\epsilon^{\alpha\beta\mu\nu}B_{\alpha\beta}, K_{\mu\nu} = \eta_{\mu\nu} + ge_0\mathcal{R}_{\mu\lambda}(g) \star F^\lambda_{\ \nu}, \tag{6.86}$$

and $\epsilon^{\mu\nu\alpha\beta}$ is Levi-Civita symbol normalized as $\epsilon^{0123} = 1$.

Differentiation with respect to θ^n in (6.85) replaces the products of the variables θ^n by the corresponding antisymmetrized products of the matrices $i\tilde{\gamma}^n$. Finally, passing to the propagator S^c and using the relations

$$\gamma^{[\lambda}\gamma^\mu\gamma^{\nu]} = \gamma^\lambda\gamma^{[\mu}\gamma^{\nu]} - 2\eta^{\lambda[\mu}\gamma^{\nu]}, \sigma^{\mu\nu} = i\gamma^{[\mu}\gamma^{\nu]} = \frac{i}{2}\left[\gamma^\mu,\gamma^\nu\right],$$

where antisymmetrization over the corresponding sets of indices is denoted by square brackets, one gets

$$S^c(x_{out},x_{in}) = \frac{i}{2}\int_0^\infty de_0\int_{x_{in}}^{x_{out}}Dx\,M(e_0)\Phi[x,e_0]\,\exp\left\{iI[x,e_0]\right\}, \tag{6.87}$$

where $I[x,e_0]$ is the action of a relativistic spinless particle,

$$I[x,e_0] = -\int_0^1\left[\frac{\dot{x}^2}{2e_0} + \frac{e_0}{2}m^2 + g\dot{x}A(x)\right]d\tau, \tag{6.88}$$

and $\Phi[x, e_0]$ is the so-called spin factor,

$$\Phi[x, e_0] = \Big[m + (2e_0)^{-1}\dot{x}^{\mu} \star K_{\mu\lambda}\left(2\eta^{\lambda\kappa} - ge_0 B^{\lambda\kappa}\right)\gamma_{\kappa} - \frac{ig}{4}\left(me_0 + \dot{x}^{\mu} \star K_{\mu\lambda}\gamma^{\lambda}\right)$$
$$\times B_{\kappa\nu}\sigma^{\kappa\nu} + m\frac{g^2 e_0^2}{16} B^{*}_{\alpha\beta} * B^{\alpha\beta}\gamma^5 \Big] \exp\left\{-\frac{e_0}{2}\int_0^g dg' \mathrm{Tr}\mathcal{R}(g') \star \mathcal{F}\right\} . \tag{6.89}$$

In even-dimensional space-time, there exists an alternative to the Berezin–Marinov model, see [344]. The alternative pseudoclassical action is similar to the one proposed in [172] for the odd-dimensional case. The quantization of the alternative model yields an alternative expression for the spin factor. It is shown that this expression is equivalent to the one (6.89).

6.4.2 Propagator in the constant uniform electromagnetic field

Here, we calculate the spinning particle propagator in the constant uniform field ($F_{\mu\nu}$ = const), following Ref. [182, 183]. In this case, the functionals $\mathcal{R}$, K and B do not depend on the trajectory x and can be calculated straightforwardly,

$$\mathcal{R}(g) = \frac{1}{2}\left(\eta\varepsilon(\tau - \tau') - \tanh\frac{ge_0 F}{2}\right)\exp\{e_0 gF(\tau - \tau')\} ,$$
$$K = \left(\eta - \tanh\frac{ge_0 F}{2}\right)\exp\left(ge_0 F\tau\right), \quad B = \frac{2}{ge_0}\tanh\frac{ge_0 F}{2} . \tag{6.90}$$

Using (6.89) in (6.89) and integrating over τ whenever possible, we obtain the spin factor in the constant uniform field,

$$\Phi[x, e_0] = \left(\det\cosh\frac{ge_0 F}{2}\right)^{1/2}\Bigg\{ m\Bigg[1 - \frac{i}{2}\left(\tanh\frac{ge_0 F}{2}\right)_{\mu\nu}\sigma^{\mu\nu}$$
$$+ \frac{1}{4}\left(\tanh\frac{ge_0 F}{2}\right)^{*}_{\mu\nu}\left(\tanh\frac{ge_0 F}{2}\right)^{\mu\nu}\gamma^5\Bigg] + \frac{1}{e_0}\left(\int_0^1 \dot{x}\exp(ge_0 F\tau)d\tau\right)$$
$$\times\left(\eta - \tanh\frac{ge_0 F}{2}\right)\left[\left(\eta - \tanh\frac{ge_0 F}{2}\right)\gamma - \frac{i}{2}\gamma\left(\tanh\frac{ge_0 F}{2}\right)_{\mu\nu}\sigma^{\mu\nu}\right]\Bigg\} . \tag{6.91}$$

We can see that in the field under consideration the spin factor is linear in the trajectory $x^{\mu}(\tau)$. This facilitates further integrations in expression (6.87).

In spite of the fact that the spin factor is a gauge invariant object, the total propagator is not. It is clear from expression (6.87) where one has to choose a particular gauge for the potentials A_μ. Using the potentials

$$A_\mu = -\frac{1}{2}F_{\mu\nu}x^\nu , \tag{6.92}$$

for the electromagnetic field under consideration, we transform the path integral (6.87) into the quasi-Gaussian one. After that we perform the shift $x \to y + x_{cl}$, with x_{cl} being a solution of the classical equations of motion

$$\frac{\delta I}{\delta x} = 0 \Leftrightarrow \ddot{x}_\mu - ge_0F_{\mu\nu}\dot{x}^\nu = 0 , \tag{6.93}$$

subjected to the boundary conditions $x_{cl}(0) = x_{\text{in}}$ and $x_{cl}(1) = x_{\text{out}}$. Then the new trajectories of integration y obey zero boundary conditions, $y(0) = y(1) = 0$. Due to the quadratic structure of the action $I[x, e_0]$ and the linearity of the spin factor in x one can make the following substitutions in the path integral:

$$\begin{aligned} &I[y + x_{cl}, e_0] \to I[x_{cl}, e_0] + I[y, e_0] + \frac{e_0}{2}m^2, \\ &\Phi[y + x_{cl}, e_0] \to \Phi[x_{cl}, e_0] = \Psi(x_{\text{out}}, x_{\text{in}}, e_0) . \end{aligned} \tag{6.94}$$

By also doing a convenient replacement: $p \to p/\sqrt{e_0}$, $y \to y\sqrt{e_0}$, we obtain

$$\begin{aligned} S^c = {} & \frac{i}{2}\int_0^\infty \frac{de_0}{e_0^2}\Psi(x_{\text{out}}, x_{\text{in}}, e_0)\exp iI[x_{cl}, e_0] \\ & \times \int_0^\infty Dy \int Dp \exp\left\{\frac{i}{2}\int_0^1 \left(p^2 - \dot{y}^2 - ge_0yF\dot{y}\right)d\tau\right\} . \end{aligned} \tag{6.95}$$

One can see that the path integral in representation (6.95) is, in fact, the kernel of the corresponding K–G propagator in the proper-time representation. This path integral can be represented as

$$\begin{aligned} &\int_0^\infty Dy \int Dp \exp\left\{\frac{i}{2}\int_0^1 \left(p^2 - \dot{y}^2 - ge_0yF\dot{y}\right)d\tau\right\} \\ &= \left[\frac{\operatorname{Det}(\eta_{\mu\nu}\partial_\tau^2 - ge_0F_{\mu\nu}\partial_\tau)}{\operatorname{Det}(\eta_{\mu\nu}\partial_\tau^2)}\right]^{-1/2} \int_0^\infty Dy \int Dp \exp\left\{\frac{i}{2}\int_0^1 \left(p^2 - \dot{y}^2\right)d\tau\right\} . \end{aligned} \tag{6.96}$$

Each determinant in the right hand side of equation (6.95) can be divided by the factor $\operatorname{Det}(-\eta_{\mu\nu})$, so that

$$\frac{\operatorname{Det}(\eta_{\mu\nu}\partial_\tau^2 - ge_0F_{\mu\nu}\partial_\tau)}{\operatorname{Det}(\eta_{\mu\nu}\partial_\tau^2)} = \frac{\operatorname{Det}(-\delta^\mu_\nu\partial_\tau^2 + ge_0F^\mu_\nu\partial_\tau)}{\operatorname{Det}(-\delta^\mu_\nu\partial_\tau^2)} . \tag{6.97}$$

We can also make the replacement

$$-\mathbb{I}\partial_\tau^2 + ge_0F\partial_\tau \to -\mathbb{I}\partial_\tau^2 + \frac{g^2e_0^2}{4}F^2\,, \tag{6.98}$$

in the right hand side of equation (6.97), because spectra of both operators coincide. Indeed,

$$-\mathbb{I}\partial_\tau^2 + ge_0F\partial_\tau = \exp\left(\frac{ge_0}{2}F\tau\right)\left(-\mathbb{I}\partial_\tau^2 + \frac{g^2e_0^2}{4}F^2\right)\exp\left(-\frac{ge_0}{2}F\tau\right), \tag{6.99}$$

and the transformation $y \to \exp(ge_0F\tau/2)y$ does not change the boundary conditions. Then, using equation (6.98) and the value of the free path integral (calculated in Ref. [147]),

$$\frac{i}{2}\int_0^0 Dy\int Dp\exp\left\{\frac{i}{2}\int d\tau(p^2 - \dot{y}^2)\right\} = \frac{1}{8\pi^2}\,,$$

related, in fact, to the definition of the measure, we obtain

$$S^c = \frac{\Theta^{-1/2}}{8\pi^2}\int_0^\infty \frac{de_0}{e_0^2}\Psi(x_{\text{out}}, x_{\text{in}}, e_0)\,\exp iI[x_{cl}, e_0]\,, \tag{6.100}$$

where

$$\Theta = \frac{\operatorname{Det}(-\mathbb{I}\partial_\tau^2 + \frac{g^2e_0^2}{4}F^2)}{\operatorname{Det}(-\mathbb{I}\partial_\tau^2)}\,. \tag{6.101}$$

The ratio of the determinants (6.101) can be written as

$$\begin{aligned}\Theta &= \exp\operatorname{Tr}\left[\ln\left(-\mathbb{I}\partial_\tau^2 + \frac{g^2e_0^2}{4}F^2\right) - \ln\left(-\mathbb{I}\partial_\tau^2\right)\right]\\ &= \exp\operatorname{Tr}\left[\frac{e_0^2}{2}F^2\int_0^g d\lambda\,\lambda\left(-\mathbb{I}\partial_\tau^2 + \frac{\lambda^2e_0^2}{4}F^2\right)^{-1}\right].\end{aligned} \tag{6.102}$$

The trace in the infinite-dimensional space in the right hand side of equation (6.102) can be calculated, so that only the trace in the four-dimensional space remains,

$$\Theta = \exp\operatorname{tr}\left[\frac{e_0^2}{2}F^2\int_0^g d\lambda\,\lambda\sum_{n=1}^\infty\left(\pi^2n^2\mathbb{I} + \frac{\lambda^2e_0^2}{4}F^2\right)^{-1}\right]. \tag{6.103}$$

For any 4×4 matrix κ the formula

$$\sum_{n=1}^\infty\left(\pi^2n^2 + \kappa^2\right)^{-1} = \frac{1}{2\kappa}\coth\kappa - \frac{1}{2\kappa^2}\,,$$

holds true. Using it and integrating in (6.103), we find

$$\Theta = \det\left(\frac{\sinh\frac{ge_0F}{2}}{\frac{ge_0F}{2}}\right).$$

Thus,

$$S^c = \frac{1}{32\pi^2}\int_0^\infty de_0\left(\det\frac{\sinh\frac{ge_0F}{2}}{gF}\right)^{-1/2}\Psi(x_{\text{out}},x_{\text{in}},e_0)\exp iI[x_{cl},e_0]\,, \tag{6.104}$$

where the matrix function $\Psi(x_{\text{out}},x_{\text{in}},e_0)$ is the spin factor on the classical trajectory x_{cl}, see equation (6.94).

To find this function, we solve equation (6.93) for the classical trajectory in the electromagnetic field under consideration:

$$x_{cl} = \left[\exp(ge_0F)-\eta\right]^{-1}\left[\exp(ge_0F\tau)(x_{\text{out}}-x_{\text{in}})+\exp(ge_0F)x_{\text{in}}-x_{\text{out}}\right]. \tag{6.105}$$

Substituting (6.105) into equations (6.94) and (6.104), we obtain

$$\begin{aligned} S^c = {} & \frac{1}{32\pi^2}\int_0^\infty de_0\left(\det\frac{\sinh\frac{ge_0F}{2}}{gF}\right)^{-1/2}\Psi(x_{\text{out}},x_{\text{in}},e_0) \\ & \times\exp\left\{\frac{ig}{2}x_{\text{out}}Fx_{\text{in}}-\frac{i}{2}e_0m^2-\frac{ig}{4}(x_{\text{out}}-x_{\text{in}})F\coth\left(\frac{ge_0F}{2}\right)(x_{\text{out}}-x_{\text{in}})\right\}, \end{aligned} \tag{6.106}$$

where

$$\begin{aligned} \Psi(x_{\text{out}},x_{\text{in}},e_0) = {} & \left[m+\frac{g}{2}(x_{\text{out}}-x_{\text{in}})F\left(\coth\frac{ge_0F}{2}-1\right)\gamma\right] \\ & \times\sqrt{\det\cosh\frac{ge_0F}{2}}\left[1-\frac{i}{2}\left(\tanh\frac{ge_0F}{2}\right)_{\mu\nu}\sigma^{\mu\nu}\right. \\ & \left.+\frac{1}{8}\epsilon^{\alpha\beta\mu\nu}\left(\tanh\frac{ge_0F}{2}\right)_{\alpha\beta}\left(\tanh\frac{ge_0F}{2}\right)_{\mu\nu}\gamma^5\right]. \end{aligned} \tag{6.107}$$

Now we are going to compare representation (6.106) with the result which Schwinger derived in the same case by means of the proper-time method, see Ref. [294]. The Schwinger representation has the form

$$\begin{aligned} S^c(x_{\text{out}},x_{\text{in}}) = {} & \frac{1}{32\pi^2}\left[\gamma^\mu\left(i\frac{\partial}{\partial x^\mu_{\text{out}}}-gA_\mu(x_{\text{out}})\right)+m\right]\int_0^\infty de_0\left(\det\frac{\sinh\frac{ge_0F}{2}}{gF}\right)^{-1/2} \\ & \times\exp\left\{\frac{i}{2}\left[gx_{\text{out}}Fx_{\text{in}}-e_0m^2-(x_{\text{out}}-x_{\text{in}})\frac{gF}{2}\coth\frac{ge_0F}{2}(x_{\text{out}}-x_{\text{in}})-\frac{ge_0}{2}F_{\mu\nu}\sigma^{\mu\nu}\right]\right\}. \end{aligned} \tag{6.108}$$

By doing the differentiation with respect to x^μ_{out}, we transform formula (6.108) to a form which is convenient for the comparison with our representation (6.104),

$$S^c = \frac{1}{32\pi^2}\int_0^\infty de_0 \left(\det \frac{\sinh \frac{ge_0F}{2}}{gF}\right)^{-1/2} \Psi_S(x_{\text{out}}, x_{\text{in}}, e_0)$$
$$\times \exp\left\{i\frac{g}{2}x_{\text{out}}Fx_{\text{in}} - \frac{i}{2}e_0m^2 - i\frac{g}{4}(x_{\text{out}} - x_{\text{in}})F\coth\left(\frac{ge_0F}{2}\right)(x_{\text{out}} - x_{\text{in}})\right\}, \tag{6.109}$$

where the function Ψ_S is given by

$$\Psi_S(x_{\text{out}}, x_{\text{in}}, e_0) = \left[m + \frac{g}{2}(x_{\text{out}} - x_{\text{in}})F(\coth\frac{ge_0F}{2} - 1)\gamma\right]\exp(-i\frac{ge_0}{4}F_{\mu\nu}\sigma^{\mu\nu}). \tag{6.110}$$

Thus one needs only to compare the functions Ψ and Ψ_S. They coincide, since equation (A.65) from Section A.2.2 holds true.

6.4.3 Propagator in a constant uniform field and a plane wave field

Here, we calculate the Dirac propagator in a superposition of the constant uniform field and a plane wave field, see [182]. Potentials A^{comb}_μ of such a field are chosen to be

$$A^{comb}_\mu = -\frac{1}{2}F_{\mu\nu}x^\nu + a_\mu(nx), \quad F_{\mu\nu} = \text{const}, \tag{6.111}$$

where $a_\mu(\phi)$ is a vector-valued function of a real variable ϕ, and n is a normalized isotropic vector $n^\mu = (1, \mathbf{n})$, $\mathbf{n}^2 = 1$, $n^2 = 0$. They produce the electromagnetic field of the form

$$F^{comb}_{\mu\nu}(nx) = F_{\mu\nu} + f_{\mu\nu}(nx), \tag{6.112}$$

which is a superposition of the constant uniform field $F_{\mu\nu}$ and the plane-wave field

$$f_{\mu\nu}(nx) = n_\mu a'_\nu(nx) - n_\nu a'_\mu(nx),$$

where the prime denotes a differentiation. Without loss of generality we can choose a_μ to be transversal,

$$n^\mu a_\mu(\phi) = 0. \tag{6.113}$$

The dependence of the spin factor $\Phi[x, e_0]$ (6.89) on the trajectory $x^\mu(\tau)$ is twofold. In addition to the direct dependence, there is an indirect one through the external field, if the latter depends on x. In the case under consideration the external field depends on $x^\mu(\tau)$ only through the scalar combination $nx(\tau)$. Replacing the latter by an auxiliary scalar trajectory $\phi(\tau)$ one obtains the new quantity $\tilde{\Phi}[x, \phi, e_0]$,

$$\tilde{\Phi}[x,\phi,e_0] = \left[m + (2e_0)^{-1}\dot{x}^\mu \star \tilde{K}_{\mu\lambda}\left(2\eta^{\lambda\kappa} - ge_0\tilde{B}^{\lambda\kappa}\right)\gamma_\kappa - \frac{ig}{4}\left(me_0 + \dot{x}^\mu \star \tilde{K}_{\mu\lambda}\gamma^\lambda\right)\tilde{B}_{\kappa\nu}\sigma^{\kappa\nu} \right.$$
$$\left. + m\frac{g^2e_0^2}{16}\tilde{B}_{\alpha\beta}\tilde{B}^{*\alpha\beta}\bar{\gamma}^5 \right] \exp\left\{ -\frac{e_0}{2}\int_0^g dg' \mathrm{Tr}\tilde{\mathfrak{R}}(g') \star \mathfrak{F}^{comb}(\phi) \right\}, \quad (6.114)$$

where $\mathfrak{F}^{comb}_{\mu\nu}(\phi|\tau - \tau') = [F_{\mu\nu} + f_{\mu\nu}(\phi(\tau))]\delta(\tau, \tau')$ and

$$\tilde{B}_{\mu\nu} = F^{comb}_{\mu\nu}(\phi) \star \tilde{K}^\lambda_\nu, \tilde{K}_{\mu\nu} = \eta_{\mu\nu} + ge_0\tilde{\mathfrak{R}}_{\mu\nu}(g) \star F^{comb}(\phi),$$
$$\left[\eta\frac{\partial}{\partial\tau} - ge_0F^{comb}(\phi(\tau))\right]\tilde{\mathfrak{R}}(q|\tau,\tau') = \eta\delta(\tau,\tau'),$$
$$\tilde{\mathfrak{R}}(g|1,\tau) = -\tilde{\mathfrak{R}}(g|0,\tau), \quad \forall\tau \in (0,1). \quad (6.115)$$

It is obvious that

$$\tilde{\mathfrak{R}}(g)\Big|_{\phi(\tau)=nx(\tau)} = \mathfrak{R}(g), \quad \tilde{K}\Big|_{\phi(\tau)=nx(\tau)} = K, \tilde{B}\Big|_{\phi(\tau)=nx(\tau)} = B, \quad (6.116)$$

and, therefore,

$$\tilde{\Phi}[x,nx,e_0] = \Phi[x,e_0]. \quad (6.117)$$

Inserting the integral of the δ-function,

$$\int D\phi D\lambda e^{i\lambda\star(\phi-nx)} = 1,$$

into the right hand side of equation (6.87) and using (6.117) one transforms the path integral (6.87) into a quasi-Gaussian one of a simple form

$$S^c(x_{\mathrm{out}},x_{\mathrm{in}}) = \frac{i}{2}\int_0^\infty de_0 \int D\phi\, D\lambda\, e^{i\lambda\star\phi} \int_{x_{\mathrm{in}}}^{x_{\mathrm{out}}} Dx\, M(e_0)\tilde{\Phi}[x,\phi,e_0]$$
$$\times \exp\{i\tilde{I}[x,\phi,e_0] - i\lambda \star (nx)\}. \quad (6.118)$$

The action functional

$$\tilde{I}[x,\phi,e_0] = -\frac{\dot{x}\star\dot{x}}{2e_0} - \frac{e_0}{2}m^2 - \frac{g}{2}x \star \bar{\mathfrak{F}} \star \dot{x} - ga(\phi) \star \dot{x} \quad (6.119)$$

(where $\bar{\mathfrak{F}}(\tau,\tau') = F\delta(\tau-\tau')$) contains only linear and bilinear terms in x (the bilinear part not depending on the wave potential a_μ). The spin factor $\tilde{\Phi}[x,\phi,e_0]$ is linear in x and (following the same way of reasoning as in the case of a constant field) one finds

$$S^c = \frac{1}{32\pi^2}\int_0^\infty de_0 \left(\det\frac{\sinh\frac{ge_0F}{2}}{gF}\right)^{-1/2} \int D\phi\, D\lambda e^{i\lambda\star(\phi-nx_q)}\tilde{\Phi}[x_q,\phi,e_0]e^{i\tilde{I}[x_q,\phi,e_0]}, \quad (6.120)$$

where x_q is the solution to the equation

$$\ddot{x}_q - ge_0F\dot{x}_q = e_0\lambda n - e_0 g a'(\phi)\dot{\phi}\,, \tag{6.121}$$

obeying the boundary conditions $x_q(0) = x_{\text{in}}, x_q(1) = x_{\text{out}}$.

Introducing an appropriate Green's function $\mathcal{G} = \mathcal{G}(\tau, \tau')$ for the second-order operator,

$$\left[\eta\frac{\partial^2}{\partial\tau^2} - ge_0F\frac{\partial}{\partial\tau}\right]\mathcal{G}(\tau,\tau') = \eta\delta(\tau-\tau')\,,$$
$$\mathcal{G}(0,\tau) = \mathcal{G}(1,\tau) = \mathcal{G}(\tau,0) = \mathcal{G}(\tau,1) = 0, \quad \forall\tau \in (0,1)\,, \tag{6.122}$$

we represent this solution in the form

$$x_q = x_{cl} + e_0\mathcal{G} \star \left(\lambda n - g a'(\phi)\dot{\phi}\right)\,. \tag{6.123}$$

The value of the action functional $\tilde{I}[x, \phi, e_0]$ on the solution x_q is given by

$$\begin{aligned}\tilde{I}[x_q, \phi, e_0] &= \bar{I}[x_{cl}, e_0] - ga(\phi) \star \dot{x}_{cl} \\ &\quad - \frac{e_0}{2}\left(ga'(\phi)\dot{\phi} - \lambda n\right) \star \mathcal{G} \star \left(ga'(\phi)\dot{\phi} - \lambda n\right) + \lambda n \star (x_q - x_{cl})\,,\end{aligned} \tag{6.124}$$

where

$$\bar{I}[x, e_0] = -\frac{1}{2e_0}\dot{x} \star \dot{x} - \frac{e_0}{2}m^2 - \frac{g}{2}x \star \bar{\mathcal{F}} \star \dot{x}$$

is the scalar particle action in the uniform constant field F.

The functional integral over λ in (6.118) is a quasi-Gaussian integral of simple form (let us remind that x_q is linear in λ, see equation (6.123)) and the integration can be done explicitly. As a result we obtain an expression for the propagator in which the only functional integration is over the scalar trajectory $\phi(\tau)$. However, the latter integration can hardly be performed explicitly in the general case (for arbitrary $a_\mu(\phi)$). Nevertheless, there exists a specific combination [81] for which the integration can be done and explicit formulae for the propagator can be derived. The latter are comparable with the corresponding Schwinger-type formulae obtained in Refs. [150–152].

Namely, let us choose the wave vector n to coincide with a real eigenvector of the matrix F (see Section 6.4.4),

$$F_{\mu\nu}n^\nu = -\mathcal{E}n_\mu, \quad n^2 = 0, \; \mathbf{n}^2 = 1\,. \tag{6.125}$$

In this case $nx_q = nx_{cl}$, and, moreover, the action functional (6.119) is "on-shell" invariant with respect to the longitudinal shifts

$$x_q(\tau) \to x_q(\tau) + \alpha(\tau)n, \alpha(0) = \alpha(1) = 0\,, \tag{6.126}$$

by virtue of (6.125) and the transversality (6.113) of the wave potential a_μ. Then the quantity $\tilde{I}[x_q, \phi, e_0]$ does not depend on λ and has the form

$$\tilde{I}[x_q, \phi, e_0] = \tilde{I}[x_{tr}, \phi, e_0],\ x_{tr} = x_{cl} - ge_0\mathcal{G} \star \left(a'(\phi)\dot{\phi}\right), \tag{6.127}$$

where x_{tr} is a solution to the equation

$$\ddot{x}_{tr} - ge_0F\dot{x}_{tr} = -ge_0a'(\phi)\dot{\phi}, \tag{6.128}$$

obeying the boundary conditions (6.121). However, the spin factor $\tilde{\Phi}[x_q, \phi, e_0]$ does not show the "on-shell" invariance and, therefore, is λ-dependent. Presenting x_q as a sum $x_q = x_{tr} + e_0\mathcal{G}n \star \lambda$, and substituting it into the expansion of the spin factor in the antisymmetrized products of γ-matrices,

$$\begin{aligned}\tilde{\Phi}[x, \phi, e_0] = &\left[e_0^{-1}\dot{x}^\mu \star \tilde{R}_{\mu\nu}\left(\gamma^\nu + \frac{ge_0}{4}\tilde{B}_{\alpha\beta}\gamma^{[\nu}\gamma^\alpha\gamma^{\beta]}\right)\right.\\ &\left.+m\left(1 + \frac{ge_0}{4}\tilde{B}_{\alpha\beta}\gamma^{[\alpha}\gamma^{\beta]} + \frac{g^2e_0^2}{32}\tilde{B}_{\alpha\beta}\tilde{B}_{\mu\nu}\gamma^{[\alpha}\gamma^\beta\gamma^\mu\gamma^{\nu]}\right)\right]\Lambda[\phi, e_0],\\ \Lambda[\phi, e_0] = &\exp\left\{-\frac{e_0}{2}\int_0^g dg'\mathrm{Tr}\tilde{\mathfrak{R}}(g') \star \mathfrak{F}^{comb}(\phi)\right\},\end{aligned} \tag{6.129}$$

we obtain,

$$\begin{aligned}\tilde{\Phi}[x_q, \phi, e_0] &= \tilde{\Phi}[x_{tr}, \phi, e_0] + \lambda \star l[\phi, e_0],\\ l[\phi, e_0] &= n^\kappa\mathcal{G}^{(r)}_{\kappa\mu} \star \tilde{K}^\mu_\nu\left(\gamma^\nu + \frac{ge_0}{4}\tilde{B}_{\alpha\beta}\gamma^{[\nu}\gamma^\alpha\gamma^{\beta]}\right)\Lambda[\phi, e_0],\\ \mathcal{G}^{(r)}(\tau, \tau') &= \frac{\partial}{\partial\tau'}\mathcal{G}(\tau, \tau').\end{aligned} \tag{6.130}$$

It turns out that $l[\phi, e_0]$ does not depend on ϕ. First, expanding $\tilde{\mathfrak{R}}$ in powers of f and using (6.125) and (6.113) one derives that $\Lambda[\phi, e_0]$ coincides with the expression

$$\bar{\Lambda}(e_0) = \exp\left\{\frac{e_0}{2}\int_0^g dg'\mathrm{Tr}\bar{\mathfrak{R}}(g) \star \bar{\mathfrak{F}}\right\}. \tag{6.131}$$

Second[6],

$$n_\mu\tilde{K}^\mu_\nu = n_\mu\bar{K}^\mu_\nu = \frac{1}{2}\left(n\bar{K}\bar{n}\right)n_\nu. \tag{6.132}$$

---

6 In this section we denote by $\bar{\mathfrak{R}}(g)$, $\bar{K}$, $\bar{B}$ the quantities given by (6.90), i.e. corresponding to the case of a constant uniform field F.

Indeed, using the definitions (6.116) one finds that $\tilde{K}$ satisfies the equation

$$\left[\frac{\partial}{\partial\tau} - ge_0(F + f(\phi(t)))\right]\tilde{K}(\tau) = 0, \quad \tilde{K}(0) + \tilde{K}(1) = 2\eta\,. \tag{6.133}$$

Multiplying (6.133) by n and using the properties (6.125), (6.113), we find

$$\left(\frac{\partial}{\partial\tau} - ge_0\mathcal{E}\right)n\tilde{K} = 0\,, \quad n\tilde{K}(0) + n\tilde{K}(1) = 2n\,. \tag{6.134}$$

At the same time $n\bar{K}$ obeys equation (6.134). Therefore, $n\tilde{K}$ and $n\bar{K}$ coincide. Then using equation (6.113) and the properties of $n, \bar{n}$ (see Section 6.4.4) one gets (6.132). Third, using the same properties of the electromagnetic field one can derive

$$\tilde{B}_{\alpha\beta} = \bar{B}_{\alpha\beta} + n_\alpha b_\beta - b_\alpha n_\beta\,, \tag{6.135}$$

where b_α depends on ϕ and $\bar{B}$ is given by (6.90). Substituting (6.135) into (6.130) and using (6.125) one finds

$$l[\phi, e_0] = \frac{1}{4}\left(n\mathcal{G}^{(r)}\bar{n}\right) \star \left(n\bar{K}\bar{n}\right)\left[n_\nu\gamma^\nu + \frac{ge_0}{4}n_\nu\left(\bar{B}_{\alpha\beta} + n_\alpha b_\beta - b_\alpha n_\beta\right)\gamma^{[\nu}\gamma^\alpha\gamma^{\beta]}\right]\bar{\Lambda}(e_0)\,,$$

and the contribution of the ϕ-dependent terms vanishes by virtue of the complete antisymmetry of the term $\gamma^{[\nu}\gamma^\alpha\gamma^{\beta]}$. Therefore $l[\phi, e_0]$ can be replaced by

$$\bar{l}(e_0) = n^\kappa\mathcal{G}^{(r)}_{\kappa\mu} \star \bar{K}^\mu_\nu\left(\gamma^\nu + \frac{ge_0}{4}\bar{B}_{\alpha\beta}\gamma^{[\nu}\gamma^\alpha\gamma^{\beta]}\right)\bar{\Lambda}(e_0)\,. \tag{6.136}$$

Substituting (6.136) into (6.130) and then into (6.120), using (6.127) and

$$\lambda(\tau)e^{i\lambda\star\phi} = -i\frac{\delta}{\delta\phi(\tau)}e^{i\lambda\star\phi}\,,$$

and integrating by parts, we find

$$\begin{aligned} S^c &= \frac{1}{32\pi^2}\int_0^\infty de_0\left(\det\frac{\sinh\frac{ge_0F}{2}}{gF}\right)^{-1/2}\int D\phi\, D\lambda e^{i\lambda\star(\phi - nx_{cl})} \\ &\times\left[\tilde{\Phi}[x_{tr}, \phi, e_0] - \left(\frac{\delta}{\delta\phi}\tilde{I}[x_{tr}, \phi, e_0]\right)\star\bar{l}(e_0)\right]\exp\left\{i\tilde{I}[x_{tr}, \phi, e_0]\right\}\,. \end{aligned} \tag{6.137}$$

Using equations (6.135), (6.130), and

$$\frac{\delta}{\delta\phi}\tilde{I}[x_{tr}, \phi, e_0] = -a'(\phi)\dot{x}_{tr}\,,$$

we transform (6.137) to the following form

$$S^c = \frac{1}{32\pi^2}\int_0^\infty de_0\left(\det\frac{\sinh\frac{ge_0F}{2}}{gF}\right)^{-1/2}\int D\phi\, D\lambda e^{i\lambda\star(\phi - nx_{cl})}\bar{\Phi}[\tilde{x}, \phi, e_0]\exp\left\{i\tilde{I}[x_{tr}, \phi, e_0]\right\},$$

where

$$\tilde{x}^{\mu} = x^{\mu}_{tr} + ge_0 n^{\mu} a'_{\kappa}(\phi)\dot{x}^{\kappa}_{tr} .$$

One can straightforwardly check that the trajectory $\tilde{x}$ satisfies the equation

$$\ddot{\tilde{x}}_{\mu} - ge_0\left(F_{\mu\nu} + n_{\mu}a'_{\nu}(\phi)\right)\dot{\tilde{x}}^{\nu} = -ge_0 a'_{\mu}(\phi)\dot{\phi} \tag{6.138}$$

and the boundary conditions (6.121). The trajectory x_{tr} in the action $\tilde{I}$ can be replaced by $\tilde{x}$ due to the invariance of the action $\tilde{I}[x, \phi, e_0]$ under the longitudinal shifts (6.126). The integration over λ and ϕ is straightforward now. One needs only to take into account that $\tilde{x}|_{\phi=nx_{cl}} \equiv x_{\text{comb}}$ is the solution (subjected to the boundary conditions (6.121)) to the equation of motion

$$\ddot{x}^{\mu}_{\text{comb}} - ge_0\left(F + f(nx_{\text{comb}})\right)^{\mu}_{\nu}\dot{x}^{\nu}_{\text{comb}} = 0 . \tag{6.139}$$

Indeed, equation (6.138) turns out to be equivalent to equation (6.139), when $\phi = nx_{cl}$ and the relation $nx_{\text{comb}} = nx_{cl}$ is taken into account. Therefore,

$$\tilde{\Phi}[\tilde{x}, \phi, e_0]\big|_{\phi=nx_{cl}} = \Phi[x_{\text{comb}}, e_0], \quad \tilde{I}[\tilde{x}, \phi, e_0]\big|_{\phi=nx_{cl}} = I[x_{\text{comb}}, e_0] . \tag{6.140}$$

Finally, we get

$$S^c = \frac{1}{32\pi^2}\int_0^{\infty} de_0 \left(\det \frac{\sinh \frac{ge_0F}{2}}{gF}\right)^{-\frac{1}{2}} \Phi[x_{\text{comb}}, e_0] e^{I[x_{\text{comb}}, e_0]} , \tag{6.141}$$

where

$$\Phi[x_{\text{comb}}, e_0] = \Big[e_0^{-1}\dot{x}^{\mu}_{\text{comb}} \star K_{\mu\nu}\left(\gamma^{\nu} + \frac{ge_0}{4}B_{\alpha\beta}\gamma^{[\nu}\gamma^{\alpha}\gamma^{\beta]}\right) + m\left(1 + \frac{ge_0}{4}B_{\alpha\beta}\gamma^{[\alpha}\gamma^{\beta]} + \frac{g^2e_0^2}{32}B_{\alpha\beta}B_{\mu\nu}\gamma^{[\alpha}\gamma^{\beta}\gamma^{\mu}\gamma^{\nu]}\right)\Big]\bar{\Lambda}(e_0) . \tag{6.142}$$

The vector $\dot{x}^{\mu}_{\text{comb}}$ satisfies equation (6.139) and can be presented as

$$\dot{x}_{\text{comb}}(\tau) = T_A \exp\left\{-ge_0\int_{\tau}^{1} F^{comb}(nx_{cl}(\tau))d\tau\right\}\dot{x}(1) , \tag{6.143}$$

where T_A denotes the antichronological T-product. On the other hand, the tensor trajectory $K(\tau)$ satisfies equation (6.133), wherein one has to replace ϕ by nx_{cl}. Therefore we have[7]

$$K(\tau) = T_A \exp\left\{-ge_0\int_{\tau}^{1} F^{comb}(nx_{cl})d\tau\right\} K(1) ,$$
$$\dot{x}_{\text{comb}} \star K = \dot{x}_{\text{comb}}(1)K(1) . \tag{6.144}$$

7 The operator $T_A \exp\{-ge_0\int_{\tau}^{1} F^{comb}(nx_{cl})d\tau\}$ preserves the scalar product due to the antisymmetry of the stress tensor.

Substituting (6.143) into (6.142) and taking into account the relation $B_{[\alpha\beta}B_{\mu\nu]} = \bar{B}_{[\alpha\beta}\bar{B}_{\mu\nu]}$, we find

$$\Phi[x_{\text{comb}}, e_0] = \left[e_0^{-1}\dot{x}^{\mu}_{\text{comb}}(1)K_{\mu\nu}(1)\left(\gamma^{\nu} + \frac{ge_0}{4}B_{\alpha\beta}\gamma^{[\nu}\gamma^{\alpha}\gamma^{\beta]}\right) + m\left(1 + \frac{ge_0}{4}B_{\alpha\beta}\gamma^{[\alpha}\gamma^{\beta]} + \frac{g^2e_0^2}{32}\bar{B}_{\alpha\beta}\bar{B}_{\mu\nu}\gamma^{[\alpha}\gamma^{\beta}\gamma^{\mu}\gamma^{\nu]}\right)\right]\bar{\Lambda}(e_0)\,. \tag{6.145}$$

A representation for the Dirac propagator in the special field combination (6.125) was given in Refs. [145, 146, 150–152, 175] in terms of proper-time integral. Another, more complicated, representation has been obtained before in [81]. In our notation the representation [145, 146, 175] can be written as

$$S^c(x_{\text{out}}, x_{\text{in}}) = \left[\gamma^{\mu}\left(i\frac{\partial}{\partial x^{\mu}_{\text{out}}} - gA^{comb}_{\mu}(nx_{\text{out}})\right) + m\right] \times \frac{1}{32\pi^2}\int_0^{\infty} de_0 \left(\det \frac{\sinh \frac{ge_0F}{2}}{gF}\right)^{-\frac{1}{2}} e^{iI[x_{\text{comb}},e_0]}\Delta[nx_{cl}, e_0]\,, \tag{6.146}$$

where

$$\Delta[nx_{cl}, e_0] = T\exp\left\{-\frac{ige_0}{4}\int_0^1 d\tau\,(F + f(nx_{cl}(\tau)))_{\mu\nu}\,\sigma^{\mu\nu}\right\} = \exp\left\{-\frac{ige_0}{4}F_{\mu\nu}\sigma^{\mu\nu}\right\} - \frac{ige_0}{4}\int_0^1 d\tau\left(e^{\frac{ge_0}{2}F(1-2\tau)}f(nx_{cl}(\tau))e^{-\frac{ge_0}{2}F(1-2\tau)}\right)_{\mu\nu}\sigma^{\mu\nu}, \tag{6.147}$$

and T denotes the chronological T-product. It was proved in Ref. [182] that the representations (6.141) and (6.146) are equivalent.

6.4.4 Propagator in a constant uniform field in 2 + 1 dimensions

In $D = 2 + 1$ dimensions the equation for the Dirac propagator has the form (6.47), where γ matrices in 2 + 1 dimensions can be taken, for example, in the form

$$\gamma^0 = \sigma^3, \quad \gamma^1 = i\sigma^2, \quad \gamma^2 = -i\sigma^1, \quad [\gamma^{\mu}, \gamma^{\nu}]_+ = 2\eta^{\mu\nu}, \quad \eta^{\mu\nu} = \text{diag}(1, -1, -1),$$
$$\mu, \nu = 0, 1, 2\,.$$

In this particular case they obey the relations

$$[\gamma^{\mu}, \gamma^{\nu}] = -2i\epsilon^{\mu\nu\lambda}\gamma_{\lambda}, \quad \gamma^{\mu} = \frac{i}{2}\epsilon^{\mu\nu\lambda}\gamma_{\nu}\gamma_{\lambda}\,. \tag{6.148}$$

The path integral representation for the Dirac propagator in 2 + 1 dimensions can be extracted from the general formula (6.71). This representation reads

$$S^c = \frac{1}{2}\exp\left(i\gamma^\mu \frac{\partial_\ell}{\partial\theta^\mu}\right)\int_0^\infty de_0 \int d\chi_0 \int_{e_0} M(e)De \int_{\chi_0} D\chi \int_{x_{in}}^{x_{out}} Dx \int D\pi \int D\nu$$
$$\times \int_{\psi(0)+\psi(1)=\theta} \mathcal{D}\psi \exp\left\{ i\int_0^1 \left[-\frac{\dot{x}^2}{2e} - \frac{e}{2}m^2 - g\dot{x}_\mu A^\mu + iegF_{\mu\nu}\psi^\mu\psi^\nu \right.\right.$$
$$\left.\left. + \chi\left(\frac{2i}{e}\epsilon_{\mu\nu\lambda}\dot{x}^\mu\psi^\nu\psi^\lambda - m\right) - i\psi_\mu\dot{\psi}^\nu + \pi\dot{e} + \nu\dot{\chi}\right] d\tau + \psi_\mu(1)\psi^\mu(0)\right\}\Bigg|_{\theta=0} , \tag{6.149}$$

where $x(\tau), p(\tau), e(\tau), \pi(\tau)$ are even and $\psi(\tau), \chi_1(\tau), \chi_2(\tau), \nu_1(\tau), \nu_2(\tau)$ are odd trajectories, obeying the boundary conditions (6.69), and the notations used are (6.70). The measure $M(e)$ is defined by the equation (6.46) in the corresponding dimensions[8], and

$$\mathcal{D}\psi = D\psi\left[\int_{\psi(0)+\psi(1)=0} D\psi \exp\left\{\int_0^1 \psi_\mu\dot{\psi}^\mu d\tau\right\}\right]^{-1} .$$

Integrating over the Grassmann variables in the same way as in the case of 3 + 1 dimensions, we obtain

$$S^c(x_{out}, x_{in}) = \frac{i}{2}\int_0^\infty de_0 M(e_0) \int_{x_{in}}^{x_{out}} Dx\, \Phi[x, e_0] \exp\{iI[x, e_0]\} , \tag{6.150}$$

where

$$\Phi[x, e_0] = \left[\left(m + \frac{i}{e_0}\int_0^1 d\tau \epsilon_{\mu\nu\lambda}\dot{x}^\mu(\tau)\mathcal{R}^{\nu\lambda}(g|\tau,\tau)\right)\left(1 + \frac{ge_0}{4}B_{\alpha\beta}\gamma^\alpha\gamma^\beta\right)\right.$$
$$\left. + \frac{i}{2e_0}\int_0^1 d\tau\, \epsilon_{\mu\nu\lambda}\dot{x}^\mu(\tau)K^\nu{}_\alpha(\tau)K^\lambda{}_\beta(\tau)\gamma^\alpha\gamma^\beta\right] \exp\left\{-\frac{e_0}{2}\int_0^g dg' \mathrm{Tr}\mathcal{R}(g) \star \mathcal{F}\right\} \tag{6.151}$$

is the spin factor in 2 + 1 dimensions and the quantities $I[x, e_0]$, $\mathcal{R}(g) \equiv \mathcal{R}(g|\tau, \tau')$, $K \equiv K(\tau)$, B, and $\mathcal{F}$ are defined by equations (6.88), (6.84), (6.86), and (6.80) respectively.

---

8 We will refer to some formulae from the previous sections without specifying that the number of dimensions is 2 + 1 now.

Because of relations (6.148) one can also represent the spin factor in the form

$$\Phi[x, e_0] = \left\{m + \frac{i}{e_0}\dot{x} \star r(g) + \left[\left(-i\frac{ge_0}{4}m + \frac{g}{4}\dot{x} \star r(g)\right)u_\alpha + \frac{1}{2e_0}(\dot{x} \star T)_\alpha\right]\gamma^\alpha\right\} \exp\left[-\frac{e_0}{2}\int_0^g dg' \mathrm{Tr}\mathcal{R}(g') \star \mathcal{F}\right], \tag{6.152}$$

where

$$r_\mu(g) \equiv r_\mu(g|\tau) = \epsilon_{\mu\nu\lambda}\mathcal{R}^{\nu\lambda}(g|\tau, \tau), \quad u^\mu = \epsilon^{\mu\alpha\beta}B_{\alpha\beta}, \quad T_\mu^\rho = \epsilon_{\mu\nu\lambda}\epsilon^{\rho\alpha\beta}K_\alpha^\nu K_\beta^\lambda .$$

In the case of a constant uniform field $F_{\mu\nu} = \text{const}$ one can calculate the path integral (6.150) by explicitly integrating over the Bosonic trajectories. Following the same way as in 3 + 1 dimensions and taking into account that in 2 + 1 dimensions we have

$$\frac{i}{2}M(e_0)\int_0^0 Dx \exp\left\{-\frac{i}{2e_0}\dot{x} \star \dot{x}\right\} = \frac{e^{i\pi/4}}{2(2\pi e_0)^{3/2}},$$

one gets for the propagator (6.150)

$$S^c = \frac{e^{i\pi/4}}{2(4\pi)^{3/2}}\int_0^\infty de_0 \left(\det \frac{\sinh \frac{ge_0F}{2}}{gF}\right)^{-1/2} e^{iI[x_{cl}, e_0]}\Phi[x_{cl}, e_0], \tag{6.153}$$

where x_{cl}, $\mathcal{R}(g)$, K, B are given by (6.105) and (6.90).

The antisymmetric matrices $F_{\mu\nu}$ can be classified by the value of the invariant φ (see Section 6.4.4). In the case $\varphi^2 > 0$ one can find a Lorentz frame in which the magnetic field vanishes. On the other hand, $\varphi^2 < 0$ implies that the electric field vanishes in an appropriate Lorentz frame. The case $\varphi^2 = 0, F \neq 0$ corresponds to nonvanishing electric and magnetic fields of "equal magnitude" (and this property is Lorentz invariant). We will consider the case $\varphi^2 \neq 0$. The case $\varphi^2 = 0$ can be easily treated, for example taking the limit $\varphi \to 0$.

In the case under consideration one can avoid the first integrations over τ in the spin factor (6.151). Indeed, due to the specific form of $\mathcal{R}(g)$ the term containing it in (6.151) vanishes. On the other hand, we have in the case under consideration

$$\dot{x}_{cl}(\tau) = e^{ge_0F(\tau-1)}\dot{x}_{cl}(1), K(\tau) = e^{ge_0F(\tau-1)}K(1) .$$

Taking into account that $e^{ge_0F(\tau-1)}$ is an operator respecting the scalar product, one can easily perform the second integration over τ in (6.151). Finally, calculating the determinants involved, by means of (6.161), one gets

$$S^c = \sqrt{\frac{i}{16(2\pi)^3}} \int_0^\infty \frac{de_0}{\sqrt{e_0}} \frac{g\varphi}{\sinh \frac{ge_0\varphi}{2}} e^{iI[x_{cl},e_0]} \Phi[x_{cl}, e_0],$$

$$\Phi[x_{cl}, e_0] = \left\{ m\left(1 + \frac{ge_0}{4} B_{\alpha\beta}\gamma^\alpha\gamma^\beta\right) + \frac{i}{2e_0} \dot{x}^\mu(1)\epsilon_{\mu\nu\lambda} \left[K^\nu_\alpha K^\lambda_\beta \gamma^\alpha\gamma^\beta \right.\right.$$
$$\left.\left. \times -\frac{ge_0}{2} B^{\nu\lambda}\left(1 + \frac{ge_0}{4} B_{\alpha\beta}\gamma^\alpha\gamma^\beta\right)\right]\right\} \cosh \frac{ge_0\varphi}{2} . \tag{6.154}$$

On the other hand one can obtain a representation for the propagator using the Schwinger proper-time method (we do not present the calculations here). Such a representation has the form

$$S^c(x_{\text{out}}, x_{\text{in}}) = \left[\gamma^\mu \left(i\frac{\partial}{\partial x^\mu_{\text{out}}} + gA_\mu(x_{\text{out}}) \right) + m \right]$$
$$\times \sqrt{\frac{i}{16(2\pi)^3}} \int_0^\infty \frac{de_0}{\sqrt{e_0}} \frac{g\varphi}{\sinh \frac{ge_0\varphi}{2}} e^{iI[c_{cl},e_0]} e^{\frac{ge_0}{4} F_{\alpha\beta}\gamma^\alpha\gamma^\beta} . \tag{6.155}$$

To compare both representations we take the derivative in (6.155) and use the relation

$$\frac{\partial}{\partial x^\mu_{\text{out}}} I[x_{cl}, e_0] = -e_0^{-1} (\dot{x}_{cl})_\mu (1) - gA_\mu(x_{\text{out}}) .$$

Then one obtains

$$S^c(x_{\text{out}}, x_{\text{in}}) = \frac{e^{i\pi/4}}{4(2\pi)^{3/2}} \int_0^\infty \frac{de_0}{\sqrt{e_0}} \frac{g\varphi}{\sinh \frac{ge_0\varphi}{2}} \Psi_S(x_{\text{out}}, x_{\text{in}}, e_0) e^{iI[x_{cl},e_0]} , \tag{6.156}$$

where

$$\Psi_S(x_{\text{out}}, x_{\text{in}}, e_0) = \left(e_0^{-1}\gamma_\mu \dot{x}^\mu_{cl}(1) + m\right) e^{\frac{ge_0}{4} F_{\alpha\beta}\gamma^\alpha\gamma^\beta} . \tag{6.157}$$

Comparing (6.154) and (6.156), with the use of the identities (see Section 6.4.4)

$$\exp\left\{\frac{ge_0}{4} F_{\alpha\beta}\gamma^\alpha\gamma^\beta\right\} = \left(1 + \frac{ge_0}{4} B_{\alpha\beta}\gamma^\alpha\gamma^\beta\right) \cosh \frac{ge_0\varphi}{2} , \tag{6.158}$$

$$\gamma_\mu \exp\left\{\frac{ge_0}{4} F_{\alpha\beta}\gamma^\alpha\gamma^\beta\right\} = \frac{i}{2}\epsilon_{\mu\nu\lambda} \left[K^\nu_\alpha(1)\, K^\lambda_\beta(1)\gamma^\alpha\gamma^\beta - \frac{8e_0}{2} B^{\nu\lambda} \right.$$
$$\left. \times \left(1 + \frac{ge_0}{4} B_{\alpha\beta}\gamma^\alpha\gamma^\beta\right)\right] \cosh \frac{ge_0\varphi}{2} , \tag{6.159}$$

one can verify that the both representations coincide.

Appendix

(I) Let $F_{\mu\nu}$ be an antisymmetric matrix in 2 + 1 dimensions. The antisymmetry implies $\text{tr}F = 0$, $\det F = 0$, so that the sum and the product of the eigenvalues vanishes. The eigenvalues are $0, \varphi, -\varphi$, where the real number φ^2 coincides with the invariant

$$\varphi^2 = \frac{1}{2}\mathrm{tr}F^2 .$$

In the case of nonvanishing φ there exist three eigenvectors of F, and F^2 is proportional to a projection operator P onto a certain two-dimensional subspace,

$$F^2 = \varphi^2 P, \quad P^2 = P, \quad \mathrm{tr}P = 2, \quad PF = FP = F . \tag{6.160}$$

Then, for an even function h,

$$h(F) = h(0)\,(1 - P) + h(\varphi)P , \tag{6.161}$$

while for an odd one

$$h(F) = \frac{F}{\varphi}h(\varphi) . \tag{6.162}$$

The case of vanishing φ (and $F \neq 0$) corresponds to a nilpotent matrix, $F^3 = 0$.

(II) To prove the identity (6.158) we introduce the new four-vector z^μ,

$$z^\mu = \epsilon^{\mu\nu\lambda}F_{\nu\lambda}, \quad z^2 = -4\phi^2$$

and transform the left hand side of (6.158) using equation (6.148),

$$\exp\left\{\frac{ge_0}{4}F_{\mu\nu}\gamma^\mu\gamma^\nu\right\} = \cosh\frac{ge_0\varphi}{2}\left(1 - \frac{iz\gamma}{2\varphi}\tanh\frac{ge_0\varphi}{2}\right) .$$

Taking into account equations (6.86), (6.162) and the relation $iz\gamma = -\gamma F\gamma$, one gets (6.158).

Multiplying (6.159) by $K(1)$ and using (6.148) one transforms (6.159) into the equivalent identity

$$K^\mu_\lambda(1)\gamma^\lambda e^{\frac{ge_0}{4}F_{\alpha\beta}\gamma^\alpha\gamma^\beta} = \gamma^\mu\cosh\frac{ge_0\varphi}{2}\det K(1) , \tag{6.163}$$

which can be easily proved. Taking into account the identity

$$\gamma^\lambda e^{\frac{ge_0}{4}F_{\alpha\beta}\gamma^\alpha\gamma^\beta} = \left(e^{-\frac{ge_0}{2}F}\right)^\lambda_{\ \rho}\gamma^\rho ,$$

which can be derived from (6.158) and using (6.86) one transforms the left hand side of (6.163) as follows:

$$K^\mu_\lambda(1)\gamma^\lambda e^{\frac{ge_0}{4}F_{\alpha\beta}\gamma^\alpha\gamma^\beta} = \left(\cosh\frac{ge_0}{2}\varphi\right)^{-1}\gamma^\mu . \tag{6.164}$$

Calculating the determinant

$$\det K(1) = \left(\cosh\frac{ge_0\varphi}{2}\right)^{-2} ,$$

one finds that the right hand side of (6.164) coincides with that of (6.163).

7 Electron interacting with a quantized electromagnetic plane wave

In this Chapter, we consider the Dirac equation in which electromagnetic potentials are a sum of two parts $A_\mu(x)$ and $\hat{A}_\mu(x)$. Here $A_\mu(x)$ are electromagnetic potentials of an external classical field and $\hat{A}_\mu(x)$ are operator-valued potentials of a quantized plane-wave field. At the same time the bispinor wave functions depend both on coordinates x of the charge and on photon variables. The operators $\hat{A}_\mu(x)$ act on the latter. In what follows, we call the above described equation the *Dirac equation with quantized plane wave*.

Solutions of such a Dirac equation were studied for the first time in Refs. [92, 194]. Various extensions, detailed investigations, and applications of this problem can be found in Refs. [1, 14–17, 21–23, 38, 41, 95, 130–133, 135, 139, 163, 173, 212, 215, 225, 298, 343].

7.1 Dirac equation with quantized plane wave

7.1.1 General

In this section, we present an interpretation of the Dirac equation with quantized plane wave in the context of QED.

Let $\hat{H}_{\mathrm{QED}}$ be the Hamiltonian of QED and $|\Psi\rangle$ be a state vector of interacting Dirac $\psi(x)$ and electromagnetic $A^\mu(x)$ fields (electron-positron and photon fields). In the Schrödinger picture, the state vectors depend on time $t = x^0$, so that $|\Psi\rangle = |\Psi(x^0)\rangle$, the time evolution of $|\Psi(x^0)\rangle$ is described by the Schrödinger equation

$$i\frac{\partial\left|\Psi(x^0)\right\rangle}{\partial x^0} = \hat{H}_{\mathrm{QED}}\left|\Psi(x^0)\right\rangle . \tag{7.1}$$

The total space of states of QED can be considered as the direct product of particle states and Fock space of photon states. It is known that the Ket-basis in the particle subspace can be taken in the following form:

$$\left\langle 0|\psi^{(-)}(\mathbf{r}_1)\ldots\psi^{(-)}(\mathbf{r}_n)\bar{\psi}_c^{(-)}(\mathbf{r}_1')\ldots\bar{\psi}_c^{(-)}(\mathbf{r}_m')\right.$$

(see, for example, Ref. [292]). Here $\langle 0|$ is the vacuum vector in the particle subspace, $\psi(\mathbf{r})$ are operators of the Dirac field in the Schrödinger picture, $\psi_c(\mathbf{r}) = C\bar{\psi}^T(\mathbf{r})$ are charge conjugated operators, and $\psi^{(-)}(\mathbf{r})$, $\bar{\psi}_c^{(-)}(\mathbf{r})$ are their parts that contain particle annihilation operators.

Consider the following amplitudes:

$$\langle 0|\,\psi^{(-)}(\mathbf{r}_1)\ldots\psi^{(-)}(\mathbf{r}_n)\bar{\psi}_c^{(-)}(\mathbf{r}_1')\ldots\bar{\psi}_c^{(-)}(\mathbf{r}_m')\left|\Psi(x^0)\right\rangle . \tag{7.2}$$

On the one hand, these amplitudes are vectors in the Fock space of photon states and, on the other hand, they are 4^{n+m} component wave functions of states with a given number of charged particles. It is obvious that fixing the infinite set of amplitudes (7.2) is equivalent to fixing the state vector $|\Psi(x^0)\rangle$. In particular, the first amplitude

$$\langle 0|\, \psi^{(-)}(\mathbf{r}) \left|\Psi(x^0)\right\rangle = \Psi(x), \; x = \left(x^0, \mathbf{r}\right) , \tag{7.3}$$

is the projection of the vector $|\Psi(x^0)\rangle$ onto the single-electron state and can be considered as a four-component wave function of an electron and of photons.

Equation (7.1) induces a set of coupled equations for amplitudes (7.2). If the electromagnetic field is quantized in the Coulomb gauge, where, in particular, $\hat{A}_0 = 0$, and $\operatorname{div} \hat{\mathbf{A}}(\mathbf{r}) = 0$, see for example [100, 174, 201, 292], then the corresponding equation written for the single-electron amplitude (7.3) has the form

$$i\frac{\partial \Psi(x)}{\partial x^0} = \hat{H}_{e,\gamma}\Psi(x) + \Delta , \tag{7.4}$$

where

$$\hat{H}_{e,\gamma} = \hat{H}_{\gamma} + \boldsymbol{\alpha}\hat{\mathbf{P}} + m\beta, \quad \hat{\mathbf{P}} = \hat{\mathbf{p}} - e\hat{\mathbf{A}}(\mathbf{r}) . \tag{7.5}$$

Here $\hat{H}_\gamma$ is the Hamiltonian of free transversal photons, $\hat{\mathbf{A}}(\mathbf{r})$ is the operator-valued vector potential of such photons in the Schrödinger picture, and all the contributions from many-particle amplitudes (7.2) are absorbed into the term Δ. The operators $\hat{H}_\gamma$ and $\hat{\mathbf{A}}(\mathbf{r})$ can be represented as combinations of the photon creation and annihilation operators $a^{+}_{\mathbf{k},\lambda}$ and $a_{\mathbf{k},\lambda}$ with a definite momentum $\mathbf{k}$ and polarization $\lambda = 1, 2$,

$$\begin{aligned} \hat{H}_\gamma &= \sum_{\mathbf{k},\lambda} |\mathbf{k}|\, a^{+}_{\mathbf{k},\lambda} a_{\mathbf{k},\lambda} , \\ \hat{\mathbf{A}}(\mathbf{r}) &= \sum_{\mathbf{k},\lambda} \frac{1}{\sqrt{2\,|\mathbf{k}|\, V}} \left[a_{\mathbf{k},\lambda} e^{i\mathbf{k}\mathbf{x}} \mathbf{e}_\lambda(\mathbf{k}) + a^{+}_{\mathbf{k},\lambda} e^{-i\mathbf{k}\mathbf{x}} \mathbf{e}^{*}_\lambda(\mathbf{k}) \right] . \end{aligned} \tag{7.6}$$

Here $V = L^3$ is the quantization box volume, and $\mathbf{e}_\lambda(\mathbf{k})$ are polarization vectors possessing the properties

$$\mathbf{e}_\lambda(\mathbf{k})\mathbf{k} = \mathbf{e}^{*}_\lambda(\mathbf{k})\mathbf{k} = 0, \quad \mathbf{e}^{*}_\lambda(\mathbf{k})\mathbf{e}_{\lambda'}(\mathbf{k}) = \delta_{\lambda,\lambda'}, \; \lambda, \lambda' = 1, 2 . \tag{7.7}$$

If the operator potential (7.6) is considered to be, for example, space-periodic with the period L along each Cartesian axis, then the vector $\mathbf{k}$ takes on the following discrete values:

$$\mathbf{k} = 2\pi L^{-1} \left(n_1, n_2, n_3\right), \quad n_s \in \mathbb{Z} . \tag{7.8}$$

The operator $\hat{H}_{e,\gamma}$ can be regarded as a Hamiltonian of a charge e interacting with the quantized electromagnetic field.

Disregarding the effects of virtual pair creation and the related appearance of the higher amplitudes in equation (7.4) (i.e. neglecting the term Δ), we are left with the

Schrödinger equation for a single charge e interacting with the transversal electromagnetic field. This equation forms the basis for many models of one-electron type. The amplitude (7.3) can be treated as a wave function of one electron interacting with the transversal electromagnetic field. Formally it is a bispinor, each component of which lives in the Fock space of the transversal photons. In what follows, we call it the *Fock bispinor*.

Solutions of the Schrödinger equation with the potential (7.6) and $\Delta = 0$ make it possible to study in a model way (i.e. disregarding the real and virtual creation of electron-positron pairs) the problems in which a single charge is involved along with an arbitrary number of photons. Unfortunately, even with this simplification, the problem has not yet any exact solutions, so that one is forced to treat the radiative interaction as a perturbation, starting with the solutions of the free equation as unperturbed. It is, nevertheless, quite evident that there exist some situations for which corrections to the solution of the free problem are not small. For instance, as long as states with a large photon density d are dealt with, one should take into account all the terms of the perturbation expansion, since the perturbative matrix elements are proportional to $e^2 d$.

Let us consider a certain possibility of taking the whole perturbation series into account. Suppose we are interested in processes where an electron in an intense plane-wave field is involved. From the quantum-mechanical point of view this means that there exist initial states with a large density of photons that move in a common direction $\mathbf{n}$ along the plane-wave. Then it is reasonable to chose expression (7.5) for the unperturbed Hamiltonian in Equation (7.4), in which $\hat{\mathbf{A}}$ is the part of the potential of the quantized electromagnetic field containing only the summation over the transversal photons that move in the common direction $\mathbf{n}$ (the photons of the plane wave). If such a problem can be solved exactly, then the interaction of a charge with the photons of the plane wave is kept exactly, while the corrections to such a solution could be small.

Below, in this Chapter we consider solutions of such a problem in detail. More specifically, we consider the Schrödinger equation of the form

$$i\frac{\partial \Psi(x)}{\partial x^0} = \hat{H}_{e,\gamma}\Psi(x), \quad \hat{H}_{e,\gamma_{\mathbf{n}}} = \hat{H}_{\gamma_{\mathbf{n}}} + \boldsymbol{\alpha}\hat{\mathbf{P}} + m\beta, \quad \hat{\mathbf{P}} = \hat{\mathbf{p}} - e\hat{\mathbf{A}}(\mathbf{r}), \tag{7.9}$$

where $\hat{H}_{e,\gamma_{\mathbf{n}}}$ is the Hamiltonian of free transversal photons that are moving in the common direction $\mathbf{n}$ (the Hamiltonian of the free plane-wave field), and $\hat{\mathbf{A}}(\mathbf{r})$ is the corresponding operator-valued vector potential of these photons. Provided that $\mathbf{n} = (0, 0, 1)$ is chosen along the axis z, we have for photons that are moving in the positive direction of z the following momenta:

$$\mathbf{k} = \mathbf{k}_s = 2\pi L^{-1}(0, 0, s) = \kappa_s \mathbf{n}, \quad \kappa_s = \kappa_0 s, \quad \kappa_0 = 2\pi L^{-1}, \quad s \in \mathbb{N},$$
$$a_{s,\lambda} = a_{\mathbf{k}_s,\lambda}, \quad a^+_{s,\lambda} = a^+_{\mathbf{k}_s,\lambda}. \tag{7.10}$$

Then

$$\hat{H}_{\gamma_n} = \sum_{s,\lambda} \kappa_s a^+_{s,\lambda} a_{s,\lambda} ,$$

$$\hat{\mathbf{A}}(\mathbf{r}) = \sum_{s,\lambda} \frac{1}{\sqrt{2\kappa_s V}} \left[a_{s,\lambda} e^{i\kappa_s(\mathbf{nx})} \mathbf{e}_\lambda(s) + a^+_{s,\lambda} e^{-i\kappa_s(\mathbf{nx})} \mathbf{e}^*_\lambda(s) \right] . \tag{7.11}$$

By passing to the interaction picture with respect to the Hamiltonian $\hat{H}_{\gamma_n}$,

$$\Psi(x) = \exp\left(-i\hat{H}_{\gamma_n} x^0\right) \boldsymbol{\Psi}(x) , \tag{7.12}$$

in equation (7.9), we can easily find that the Fock bispinor $\Psi(x)$ satisfies the Dirac equation of the form

$$\left(\gamma^\mu \hat{P}_\mu - m\right) \Psi(x) = 0, \quad \hat{P}_\mu = i\partial_\mu - e\hat{A}^\mu(x) , \tag{7.13}$$

with the operator-valued electromagnetic potential

$$\hat{A}^\mu(x) = \left(0, \hat{\mathbf{A}}(x)\right), \quad \hat{\mathbf{A}}(x) = \exp\left(i\hat{H}_{\gamma_n} x^0\right) \hat{\mathbf{A}}(\mathbf{r}) \exp\left(-i\hat{H}_{\gamma_n} x^0\right) .$$

This potential can be written as

$$\hat{A}^\mu(x) = \hat{A}_\mu(u_0) = \sum_{s,\lambda} \frac{1}{|e|} \sqrt{\frac{\delta}{\kappa_s}} \left[a_{s,\lambda} e^{-i\kappa_s u_0} e^\mu_\lambda(s) + a^+_{s,\lambda} e^{i\kappa_s u_0} e^{\mu *}_\lambda(s) \right] . \tag{7.14}$$

The variable u_0 is just the light-cone time $u_0 = nx$ which takes the form $u_0 = x^0 - x^3$ provided that $\mathbf{n}$ is chosen along the axis x, i.e. $\mathbf{n} = (0, 0, 1)$ and $\delta = e^2/2L^3$. The quantity δ characterizes a strength of the interaction between the charge and the plane-wave field. If we interpret $V^{-1} = L^{-3}$ as the electron density ρ then $\delta \sim e^2\rho$. The dimensionality of δ is $[\delta] = l^{-3}$ (where l is the dimensionality of length). Being written with $\hbar$ and c restored it has the form

$$\delta = \frac{\alpha\rho}{2} = \frac{\alpha\kappa_0^3}{16\pi^3} , \quad \alpha = \frac{e^2}{\hbar c} , \tag{7.15}$$

where α is fine structure constant, $\alpha \simeq 1/137$.

The polarization vectors are now labelled by integers s and by λ and obey the conditions

$$e^\mu_\lambda(s) = \left(0, \mathbf{e}_\lambda(\mathbf{k})\right) , \quad \mathbf{k} = \kappa_s \mathbf{n} ,$$

$$n e_\lambda(s) = n e^*_\lambda(s) = 0, \quad e^*_\lambda(s) e_{\lambda'}(s) = -\delta_{\lambda,\lambda'}, \quad n^\mu = (1, \mathbf{n}) , \quad n^2 = 0 . \tag{7.16}$$

The photon creation and annihilation operators $a^+_{s,\lambda}$ and $a_{s,\lambda}$ are also labeled by the integers s and by λ and obey the Bose commutation relations

$$a_{s',\lambda'} a^+_{s,\lambda} - a^+_{s,\lambda} a_{s',\lambda'} = \delta_{s,s'} \delta_{\lambda,\lambda'} ,$$

$$a_{s',\lambda'} a_{s,\lambda} - a_{s,\lambda} a_{s',\lambda'} = a^+_{s',\lambda'} a^+_{s,\lambda} - a^+_{s,\lambda} a^+_{s',\lambda'} = 0 . \tag{7.17}$$

Note that if the scalar QED is considered, the approximations made result in the K–G equation with the operator-valued potential (7.14), see below.

Finally, if an external classical electromagnetic field $A^{\mu}_{\text{ext}}(x)$ is acting on the system under consideration (electron interacting with the quantized plane-wave), its potentials should be combined with the potential (7.14) in an additive way, so that the equation for the Fock-bispinor $\Psi(x)$ takes the form of the Dirac (or K–G) equation with the potentials consisting of two terms

$$A^{\mu} = \hat{A}^{\mu}(x) + A^{\mu}_{\text{ext}}(x)\,. \tag{7.18}$$

7.1.2 Separation of variables

Here, we study the Dirac equation considering the potential involved in it as (7.18) where $\hat{A}^{\mu}(x)$ is the operator-valued potential of a quantized plane-wave, and $A^{\mu}_{\text{ext}}(x)$ is the potential of an external classical field of the type (2.103), i.e. of the field that is a combination of the crossed and longitudinal fields. Evidently, the whole combination (7.18) gives the (already operator-valued) field of the same structure (2.103). We separate spin variables in equation (7.13) by the standard trick (2.107), i.e. we represent the Fock-bispinor as follows:

$$\Psi(x) = \mathcal{K}\left\{[1 + (\boldsymbol{\sigma}\mathbf{n})]\,\psi_1(x) + [1 - (\boldsymbol{\sigma}\mathbf{n})]\,\psi_{-1}(x)\right\} v\,, \tag{7.19}$$

where the operator $\mathcal{K}$ has the form

$$\begin{aligned} \mathcal{K} &= \begin{pmatrix} m + (n\hat{P}) + (\boldsymbol{\sigma}\mathbf{n})\,(\boldsymbol{\sigma}\hat{\mathbf{P}}_{\perp}) \\ \left[m - (n\hat{P})\right](\boldsymbol{\sigma}\mathbf{n}) + (\boldsymbol{\sigma}\hat{\mathbf{P}}_{\perp}) \end{pmatrix}, \\ \hat{\mathbf{P}}_{\perp} &= \hat{\mathbf{P}} - \mathbf{n}(\boldsymbol{\sigma}\hat{\mathbf{P}})\,. \end{aligned} \tag{7.20}$$

The spinor v is constant, while functions ψ_{ζ} ($\zeta = \pm 1$) satisfy the independent equations

$$\left(\hat{P}_0^2 - \hat{\mathbf{P}}^2 - m^2 + ieE_3 - \zeta e\bar{H}_3\right)\psi_{\zeta}(x) = 0\,, \tag{7.21}$$

where the designations

$$E_3 = (\mathbf{nE})\,, \quad \bar{H}_3 = (\mathbf{nH}) + \bar{b}\,, \tag{7.22}$$

$$\bar{b} = \delta \sum_{k,\lambda} \frac{\xi_2(k,\lambda)}{\kappa_k}, \quad \xi_2(k,\lambda) = i\left(\mathbf{n}\left[\mathbf{e}_{\lambda}(k) \times \mathbf{e}^{*}_{\lambda}(k)\right]\right) \tag{7.23}$$

are used.

Evidently, $(\mathbf{nE})$ and $(\mathbf{nH})$ can only be different from zero provided that a classical field A^{μ}_{ext} is present. The quantity $\bar{b}$ originates from the quantized plane wave, but manifests itself physically as a constant and homogeneous longitudinal classical magnetic field (not induced by a classical potential). It is known (for example, see Ref. [86])

that the quantity $\xi_2(k,\lambda)$ determines the extent of the circular polarization of a photon labelled by k and λ. If the sum in (7.23) includes, for each k, both the values $\lambda = 1, 2$, we have $\bar{b} = 0$ due to the property $\xi_2(k, 1) = -\xi_2(k, 2)$. In this case, no extra magnetic field appears. If each wave mode is decomposed using the vectors of linear polarization, the same situation occurs, since in this case $\mathbf{e}_\lambda(k) = \mathbf{e}^*_\lambda(k) \Longrightarrow \xi_2(k,\lambda) = 0$. If, however, not every value of λ is included in the sum (7.23), and the decomposition in elliptical polarization is used, then $\bar{b} \neq 0$, and it will be shown in the next Section that in this case the energy of the system depends on the particle spin and on the wave polarization. This conclusion, drawn first in Refs. [15, 16], is intriguing, but hardly acceptable physically. Indeed, it would be most consistent to take into account all the photons propagating along the given direction $\mathbf{n}$, and to include thereby both the terms $\lambda = 1$ and $\lambda = 2$ for each k in the sum (7.23). This, however, implies $\bar{b} = 0$. In what follows, we shall assume this to be the case, everywhere except in Section 7.2.

In Sections 7.1.2, 7.2, 7.3, 7.4, 7.5, and 7.6 we shall be dealing with particles moving in the quantized plane wave alone (without any classical field), and we shall set everywhere $\bar{b} = 0$ in equation (7.22) except for Section 7.2. Then $E_3 = \bar{H}_3 = 0$ and the Fock-bispinor can be written as

$$\Psi(x) = \mathcal{K}\Phi(x)v\,, \tag{7.24}$$

where $\Phi(x)$ is a solution of the corresponding K–G equation, it can be called the *Fock-scalar*. It is readily seen that in this case the quantity $(n\hat{P}) = (n\hat{p})$ is an integral of motion. We shall consider the Fock-bispinor to be an eigenfunction of the operator $(n\hat{p})$ with the eigenvalue equal to (np). Setting

$$\begin{aligned}\Phi(x) &= \exp\left[-\frac{i}{2}(np)u_3\right]\varphi(u_0,\mathbf{r}_\perp)\,,\\ u_3 &= (\bar{n}x),\quad \bar{n}^\mu = (1,-\mathbf{n}),\quad \mathbf{r}_\perp = \mathbf{r}-\mathbf{n}\,(\mathbf{nr})\,,\end{aligned} \tag{7.25}$$

we obtain in this case that the function $\varphi(u_0,\mathbf{r}_\perp)$ satisfies the Schrödinger equation:

$$i\partial_0\varphi(u_0,\mathbf{r}_\perp) = \mathcal{H}\varphi(u_0,\mathbf{r}_\perp),\quad \mathcal{H} = \frac{\hat{\mathbf{P}}_\perp^2 + m^2}{2(np)}\,, \tag{7.26}$$

which will be assumed as a basis for the further study of the problem.

Consider the operators

$$G_\mu = i\partial_\mu + n_\mu \sum_{s,\lambda}\kappa_s a^+_{s,\lambda}a_{s,\lambda}\,. \tag{7.27}$$

They commute among themselves $[G_\nu, G_\mu]_- = 0$, and with all the operators $\hat{P}_\mu$, $[\hat{P}_\mu, G_\nu]_- = 0$. One can see that both the Fock-bispinor Ψ and the Fock-scalar Φ can be chosen to be eigenfunctions of the operators G_μ,

$$G_\mu\Psi = g_\mu\Psi,\quad G_\mu\Phi = g_\mu\Phi\,. \tag{7.28}$$

It is seen from equation (7.28) that the derivatives ∂_μ can be eliminated from equation (7.26). Thus, we are left with equations that only contain photon variables. The operator G_μ has evidently the meaning of the energy-momentum operator for the system composed of the particle and the quantized plane wave. It is also evident by virtue of the property $n^2 = 0$ that $(nG) = (n\hat{p})$.

For the K–G equation, it is convenient to write the operator G_μ as

$$G^\mu = p_K^\mu + n^\mu \mathcal{H}_K, \quad p_K^\mu = i\partial^\mu - \left[\frac{e(gA)}{(ng)} - \frac{e^2A^2}{2(ng)}\right] n^\mu,$$
$$\mathcal{H}_K = \sum_{s,\lambda} \kappa_s a_{s,\lambda}^{+} a_{s,\lambda} + \frac{e(gA)}{(ng)} - \frac{e^2A^2}{2(ng)}, \tag{7.29}$$

where the operators p_K^μ and $\mathcal{H}_K$ are each integrals of motion. When the interaction between the plane wave and the particle is switched off, the operator $\hat{p}_K^\mu$ becomes the energy-momentum operator of the particle, while the operator $n^\mu \mathcal{H}_K$ corresponds to the energy-momentum of free photons. It is therefore appropriate to refer to p_K^μ as the quasiparticle energy-momentum, and to $n^\mu \mathcal{H}_K$ as the energy-momentum of *quasiphotons*. The meaning of these terms will be made more precise below.

A similar division can also be done in the case of the Dirac equation,

$$G^\mu = p_D^\mu + n^\mu \mathcal{H}_D, \quad p_D^\mu = p_K^\mu + R_0 n^\mu,$$
$$\mathcal{H}_D = \mathcal{H}_K - R_0, \quad R_0 = \frac{e\sigma^{\alpha\beta}F_{\alpha\beta}}{4(ng)},$$
$$F_{\alpha\beta} = \partial_\alpha A_\beta - \partial_\beta A_\alpha = n_\alpha A'_\beta - n_\beta A'_\alpha, \tag{7.30}$$

where the prime denotes the derivative with respect to u_0. Here, again, p_D^μ and $\mathcal{H}_D$ are integrals of motion.

It might seem that for separating the particle and photon variables the requirement that equations (7.28) be fulfilled is necessary. This is not the case, however. Let us transform the Fock-bispinors or Fork-scalars using the unitary operator

$$U_1 = \exp\left(iu_0 \sum_{s,\lambda} \kappa_s a_{s,\lambda}^{+} a_{s,\lambda}\right). \tag{7.31}$$

The following relations are readily obtained:

$$U_1^{+} G_\mu U_1 = i\partial_\mu, U_1^{+} \hat{P}_\mu U_1 = i\partial_\mu + Q_\mu - n_\mu \sum_{s,\lambda} \kappa_s a_{s,\lambda}^{+} a_{s,\lambda}$$
$$Q^\mu = (0, \mathbf{Q}), \quad \mathbf{Q} = \sum_{k,\lambda} \sqrt{\frac{\delta}{\kappa_k}} \left[a_{k,\lambda} \mathbf{e}_\lambda(k) + a_{k,\lambda}^{+} \mathbf{e}_\lambda^{*}(k)\right]. \tag{7.32}$$

With their help, one finally finds that the Fock-bispinors or Fork-scalars can be written as

$$\Psi = \exp\left[-\frac{i}{2}(np)u_3\right]U_1\overline{\mathcal{K}}\psi_0 v, \quad \Phi = \exp\left[-\frac{i}{2}(np)u_3\right]U_1\psi_0,$$
$$\overline{\mathcal{K}} = \begin{pmatrix} m + (np) + (\boldsymbol{\sigma}\mathbf{n})(\boldsymbol{\sigma}\mathbf{F}) \\ [m-(np)](\boldsymbol{\sigma}\mathbf{n}) + (\boldsymbol{\sigma}\mathbf{F}) \end{pmatrix}, \quad \mathbf{F} = -i\boldsymbol{\nabla}_\perp + \mathbf{Q}, \quad \boldsymbol{\nabla}_\perp = \boldsymbol{\nabla} - \mathbf{n}(\mathbf{n}\boldsymbol{\nabla}), \tag{7.33}$$

where the function ψ_0 is a solution of the Schrödinger equation

$$i\partial_0\psi_0 = \mathcal{H}\psi_0, \quad \mathcal{H} = \frac{\mathbf{F}^2 + m^2}{2(np)} + \sum_{s,\lambda}\kappa_s a^+_{s,\lambda}a_{s,\lambda}\,. \tag{7.34}$$

The Hamiltonian $\mathcal{H}$ in (7.34) does not explicitly depend on the space-time variables. Note the validity of the following relations:

$$\mathcal{H} = U_1^+\mathcal{H}_K U_1 = U_2^{-1}U_1^+\mathcal{H}_D U_1 U_2,$$
$$U_2 = 1 - \frac{(\gamma^\mu n_\mu)(\gamma^\nu Q_\nu)}{2(np)}, \quad U_2^{-1} = 1 + \frac{(\gamma^\mu n_\mu)(\gamma^\nu Q_\nu)}{2(np)},$$
$$\hat{p}^\mu = U_1^+ p_K^\mu U_1 = U_2^{-1}U_1^+ p_D^\mu U_1 U_2^{-1} = i\partial^\mu - n^\mu\mathcal{H}\,. \tag{7.35}$$

Now we are in a position to separate the particle variables. This corresponds to choosing the functions Ψ and Φ to be eigenfunctions of the operators p_D^μ and p_K^μ with the eigenvalues p^μ (the eigenvalues p^μ are, according to (7.35), the same for both the operators p_D^μ and p_K^μ). Then we write, instead of (7.33),

$$\Phi = \exp\left[-i(px)\right]U_1\psi, \quad \Psi = \exp\left[-i(px)\right]U_1\overline{\mathcal{K}}\psi v\,, \tag{7.36}$$

where the operator $\overline{\mathcal{K}}$ is again defined by equation (7.33) wherein one should set

$$\mathbf{F} = \mathbf{p}_\perp + \mathbf{Q}, \quad \mathbf{p}_\perp = \mathbf{p} - \mathbf{n}(\mathbf{n}\mathbf{p}), \tag{7.37}$$

and the function ψ is a solution of the Schrödinger equation

$$i\partial_0\psi = \hat{H}\psi, \quad \hat{H} = \sum_{s,\lambda}\kappa_s a^+_{s,\lambda}a_{s,\lambda} + \frac{(\mathbf{p}\mathbf{Q})}{(np)} + \frac{\mathbf{Q}^2}{2(np)}\,. \tag{7.38}$$

The quasiparticle energy-momentum satisfies the natural condition

$$p^2 = m^2 \Longrightarrow p_0^2 = \mathbf{p}^2 + m^2\,. \tag{7.39}$$

Finally, we can now easily write the solutions, satisfying equations (7.28)

$$\Phi = \exp\left[-i(gx)\right]U_1\psi_R, \quad \Psi = \exp\left[-i(gx)\right]U_1\overline{\mathcal{K}}\psi_R v\,, \tag{7.40}$$

where $\overline{\mathcal{K}}$ is defined in (7.33) under the condition (7.37), and ψ_R depends on the photon variables alone and is a solution of the stationary Schrödinger equation

$$\hat{H}\psi_R = R\psi_R\,. \tag{7.41}$$

The energy-momentum vector g^μ of the system has the form

$$g^\mu = p^\mu + n^\mu R\,. \tag{7.42}$$

It is the sum of the energy-momentum vector p^μ of the quasiparticle and that of the quasiphotons, Rn^μ.

Concluding this section, note that if we had started with equation (7.21) in the case of a spinor particle with $\bar{b} \neq 0$ we would have obtained, instead of (7.40) and (7.42),

$$\Psi = U_1 \mathcal{K} \psi_R \sum_{\xi=\pm 1} \left[1 + \zeta(\boldsymbol{\sigma}\mathbf{n})\right] \exp\left[-i(g_\xi x)\right] v\,, \tag{7.43}$$

with the energy-momentum vector g_ζ now depending upon the particle spin

$$g^\mu_\xi = p^\mu + (R + \zeta b)\, n^\mu, \quad b = \frac{\bar{b}}{2(np)}, \quad p^2 = m^2\,, \tag{7.44}$$

and the function ψ_R being again a solution of equation (7.41). Since the condition $\bar{b} \neq 0$ corresponds to nonzero average circular polarization of photons, the dependence of energy (7.44) on the spin ζ and on the quantity b may be recognized as a correlation between the photon and electron spins. This interpretation is, however, of limited value, since, as it was discussed above, the consistent treatment requires $b = 0$.

7.2 Quantized monochromatic plane wave with arbitrary polarization

In Refs. [92, 93, 163, 212] a monochromatic quantized plane wave with a given polarization was taken, although the polarization was always chosen to be linear. This circumstance proved to be crucial in the sense that an essential dependence of the energy and of the wave functions on the particle spin and on the polarization of the wave was missed. The study of solutions for a monochromatic wave of a given arbitrary polarization was accomplished in Ref. [16].

Let only a single term with fixed s and λ be present in the sums (7.14) and (7.32) (that is why we shall omit the indices s and λ throughout this section).

It then follows from (7.32) that

$$Q^\mu = (0, \mathbf{Q}), \quad \mathbf{Q} = \sqrt{\frac{\delta}{\kappa}}\left(a\mathbf{e} + a^+\mathbf{e}^*\right), \quad (\mathbf{e}\mathbf{e}^*) = 1, \quad (\mathbf{n}\mathbf{e}) = (\mathbf{n}\mathbf{e}^*) = 0\,, \tag{7.45}$$

and the Hamiltonian (7.38) can be written in the form of expression (7.232) of Section 7.8, with coefficients A, B, D, H_0 equal to

$$A = \kappa + \frac{\delta}{\kappa(np)}, \quad B = \frac{\delta \mathbf{e}^2}{\kappa(np)}, \quad D = \sqrt{\frac{\delta}{\kappa}}\frac{(\mathbf{p}\mathbf{e}^*)}{(np)}, \quad H_0 = \frac{\delta}{2\kappa(np)}\,. \tag{7.46}$$

Note that the quantity $\bar{b}$ in equation (7.44) disappears only for a linearly polarized photon. Therefore, in the case of arbitrary polarization (the complex vector **e**) one should use expressions (7.43) and (7.44) with

$$b = \frac{i\delta(\mathbf{n}\,[\mathbf{e} \times \mathbf{e}^*])}{2\kappa(np)} \tag{7.47}$$

for the Fock-bispinors.

Let us try to reduce the Hamiltonian to the form (7.246)

$$\hat{H} = rc^+c + \overline{H}_0 \tag{7.48}$$

by means of the linear canonical transformation (7.235) from the photon creation and annihilation operators to new creation and annihilation operators of the quasiphotons. Here u, v and r are now numbers, and it is necessary and sufficient to require that

$$uu^* - vv^* = 1 \tag{7.49}$$

for relations (7.237) to hold true. By solving equations (7.241) together with (7.49) we find

$$\begin{gathered} r^2 = A^2 - BB^*, \quad u = \sqrt{\frac{1}{2}\left(1 + \frac{A}{r}\right)}, \quad v = \frac{A - r}{B}u^*, \\ \gamma = \frac{B^*D^* - AD}{A^2 - BB^*}, \quad \overline{H}_0 = H_0 + \frac{BD^2 + B^*D^{*2} - 2ADD^*}{2r^2} + \frac{r - A}{2}\,. \end{gathered} \tag{7.50}$$

It is obvious that in this case it is sufficient to consider $B \neq 0$, since otherwise $r = A, u = 1, v = 0$. After substituting the values (7.46) into (7.50) we get

$$\begin{gathered} r = \left\{\left[\kappa + \frac{\delta}{\kappa(np)}\right]^2 - \frac{\mu^2\delta^2}{\kappa^2(np)^2}\right\}^{1/2}, \quad \mu = |\mathbf{e}^2|, \quad 0 \le \mu \le 1, \\ \overline{H}_0 = \frac{r - \kappa}{2} - \frac{\delta}{\kappa(np)^2 r^2}\left[\frac{\delta \mathbf{p}_\perp^2(1 - \mu^2)}{2\kappa(np)} + \kappa|(\mathbf{ep})|^2\right]. \end{gathered} \tag{7.51}$$

The eigenvalues R and the corresponding eigenfunctions in equation (7.41) are

$$R = rN + \overline{H}_0, \quad N \in \mathbb{Z}_+\,, \tag{7.52}$$

$$\psi_R = |N\rangle = \frac{(c^+)^N}{\sqrt{N!}}\,|0\rangle\,, \quad c\,|0\rangle = 0\,. \tag{7.53}$$

Here, $|0\rangle$ is, as usual, the vacuum vector for quasiphoton annihilation operators. The spectrum of the operator H depends on the polarization of the wave. This dependence is present both in the expression for $\overline{H}_0$ and in the quasiphoton frequency r. Equations (7.48) and (7.51) hold true provided that r is real and positive. It is seen from (2.82) that this condition is not fulfilled if

$$-(1 + \mu)\delta\kappa^{-2} \le (np) \le -(1 - \mu)\delta\kappa^{-2}\,. \tag{7.54}$$

This means that the range (7.54) is forbidden for (np). Note that in quantum theory the values $(np) < 0$ are admissible. If $\mu = 0$ (circularly polarized photon), the interval (7.54) degenerates to a point, $(np) = -\delta\kappa^{-2}$, with $r = 0$ in this point. If $\mu = 1$ (linearly polarized photon) the interval (7.54) becomes the largest

$$-2\delta\kappa^{-2} \le np \le 0\,. \tag{7.55}$$

If (up) lies inside the interval (7.54), the operator $\hat{H}$ cannot be reduced to the form (7.48). Let us try in this case to reduce it to the form (7.241),

$$\hat{H} = r(c^{+2} + c^2) + \overline{H}\,. \tag{7.56}$$

By requiring that equations (7.247) be obeyed, we obtain (as distinct from (7.51))

$$\begin{aligned}
r &= \sqrt{BB^* - A^2} = \left\{\frac{\mu^2\delta^2}{\kappa^2(np)^2} - \left[\kappa + \frac{\delta}{\kappa(np)}\right]^2\right\}^{1/2},\\
\overline{H} &= \frac{\delta}{\kappa(np)^2 r^2}\left[\frac{\delta \mathbf{p}_\perp^2(1-\mu^2)}{2\kappa(np)} + \kappa|(\mathbf{ep})|^2\right] - \kappa - \frac{\delta}{2\kappa(np)},\\
u &= \sqrt{\frac{1}{2}\left(\frac{|B|}{r}+1\right)}e^{-i\alpha/2}, \quad v = \frac{A}{|A|}\sqrt{\frac{1}{2}\left(\frac{|B|}{r}-1\right)}e^{-i\alpha/2}, \quad B = |B|e^{i\alpha}\,.
\end{aligned} \tag{7.57}$$

Now (7.56) holds true and γ is again defined by equation (7.50).

Thus, there are two domains for (np): if (np) does not belong to the interval (7.54) (we refer to this case as normal), then the Hamiltonian is reduced to the form (7.48) by a linear canonical transformation; if (np) lies inside the interval (7.54) (we refer to this case as abnormal), then a linear canonical transformation reduces the Hamiltonian to the form (7.56). Solution of the stationary Schrödinger equation with the Hamiltonian like (7.56) will be considered in Section 7.5 below.

7.3 Quantized plane wave of general form

Consider the sum over the two possible values of λ for a fixed number s in (7.32). Without indicating the s-dependence explicitly, write the s-th mode of the vector Q,

$$Q^\mu = \sqrt{\frac{\delta}{\kappa}}\sum_\lambda \left(a_\lambda e_\lambda^\mu + a_\lambda^+ e_\lambda^{*\mu}\right), \tag{7.58}$$

and consider two real vectors e_ζ ($\zeta = 1, 2$) subject to the conditions

$$(e_\zeta, e_{\zeta'}) = -\delta_{\zeta,\zeta'}, \quad (ne_\zeta) = 0, \quad e_\zeta^* = e_\zeta\,. \tag{7.59}$$

Then we decompose each vector e_λ into a sum over the vectors e_ζ

$$e_\lambda = \sum_\zeta \alpha_{\zeta\lambda} e_\zeta, \quad e_\lambda^* = \sum_\zeta \alpha_{\zeta\lambda}^* e_\zeta\,, \tag{7.60}$$
$$\alpha_{\zeta\lambda} = -\left(e_\zeta e_\lambda\right), \quad \alpha_{\zeta\lambda}^* = -\left(e_\zeta e_\lambda^*\right).$$

Evidently, each vector e_ζ can be decomposed over the vectors e_λ,

$$e_\zeta = \sum_\lambda \beta_{\zeta\lambda} e_\lambda, \quad \beta_{\zeta\lambda} = -(e_\zeta e_\lambda^*) = \alpha_{\zeta\lambda}^*\,. \tag{7.61}$$

Relations (7.60) and (7.61) imply

$$e_\lambda = \sum_\zeta \alpha_{\zeta\lambda} e_\zeta = \sum_\zeta \alpha_{\zeta\lambda} \sum_{\lambda'} \alpha_{\zeta\lambda'}^* e_{\lambda'} = \sum_{\lambda'} e_{\lambda'} \left(\sum_\zeta \alpha_{\zeta\lambda} \alpha_{\zeta\lambda'}^* \right),$$
$$e_\zeta = \sum_\lambda \alpha_{\zeta\lambda}^* e_\lambda = \sum_\lambda \alpha_{\zeta\lambda}^* \sum_{\zeta'} \alpha_{\zeta'\lambda} e_{\zeta'} = \sum_{\zeta'} e_{\zeta'} \left(\sum_\lambda \alpha_{\zeta'\lambda} \alpha_{\zeta\lambda}^* \right).$$

From this the obvious relations for the 2×2 matrix α whose elements are $\alpha_{\zeta\lambda}$ follow

$$\alpha\alpha^+ = \alpha^+\alpha = 1\,, \tag{7.62}$$

which mean that α is a Hermitian matrix. By substituting the decomposition (7.60) into (7.58), we find

$$Q = \sqrt{\frac{\delta}{\kappa}} \sum_{\lambda,\zeta} e_\zeta \left(\alpha_{\zeta\lambda} a_\lambda + \alpha_{\zeta\lambda}^* a_\lambda^+ \right) = \sqrt{\frac{\delta}{\kappa}} \sum_\zeta e_\zeta \left(b_\zeta + b_\zeta^+ \right),$$
$$b_\zeta = \sum_\lambda \alpha_{\zeta\lambda} a_\lambda, \quad b_\zeta^+ = \sum_\lambda \alpha_{\zeta\lambda}^* a_\lambda^+; \quad a_\lambda = \sum_\zeta \alpha_{\zeta\lambda}^* b_\zeta, \quad a_\lambda^+ = \sum_\zeta \alpha_{\zeta\lambda} b_\zeta^+\,. \tag{7.63}$$

By comparing (7.63) with (7.235), we find that (7.63) is a linear canonical transformation of the operators a_λ^+ and a_λ provided that one sets $u = a$, $v = \gamma = 0$ in (7.235), with equations (7.237) being fulfilled by virtue of (7.62). It also follows from (7.62) that

$$\sum_\lambda a_\lambda^+ a_\lambda = \sum_\zeta b_\zeta^+ b_\zeta\,. \tag{7.64}$$

Consequently, the operator analogous to U_1 (7.31) can be written as the same expression (7.31), but with the operators a_s and a_s^+ replaced by b_s, b_s^+, respectively.

Thus, if the summation is carried out over the two possible polarizations λ for every fixed s in equations (7.14) and (7.32), one can always pass to the decomposition over real polarization vectors with the help of a linear canonical transformation. In other words, in this case there exists a linear canonical transformation (independent for each mode) that brings about the transition from the decomposition over arbitrary polarizations to that over two possible linear polarizations. In this case the parameter $\bar{b}$ in equations (7.22), (7.23) and (7.44) is zero. The transition from arbitrary polarizations to linear ones was first accomplished for this problem in Ref. [15]. We stress once again that this transformation cannot be performed in the Hamiltonian of the previous Section, since the summation over polarizations is absent from it.

In what follows, we shall always assume that the summation over all λ's is present and that the linear canonical transformation leading to the decomposition over linear polarizations has been already performed. Real e_λ can be chosen to be the same for every s. Then one obtains for the vector $\mathbf{Q}$ in (7.32)

$$\mathbf{Q} = \sum_{s,\lambda} \sqrt{\frac{\delta}{\kappa_s}} \left(a_{s,\lambda} + \overset{+}{a}_{s,\lambda}\right) \mathbf{e}_\lambda \,, \tag{7.65}$$

where $\mathbf{e}_\lambda = \mathbf{e}^*_\lambda$, $(\mathbf{e}_\lambda \mathbf{e}_{\lambda'}) = \delta_{\lambda,\lambda'}$, $(\mathbf{n}\mathbf{e}_\lambda) = 0$. The Hamiltonian (7.38) is in this case diagonal with respect to λ and can be written as equation (7.232),

$$\begin{aligned} \hat{H} = & \sum_{k,s,\lambda} \left[A_{ks} \overset{+}{a}_{k,\lambda} a_{s,\lambda} + \frac{1}{2} B_{ks} \left(a_{k,\lambda} a_{s,\lambda} + \overset{+}{a}_{s,\lambda} \overset{+}{a}_{\zeta,\lambda} \right) \right] \\ & + \sum_{s,\lambda} D_{s,\lambda} \left(a_{s,\lambda} + \overset{+}{a}_{s,\lambda} \right) + H_0 \,, \end{aligned} \tag{7.66}$$

with the matrices A and B, the column D, and a real constant H_0 set as

$$\begin{aligned} A_{ks} &= \kappa_k \delta_{ks} + \frac{\varsigma}{2\sqrt{\kappa_k \kappa_s}} \,, \qquad B_{ks} = \frac{\varsigma}{2\sqrt{\kappa_k \kappa_s}} \,, \\ D_{s,\lambda} &= \sqrt{\frac{\varsigma(np)}{2\kappa_s}} \frac{(\mathbf{p}\mathbf{e}_\lambda)}{(np)} \,, \quad H_0 = \frac{\varsigma}{2} \sum_s \kappa_s^{-1} \,, \quad \varsigma = \frac{2\delta}{(np)} \,. \end{aligned} \tag{7.67}$$

One readily finds the matrix $(A+B)^{-1}$. By direct verification it is established that

$$(A+B)^{-1}_{ks} = \frac{\delta_{ks}}{\kappa_k} - \frac{\varsigma}{(1+\varsigma\nu)\sqrt{\kappa_k^3 \kappa_s^3}} \,, \qquad \nu = \sum_s \kappa_s^{-2} \,. \tag{7.68}$$

Therefore, the inverse matrix $(A+B)^{-1}$ exists provided that there exists the quantity ν, and that $\varsigma \neq -\nu^{-1}$. For example, if the condition (7.10) holds true, the quantity ν does exist and is equal to

$$\nu = \kappa_0^{-2} \sum_{s=1}^{\infty} s^{-2} = \frac{\pi^2}{6\kappa_0^2} \,. \tag{7.69}$$

If the summation over s covers an infinite set of values, the quantity H_0 in (7.67) diverges. This is the known logarithmic divergence of QED. If the operators require to be written in the normal form, the term H_0 can be merely eliminated from the Hamiltonian (7.66). If, however, this requirement is not imposed, the divergence in H_0, as we shall clarify below, is removed by the mass renormalization.

7.4 Canonical forms for Hamiltonian of quasiphotons

Various transformations of the Hamiltonian (7.66) have been considered in Refs. [1, 94] and [14, 15, 38, 95, 130, 132, 213]. Here we derive canonical forms of this operator using linear canonical transformations of creation and annihilation operators, see Section 7.8.

Let us try first to reduce the Hamiltonian (7.66) to the following canonical form:

$$\hat{H} = \sum_{k,\lambda} r_k c^+_{k,\lambda} c_{k,\lambda} + \overline{H}_0 \,, \tag{7.70}$$

where r_k are real positive numbers. The number of modes in the sum (7.70) can be taken as both finite $N < \infty$ and infinite $N = \infty$. In the latter case the sum (7.70) must exist. For instance, if conditions (7.16) and (7.15) are imposed, the sum (7.70) will be seen below to make sense. It follows from equations (7.241) and from the special form of the matrices A and B (7.67) that the transformation matrices u and v in equations (7.235) should obey the relations

$$u_{ks}(r_s - \kappa_k) = \frac{\varsigma}{2\sqrt{\kappa_k}} \sum_l \frac{u_{ls} - v^*_{ls}}{\sqrt{\kappa_l}} \,,$$
$$v^*_{ks}(r_s + \kappa_k) = \frac{\varsigma}{2\sqrt{\kappa_k}} \sum_l \frac{u_{ls} - v^*_{ls}}{\sqrt{\kappa_l}} \,. \tag{7.71}$$

By simple transformations we can establish that solutions of equations (7.71) should be sought for in the form

$$u_{ks} = \frac{1}{2}\left(\sqrt{\frac{r_s}{\kappa_k}} + \sqrt{\frac{\kappa_k}{r_s}}\right) q_{ks} \,,$$
$$v^*_{ks} = \frac{1}{2}\left(\sqrt{\frac{r_s}{\kappa_k}} - \sqrt{\frac{\kappa_k}{r_s}}\right) q_{ks} \,, \tag{7.72}$$

where the matrix q, whose elements are q_{ks}, has the structure

$$q_{ks} = \frac{q_s}{r_s^2 - \kappa_k^2} \,, \tag{7.73}$$

and the numbers r_s are roots of the dispersion equation

$$\sum_k \frac{1}{r_s^2 - \kappa_k^2} = \frac{1}{\varsigma} \,, \tag{7.74}$$

that determines r_s as functions of ς, i.e. $r_s = r_s(\varsigma)$. After the interaction is switched off ($\delta = \varsigma = 0$) the Hamiltonian (7.66) already has the form (7.70), and $r_s = \kappa_s$. Let us enumerate the roots of equation (7.74), i.e. the functions $r_s(\varsigma)$ in such a way that

$r_s(0) = \kappa_s$, which is in agreement with equation (7.74) itself. The numbers q_s in (7.73) are arbitrary and are not determined by equations (7.71). To fix them we require that the conditions (7.237) should be fulfilled. These conditions can be satisfied provided that the matrix q_{ks} is chosen to be real and orthogonal

$$q^T q = 1 \Longrightarrow \sum_l q_{ls} q_{lk} = \delta_{ks} , \tag{7.75}$$

$$q q^T = 1 \Longrightarrow \sum_l q_{kl} q_{sl} = \delta_{ks} . \tag{7.76}$$

It is well known that condition (7.76) is a consequence of (7.75). In our case, however, the verification and the use of these two conditions very much differ. After substituting (7.73) into (7.75) we obtain

$$\sum_l \frac{q_s^2}{(r_s^2 - \kappa_l^2)(r_k^2 - \kappa_l^2)} = \delta_{ks} . \tag{7.77}$$

If $k \neq s$, we have

$$\sum_l \frac{1}{(r_s^2 - \kappa_l^2)(r_k^2 - \kappa_l^2)} = \frac{1}{r_k^2 - r_s^2} \sum_l \left(\frac{1}{r_s^2 - \kappa_l^2} - \frac{1}{r_k^2 - \kappa_l^2} \right) = 0 ,$$

due to equation (7.74). At $k = s$ we have the possibility of fixing q_s as

$$q_s = \left[\sum_l (r_s^2 - \kappa_l^2)^{-2} \right]^{-1/2} . \tag{7.78}$$

From (7.73) and (7.74) it follows that $q_s = \varsigma \sum_k q_{ks}$. Thus, we see that for the Hamiltonian to be reduced to the form (7.70) it is necessary that the roots $r_s^2(\varsigma)$ of the dispersion equation (7.74) should be positive.

Now we are going to study the properties of the roots of the dispersion equation (7.74). The notation $x_s(\varsigma) = r_s^2(\varsigma)$ will be used, where the function $x_s(\varsigma)$ is determined by the condition

$$\sum_k \frac{\varsigma}{x_s(\varsigma) - \kappa_k^2} = 1, \quad x_s(0) = \kappa_s^2 . \tag{7.79}$$

Equation (7.79) will also be referred to as the dispersion equation.

By differentiating (7.79) with respect to ς we find

$$\frac{dx_s(\varsigma)}{d\varsigma} = x_s'(\varsigma) = \varsigma^{-2} \left\{ \sum_l \left[x_s(\varsigma) - \kappa_l^2 \right]^{-2} \right\}^{-1} . \tag{7.80}$$

Hence $x_s'(\varsigma) > 0$ and $x_s(\varsigma)$ are monotonically increasing functions of ς. By comparing (7.80) with (7.78) we find

$$q_s = \varsigma \sqrt{x_s'(\varsigma)}, \quad q_{ks} = \frac{\varsigma \sqrt{x_s'(\varsigma)}}{x_s(\varsigma) - \kappa_k^2} . \tag{7.81}$$

Assuming ς to be small it is an easy matter to find, from equation (7.79), the following expansions

$$x_s \approx \kappa_s^2 + \varsigma + \varsigma^2 A_s + \varsigma^3 \left[A_s^2 - B_s\right] + \dots,$$
$$q_s \approx \varsigma + \varsigma^2 A_s + \varsigma^3 \left[A_s^2 - \frac{3}{2} B_s\right] + \dots, \tag{7.82}$$

where

$$A_s = \sum_{l \neq s} \left(\kappa_s^2 - \kappa_l^2\right)^{-1}, \quad B_s = \sum_{l \neq s} \left(\kappa_s^2 - \kappa_l^2\right)^{-2}.$$

It follows, in particular, from (7.82) that $x_s'(0) = 1$. It can be directly checked that the matrix q diagonalizes the matrix V

$$q^T V q = X(\varsigma), \quad V_{ks} = \kappa_k^2 \delta_{k,s} + \varsigma, \tag{7.83}$$

where the elements of the diagonal matrix $X(\varsigma)$ are $x_s(\varsigma)$. From (7.83), (7.75) and (7.76) we have

$$V = q X(\varsigma) q^T, \quad V_{ks} = \sum_l x_l(\varsigma) q_{kl} q_{sl}. \tag{7.84}$$

Equations (7.84) and (7.76) give rise to many interesting relations. For instance, by setting $k = s$ in (7.76) we find with the use of (7.81)

$$\sum_l \frac{x_l'(\varsigma)}{\left[x_l(\varsigma) - \kappa_k^2\right]^2} = \frac{1}{\varsigma^2}. \tag{7.85}$$

From (7.81) it follows that $x_s'(\varsigma) = \varsigma^{-2}[x_s(\varsigma) - \kappa_k^2]^2 q_{ks}^2$. Let us substitute this expression into the sum

$$\sum_l \frac{x_l'(\varsigma)}{x_l(\varsigma) - \kappa_k^2} = \varsigma^{-2} \sum_l \left[x_l(\varsigma) - \kappa_k^2\right] q_{kl}^2,$$

and take into account that equation (7.84) at $k = s$ implies

$$\sum_l x_l(\varsigma) q_{kl}^2 = \kappa_k^2 + \varsigma, \tag{7.86}$$

and equation (7.76) at $k = s$ implies $\sum_l q_{kl}^2 = 1$. Taking into account this result and equation (7.86), we finally find

$$\sum_l \frac{x_l'(\varsigma)}{x_l(\varsigma) - \kappa_k^2} = \frac{1}{\varsigma}. \tag{7.87}$$

It is an easy matter to find the matrices V^{-1}, V^{-2}, and so on. One readily checks that

$$(V^{-1})_{ks} = \kappa_k^{-2} \delta_{k,s} - \frac{\varsigma}{(1 + \varsigma\nu)\, \kappa_k^2 \kappa_s^2}, \tag{7.88}$$

$$(V^{-2})_{ks} = \frac{1}{\kappa_k^2 \kappa_s^2} \left[\delta_{ks} - \frac{\varsigma}{1 + \varsigma\nu} \left(\frac{1}{\kappa_k^2} + \frac{1}{\kappa_s^2}\right) + \frac{\varsigma^2 \nu_1}{(1 + \varsigma\nu)^2}\right]. \tag{7.89}$$

The designation

$$\nu_n = \sum_k \kappa_k^{-2-2n}, \quad n \in \mathbb{Z}_+, \quad \nu_0 = \nu \tag{7.90}$$

is henceforth used.

Let us note, in passing, a remarkable fact. Relations (7.88) and (7.90) are valid both for finite and infinite N (the dimension of the matrix V is $N \times N$). The calculation of the matrix V^2 leads, however, to the result

$$(V^2)_{ks} = \kappa_k^2 \delta_{ks} + \varsigma\left(\kappa_k^2 + \kappa_s^2\right) + \varsigma N^2 . \tag{7.91}$$

Hence the matrix V^2 does not exist in the limit $N \to \infty$. It can be easily established by direct verification that the matrix (7.89) is indeed inverse to the matrix (7.91) for any N, but (7.89) does exist for $N \to \infty$, too, provided that the quantity ν exists (the existence of ν implies that of ν_n, $n > 0$) and $\varsigma \neq -\nu^{-1}$.

From (7.83) we find with the use of (7.75) and (7.76)

$$X^{-1}(\varsigma) = q^T V^{-1} q, \quad X^{-n}(\varsigma) = q^T V^{-n} q , \tag{7.92}$$

whence we establish with the help of (7.88) and (7.89) that

$$\sum_l x_l^{-n}(\varsigma) = \operatorname{tr} V^{-n}, \quad \sum_l x_l^{-1}(\varsigma) = \nu - \frac{\varsigma \nu_1}{1 + \varsigma \nu},$$
$$\sum_l x_l^{-2}(\varsigma) = \nu_1 - \frac{2\varsigma \nu_2}{1 + \varsigma\nu} + \frac{\varsigma^2 \nu_1^2}{(1 + \varsigma\nu)^2} . \tag{7.93}$$

Equations (7.81) and $q_s = \varsigma \sum_k q_{ks}$ imply that $x_l'(\varsigma) = \varsigma^{-2} q_l^2 = \sum_{k,s} q_{kl} q_{sl}$. Then after the use of (7.84) we get

$$\sum_l x_l^{-1}(\varsigma) x_l'(\varsigma) = \sum_{k,s,l} q_{kl} x_l^{-1}(\varsigma) q_{sl} = \sum_{k,s} \left(q X^{-1}(\varsigma) q^T\right)_{ks} = \sum_{ks} \left(V^{-1}\right)_{ks} . \tag{7.94}$$

The substitution of equation (7.88) into (7.94) results in

$$\sum_s x_s^{-1}(\varsigma) x_s'(\varsigma) = \frac{\nu}{1 + \varsigma\nu} . \tag{7.95}$$

Similarly one obtains

$$\sum_s \frac{x_s'(\varsigma)}{x_s^n(\varsigma)} = \sum_{ks} (V^{-n})_{ks} \quad \sum_s \frac{x_s'(\varsigma)}{x_s^2(\varsigma)} = \frac{\nu_1}{(1 + \varsigma\nu)^2} . \tag{7.96}$$

Equation (7.95) implies

$$\frac{d}{d\varsigma} \sum_s \ln|x_s(\varsigma)| = \frac{d}{d\varsigma} \ln|1 + \varsigma\nu| \Longrightarrow \prod_s x_s(\varsigma) \kappa_s^{-2} = 1 + \varsigma\nu . \tag{7.97}$$

The summation over l in equation (7.94) gives $\sum_s x_s'(\varsigma) = N$. In the limit $N \to \infty$ this sum does not exist. Equation (7.85) implies that

$$\frac{d}{d\varsigma}\sum_l \frac{1}{x_l(\varsigma) - \kappa_k^2} = \frac{d}{d\varsigma}\frac{1}{\varsigma} \Longrightarrow \sum_l \frac{1}{x_l(\varsigma) - \kappa_k^2} = \frac{1}{\varsigma} - 2\sum_{l \neq k} \frac{1}{\kappa_k^2 - \kappa_l^2} . \tag{7.98}$$

In a similar way, integration over ς of the sum (7.87) results in

$$\prod_l \frac{x_l(\varsigma) - \kappa_k^2}{\delta_{kl} + \kappa_l^2 - \kappa_k^2} = \varsigma . \tag{7.99}$$

Clearly, by similar means one can obtain many other relations characterizing the properties of the roots of the dispersion equation (7.79).

If $N = 1$, one has $x_1(\varsigma) = x(\varsigma) = \kappa^2 + \varsigma$. For $N = 2$,

$$x_{1,2}(\varsigma) = \frac{1}{2}\left(\kappa_1^2 + \kappa_2^2 + 2\alpha \mp \sqrt{(\kappa_2^2 - \kappa_1^2)^2 + 4\alpha^2}\right) . \tag{7.100}$$

The validity of the properties obtained above can be easily checked for these special cases.

For $N = \infty$ and with the periodic conditions (7.8), by taking into account summation formulas (see [191])

$$\begin{aligned} \sum_{n=1}^{\infty}(x^2 - n^2)^{-1} &= \frac{1}{2x^2}(\pi x \cot \pi x - 1) , \\ \sum_{n=1}^{\infty}(x^2 + n^2)^{-1} &= \frac{1}{2x^2}(\pi x \coth \pi x - 1) , \end{aligned} \tag{7.101}$$

we can write (7.79) in the form of the following transcendental equation:

$$\gamma = \frac{\rho_s^2}{\rho_s \cot \rho_s - 1}, \quad \rho_s^2 > 0, \quad \rho_s(0) = \pi s, \quad s \in \mathbb{N} , \tag{7.102}$$

$$\gamma = -\frac{|\rho_s|^2}{|\rho_s| \coth |\rho_s| - 1}, \quad \rho_s^2 < 0 , \tag{7.103}$$

where $\gamma = \pi\varsigma/2\kappa_0^2$ and $\rho_s^2(\gamma) = \pi^2 x_s(\varsigma)/\kappa_0^2$.

From equation (7.103), however, it follows that there exists a single root $\rho_1(\gamma)$ satisfying this equation, and $\gamma < -3$, $\rho_1(-3) = 0$. This root is continued to the region $\gamma > -3$ to become the positive root $\rho_1(\gamma)$ of equation (7.102). At $\rho_1 \to +0$ it also follows from equation (7.102) that $\gamma \to -3$.

Equations (7.102) or (7.103) imply that

$$x_s'(\varsigma) = \rho_s \frac{d\rho_s}{d\gamma} = \frac{\rho_s^2}{\rho_s^2 + \gamma^2 + 3\gamma} . \tag{7.104}$$

Equation (7.104) is valid for any sign of ρ_s^2. Equation (7.81) then implies

$$q_{ks} = \frac{2\gamma}{\rho_s^2 - \pi^2 k^2} \sqrt{\frac{\rho_s^2}{\rho_s^2 + \gamma^2 + 3\gamma}} . \tag{7.105}$$

The quantities ν_n are

$$\nu = \nu_0 = \frac{\pi^2}{6\kappa_0^2}, \quad \nu_1 = \frac{\pi^4}{90\kappa_0^4}, \quad \dots, \quad \nu_n = \left(\frac{2\pi}{\kappa_0}\right)^{2n+2} \frac{|B_{2n+2}|}{2(2n+2)!}, \tag{7.106}$$

where B_n are the Bernoulli numbers (see [191]).

It follows from (7.102) that when $\gamma \to +\infty$ ($\varsigma \to +\infty$) the expansion

$$\rho_s(\gamma) \approx \overline{\varphi}_s \left[1 - \frac{\overline{\varphi}_s^2}{(1+\overline{\varphi}_s^2)\gamma} + \dots\right], \quad s \in \mathbb{N} \tag{7.107}$$

takes place, where $\overline{\varphi}_s$ are positive roots of the equation $\overline{\varphi}_s = \tan\overline{\varphi}_s$, enumerated in order of increasing magnitude: $\overline{\varphi}_s > \overline{\varphi}_{s'}$ if $s > s'$. When $\gamma \to -\infty$, one has

$$\rho_{s+1}(\gamma) \approx \overline{\varphi}_s \left[1 - \frac{\overline{\varphi}_s^2}{(1+\overline{\varphi}_s^2)\gamma} + \dots\right], \quad s \in \mathbb{N} . \tag{7.108}$$

For the root ρ_1, $\rho_1^2 < 0$, we find from (7.103) that $|\rho_1| \approx -\gamma - 1 + \gamma^{-1} + \dots$ as $\gamma \to -\infty$. Thus, when $\gamma \to +\infty$ the root ρ_s tends to $\overline{\varphi}_s$ from below, whereas for $\gamma \to -\infty$ the root ρ_s tends to φ_{s-1} ($s = 2, 3, \dots$) from above, and $|\rho_1|$ grows infinitely.

Now we can describe in general the behavior of the roots of the dispersion equation (7.79) with any number N of modes.

The number of roots of equation (7.79) is equal to that of the modes, N. If $N > 1$, all the roots $x_s(\varsigma)$ are positive for $s > 1$, the root $x_1(\varsigma)$ is positive for $\varsigma > -\nu^{-1}$ and negative for $\varsigma < -\nu^{-1}$. With the value of ς given by (7.67), we find that $x_1(\varsigma) < 0$ for

$$-2\nu\delta < (np) < 0 . \tag{7.109}$$

The domain (7.109) of the values of (np) will be called abnormal, while the rest of the values of (np) will be referred to as belonging to the normal domain. The last of the roots, $x_N(\varsigma)$ (for finite N), grows infinitely as $\varsigma \to +\infty$. For N infinite all the roots are bounded as $\varsigma \to +\infty$, i.e. one has

$$x_s(\varsigma) < \varphi_s, \quad s < N; \; x_s(\varsigma) > \varphi_{s-1}, \quad s > 1 ,$$

where φ_s are roots of the equation $\sum_k (\varphi_s - \kappa_k^2)^{-1} = 0$, enumerated in the order of magnitude. All the roots of this equation are evidently positive, their number being equal to $N - 1$ and $\kappa_s^2 < \varphi_s < \kappa_{s+1}^2$, $1 \le s < N$. The dependence of the roots x_s upon ς is sketched on the Figure 7.1. The asymptotic behavior of the roots for $\varsigma \to \pm\infty$

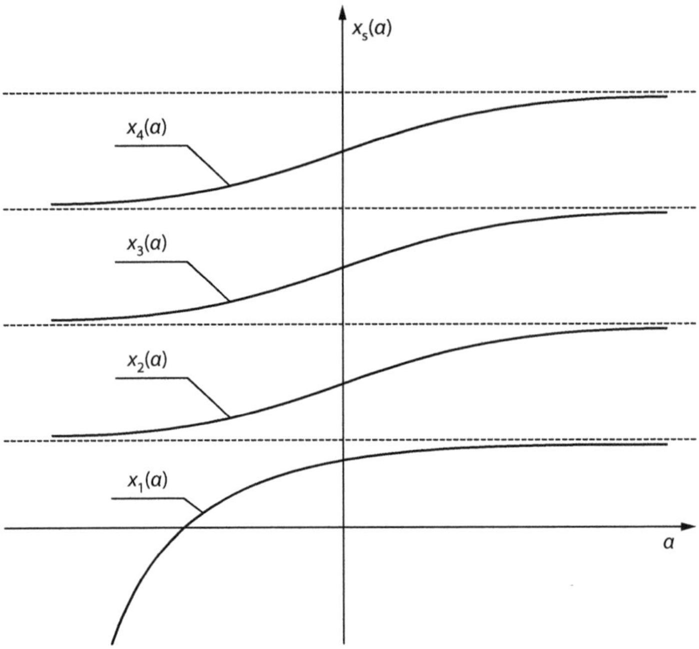

Fig. 7.1. The roots of the dispersion equation.

is shown by the dashed lines that intersect the axis x_s in the points φ_s. The first root $x_1(\varsigma)$ tends to zero at $\varsigma = -\nu^{-1}$, and its derivative is

$$\frac{dx_1(\varsigma)}{d\varsigma} = \frac{\nu^2}{\nu_1}, \quad \varsigma = -\nu^{-1}; \frac{\nu^2}{\nu_1} = \frac{5}{2}, \quad N = \infty, \quad \kappa_s = \kappa_0 s\,.$$

We have thus established that the Hamiltonian (7.66) and (7.67) can only be reduced to the form (7.70) within the normal domain.

To determine the quantity $\overline{H}_0$ it remains to find the quantities $\gamma_{k,\lambda}$. With the help of equation (7.244) they are readily found in our case to be equal to

$$\gamma_{k,\lambda} = -\sqrt{\frac{\varsigma(np)}{2\kappa_k^3}} \frac{(\mathbf{pe}_\lambda)}{(1+\varsigma\nu)(np)}\,. \tag{7.110}$$

Taking this into account we can calculate the quantity

$$\Omega_1 = \sum_{k,\lambda} D_{k,\lambda}\gamma_{k,\lambda} = -\frac{\varsigma\nu\mathbf{p}_\perp^2}{2(1+\varsigma\nu)(np)}\,. \tag{7.111}$$

It is straightforward to make sure that Ω_0,

$$\Omega_0 = -\mathrm{tr}v r v^+ = -\frac{1}{4}\sum_{k,s,\lambda}\left(\frac{r_s}{\kappa_k} + \frac{\kappa_k}{r_s} - 2\right) r_s q_{ks}^2$$
$$= -\sum_k(\kappa_k - r_k) - \frac{1}{2}\sum_{k,s}\frac{q_s^2}{\kappa_k(r_s^2 - \kappa_k^2)} = \sum_k\left(r_k - \kappa_k - \frac{\varsigma}{2\kappa_k}\right)$$

is finite for any N (including infinite N if the periodicity condition is imposed).

Thus, the Hamiltonian is reduced in the normal domain to the form

$$\hat{H} = \sum_{k,\lambda}\sqrt{x_k(\varsigma)}c_{k,\lambda}^+ c_{k,\lambda} + \overline{H}_0,$$
$$\overline{H}_0 = \Omega_0 + \Omega_1 + H_0, \quad H_0 = \frac{\varsigma}{2}\sum_s \kappa_s^{-1}. \tag{7.112}$$

As for the abnormal domain the first mode cannot be represented as (7.93) there. Let us try to reduce the Hamiltonian in this case to the form

$$\hat{H} = -\frac{r_1}{2}\sum_\lambda(c_{1,\lambda}^2 + c_{1,\lambda}^{+2}) + \sum_{k=2,\lambda} r_k c_{k,\lambda}^+ c_{k,\lambda} + \overline{H}_0. \tag{7.113}$$

Then for the value $s = 1$ (and only for this value) we obtain in place of equation (7.71) the following equation:

$$-r_1 u_{k1}^* + \kappa_k v_{k1}^* = \frac{\varsigma}{2\sqrt{\kappa_k}}\sum_l \frac{u_{l1} - v_{l1}^*}{\sqrt{\kappa_l}},$$
$$-r_1 v_{k1} - \kappa_k u_{k1} = \frac{\varsigma}{2\sqrt{\kappa_k}}\sum_l \frac{u_{l1} - v_{l1}^*}{\sqrt{\kappa_l}}. \tag{7.114}$$

The solution of these equations subject to the conditions (7.237) is

$$u_{k1} = \frac{1}{2}\left(\sqrt{\frac{r_1}{\kappa_k}} + \sqrt{\frac{\kappa_k}{r_1}}\right) q_{k1},$$
$$v_{k1}^* = \frac{1}{2}\left(\sqrt{\frac{r_1}{\kappa_k}} - \sqrt{\frac{\kappa_k}{r_1}}\right) q_{k1}, \tag{7.115}$$

where q_{k1} are, as before, defined by equation (7.81) with $x_1(\varsigma)$ being the first (negative in the normal domain) root of the dispersion equation (7.79). The quantity r_1 reads $r_1 = r_1(\varsigma) = \sqrt{|x_1(\varsigma)|}$, i.e. it is real and positive. Clearly, the quantities $\gamma_{k,\lambda}$ in the abnormal domain are, as before, defined by equation (7.110). The quantity $\overline{H}_0$ coincides with the one defined in (7.112).

It remains to test whether the linear canonical transformations performed are indeed proper transformations. As is known, transformations (7.235) are proper provided that

$$J_1 = \sum_{k,\lambda}\gamma_{k,\lambda}^2 < \infty, \quad J_2 = \sum_{ks} v_{ks}^2 < \infty. \tag{7.116}$$

From (7.110) one obtains

$$J_1 = \frac{\varsigma \mathbf{p}_\perp^2}{2(np)(1+\nu\varsigma)^2} \sum_k \kappa_k^{-3}, \sum_k \kappa_k^{-3} < \sum_k \kappa_1^{-1}\kappa_k^{-2} = \frac{\nu}{\kappa_1}.$$

This holds true provided that we have enumerated κ_k in the growing order, consistent with our convention. Hence

$$J_1 < \frac{\varsigma\nu \mathbf{p}_\perp^2}{2(np)\kappa_1(1+\varsigma\nu)^2}, \tag{7.117}$$

and J_1 exists only for $\varsigma \neq -\nu^{-1}$ and $(np) \neq 0$, i.e. singularities only arise in the boundary points of the normal domain. The quantity J_2 is estimated as

$$|J_2| = \left|\sum_{k,s} v_{ks}^2\right| \le \sum_{k,s} \left|v_{ks}^2\right| \left|\frac{r_s}{r_1}\right| = \frac{1}{|r_1|}|\mathrm{tr} v r v^+| = \left|\frac{\Omega_0}{r_1}\right|, \quad J_2 \le \left|\frac{\Omega_0}{r_1}\right|.$$

Therefore, the canonical transformation used is proper, except on the boundary of the normal domain.

Thus, we have shown that there only exist two nonequivalent canonical forms of the Hamiltonian (7.66) and (7.67). These are equation (7.112) in the normal domain and equation (7.113) in the abnormal domain.

In principle, the use of linear canonical transformations is sufficient for answering every question about a particle placed in a quantized plane-wave field, and there is no need to exploit an explicit representation for the creation and annihilation operators. Nonetheless, it sometimes turns out to be convenient for practical purposes to choose a special representation for the creation and annihilation operators. This was just the case in the investigations of Refs. [1, 14, 92–95, 130, 163, 194, 212, 213] and [52, 132, 133, 215, 298, 343]. Here we are going to consider the problem of reduction of the Hamiltonian (7.66) and (7.67) to the canonical form in the coordinate representation.

Assuming that the operators $a_{k,\lambda}$ act on functions of the variable $\xi_{k,\lambda}$, let us define them in agreement with equations (7.231) and the commutation relations (7.17), as

$$a_{k,\lambda} = \frac{1}{\sqrt{2}}\left(\xi_{k,\lambda} + \frac{\partial}{\partial \xi_{k,\lambda}}\right), \quad a_{k,\lambda}^+ = \frac{1}{\sqrt{2}}\left(\xi_{k,\lambda} - \frac{\partial}{\partial \xi_{k,\lambda}}\right). \tag{7.118}$$

Then the Hamiltonian (7.66) and (7.67) takes the form

$$\hat{H} = \sum_{k,\lambda}\left[\frac{\kappa_k}{2}\left(\xi_{k,\lambda}^2 - \frac{\partial^2}{\partial \xi_{k,\lambda}^2} - 1\right) + \sqrt{\frac{\varsigma(np)}{\kappa_k}}\frac{(\mathbf{p}\mathbf{e}_\lambda)}{(np)}\xi_{k,\lambda}\right] + \frac{\varsigma}{2}\sum_{k,s,\lambda}\frac{\xi_{k,\lambda}\xi_{s,\lambda}}{\sqrt{\kappa_k\kappa_s}}. \tag{7.119}$$

One should change the variables $\xi_{k,\lambda}$ in (7.119) in such a way that, in the new variables, the Hamiltonian (7.119) be reduced to a sum of operators acting each on its own variable. One sees from (7.119) that the Hamiltonian is already diagonal in the index λ.

It is also diagonal with respect to the second partial derivatives, and this diagonality should not be violated. The nondiagonality of (7.119) relates only to the variables $\xi_{k,\lambda}$. Thus, we are to perform a linear change of variables $\xi_{k,\lambda}$ that preserves the diagonality of the quadratic form of the second derivatives and diagonalizes the quadratic form of the variables $\xi_{k,\lambda}$. Using the properties of the roots of the dispersion equation (7.79), it can be directly verified that the required change of variables has the form ($y_{k,\lambda}$ are the new variables)

$$\xi_{k,\lambda} = \sqrt{\kappa_k} \sum_s q_{ks} |x_s(\varsigma)|^{-1/4} y_{s,\lambda} + \sqrt{2}\gamma_{k,\lambda} , \tag{7.120}$$

where q_{ks} and $\gamma_{k,\lambda}$ are defined by equations (7.81) and (7.110), respectively. Relation (7.120) is readily inverted to give

$$y_{k,\lambda} = |x_k(\varsigma)|^{1/4} \sum_s \frac{q_{sk}}{\sqrt{\kappa_s}} \xi_{s,\lambda} + \epsilon_k \frac{(\mathbf{pe}_\lambda)}{(np)} \sqrt{\frac{\varsigma(np)x_k'(\varsigma)}{|x_k(\varsigma)|^{3/2}}} ,$$

where $\epsilon_k = 1, \quad k > 1; \quad \epsilon_1 = \operatorname{sgn} x_1(\varsigma)$. After the change of variables the Hamiltonian (7.119) takes the form

$$\hat{H} = \frac{1}{2} \sum_{k,\lambda} \sqrt{|x_k(\varsigma)|} \left(\epsilon_k y_{k,\lambda}^2 - \frac{\partial^2}{\partial y_{k,\lambda}^2} - 1 \right) + \overline{H}_0 , \tag{7.121}$$

where $\overline{H}_0$ is given as (7.112).

Finding the Jacobian of the transformation from the variables $\xi_{k,\lambda}$ to the variables $y_{k,\lambda}$ is an easy matter, since q_{sk} is an orthogonal matrix, and hence $\det q_{sk} = 1$. Now, it follows from (7.120) that

$$J = \frac{D(\xi_{k,\lambda})}{D(y_{s,\lambda})} = \det \left\| \left(\frac{\kappa_k^2}{|x_s(\varsigma)|} \right)^{1/4} q_{ks} \right\| = \left(\prod_k \frac{\kappa_k^2}{|x_s(\varsigma)|} \right)^{1/4} .$$

Taking into account (7.97) we finally obtain

$$J = |I|, \quad I = (1 + \varsigma v)^{-1/4} . \tag{7.122}$$

7.5 Stationary and coherent states

7.5.1 Stationary states

Here we concern ourselves with finding the stationary states of the system under consideration, i.e. with finding stationary solutions to equation (7.41). In the normal domain, the Hamiltonian $\hat{H}$ takes the form (7.112), and the stationary states of the system, in fact, are already given by equation (7.41). We need to determine the function ψ_R. Let us represent this function as follows:

$$\psi_R = \prod_{k,\lambda} |N_{k,\lambda}\rangle , \tag{7.123}$$

where $|N_{k,\lambda}\rangle$ are eigenfunctions of the operators $c^{+}_{k,\lambda}c_{k,\lambda}$ with the integers $N_{k,\lambda} = 0, 1, 2, \ldots$ as the eigenvalues:

$$c^{+}_{k,\lambda}c_{k,\lambda}\,|N_{k,\lambda}\rangle = N_{k,\lambda}\,|N_{k,\lambda}\rangle\,, \quad \langle N_{k,\lambda}|N'_{k,\lambda}\rangle = \delta_{N_{k,\lambda},N'_{k,\lambda}}\,,$$
$$|N_{k,\lambda}\rangle = (N_{k,\lambda})^{-1/2}\,(c^{+}_{k,\lambda})^{N_{k,\lambda}}\,|0\rangle_{k,\lambda}\,, \quad c_{k,\lambda}\,|0\rangle_{k,\lambda} = 0\,. \tag{7.124}$$

Here $|0\rangle_{k,\lambda}$ is the vacuum vector for the annihilation operator $c_{k,\lambda}$ of the quasiphoton. As follows from (7.118), in the coordinate representation we have

$$c_{k,\lambda} = \frac{1}{\sqrt{2}}\left(y_{k,\lambda} + \frac{\partial}{\partial y_{k,\lambda}}\right), \quad c^{+}_{k,\lambda} = \frac{1}{\sqrt{2}}\left(y_{k,\lambda} - \frac{\partial}{\partial y_{k,\lambda}}\right),$$
$$c^{+}_{k,\lambda}c_{k,\lambda} = \frac{1}{2}\left(y^2_{k,\lambda} - \frac{\partial}{\partial y^2_{k,\lambda}} - 1\right), \quad c^{+2}_{k,\lambda} + c^{2}_{k,\lambda} = y^2_{k,\lambda} + \frac{\partial^2}{\partial y^2_{k,\lambda}}\,. \tag{7.125}$$

For $|N_{k,\lambda}\rangle$ one obtains in the coordinate representation

$$|N_{k,\lambda}\rangle = U_{N_{k,\lambda}}(y_{k,\lambda})\,,$$
$$U_n(x) = \left(2^n n!\sqrt{\pi}\right)^{-1/2} e^{-x^2/2} H_n(x), \quad n = 0, 1, 2, \ldots\,, \tag{7.126}$$

where $H_n(x)$ are the Hermite polynomials [191]. The quantity R from (7.41) is obtained in accordance with (7.112) to be

$$R = \sum_{k,\lambda} \sqrt{\varkappa_k(\varsigma)N_{k,\lambda}} + \overline{H}_0\,. \tag{7.127}$$

In the abnormal domain one should take

$$\psi_R = \prod_{\lambda} f^{\pm}_{\lambda} \prod_{k=2} |N_{k,\lambda}\rangle\,, \tag{7.128}$$

where $|N_{k,\lambda}\rangle$ are, as before, the eigenfunctions defined by (7.124) for $k > 1$, and $f^{\pm}_{\lambda}$ are eigenfunctions of the operator $c^{+2}_{k,\lambda} + c^{2}_{k,\lambda}$ with the eigenvalues ω_λ

$$(c^{2}_{1,\lambda} + c^{+2}_{1,\lambda})f^{\pm}_{\lambda} = \omega_\lambda f^{\pm}_{\lambda}\,. \tag{7.129}$$

Let us consider the functions $f^{\pm}(\omega, x)$ that are, in the coordinate representation, solutions of the equation

$$\left(\frac{d^2}{dx^2} + x^2\right) f^{\pm}(\omega, x) = \omega f^{\pm}(\omega, x)\,. \tag{7.130}$$

For any real ω there are two linearly independent solutions of equation (7.130). For instance, the even, $f^{+}(\omega, x)$, and odd, $f^{-}(\omega, x)$, solutions, with respect to x, are

$$f^{+}(\omega, x) = e^{-ix^2}\Phi\left(\frac{1 - i\omega}{4}, \frac{1}{2}; ix^2\right),$$
$$f^{-}(\omega, x) = xe^{-ix^2}\Phi\left(\frac{3 - i\omega}{4}, \frac{3}{2}; ix^2\right), \tag{7.131}$$

where $\Phi(\alpha, \beta; x)$ is the confluent hypergeometric function. Other fundamental solutions of equation (7.130) can be chosen as the parabolic cylinder functions

$$f^{\pm}(\omega, x) = D_{\frac{i\omega-1}{2}}[\pm(1+i)x] . \qquad (7.132)$$

The eigenfunctions $f_\lambda^{\pm}$ in the coordinate representation are $f_\lambda^{\pm} = f^{\pm}(\omega_\lambda, y_{1,\lambda})$. Every real value ω_λ is two-fold degenerate, denoted by the superscripts $\pm$. The quantity R in the abnormal domain is found to be

$$R = -\sum_\lambda \sqrt{|x_1(\varsigma)|}(\omega_\lambda + 1) + \sum_{k=2,\lambda} \sqrt{x_k(\varsigma)} N_{k,\lambda} + \overline{H}_0 . \qquad (7.133)$$

Let us now discuss the question of the complete set of quantum numbers for the stationary states. As far as the scalar particle is concerned, for this set one can take the three-dimensional momentum **g** of the whole system, and the occupation numbers $N_{k,\lambda}$ in the normal domain ($k \geq 1$) or the numbers ω_λ and $N_{k,\lambda}$ ($k > 1$) in the abnormal domain. For a spinor particle the spin quantum number ζ should be added which only determines the structure of the spinor v in (7.40). Relation (7.39) fixes the energy g^0 as a function of **g** and of the occupation numbers $N_{k,\lambda}$ (or $N_{k,\lambda}$ and ω_λ in the abnormal domain). Indeed, equation (7.42) implies

$$g^0 = \frac{1}{2(ng)}\left[m^2 + 2(ng)R + \mathbf{g}_\perp^2 + (ng)^2\right], \quad \mathbf{g}_\perp = \mathbf{g} - \mathbf{n}\,(\mathbf{ng}) . \qquad (7.134)$$

Taking into account that $(ng) = (np)$, and that R is, according to (7.127) or (7.133) and to the dispersion equation (7.79), also a function of (ng), we conclude that (7.134) is a transcendental equation determining g^0 as a function of **g** and of the quantum numbers $N_{k,\lambda}$. However, we cannot in practice solve this equation explicitly.

The situation is essentially simplified by choosing the quantities $\mathbf{g}_\perp = \mathbf{p}_\perp$, $(ng) = (np)$, $N_{k,\lambda}$ (or ω_λ and $N_{k,\lambda}$ in the abnormal domain) as independent quantum numbers. Then equation (7.134) gives the energy g^0 of the system as an explicit function of the independent quantum numbers.

In case the periodic conditions are imposed on the potentials and the relation (7.10) is fulfilled, the choice of quantum numbers must be made in agreement with these conditions. Namely, if we demand that the wave functions obey the property

$$\Psi(x^0, \mathbf{r}) = \Psi(x^0, \mathbf{r} + \mathbf{L}) , \qquad (7.135)$$

this leads to independent quantization of the components of **g**. Independent quantization of the integrals of motion $\mathbf{g}_\perp$ and (ng) is possible provided that the periodicity conditions are imposed in the light-cone variables (2.37)

$$\Psi(u^0, \mathbf{u}) = \Psi(u^0, \mathbf{u} + \mathbf{L}) . \qquad (7.136)$$

Then, in place of the relation (7.135), we shall have

$$\Psi(x^0, \mathbf{r} + \mathbf{L}) = e^{-ig^0 L}\Psi(x^0, \mathbf{r}) , \qquad (7.137)$$

which means that the physical pattern repeats itself in every three-dimensional block of volume L^3. Under the conditions (7.136) the independent integral of motion $(ng) = (np)$ is quantized as follows:

$$(ng) = (np) = \frac{4\pi}{L} l_3, \quad l_3 \in \mathbb{Z} . \tag{7.138}$$

Using (7.15), (7.69) and (7.109) we obtain that in this case the abnormal domain of (np) is

$$-\frac{1}{96\pi} \frac{e^2}{\hbar c} \frac{4\pi}{L} < (np) < 0 . \tag{7.139}$$

Taking into account that for the electron

$$\frac{1}{96\pi} \frac{e^2}{\hbar c} \approx 2.4 \times 10^{-5} < 1 , \tag{7.140}$$

and that (np) is quantized, we see that the abnormal domain of (np) does not exist.

As noted above, when the number of modes is infinite, the quantity H_0 in (7.112) diverges. According to QED, divergence of this type can be removed by mass renormalization. This divergence only contributes to the energy g^0 through (see equation (7.134))

$$m^2 + 2(np)H_0 = m^2 + 2\delta \sum_k \kappa_k^{-1} . \tag{7.141}$$

Let us define the physical mass m^* by the relation

$$m^2 = m^{*2} - 2\delta \sum_k \kappa_k^{-1} , \tag{7.142}$$

and transform equation (7.134) to

$$g^0 = \frac{1}{2(ng)} \left[m^{*2} + 2(ng)R^* + \mathbf{g}_\perp^2 + (ng)^2 \right] , \quad R^* = R - H_0 . \tag{7.143}$$

The quantity R^* no longer contains the divergency even when the number of modes in the quantized wave is infinite. By recognizing m^* as a finite quantity coinciding with the experimentally observed mass of the particle we obtain the finite value (7.143) for the energy g^0 of the system. Note that the value of m determined by (7.142) is in this case infinite and even purely imaginary. Thus, the mass renormalization provides, indeed, the finiteness of the physical characteristics of the system under study.

7.5.2 Relations of orthogonality, normalization and completeness

In establishing the orthogonality relations for the wave functions of the stationary states, we confine ourselves to consideration of only one mode of the wave ($N = 1$).

This assumption appreciably simplifies calculations without introducing any important limitations. The Fock-bispinor wave functions of the stationary states are given by equation (7.40) under the choice of the function ψ_R in the form (7.123). We will take (np), $\mathbf{p}_\perp$, N_λ, ζ as independent quantum numbers.

Consider the scalar product of the function (7.40) on the hyperplane $x^0 = \text{const}$. It is easily found that

$$\begin{aligned}\langle\Psi'|\Psi\rangle &= 2L^3 v^+ v \delta_{\mathbf{p}'_\perp,\mathbf{p}_\perp} \delta_{(\mathbf{ng}'),(\mathbf{ng})} \exp\left[i\left(g_0^1 - g_0\right)x^0\right] \\ &\times \left\{\left[m^2 + (np)(np') + I^8 \mathbf{p}_\perp + \frac{\varsigma(np)I^2}{\kappa}\sum_\lambda (N_\lambda + 1/2)\right] J_{N_1,N_1'} J_{N_2,N_2'}\right. \\ &- I^5\sqrt{\frac{2\varsigma(np)}{\kappa}}\left[(\mathbf{pe}_1)\left(\sqrt{N_1+1}J_{N_1+1,N_1'} + \sqrt{N_1}J_{N_1-1,N_1'}\right)J_{N_2,N_2'}\right. \\ &\left.+(\mathbf{pe}_2)\left(\sqrt{N_2+1}J_{N_2+1,N_1'} + \sqrt{N_2}J_{N_2-1,N_2'}\right)J_{N_1,N_1'}\right] \\ &+ \frac{\varsigma(np)I^2}{2\kappa}\left[\left(\sqrt{(N_1+1)(N_1+2)}J_{N_1+2,N_1'} + \sqrt{N_1(N_1-1)}J_{N_1-2,N_1'}\right)J_{N_2,N_2'}\right. \\ &\left.\left.+ \left(\sqrt{(N_2+1)(N_2+2)}J_{N_2+2,N_2'} + \sqrt{N_2(N_2-1)}J_{N_2-2,N_2'}\right)J_{N_1,N_1'}\right]\right\},\end{aligned} \tag{7.144}$$

where the quantity I is defined as (7.122). In (7.144) the quantum numbers of the primed wave function Ψ' are $\mathbf{p}'_\perp$, N'_λ, ζ', and the designation

$$J_{N_\lambda,N'_\lambda} = {}_{c'}\langle 0'|\left(c'_\lambda\right)^{N'_\lambda}(\overset{+}{c}_\lambda)^{N_\lambda}|0\rangle_c \left(N_\lambda! N'_\lambda!\right)^{-1/2} \tag{7.145}$$

is used. If $(np) = (np')$ one has $c_\lambda = c'_\lambda$ and the orthogonality of the functions (7.144) is evident. If, however, $(np) \neq (np')$ the following recurrence relation can be established:

$$\begin{aligned}&\sqrt{(N+1)(N+2)}J_{N+2,N'} + \sqrt{N(N-1)}J_{N-2,N'} \\ &= \frac{2(\mathbf{pe})I^3}{\kappa}\sqrt{\frac{2}{\varsigma(np)\kappa}}\left(\sqrt{N+1}J_{N+1,N'} + \sqrt{N}J_{N-1,N'}\right) \\ &+ \left\{\frac{2[(np')-(np)](\mathbf{pe})^2 I^6 I'^6}{\kappa(np)(np')} + 2N + 1 + \frac{2I^2 I'^2}{I^4 - I'^4}\left(2N'+1\right)\right\}J_{N,N'},\end{aligned}$$

with the use of which we can reduce equation (7.144) to the form

$$\begin{aligned}\langle\Psi'|\Psi\rangle &= 4L^3 v^{+'} v \delta_{p'_\perp,p_\perp}\delta_{(\mathbf{ng}'),(\mathbf{pg})}\left[(\mathbf{ng}) - (\mathbf{ng}')\right] \\ &\times (np)(np')\left[(np') - (np)\right]^{-1}\exp\left[i\left(g'_0 - g_0\right)t\right] J_{N_1,N_1'} J_{N_2,N_2'},\end{aligned}$$

whence it is seen that this expression is always equal to zero because $[(\mathbf{ng}) - (\mathbf{ng}')] \times \delta_{(\mathbf{ng}'),(\mathbf{ng})} = 0$. The orthogonality of the functions (7.40) relative to the scalar product on the light-cone plane can be checked more easily. One obtains from (7.40)

$$\left\langle \Psi' \right| [1 - (\boldsymbol{\alpha}\mathbf{n})] \left|\Psi\right\rangle = 4L^3 v^{+'} v(np)^2 \delta_{\mathbf{p}_\perp,\mathbf{p}'_\perp} \delta_{(np),(np')} \prod_\lambda \delta_{N_\lambda,N'_\lambda} . \tag{7.146}$$

The normalization condition for the spinor v is determined by (7.146) to be $v^+ v = [4L^3(np)^2]^{-1}$. Expression (7.146) remains valid with any number of modes. It is equally easy to check the completeness condition on the light-cone plane:

$$\sum_{(np),\mathbf{p}_\perp,\{N_{k,\lambda}\},\zeta} P_{(-)} \left|\Psi_{(np),\mathbf{p}_\perp,N_{k,\lambda,\zeta}}(u^0,\mathbf{u})\right\rangle \left\langle \Psi_{(np),\mathbf{p}_\perp,N_{k,\lambda,\zeta}}(u^0,\mathbf{u}')\right| P_{(-)} = P_{(-)}\delta(\mathbf{u}-\mathbf{u}'),$$

where $P_{(-)}$ is given by (2.70). Thus, in the present case the use of the scalar product on the null plane is technically very convenient.

7.6 Reduction to Volkov solutions

One can see that the Volkov solutions for an electron moving in the classical plane-wave considered in Section 3.2 can be obtained as a certain limit from the Fock-bispinors studied in the present chapter.

Let us take the Fock-bispinor (7.36) and rewrite it as follows:

$$\Psi = \exp\left[-i(px)\right] \overline{\mathcal{K}}\psi(u_0)v, \quad \Psi = \exp\left[-i(px)\right] \psi(u_0) , \tag{7.147}$$

where the operator $\overline{\mathcal{K}}$ is given by equation (7.33) with $\mathbf{F} = \mathbf{p}_\perp - e\mathbf{A}(u_0)$, and the operator-valued quantized potential $\mathbf{A}(u_0)$ is defined by equations (7.14) and (7.16). Now the equation for the Fock-bispinor ψ reads

$$\left[i\partial_{u_0} - \frac{e\,(A(u_0)p)}{(np)} + \frac{e^2A^2(u_0)}{2\,(np)} \right] \psi(u_0) = 0 . \tag{7.148}$$

The general solution of equation (7.148) can be written using the evolution operator as

$$\psi(u_0) = U(u_0)\psi(0) , \tag{7.149}$$

where $\psi(0)$ is an arbitrary u_0-independent vector in the Fock space of the plane-wave photons. The evolution operator $U(u_0)$ can be written as the product ordered with respect to the variable u_0 (the chronological T_{u_0}-product)

$$U(u_0) = T_{u_0} \exp\left\{ -i \int_0^{u_0} \left[-\frac{e\,(pA\,(u_0))}{(np)} + \frac{e^2A^2\,(u_0)}{2\,(np)} \right] du_0 \right\} . \tag{7.150}$$

Solutions (7.149) also include stationary states (7.123), provided $\psi(0)$ is chosen to be an eigenvector for the operator $\hat{H}$ defined by equation (7.38).

States (7.147) are characterized by the fact that the quasiparticle energy – momentum vector (7.29) is taken as an integral of motion, which also holds true for the Volkov solution.

Let us choose for $\psi(0)$ the Fock-bispinors subject to the conditions

$$\langle\psi(0)|\, a_{k,\lambda}\, |\psi(0)\rangle = z_{k,\lambda}, \quad \langle\psi(0)|\, \overset{+}{a}_{k,\lambda} a_{k,\lambda}\, |\psi(0)\rangle = |z_{k,\lambda}|^2 \,. \tag{7.151}$$

The only class of Fock-bispinors that can satisfy conditions (7.151) are coherent states:

$$a_{k,\lambda}\,|z\rangle = z_{k,\lambda}\,|z\rangle\,, \quad |\psi(0)\rangle = |z\rangle \,. \tag{7.152}$$

It is evident from the physical point of view that when going to the limiting case of the classical electromagnetic field, the quantity $\epsilon_{k,\lambda} = |z_{k,\lambda}|^2 L^{-3}$ should be understood as the density of the number of photons of the sort k, λ in a cube of volume L^3, while the density of particles in the same volume is $\rho = L^{-3}$. The possibility of treating the electromagnetic field as a fixed (classical) is provided by the condition $\rho \ll \epsilon_{k,\lambda}$ which means the absence of the back reaction of the particles on the electromagnetic field. Noting the equality

$$L^{-3/2}\overset{+}{a}_{k,\lambda}\psi(0) = \sqrt{\epsilon_{k,\lambda}}\, \exp\left(-i \arg z_{k,\lambda}\right) \psi(0) + \sqrt{\rho}\varphi \,, \tag{7.153}$$

where φ is a unit-norm vector orthogonal to the vector $\psi(0)$, we conclude that, if $\epsilon_{k,\lambda} \gg \rho$, the approximate equality

$$L^{-3/2}\overset{+}{a}_{k,\lambda}\psi(0) \approx \overset{*}{z}_{k,\lambda} L^{-3/2}\psi(0)$$

holds true and that, in this case, equations (7.149) and (7.150) imply that

$$\psi(u_0) \approx \exp\left\{-i\int\limits_0^{u_0}\left[-\frac{e(A_{\mathrm{cl}}(u_0)p)}{(np)} + \frac{e^2A^2_{\mathrm{cl}}(u_0)}{2(np)}\right]du_0\right\}\psi(0)\,, \tag{7.154}$$

where $A_{\mathrm{cl}}(u_0)$ is the classical potential of the external plane-wave field given as the Fourier series

$$|e|A^{\mu}_{\mathrm{cl}}(u_0) = \sum_{s,\lambda}\sqrt{\frac{\delta}{\kappa_s}}\left[z_{s,\lambda}e^{\mu}_{\lambda}(s)\exp(-i\kappa_s u_0) + \overset{*}{z}_{s,\lambda}e^{*\mu}_{\lambda}(s)\exp(-i\kappa_s u_0)\right]\,, \tag{7.155}$$

with arbitrary coefficients $z_{s,\lambda}$. The polarization vectors $e^{\mu}_{\lambda}(s)$ have, as before, the properties (7.16). Finally, one obtains the approximate expression for the solutions (7.147) in this case

$$\Psi \approx \exp(-iS)\overline{\mathcal{K}}v\psi(0), \quad \Phi \approx \exp(-iS)\psi(0)\,,$$

$$S = (px) + \int\limits_0^{u_0}\left[-\frac{e(A_{\mathrm{cl}}(u_0)p)}{(np)} + \frac{e^2A^2_{\mathrm{cl}}(u_0)}{2(np)}\right]du_0 \,. \tag{7.156}$$

The operator $\overline{\mathcal{K}}$ is given by equation (7.33), in which one should now set $\mathbf{F} = \mathbf{p}_\perp - e\mathbf{A}_{\text{cl}}(u_0)$. Solutions (7.156) are the Volkov functions studied in Section 3.2, multiplied by the vector $\psi(0)$ in the Fock space of plane-wave photons, which is interpreted in the sense that the particles do not affect the field in which they move.

7.7 Electron interacting with quantized plane-wave and with external electromagnetic background

In the papers [1, 93, 130, 212, 213] the problem considered in this Chapter was extended to include, together with the quantized plane-wave field, the homogeneous and constant external classical magnetic field. In Refs. [17, 23] additional fields were taken, such as the classical plane-wave field and the running electric field. The authors of Ref. [298] formulated and solved the problem of finding all classical fields in the K–G equation which allow separation of variables when combined with the quantized plane-wave field. The results of that work made it possible to find, in Ref. [20], the wave functions of the system when complicated external electromagnetic fields are present.

The common feature of all the classical external fields which allow exact solutions when combined with the quantized plane wave is that the problem for all of them is reduced after some transformations to studying a system whose Hamiltonian is quadratic in the creation and annihilation operators with, generally, time-dependent coefficients.

7.7.1 Classical plane wave along the quantized field

Assume that the potentials $A^\mu(x)$ involved in the Dirac and K–G equations have now the form (7.18) where $A^\mu_{\text{ext}}(x)$ are the potentials of the external classical field with the structure of (2.103). Besides, we always mean that the wave functions (Fock-bispinors) correspond to the states with a given integral of motion (np), and that in the light-cone variables (2.37) the potential component $(\tilde{A}_{\text{ext}})_3$ of the classical field is of the form (3.136) and equation (3.137) holds true, i.e.

$$|e|(\tilde{A}_{\text{ext}})_3 = \frac{1}{2}g(u_0), \quad |e|E_z = |e|(\mathbf{nE}) = |e|E_3 = g'(u_0)\,.$$

Solutions of the Dirac and K–G equations will be sought in the same way as (4.142), (4.143) and (7.33), in the form

$$\Phi = P^{-1/2}\exp\left[-\frac{i}{2}(np)u_3\right]U_1\psi,\ P = (np) + g(u_0)\,,$$

$$\Psi = P^{-1}\exp\left[-\frac{i}{2}(np)u_3\right]U_1\mathcal{K}\left\{\sum_{\zeta=\pm1}[1+\zeta(\sigma\mathbf{n})]\,e^{i\zeta t}\right\}\psi v\,. \tag{7.157}$$

The operator U_1 is given as (7.31), while for the operator $\mathcal{K}$ equation (4.143) is valid, wherein one should set

$$\mathbf{F} = i\nabla_\perp - \mathbf{Q} + e(\mathbf{A}_{\text{ext}})_\perp, \quad (\mathbf{A}_{\text{ext}})_\perp = \mathbf{A}_{\text{ext}} - \mathbf{n}\left(\mathbf{n}\mathbf{A}_{\text{ext}}\right) . \tag{7.158}$$

Here the operator $\mathbf{Q}$ is given by equation (7.65). The function $t = t(u_0)$ is to be determined separately for each classical field.

If condition $(\mathbf{nH}) = 0$ is fulfilled for the classical electromagnetic magnetic field the Fock-bispinor is given by an expression simpler than (7.157)

$$\Psi = P^{-1} \exp\left[-\frac{i}{2}(np)u_3\right] U_1 \mathcal{K}\psi v . \tag{7.159}$$

This follows from equation (7.157) at $t = 0$.

The function ψ in equation (7.157) satisfies the Schrödinger equation

$$i\partial_0\psi = \mathcal{H}\psi, \quad \mathcal{H} = \frac{\mathbf{F}^2 + m^2}{2P} + \sum_{k,\lambda} \kappa_k a^+_{k,\lambda} a_{k,\lambda} . \tag{7.160}$$

Following Ref. [17] we shall consider solutions of equation (7.160) in the case when the plane-wave is chosen as an external classical field, that runs in the same direction $\mathbf{n}$ as the quantized wave. Then

$$|e|(\mathbf{A}_{\text{ext}})_\perp = \mathbf{A}(u_0), \quad (\mathbf{nA}) = 0, \quad g(u_0) = 0 . \tag{7.161}$$

The dependence of the vector $\mathbf{A}(u_0)$ on u_0 is otherwise arbitrary. In this case it is more convenient to make use of the Ansatz

$$\Phi = \exp\left[-i(gx)\right] U_1\psi, \quad \Psi = \exp[-i(gx)] U_1 \mathcal{K}\psi v , \tag{7.162}$$

instead of (7.157). We have set

$$g^\mu = p^\mu + Rn^\mu, \quad p^2 = m^2 \tag{7.163}$$

in (7.162). The operator $\mathcal{K}$ has again the form (3.144) where one should set

$$\mathbf{F} = -\left[\mathbf{p}_\perp + \mathbf{Q} + \mathbf{A}(u_0)\right] , \tag{7.164}$$

so far as the present problem is considered. The function ψ, in (7.162) obeys the equation

$$\begin{aligned} &i\partial_0\psi = (\bar{H} - R)\psi, \\ &\bar{H} = \frac{(\mathbf{pQ})}{(np)} + \frac{\mathbf{Q}^2}{2(np)} + \frac{(\mathbf{QA})}{(np)} + \frac{(\mathbf{pA})}{(np)} + \frac{\mathbf{A}^2}{2(np)} + \sum_{k,\lambda} \kappa_k a^+_{k,\lambda} a_{k,\lambda} , \end{aligned} \tag{7.165}$$

that follows from (7.160) after equations (7.163) and (7.164) are taken into account. If $\mathbf{A} = 0$ the operator $\bar{H}$ coincides with (7.38). The operator $\bar{H}$ contains not only the photon variables and a term depending on u_0 alone, but also crossed terms like $\sim$ (**QA**) which mix the photon variables and u_0.

Let us change to new creation, $c^+_{k,\lambda}$, and annihilation, $c_{k,\lambda}$, operators in $\bar{H}$, using equations (7.235) with the matrices u_{ks} and v_{ks} and the column $\gamma_{k,\lambda}$ chosen as (7.72), (7.81) and (7.110) respectively. We confine ourselves to considering the normal domain of (np). After a transformation, the operator **Q** becomes

$$\mathbf{Q} = \sum_{k,\lambda} \left[\frac{\varsigma(np)x'_k(\varsigma)}{2\sqrt{x_k(\varsigma)}} \right]^{1/2} \left(c_{k,\lambda} + c^+_{k,\lambda} \right) \mathbf{e}_\lambda - \frac{\varsigma\nu}{1+\varsigma\nu} \mathbf{p}_\perp . \tag{7.166}$$

The operator $\bar{H}$ in (7.165) is now

$$\bar{H} = \hat{H} + \frac{(\mathbf{QA})}{(np)} + \frac{(\mathbf{pA})}{(np)} + \frac{\mathbf{A}^2}{2(np)} , \tag{7.167}$$

where $\hat{H}$ is given by (7.93). By substituting here the special value (7.166) of **Q** it is found that

$$\begin{aligned} \bar{H} = \hat{H} &+ \sum_{k,\lambda} \left[\frac{\varsigma(np)x'_k(\varsigma)}{2\sqrt{x_k(\varsigma)}} \right]^{1/2} \frac{\left(c_{k,\lambda} + c^+_{k,\lambda} \right)}{(np)} (\mathbf{A}\mathbf{e}_\lambda) \\ &+ \frac{(\mathbf{pA})}{(np)(1+\varsigma\nu)} + \frac{\mathbf{A}^2}{2(np)} . \end{aligned} \tag{7.168}$$

Let us transform the Fock-bispinor ψ in equation (7.165) as follows:

$$\psi = U_3 \Phi, \quad U_3 = \exp \left\{ \sum_{k,\lambda} \left[z_{k,\lambda}(u_0) c^+_{k,\lambda} - z^*_{k,\lambda}(u_0) c_{k,\lambda} \right] \right\} , \tag{7.169}$$

and select the quantities $z_{k,\lambda}(u_0)$ lest the transformed equation contains terms linear in $c_{k,\lambda}$ and $c^+_{k,\lambda}$. When carrying out the transformation one should bear in mind that

$$\partial_0 U_3 = U_3 \left[z' a^+ - z^{*\prime} a + \frac{1}{2} \left(z' z^* - z z^{*\prime} \right) \right] . \tag{7.170}$$

Making use of (7.170) and setting

$$\begin{aligned} z_{k,\lambda}(u_0) = &- \frac{i}{(np)} \left[\frac{\varsigma(np)x'_k(\varsigma)}{2\sqrt{x_k(\varsigma)}} \right]^{1/2} \\ &\times \int^{u_0} \left(\mathbf{A}(u'_0)\mathbf{e} \right) \exp \left[i \left(u'_0 - u_0 \right) \sqrt{x_k(\varsigma)} \right] du'_0 , \end{aligned} \tag{7.171}$$

we find that the function Φ is a solution to the equation

$$i\partial_0 \Phi = \left[\hat{H} + \partial_0 S(u_0) - R \right] \Phi , \tag{7.172}$$

where the function $S(u_0)$ is given in the following way:

$$S(u_0) = \int^{u_0} \left[\frac{(\mathbf{A}(u_0')\mathbf{p})}{(np)(1+\varsigma\nu)} + \frac{\mathbf{A}^2(u_0')}{2(np)} + \int^{u_0'} \frac{\mathcal{D}(u_0'' - u_0')}{2(np)} \left(\mathbf{A}(u_0')\mathbf{A}(u_0'')\right) du_0'' \right] du_0' \,. \tag{7.173}$$

The function $\mathcal{D}(x)$ is only determined by the properties of the quantized plane wave. It is

$$\mathcal{D}(x) = \varsigma \sum_k \frac{x_k'(\varsigma)}{\sqrt{x_k(\varsigma)}} \sin\left(x\sqrt{x_k(\varsigma)}\right) \,. \tag{7.174}$$

Now, solutions of equation (7.172) are obtained without further effort. These are

$$\Phi = \exp[-iS(u_0)]\psi_R, \quad \psi = \exp[-iS(u_0)]U_3\psi_R \,, \tag{7.175}$$

where ψ_R is the solution of the stationary equation (7.41), which has the form (7.123), and eigenvalue R is (7.127). If the interaction with the quantized wave is switched off ($\varsigma \to 0$), from (7.162) and (7.175) we obtain the Volkov solutions studied in Section 3.2, multiplied by the stationary wave function of free photons; if $\mathbf{A}(u_0) = 0$, the stationary states (7.40), (7.123) are obtained.

Our treatment does not impose any limitations on the number of modes in the quantized wave, although the most interesting situation arises when an infinite number of modes is considered and, moreover, the frequencies κ_k are so chosen that the following expansion takes place:

$$\mathbf{A}(u_0) = \sum_{k,\lambda} \left[b_{k,\lambda} \exp(-i\kappa_k u_0) + b_{k,\lambda}^* \exp(i\kappa_k u_0) \right] \mathbf{e}_\lambda \,, \tag{7.176}$$

where the $b_{k,\lambda}$ are constants. In this case, one easily finds with the use of (7.174) and (7.79) that

$$\int^{u_0} \mathcal{D}\left(u_0' - u_0\right) \mathbf{A}(u_0') du_0' = -\mathbf{A}(u_0) \,. \tag{7.177}$$

Simultaneously, an especially simple expression is obtained for $S(u_0)$

$$S(u_0) = \int^{u_0} \frac{(\mathbf{A}(u_0')\mathbf{p})}{(np)(1+\varsigma\nu)} du_0' \,. \tag{7.178}$$

The calculation of the mean value of the potential using the functions (7.162) and (7.175) leads to the result

$$|e| \langle \mathbf{A} \rangle = -\mathbf{A}(u_0) - \frac{\varsigma\nu}{1+\varsigma\nu} \mathbf{p}_\perp \,. \tag{7.179}$$

The constant part of the potential, the one proportional to $\mathbf{p}_\perp$, is inessential. Hence equation (7.179) means that we have found such states of the system in which the classical component of the wave field is entirely cancelled owing to the photons re-emitted by the electron in the direction $\mathbf{n}$. This by no means is the total elimination of the wave field by the electron, since the state of the radiation field is not generally characterized completely by the mean value of the potential. For instance, the potential mean value is zero in stationary states with a fixed number of photons.

7.7.2 Classical magnetic field directed along the quantized plane wave

Let us consider a particle in the quantized plane-wave field combined with the constant and homogeneous classical magnetic field directed along the vector $\mathbf{n}$. Then $g(u_0) = 0$ and one should set in (7.157)

$$t(u_0) = bu_0, \quad b = \frac{\gamma}{2(np)}, \quad \gamma = |eH| . \tag{7.180}$$

Then the function ψ from equation (7.157) obeys equation (7.160) with the classical electromagnetic field potential of the form

$$|e|(\mathbf{A}_{\text{ext}})_\perp = -\gamma u_2 \mathbf{e}_1 . \tag{7.181}$$

Let us work in the light-cone variables (2.37). The following Ansatz can be used for finding the solution of the original equations:

$$\begin{aligned} \Phi &= \exp[-i(px)]U_1\psi, \\ \Psi &= U_1\mathcal{K}\left\{\sum_{\zeta=\pm 1}[1+\zeta(\boldsymbol{\sigma}\mathbf{n})]\exp(ip_\zeta u_0)\right\}\psi v , \end{aligned} \tag{7.182}$$

where

$$p_\zeta^\mu = p^\mu + \zeta b n^\mu, \quad (np_\zeta) = (np), \quad (\mathbf{p}_\zeta \mathbf{e}_1) = (\mathbf{p}\mathbf{e}_1), \quad (\mathbf{p}\mathbf{e}_2) = 0 . \tag{7.183}$$

Now equation (7.160) becomes the stationary Schrödinger equation

$$\begin{aligned} p_0\psi &= \left[\mathcal{H} + \frac{(np)}{2}\right]\psi, \quad \mathbf{F} = \sum_\lambda \mathcal{F}_\lambda \mathbf{e}_\lambda, \\ \mathcal{F}_1 &= \gamma u_2 - (\mathbf{p}\mathbf{e}_1) - (\mathbf{Q}\mathbf{e}_1), \quad \mathcal{F}_2 = i\partial_2 - (\mathbf{Q}\mathbf{e}_2) , \end{aligned} \tag{7.184}$$

where the operator $\mathcal{H}$ is given by equation (7.160) with the operator $\boldsymbol{F}$ indicated in (7.184).

Let us consider the dimensionless variable

$$\eta = \sqrt{\gamma}u_2 - \frac{(\mathbf{p}\mathbf{e}_1)}{\sqrt{\gamma}} , \tag{7.185}$$

and the new creation, a_0^+, and annihilation, a_0, Bose operators, introduced according to (7.231)

$$a_0^+ = \frac{1}{\sqrt{2}}\left(\eta - \frac{\partial}{\partial\eta}\right), \quad a_0 = \frac{1}{\sqrt{2}}\left(\eta + \frac{\partial}{\partial\eta}\right), \tag{7.186}$$

in terms of which

$$\gamma u_2 - (\mathbf{pe}_1) = \sqrt{\frac{\gamma}{2}}\,(a_0 + a_0^+), \quad \partial_2 = \sqrt{\frac{\gamma}{2}}\,(a_0 - a_0^+)\,. \tag{7.187}$$

The operators a_0 and a_0^+ commute with every operator $a_{k,\lambda}$ and $a_{k,\lambda}^+$. Taking into account equations (7.187), the Hamiltonian $\mathcal{H}$ in (7.160) represents a quadratic form in the creation $a_{k,\lambda}^+$ and annihilation $a_{k,\lambda}$ operators where the index k now takes the values $k = 0, 1, 2, \ldots$ with $a_0 = a_{0,2}$ and the numbers $k \geq 1$ designating again the photon operators,

$$\begin{aligned}\mathcal{H} = &\sum_{s,k;\lambda,\lambda'}\left[A_{s\lambda,k\lambda'}a_{s,\lambda}^+ a_{k,\lambda} + \frac{1}{2}\left(B^*_{s\lambda,k\lambda'}a_{s,\lambda}a_{k,\lambda'} + B_{s\lambda,k\lambda'}a_{s,\lambda}^+ a_{k,\lambda'}^+\right)\right]\\ &- b + \frac{m^2}{2(np)} + \frac{\varsigma}{2}\sum_{s=1}\kappa_s^{-1}\,,\end{aligned} \tag{7.188}$$

where

$$\begin{aligned}A_{s\lambda,k\lambda'} &= \left[2b(\lambda-1)\delta_{0,s} + \kappa_s(1-\delta_{0,s})\right]\delta_{s,k}\delta_{\lambda,\lambda'} + B_{s\lambda,k\lambda'},\\ B_{s\lambda,k\lambda'} &= \frac{\varsigma(1-\delta_{0,s})(1-\delta_{0,k})\delta_{\lambda,\lambda'}}{2\sqrt{\kappa_s\kappa_k}} + \frac{|(np)|}{2(np)}\sqrt{2\varsigma b}\\ &\times\left[\frac{(-i)^{\lambda'-1}}{\sqrt{\kappa_k}}(1-\lambda)(1-\delta_{0,k})\delta_{0,s} + \frac{(i)^{\lambda-1}}{\sqrt{\kappa_s}}(1-\lambda')(1-\delta_{0,s})\delta_{0,k}\right].\end{aligned} \tag{7.189}$$

Let us now perform the canonical transformation (7.235) with the parameters taken as

$$\begin{aligned}u_{s\lambda,k\lambda'} &= \left[\left(\sqrt{\frac{r_{k\lambda'}}{\kappa_s}} + \sqrt{\frac{\kappa_s}{r_{k\lambda'}}}\right)\frac{(3-2\lambda')(2-\lambda) - i(\lambda-1)}{2(r_{k\lambda'}^2 - \kappa_s^2)}(1-\delta_{0,s})\right.\\ &\left.+\ \delta_{0,s}(\lambda-1)(\lambda'-1)\sqrt{\frac{2b}{\varsigma r_{k\lambda'}^3}}\right]\left[1 + (\lambda'-2)\delta_{0,k}\right]q_{k\lambda'}\\ v_{s\lambda,k\lambda'} &= \left[\left(\sqrt{\frac{r_{k\lambda'}}{\kappa_s}} - \sqrt{\frac{\kappa_s}{r_{k\lambda'}}}\right)\frac{(3-2\lambda')(2-\lambda) - i(\lambda-1)}{2(r_{k\lambda'}^2 - \kappa_s^2)}(1-\delta_{0,s})\right.\\ &\left.+\delta_{0,s}(1-\lambda)(2-\lambda')\sqrt{\frac{2b}{\varsigma r_{k\lambda'}^3}}\right]\left[1 + (\lambda'-2)\delta_{0,k}\right]q_{k\lambda'}\\ q_{k\lambda} &= \left[\frac{2(2\lambda-3)b}{\varsigma r_{k\lambda}^3} + 2\sum_{s=1}(r_{k\lambda}^2 - \kappa_s^2)^{-2}\right]^{-1/2}, \quad \gamma_{k\lambda} = 0\,.\end{aligned} \tag{7.190}$$

The quantities $r_{k\lambda}$ are positive roots of the equation

$$\sum_{s=1} \frac{\varsigma}{r_{k\lambda}^2 - \kappa_s^2} = 1 + \frac{2(3-2\lambda)b}{r_{k\lambda}}, \quad r_{01} = 0\,. \tag{7.191}$$

The transformed Hamiltonian is

$$\mathcal{H} = \sum_{k,\lambda} r_{k\lambda} c_{k,\lambda}^{+} c_{k,\lambda} + \tilde{H}_0,$$

$$\tilde{H}_0 = \frac{m^2}{2(np)} - b - \sum_{s,k=0,1,2} \sum_{\lambda,\lambda'=1,2} r_{k\lambda'} |v_{s\lambda,k\lambda'}|^2 + \frac{\varsigma}{2} \sum_{s=1} \kappa_s^{-1}\,. \tag{7.192}$$

The eigenvectors ψ in (7.184) are

$$\psi = \prod_{k=0,1,2;\lambda=1,2} |N_{k,\lambda}\rangle\,, \quad |N_{k,\lambda}\rangle = \frac{(c_{k,\lambda}^{+})^{N_{k,\lambda}}}{\sqrt{N_{k,\lambda}!}} |0\rangle_c\,;$$

$$N_{k,\lambda} = 0, 1, 2, \ldots; \quad c_{k,\lambda} |0\rangle_c = 0\,. \tag{7.193}$$

The corresponding energy spectrum is given as

$$p^0 = \frac{np}{2} + \sum_{s,k=0,1,2} \sum_{\lambda,\lambda'=1,2} r_{k\lambda} N_{k,\lambda} + \tilde{H}_0, \quad p^3 = p^0 - (np)\,. \tag{7.194}$$

A characteristic feature of the solutions obtained is the removal, due to the magnetic field, of the degeneracy of the spectrum with respect to the polarizations λ. Clearly, the fact that the magnetic field classically determines the rotation direction of a charged particle is responsible for this phenomenon. The photon quasifrequencies themselves depend on λ, as follows from the dispersion equation (7.181). This dependence has the following character:

$$r_{k1} = r_{k1}(\varsigma, b), \quad r_{k2} = r_{k1}(\varsigma, -b), \quad k \geq 1\,. \tag{7.195}$$

The roots of the dispersion equation can be chosen in such a way that, as $\varsigma \to 0$ with b bounded, the initial conditions

$$r_{k\lambda}(\varsigma = 0, b) = \kappa_k, \quad k \geq 1; \quad r_{02}(\varsigma = 0, b) = 2b \tag{7.196}$$

should be satisfied. For small ς one obtains from equation (7.191)

$$r_{02} \approx 2b \left[1 + \varsigma \sum_{s=1} \frac{1}{4b^2 - \kappa_s^2} + \ldots \right],$$

$$r_{k\lambda} \approx \kappa_k + \frac{\varsigma}{2\kappa_k + 4(3-2\lambda)b} + \ldots, \quad k \geq 1\,. \tag{7.197}$$

Expansion (7.197) is valid provided that there is no resonance. The resonance occurs whenever $\kappa_s = 2b$ for a fixed s. In this case, one obtains for $\lambda = 2$ (for $(np) > 0$)

$$r_{02} \approx r_{s2} \approx \kappa_s + \sqrt{\frac{\varsigma}{2}} + \ldots\,,$$

which means that the analytical properties of the functions $r_{s\lambda}(\varsigma)$ have changed.

Unfortunately, no detailed study of the properties of the roots of the dispersion equations has yet been fulfilled. Those interested can find some simple properties in the original papers [1, 93, 130, 173, 212, 213].

7.8 Linear and quadratic combinations of creation and annihilation operators

7.8.1 Linear combinations

Let c_k^+ and c_k, $k = 1, \ldots, n$ be creation and annihilation operators subject to the *Bose commutation relations*,

$$[c_k, c_s^+] = \delta_{k,s}, \quad [c_k, c_s] = [c_k^+, c_s^+] = 0, \quad k, s = 1, \ldots, n\,. \tag{7.198}$$

Here we shall consider linear combinations[1]

$$F_k = \alpha_{ks} c_s + \beta_{ks} c_s^+, \quad k, s = 1, \ldots, n\,, \tag{7.199}$$

of these operators and the eigenvalue problem for these combinations following the work [24].

In what follows, the consideration will be restricted to linearly independent and commuting operators F_k. It can be shown that a sufficient condition for the linear independence of the operators F_k is the nondegeneracy of one of the matrices α or β, whereas the necessary and sufficient condition for the mutual commutativity of F_k is the matrix equality

$$\alpha\beta^T = \beta\alpha^T\,. \tag{7.200}$$

In this case, one can set the problem of finding the common system of eigenvectors for the operators F_k,

$$F_k\,|f\rangle = f_k\,|f\rangle\,, \quad f \equiv \{f_k\}\,. \tag{7.201}$$

If, in addition to (7.200), the matrices α and β obey the conditions

$$\alpha\alpha^+ - \beta\beta^+ = 1, \quad \beta^T\alpha^* = \alpha^+\beta, \quad \alpha^+\alpha - \beta^T\beta^* = 1\,, \tag{7.202}$$

where the symbols $*$ and $+$ stand for the complex and Hermitian conjugation, respectively, the relation (7.199) and its Hermitian conjugation determine linear canonical transformations from the operators c_k^+ and c_k to new creation and annihilation operators F_k^+ and F_k, see [89]. Since F_k are annihilation operators, the solution of the eigenvalue problem (7.201) is well known, [184, 216]. Namely, this solution is given by coherent states (CS).

1 In this section, we assume summation over repeated indices.

If the matrix relations (7.202) are not fulfilled, then solutions of equations (7.201) are also possible; however, in this case it is the matrix α that should be nondegenerate. Indeed, if α is degenerate, $\det\alpha = 0$, the equations $x_k\alpha_{kj} = 0$ admit nontrivial solutions for x_k. Contracting (7.201) with one of them, we obtain $x_k\beta_{kj}c_j^+|f\rangle = x_k f_k|f\rangle$. Then introducing the new creation and annihilation operators

$$a^+ = \frac{x_k\beta_{kj}c_j^+}{\sqrt{x_k\left(\beta\beta^+\right)_{ks}x_s^*}}, \quad a = \frac{x_k^*\beta_{kj}^*c_j}{\sqrt{x_k\left(\beta\beta^+\right)_{ks}x_s^*}},$$
$$\left[a_k\,, a_s^+\right] = \delta_{k,s}, \quad \left[a_k\,, a_s\right] = \left[a_k^+\,, a_s^+\right] = 0\,,$$

we can see that in this case the vectors $|f\rangle$ are eigenvectors of the new creation operator a^+. On the other hand, it is well known that the eigenvectors of creation operators do not belong to the Fock space.

With this in mind, the matrix α is assumed to be nondegenerate. In this case, we shall look for solutions of (7.201) in the form

$$|f\rangle = e^{-S}\,|z(f)\rangle\,, \quad 2S = c_k^+\gamma_{ks}c_s^+, \quad \gamma = \alpha^{-1}\beta\,, \tag{7.203}$$

where the matrix γ is symmetric (this is easily seen from (7.200)). Substituting (7.203) into (7.201) and using the well-known operator identity

$$\exp(A)B\exp(-A) = B + [A,B] + \frac{1}{2!}[A,[A,B]] + \dots\,,$$

as well as the commutation relations (7.198), we obtain

$$c_k\,|z(f)\rangle = z_k(f)\,|z(f)\rangle\,, \quad z_k(f) = \left(\alpha^{-1}\right)_{ks}f_s\,. \tag{7.204}$$

We can see that $|z(f)\rangle$ is a coherent state:

$$|z(f)\rangle = \exp\left\{z_k(f)c_k^+ - z_k^*(f)c_k\right\}|0\rangle\,, \quad c_k\,|0\rangle = 0, \quad \forall k\,. \tag{7.205}$$

Hence the eigenvectors of the operators F_k are related to CS by an invertible operator $\exp(-S)$. Using (7.203) and (7.204), it is straightforward to find the explicit form of eigenvectors of the operators F_k in the z-representation

$$\langle z|f\rangle = \exp\left\{z_k^*z_k(f) - \frac{z_k^*\gamma_{ks}z_s^* + z_k^*z_k + z_k^*(f)z_k(f)}{2}\right\}\,. \tag{7.206}$$

Let us now examine the orthogonality and normalization of eigenvectors corresponding to the operators F_k. To this end, we shall calculate the scalar product $\langle f'|f\rangle$, using the coherent state representation

$$\langle f'|f\rangle = \int\langle f'|z\rangle\langle z|f\rangle\prod_k\frac{d^2z_k}{\pi}, \quad d^2z_k = d\left(\operatorname{Re}z_k\right)d\left(\operatorname{Im}z_k\right)\,. \tag{7.207}$$

Substituting (7.206) into (7.207), we obtain

$$\langle f'|f\rangle = \exp\left\{-\frac{1}{2}z_k^*\left(f'\right)z_k\left(f'\right) - \frac{1}{2}z_k^*(f)z_k(f)\right\} I_{ff'} ,$$
$$I_{ff'} = \int \exp\left\{-x_j B_{jj'} x_{j'} + x_j M_j\left(ff'\right)\right\} \prod_j \frac{dx_j}{\sqrt{\pi}} , \tag{7.208}$$

where the set of integration variables is denoted as $\{x_j\} = \{\mathrm{Re}\, z_k, \mathrm{Im}\, z_k\}$, B stands for the block matrix

$$B = \begin{pmatrix} 1 + \mathrm{Re}\,\gamma & \mathrm{Im}\,\gamma \\ \mathrm{Im}\,\gamma & 1 - \mathrm{Re}\,\gamma \end{pmatrix} , \tag{7.209}$$

and

$$\left\{M_j\left(ff'\right)\right\} \equiv \left\{z_k^*\left(f'\right) + z_k(f),\quad i\left[z_k^*\left(f'\right) - z_k(f)\right]\right\} .$$

Provided that at least one of the eigenvalues of the matrix B is negative, the integral (7.208) is infinite for any f and f'. Consequently, in this case all the eigenvectors $|f\rangle$ are not orthogonal and have infinite norms. The latter property is characteristic of eigenvectors of some creation operators. In view of this, we shall refer to the corresponding operators F_k as creation-like operators.

In case every eigenvalue of the matrix B is nonnegative, and $\det B \neq 0$, one has

$$I_{ff'} = (\det B)^{-1/2} \exp\left\{\frac{1}{4}M_j\left(ff'\right)\left(B^{-1}\right)_{jj'} M_{j'}\left(ff'\right)\right\} , \tag{7.210}$$

and thus the eigenvectors $|f\rangle$ have finite norms but are not mutually orthogonal, which is characteristic of eigenvectors of some annihilation operators. With this in mind, such operators F_k will be called annihilation-like operators.

Let $\det B = 0$, with every nonzero eigenvalue b_λ of the matrix B being positive,

$$B_{jj'}u_{j'}^\lambda = b_\lambda u_j^\lambda,\quad b_\lambda > 0;\quad B_{jj'}v_{j'}^a = 0 ,$$

and, in addition, let $\mathrm{Re}[v_j^a M_j(ff')] = 0$ for any zero vector v_j^a. Then

$$I_{ff'} = \prod_a 2\sqrt{\pi}\delta\left(\mathrm{Im}\, v_j^a M_j\left(ff'\right)\right)\prod_\lambda b_\lambda^{-1/2} \exp\left\{\frac{1}{4b_\lambda}\left[u_j^\lambda M_j\left(ff'\right)\right]^2\right\} . \tag{7.211}$$

The most important case under consideration is the one with the number of vanishing eigenvalues of the matrix B being equal to the number of nonvanishing eigenvalues. Then the eigenvectors are normalized by the δ-function. The corresponding operators will be referred to as normal-like operators. Indeed, an example of such operators is a normal set of the operators F_k, which obeys the relations

$$F_k F_s^+ - F_s^+ F_k = 0 . \tag{7.212}$$

It is easy to see that the relations (7.212) hold true iff

$$\alpha^+\alpha - \beta\beta^+ = 0 \Longleftrightarrow \left\{\begin{array}{l} (\mathrm{Re}\,\gamma)^2 + (\mathrm{Im}\,\gamma)^2 = 1 \\ \mathrm{Re}\,\gamma\ \mathrm{Im}\,\gamma - \mathrm{Im}\,\gamma\ \mathrm{Re}\,\gamma = 0 \end{array}\right. . \tag{7.213}$$

Relations (7.213) imply that the block matrix (7.209) has two eigenvalues: 0 and 2, and that $\det B = 0$.

For a normal set of operators, the eigenvectors form an orthogonal system. Indeed, let us represent F_k in the form $F_k = A_k + iB_k$, where $A_k = (F_k + F_k^+)/2$ and $B_k = i(F_k^+ - F_k)/2$ are self-adjoint operators. With allowance for (7.212), these operators commute, and hence they possess a common orthogonal set of eigenvectors. At the same time, these vectors are eigenvectors of the operators F_k.

We present below some examples of normal sets of operators.

Consider the operators $F_k = \hat{\xi}_k$,

$$\hat{\xi}_k = \frac{1}{\sqrt{2}}\left(c_k + c_k^+\right) = \hat{\xi}_k^+ . \tag{7.214}$$

In this case, $\alpha = \beta = (\sqrt{2})^{-1}$, $\gamma = 1$, and $F_k|f\rangle = f_k|f\rangle$, hence $\hat{\xi}_k|\xi\rangle = \xi_k|\xi\rangle$. According to (7.203), the eigenvectors $|\xi\rangle$ are

$$|\xi\rangle = N_\xi \exp\left(-1/2c_k^+c_k^+\right)|z(\xi)\rangle , \quad z_k(\xi) = \sqrt{2}\xi_k = \sqrt{2}\xi_k^* , \tag{7.215}$$

where N_ξ is a normalization factor. In the z-representation, the eigenvectors (7.215) have the form

$$\langle z|\xi\rangle = N_\xi \exp\left\{\sqrt{2}z_k^*\xi_k - \xi_k\xi_k - 2^{-1}\left(z_k^*z_k^* + z_k^*z_k\right)\right\} . \tag{7.216}$$

It is easy to see that $\langle\xi'|\xi\rangle = \prod_k \delta(\xi_k' - \xi_k)$, with the normalization factor chosen as

$$N_\xi = \prod_k (2\pi)^{-1/4} \exp\left(\xi_k^2/2\right) .$$

To prove the relation of completeness $\int |\xi\rangle\langle\xi|\prod_k d\xi_k = 1$, it is sufficient to show that

$$\int \langle z|\xi\rangle\langle\xi|z'\rangle \prod_k d\xi_k = \langle z|z'\rangle = \prod_k \exp\left\{-\frac{1}{2}|z_k|^2 + z_k^*z_k' - \frac{1}{2}|z_k'|^2\right\} . \tag{7.217}$$

Using (7.216), it is straightforward to check the relation (7.217). Thus, the eigenvectors of the operators ξ_k form a complete orthogonal system. One can show that the action of the creation and annihilation operators on the vectors $|\xi\rangle$ can be written as

$$c_k = \frac{1}{\sqrt{2}}\left(\xi_k + \frac{\partial}{\partial\xi_k}\right), \quad c_k^+ = \frac{1}{\sqrt{2}}\left(\xi_k - \frac{\partial}{\partial\xi_k}\right) . \tag{7.218}$$

Consider the operators $F_k = \hat{\eta}_k$,

$$\hat{\eta}_k = \frac{i}{\sqrt{2}}\left(c_k^+ - c_k\right) = \hat{\eta}_k^+ . \tag{7.219}$$

In this case, $\alpha = -\beta = (\sqrt{2i})^{-1}$, $\gamma = -1$, and $F_k|f\rangle = f_k|f\rangle$, hence $\hat{\eta}_k|\eta\rangle = \eta_k|\eta\rangle$. The eigenvectors $|\eta\rangle$ are

$$|\eta\rangle = N_\eta \exp\left(1/2c_k^+c_k^+\right)|z_k(\eta)\rangle\,,\quad z_k(\eta) = i\sqrt{2}\eta_k\,,$$
$$\langle z|\eta\rangle = N_\eta \exp\left\{i\left[\sqrt{2}z_k^*\eta_k - \eta_k\eta_k + 2^{-1}\left(z_k^*z_k^* - z_kz_k^*\right)\right]\right\}\,. \tag{7.220}$$

Provided that $N_\eta = \prod_k (2\pi)^{-1/4}\exp(\eta_k^2/2)$, one can show that the set of eigenvectors $|\eta\rangle$ is normed to the δ-functions and obeys the completeness relation:

$$\langle\eta'|\eta\rangle = \prod_k \delta\left(\eta_k' - \eta_k\right)\,,\quad \int |\eta\rangle\langle\eta|\prod_k d\eta_k = 1\,.$$

Below, we consider some applications of the above results to quantum theory of electromagnetic fields.

Let us take the vector potential of a quantized electromagnetic field in the Coulomb gauge (div $\hat{\mathbf{A}}(\mathbf{r}) = 0$, $A^0 = 0$) and in the Schrödinger picture, given by equation (7.6). It can be represented as follows:

$$\mathbf{A}(\mathbf{r}) = \sum_{\mathbf{k}\lambda}\frac{e^{i\mathbf{k}\mathbf{r}}\mathbf{e}_{\mathbf{k}\lambda}}{\sqrt{2|\mathbf{k}|V}}Q_{\mathbf{k}\lambda}\,,\quad Q_{\mathbf{k}\lambda} = c_{\mathbf{k}\lambda} + c_{-\mathbf{k}\lambda}^+\,,\quad \lambda = 1,2\,, \tag{7.221}$$

where $c_{\mathbf{k}\lambda}^+$ and $c_{\mathbf{k}\lambda}$ are creation and annihilation operators of photons with momentum $\mathbf{k}$ and polarization λ,

$$\left[c_{\mathbf{k}\lambda}, c_{\mathbf{k}'\lambda'}^+\right] = \delta_{\mathbf{k}\mathbf{k}'}\delta_{\lambda\lambda'}\,,\quad \left[c_{\mathbf{k}\lambda}, c_{\mathbf{k}'\lambda'}\right] = \left[c_{\mathbf{k}\lambda}^+, c_{\mathbf{k}'\lambda'}^+\right] = 0\,, \tag{7.222}$$

$\mathbf{e}_{\mathbf{k}\lambda} = \mathbf{e}_{-\mathbf{k}\lambda}$ are real polarization vectors, $(\mathbf{k}\mathbf{e}_{\mathbf{k}\lambda}) = 0$, and V is the quantization volume. The corresponding operators of electric and magnetic fields are

$$\mathbf{E}(\mathbf{r}) = i\sum_{\mathbf{k}\lambda}\frac{e^{-i\mathbf{k}\mathbf{r}}\mathbf{e}_{\mathbf{k}\lambda}}{\sqrt{2|\mathbf{k}|V}}P_{\mathbf{k}\lambda}\,,\quad P_{\mathbf{k}\lambda} = c_{\mathbf{k}\lambda} - c_{-\mathbf{k}\lambda}^+\,, \tag{7.223}$$

$$\mathbf{H}(\mathbf{r}) = i\sum_{\mathbf{k}\lambda}\frac{e^{i\mathbf{k}\mathbf{r}}\left[\mathbf{k}\times\mathbf{e}_{\mathbf{k}\lambda}\right]}{\sqrt{2|\mathbf{k}|V}}Q_{\mathbf{k}\lambda}\,. \tag{7.224}$$

We see that the operators of the vector potential, electric, and magnetic fields are expressed in terms of two sets of the operators $Q_{\mathbf{k}\lambda}$ and $P_{\mathbf{k}\lambda}$. Each set is a normal set of commuting operators. The eigenvalue problem for the operators of the vector potential, electric, and magnetic fields is, thus, reduced to the eigenvalue problem for the normal sets $Q_{\mathbf{k}\lambda}$ or $P_{\mathbf{k}\lambda}$:

$$Q_{\mathbf{k}\lambda}|q\rangle = q_{\mathbf{k}\lambda}|q\rangle\,,\quad q_{\mathbf{k}\lambda}^* = q_{-\mathbf{k}\lambda}\,,$$
$$P_{\mathbf{k}\lambda}|p\rangle = p_{\mathbf{k}\lambda}|p\rangle\,,\quad p_{\mathbf{k}\lambda}^* = -p_{-\mathbf{k}\lambda}\,.$$

For the operators $Q_{\mathbf{k}\lambda}$ and $P_{\mathbf{k}\lambda}$ (considered as operators F from (7.199)), the matrix α equals to unity, and the matrices $\beta = \gamma$ have the form

$$\gamma_{\mathbf{k}\lambda,\mathbf{k}'\lambda'} = \pm\delta_{\lambda\lambda'}\delta_{\mathbf{k},-\mathbf{k}'} \,.$$

Given this, in view of (7.203), the eigenvectors $|q\rangle$ and $|p\rangle$ are

$$|q\rangle = N_q e^{-1/2c^+_{\mathbf{k}\lambda}c^+_{-\mathbf{k}\lambda}}|z(q)\rangle \,, \quad z_{\mathbf{k}\lambda}(q) = q_{\mathbf{k}\lambda} = q^*_{-\mathbf{k}\lambda} \,, \tag{7.225}$$

$$|p\rangle = N_p e^{1/2c^+_{\mathbf{k}\lambda}c^+_{-\mathbf{k}\lambda}}|z(p)\rangle \,, \quad z_{\mathbf{k}\lambda}(p) = p_{\mathbf{k}\lambda} = -p^*_{-\mathbf{k}\lambda} \,, \tag{7.226}$$

where N_p, N_q are normalization factors, and $|z(f)\rangle$ are CS normalized to unity.

One can find various representations for eigenvectors (7.225) and (7.226). The most interesting representations of this kind are the z-, ξ- and n-representation. The basis in the z-representation is composed of CS. The basis in the ξ-representation is composed of eigenvectors of the self-adjoint operators (7.214). The basis of the n-representation consists of eigenvectors of the photon number operators.

(1) z-representation. According to (7.206), the eigenvectors $|q\rangle$ and $|p\rangle$ in the z-representation can be written as

$$\langle z|q\rangle = N_q \exp \sum_{\mathbf{k}\lambda}{}^{+} \left\{ -\frac{|z_{\mathbf{k}\lambda}|^2}{2} - \frac{|z_{-\mathbf{k}\lambda}|^2}{2} - (z^*_{\mathbf{k}\lambda} - q^*_{\mathbf{k}\lambda})(z^*_{-\mathbf{k}\lambda} - q^*_{-\mathbf{k}\lambda}) \right\} ,$$

$$\langle z|p\rangle = N_p \exp \sum_{\mathbf{k}\lambda}{}^{+} \left\{ -\frac{|z_{\mathbf{k}\lambda}|^2}{2} - \frac{|z_{-\mathbf{k}\lambda}|^2}{2} - (z^*_{\mathbf{k}\lambda} - p^*_{\mathbf{k}\lambda})(z^*_{-\mathbf{k}\lambda} - p^*_{-\mathbf{k}\lambda}) \right\} , \tag{7.227}$$

where the symbol + in the sum implies summation over half of the **k**-vectors in the **k**-space (if we take **k** then we do not take −**k**).

It should be noted that the eigenfunctions of the vector potential operator $\mathbf{A}(\mathbf{r})$ were found in [190] as expansions in CS. The corresponding expansion coefficients are identical with the z-representation of the functions $|q\rangle$.

(2) ξ-representation

$$\langle \xi|q\rangle = N_q \exp \prod_{\mathbf{k}\lambda}{}^{+} \delta\left(\xi_{\mathbf{k}\lambda} + \xi_{-\mathbf{k}\lambda} - \sqrt{2}\,\mathrm{Re}\, q_{\mathbf{k}\lambda}\right)$$
$$\times \exp\left\{ i\,\mathrm{Im}\, q_{\mathbf{k}\lambda}\frac{\xi_{\mathbf{k}\lambda} - \xi_{-\mathbf{k}\lambda}}{\sqrt{2}} - \frac{|q_{\mathbf{k}\lambda}|^2}{2} \right\} ,$$
$$\langle \xi|p\rangle = N_p \exp \prod_{\mathbf{k}\lambda}{}^{+} \delta\left(\xi_{\mathbf{k}\lambda} - \xi_{-\mathbf{k}\lambda} - \sqrt{2}\,\mathrm{Re}\, p_{\mathbf{k}\lambda}\right)$$
$$\times \exp\left\{ i\,\mathrm{Im}\, p_{\mathbf{k}\lambda}\frac{\xi_{\mathbf{k}\lambda} + \xi_{-\mathbf{k}\lambda}}{\sqrt{2}} - \frac{|p_{\mathbf{k}\lambda}|^2}{2} \right\} . \tag{7.228}$$

(3) n-representation

$$\langle n|q\rangle = N_q \prod_{\mathbf{k}\lambda}^{+} \frac{q_{\mathbf{k}\lambda}^{n_{\mathbf{k}\lambda}} q_{-\mathbf{k}\lambda}^{n_{-\mathbf{k}\lambda}}}{\sqrt{n_{\mathbf{k}\lambda}! n_{-\mathbf{k}\lambda}!}} \Phi_{n_{\mathbf{k}\lambda} n_{-\mathbf{k}\lambda}}\left(|q_{\mathbf{k}\lambda}|^2\right) ,$$

$$\langle n|p\rangle = N_p \prod_{\mathbf{k}\lambda}^{+} \frac{p_{\mathbf{k}\lambda}^{n_{\mathbf{k}\lambda}} p_{-\mathbf{k}\lambda}^{n_{-\mathbf{k}\lambda}}}{\sqrt{n_{\mathbf{k}\lambda}! n_{-\mathbf{k}\lambda}!}} \Phi_{n_{\mathbf{k}\lambda} n_{-\mathbf{k}\lambda}}\left(|p_{\mathbf{k}\lambda}|^2\right) , \tag{7.229}$$

where

$$\Phi_{n_1 n_2}(x) = \sum_{j=0}^{\min(n_1,n_2)} \frac{(-1)^j n_1! n_2! x^{-j}}{j!\,(n_1 - j)!\,(n_2 - j)!} \exp(-x) .$$

Let us now examine the normalization and completeness of the eigenvector systems (7.225) and (7.226). Using any of the above representations, it is straightforward to obtain

$$\langle q'|q\rangle = \prod_{\mathbf{k}\lambda}^{+} \delta\left[\mathrm{Re}\left(q'_{\mathbf{k}\lambda} - q_{\mathbf{k}\lambda}\right)\right] \delta\left[\mathrm{Im}\left(q'_{\mathbf{k}\lambda} - q_{\mathbf{k}\lambda}\right)\right] ,$$

$$\langle p'|p\rangle = \prod_{\mathbf{k}\lambda}^{+} \delta\left[\mathrm{Re}\left(p'_{\mathbf{k}\lambda} - p_{\mathbf{k}\lambda}\right)\right] \delta\left[\mathrm{Im}\left(p'_{\mathbf{k}\lambda} - p_{\mathbf{k}\lambda}\right)\right] ,$$

with the normalization multipliers chosen as

$$N_q = \prod_{\mathbf{k}\lambda}^{+} \pi^{-1/2} \exp\left\{|q_{\mathbf{k}\lambda}|^2/2\right\}, \quad N_p = \prod_{\mathbf{k}\lambda}^{+} \pi^{-1/2} \exp\left\{|p_{\mathbf{k}\lambda}|^2/2\right\} .$$

In order to prove the completeness of the sets $|q\rangle$ or $|p\rangle$ it is evidently sufficient to show that

$$\int \langle a|q\rangle \langle q|a'\rangle \prod_{\mathbf{k}\lambda}^{+} d\left(\mathrm{Re}\, q_{\mathbf{k}\lambda}\right) d\left(\mathrm{Im}\, q_{\mathbf{k}\lambda}\right) = \langle a|a'\rangle ,$$

$$\int \langle a|p\rangle \langle p|a'\rangle \prod_{\mathbf{k}\lambda}^{+} d\left(\mathrm{Re}\, p_{\mathbf{k}\lambda}\right) d\left(\mathrm{Im}\, p_{\mathbf{k}\lambda}\right) = \langle a|a'\rangle ,$$

where $|a\rangle$ is an arbitrary complete set of vectors. Using, for instance, the ξ-representation (7.228) of wave functions, one can easily show that

$$\int \langle \xi|q\rangle \langle q|\xi'\rangle \prod_{\mathbf{k}\lambda}^{+} d\left(\mathrm{Re}\, q_{\mathbf{k}\lambda}\right) d\left(\mathrm{Im}\, q_{\mathbf{k}\lambda}\right) = \prod_{\mathbf{k}\lambda} \delta\left(\xi_{\mathbf{k}\lambda} - \xi'_{\mathbf{k}\lambda}\right) ,$$

$$\int \langle \xi|p\rangle \langle p|\xi'\rangle \prod_{\mathbf{k}\lambda} d\left(\mathrm{Re}\, p_{\mathbf{k}\lambda}\right) d\left(\mathrm{Im}\, p_{\mathbf{k}\lambda}\right) = \prod_{\mathbf{k}\lambda} \delta\left(\xi_{\mathbf{k}\lambda} - \xi'_{\mathbf{k}\lambda}\right) .$$

Thus the sets of eigenvectors $|p\rangle$ or $|q\rangle$ under consideration are not only orthogonal but also complete. Consequently, they can be used as basis vectors. The corresponding representations will be called the p- or q-representations. In particular, in the q-representation the creation and annihilation operators have the form

$$c_{\mathbf{k}\lambda}^{+} = \frac{q_{\mathbf{k}\lambda}^{*}}{2} - \frac{\partial}{\partial q_{\mathbf{k}\lambda}}, \quad c_{\mathbf{k}\lambda} = \frac{q_{\mathbf{k}\lambda}^{*}}{2} + \frac{\partial}{\partial q_{\mathbf{k}\lambda}^{*}} ,$$

while the operators of the vector potential and magnetic field act as multiplication operators.

It should be noted that for any state $|q\rangle$ or $|p\rangle$ the mean value of the number of photons $n_{\mathbf{k}\lambda}$ is infinite, as can be shown by straightforward calculations. This implies that the states under consideration are idealized states, similar to the states of absolute localization, or the states of absolutely free motion in usual quantum mechanics. As usual, realistic states can be constructed from normalized wave packets.

7.8.2 Quadratic combinations

In this Section, we consider quadratic combinations of creation and annihilation operators. Such combinations are identified by us with some quadratic Hamiltonians, that is why we call them in what follows *quadratic Hamiltonians*. A usual way to solve the spectral problem for a quadratic Hamiltonian consists of reducing it to the so-called canonical form. Below, we discuss in detail such a reduction. In addition, we determine integrals of motion and CS for general quantum systems with quadratic Hamiltonians.

Let H be a self-adjoint quantum Hamiltonian, quadratic in coordinates x^k, $k = 1, \ldots, n$ and momenta $\hat{p}_k = i\partial_k$,

$$H = \hat{p}Q\hat{p} + xR\hat{p} + \hat{p}R^T x + xMx + \hat{p}J + xG + H_1 \,, \tag{7.230}$$

where the matrices Q, R, M, the columns J, G and the number H_1 are real. The quantities Q, R, M, J, G and H_1 can depend on time t in an arbitrary way.

Consider the creation and annihilation Bose operators a_k^+, a_k, $k = 1, \ldots, n$,

$$\begin{aligned}
a_k^+ &= \frac{1}{\sqrt{2}}\left(x^k + i\hat{p}_k\right), \quad a_k = \frac{1}{\sqrt{2}}\left(x^k - i\hat{p}_k\right), \\
x^k &= \frac{1}{\sqrt{2}}\left(a_k + a_k^+\right), \quad \hat{p}_k = \frac{i}{\sqrt{2}}\left(a_k - a_k^+\right), \\
\left[a_k, a_s^+\right]_- &= \delta_{k,s}, \quad \left[a_k, a_s\right]_- = \left[a_k^+, a_s^+\right]_- = 0 \,.
\end{aligned} \tag{7.231}$$

When written in terms of the operators a_k^+ and a_k, the Hamiltonian H acquires the form

$$H = a^+ Aa + \frac{1}{2}aBa + \frac{1}{2}a^+ B^+ a^+ + D^* a + a^+ D + H_0 \,, \tag{7.232}$$

where

$$\begin{aligned}
A &= M + Q + i(R - R^T), \quad D = \frac{1}{\sqrt{2}}(G + iJ) \,, \\
B &= M - Q + i(R + R^T), \quad H_0 = H_1 + \frac{1}{2}\mathrm{tr}(Q + M) \,.
\end{aligned} \tag{7.233}$$

The relations inverse to (7.233) have the form

$$Q = \frac{1}{4}(A + A^T - B - B^+), \quad R = \frac{i}{4}(B^+ - B + A^T - A),$$
$$M = \frac{1}{4}(A + A^T + B + B^+), \quad G = \frac{1}{\sqrt{2}}(D + D^*), \quad J = \frac{i}{\sqrt{2}}(D - D^*). \tag{7.234}$$

Then the self-adjointness of the Hamiltonian, $H^+ = H$, implies that

$$Q^T = Q, \quad M^T = M, \quad A^+ = A, \quad B^T = B, \quad H_0^* = H_0.$$

Let us consider new creation and annihilation operators c_k^+ and c_k connected with the operators a_k^+ and a_k by a linear canonical transformation

$$a = uc - vc^+ + \gamma, \quad a^+ = c^+u^+ - cv^+ + \gamma^*, \tag{7.235}$$
$$[c_k, c_s^+]_- = \delta_{k,s}, \quad [c_k, c_s]_- = [c_k^+, c_s^+]_- = 0. \tag{7.236}$$

The fact that the transformation is canonical and, in particular, the requirement that the new operators c_k^+ and c_k be again creation and annihilation operators subject to the Bose commutation relations (7.236), imposes the following conditions on the matrices u and v, and the column γ:

$$uu^+ - vv^+ = 1, \quad vu^T - uv^T = 0,$$
$$u^+u - v^Tv^* = 1, \quad v^Tu^* - u^+v = 0, \tag{7.237}$$

see [89]. Relations (7.235) can be solved for the operators c_k^+ and c_k as follows:

$$c = u^+a + v^Ta^+ - u^+\gamma - v^T\gamma^*,$$
$$c^+ = a^+u + av^* - \gamma^*u - \gamma v^*. \tag{7.238}$$

The Hamiltonian H, in terms of the new operators c_k^+ and c_k, looks like

$$H = c^+\overline{A}c + \frac{1}{2}c\overline{B}c + \frac{1}{2}c^+\overline{B}^+c^+ + \overline{D}^*c + c^+\overline{D} + \overline{H}_0, \tag{7.239}$$

with

$$\overline{A} = u^+Au + v^TA^Tv^* - v^TBu - u^+B^+v^*,$$
$$\overline{B} = u^TBu + v^+Bv^* - v^+Au - u^TA^Tv^*,$$
$$\overline{D} = (u^+A - v^TB)\gamma + (u^+B^+ - v^TA^T)\gamma^* + u^+D - v^TD^*,$$
$$\overline{H}_0 = \gamma^*A\gamma + \frac{1}{2}\gamma B\gamma + \frac{1}{2}\gamma^*B^+\gamma^* + D^*\gamma + \gamma^*D$$
$$- \frac{1}{2}\mathrm{tr}(u^TBv - v^+B^+u^* - 2v^+Av) + H_0. \tag{7.240}$$

It can be readily established that $\overline{A}^{+} = \overline{A}$, $\overline{B}^{T} = \overline{B}$, $\overline{H}_0^{*} = \overline{H}_0$, which implies that the self-adjointness property of H survives the linear canonical transformations, as it should.

Let the matrices u and v obey the equations

$$ur = Au - B^{+}v^{*}, \quad v^{*}r = Bu - A^{T}v^{*}, \tag{7.241}$$

where r is a nonsingular matrix. If this matrix exists, we obtain from (7.241) with the use of (7.237) that

$$r = u^{+}Au + v^{T}A^{T}v^{*} - u^{+}B^{+}v^{*} - v^{T}Bu, \tag{7.242}$$

whence it follows that the matrix r is Hermitian ($r^{+} = r$). The column γ is taken as

$$\gamma = \left(ur^{-1}v^{T} + v\left(r^{T}\right)^{-1}u^{T}\right)D^{*} - \left(ur^{-1}u^{+} + v\left(r^{T}\right)^{-1}v^{+}\right)D. \tag{7.243}$$

It follows from (7.243) and (7.242) that γ satisfies the equation

$$A\gamma + B^{+}\gamma^{*} + D = 0. \tag{7.244}$$

The fulfilment of equation (7.244) provides the equality $\overline{D} = 0$, irrespective of whether (7.243) holds true. Using (7.241), (7.244) and (7.237) we find

$$\overline{A} = r, \quad \overline{B} = 0, \quad \overline{D} = 0, \overline{H}_0 = H_0 + \frac{1}{2}(D^{*}\gamma + \gamma^{*}D) - \operatorname{tr} vr^{T}v^{+}, \tag{7.245}$$

so that $H = c^{+}rc + \overline{H}_0$. If we further require that the matrix r be diagonal, the Hamiltonian H is reduced to the form

$$H = \sum_k r_k c_k^{+} c_k + \overline{H}_0, \tag{7.246}$$

where r_k are real diagonal elements of the matrix r. Expression (7.246) is called *the first canonical form for a quadratic combination of creation and annihilation operators.*

It may happen that there exist no such matrices r, u and v that satisfy equations (7.241). In this case, we subject the matrices u and v to the equations

$$ru^{+} = u^{T}B - v^{+}A, \quad rv^{T} = u^{T}A^{T} - v^{+}B^{+}, \tag{7.247}$$

where r is a matrix, in general, different from the one defined by equation (7.242). If this matrix exists, we obtain from (7.247) with the use of (7.237)

$$r = u^{T}Bu + v^{+}B^{+}v^{*} - v^{+}Au - u^{T}A^{T}v^{*}, \tag{7.248}$$

whence it follows that $r^{T} = r$. If the matrix r is nonsingular, with the choice

$$\gamma = \left(ur^{-1}v^{+} + v\left(r^{+}\right)^{-1}u^{+}\right)D - \left(ur^{-1}u^{*} + v\left(r^{+}\right)^{-1}v^{T}\right)D^{*}, \tag{7.249}$$

we have

$$\overline{A} = 0, \quad \overline{B} = r\,, \overline{D} = 0\,,$$

$$\overline{H}_0 = H_0 + \frac{1}{2}(D^*\gamma + \gamma^* D) - \mathrm{tr}\left(vru^+ + ur^+v^+\right)\,,$$

so that $H = 1/2\,(crc + c^+r^+c^+) + \overline{H}_0$. Finally, if we impose the requirement of diagonality on the matrix r, then the Hamiltonian H is reduced to the form

$$H = \frac{1}{2}\sum_k \left(r_k c_k c_k + r_k^* c_k^+ c_k^+\right) + \overline{H}_0\,, \tag{7.250}$$

where r_k are diagonal elements of the matrix r. Expression (7.250) is called *the second canonical form for a quadratic combination of creation and annihilation operators.*

It should be noted that the quadratic form (7.250) cannot be reduced to the form (7.246) by using transformations such as (7.235) and vice verca. Thus, expressions (7.246) and (7.250) are two independent canonical forms of Hamiltonians quadratic in coordinates and momenta, or in creation and annihilation operators.

In the general case, the Hamiltonian H can, evidently, be reduced to the canonical form in which some modes appear in the first canonical form (7.246), while others appear in the second canonical form (7.250)

$$H = \sum_\alpha r_\alpha c_\alpha^+ c_\alpha + \frac{1}{2}\sum_i \left(r_i c_i c_i + r_i^* c_i^+ c_i^+\right) + \overline{H}_0\,, \tag{7.251}$$

where the subscripts α and i range over the set of indices $(\alpha, i) = (k)$.

8 Spin equation and its solutions

8.1 Introduction

We refer as *the spin equation* (*SE*) to the following set of two ordinary linear differential equations of first order for two functions $v_1(t)$ and $v_2(t)$:

$$i\dot{V} = (\boldsymbol{\sigma}\mathbf{F})\,V, \quad V = \begin{pmatrix} v_1(t) \\ v_2(t) \end{pmatrix}, \quad \dot{V} = dV/dt\,. \tag{8.1}$$

Here, $\boldsymbol{\sigma} = (\sigma_1, \sigma_2, \sigma_3)$ are the Pauli matrices and $\mathbf{F}$ is a given time-dependent (in general, complex) three-vector,

$$\mathbf{F} = \left(F_k(t),\ k = 1, 2, 3\right)\,. \tag{8.2}$$

Sometimes, we represent the three-vector $\mathbf{F}$ as

$$\begin{aligned} &\mathbf{F} = \mathbf{K} + i\mathbf{G}, \quad \mathbf{K} = \operatorname{Re}\mathbf{F}, \quad \mathbf{G} = \operatorname{Im}\mathbf{F}\,, \\ &\mathbf{K} = \left(K_k\right), \quad \mathbf{G} = \left(G_k\right), \quad k = 1, 2, 3\,, \end{aligned} \tag{8.3}$$

where $\mathbf{K}$ and $\mathbf{G}$ are real vectors.

In what follows, the column V and the vector $\mathbf{F}$ are called the spinor and the external field, respectively.

The equation conjugate to the SE has the form

$$i\dot{V}^+ = -V^+\left(\boldsymbol{\sigma}\mathbf{F}^*\right), \quad V^+ = \begin{pmatrix} v_1^*(t) & v_2^*(t) \end{pmatrix}. \tag{8.4}$$

The SE and its conjugate equation can be written in the component form as follows:

$$i\dot{v}_1 = F_3 v_1 + \left(F_1 - iF_2\right)v_2, \quad i\dot{v}_2 = -F_3 v_2 + \left(F_1 + iF_2\right)v_1\,, \tag{8.5}$$

$$i\dot{v}_1^* = -F_3^* v_1^* - \left(F_1^* + iF_2^*\right)v_2^*, \quad i\dot{v}_2^* = F_3^* v_2^* - \left(F_1^* - iF_2^*\right)v_1^*\,. \tag{8.6}$$

The set (8.6) can be written in the form of the SE for the *anticonjugate spinor* $\bar{V} = -i\sigma_2 V^*$ (see (A.13)) with the external field $\mathbf{F}^*$,

$$i\dot{\bar{V}} = \left(\boldsymbol{\sigma}\mathbf{F}^*\right)\bar{V}\,. \tag{8.7}$$

The SE with a real external field can be treated as a reduction of the Pauli equation [261] to the $(0 + 1)$-dimensional case. Such an equation is commonly used to describe a (frozen in the space) spin-$1/2$ particle of magnetic momentum μ immersed in a magnetic field $\mathbf{B}$ (in this case, $\mathbf{F} = -\mu\mathbf{B}$). The equation was intensely studied in connection with the problem of magnetic resonance [98, 270]. Besides, complex quantum systems with a discrete energy spectrum can be placed in a special dynamic configuration in which only two stationary states are important. In those cases, it is possible to reduce the Hilbert space of the system to a two-dimensional space. Such *two-level systems* can also be described by the SE. The SE with an external field of the form $\mathbf{F} = (F_1, 0, F_3)$, $F_1 = \epsilon$, where ϵ is a constant, describes two-level systems with unperturbed energy levels $\pm\epsilon$ ($F_3 \equiv 0$) submitted to an external time-dependent interaction $F_3(t)$, inducing a transition between the unperturbed eigenstates. Two-level systems possess a wide range of applications, for example, the semi-classical theory of laser beams [256], the absorption resonance and nuclear induction experiments [270], the behavior of a molecule in a cavity immersed in electric or magnetic fields [142], and so on. Recently, this subject has attracted even more attention due to its relation to quantum computation [141], where the state of each bit of conventional computation is permitted to be any quantum-mechanical state of a *qubit* (quantum bit), which can be treated as a two-level system. The SE with complex external fields describes a possible damping of two-level systems. There exist various equations that are equivalent, or (in a sense) related to the SE. For example, the well-known rigid rotator (top) equation, which appears in the gyroscope theory, in the theory of precession of a classical gyromagnet in a magnetic field (see [142]), and so on. For periodic, or quasiperiodic, external fields, the equations of a two-level system have been studied by many authors using different approximation methods, for example perturbative expansions [74], see also [75].

When the external field $\mathbf{F}$ is neither periodic, nor quasiperiodic, no regular approach exists to finding exact solutions of the SE. One ought to stress that exact solutions are very important in view of the numerous physical applications of the SE.

The first exact solution of the SE was found by Rabi [269], for the external field of the form

$$\mathbf{F} = \left(f_1 \cos \Omega t, f_2 \sin \Omega t, F_3\right) , \tag{8.8}$$

where $f_{1,2}$, Ω, and F_3 are some constants. For the external field of the form

$$\mathbf{F} = \left(F_1, 0, F_3\right), \quad F_1 = \text{const} , \tag{8.9}$$

exact solutions for two different functions F_3 were found in [49]. These functions are

$$F_3 = c_0 \tanh t + c_1, \quad F_3 = \frac{c_0}{\cosh t} + c_1 , \tag{8.10}$$

where $c_{0,1}$ are arbitrary real constants. In the work [50], exact solutions for three sufficiently complicated functions F_3 were found. One (the simplest) of these functions is

$$\begin{aligned} & F_3 = c_0 + \frac{2\left(c_1^2 - c_0^2\right)}{Q + c_0}, \quad Q = \begin{cases} c_1 \cosh \varphi, & c_1^2 > c_0^2 \\ c_1 \cos \varphi, & c_1^2 < c_0^2 \end{cases} , \\ & \varphi = 2\left(t\sqrt{|c_1^2 - c_0^2|} + c_2\right) , \end{aligned} \tag{8.11}$$

where $c_{0,1,2}$ are arbitrary real constants.

The detailed study of the SE and its exact solution was done in [51]. These exact solutions are described below.

8.2 Associated equations

8.2.1 Associated Schrödinger equations

Consider the Schrödinger equation in 0 + 1 dimensions for a time-dependent two-component complex spinors Ψ. In the general case, the equation has the form

$$i\dot{\Psi} = H\Psi\,, \tag{8.12}$$

where the Hamiltonian H is a 2×2 complex time-dependent matrix and Ψ a spinor. The matrix H can always be decomposed in the basic matrices, $H = hI + (\boldsymbol{\sigma}\mathbf{F})$, where $h = h(t)$ and $\mathbf{F} = (F_k(t), k = 1, 2, 3)$. Making the transformation $\Psi = U \exp(-i \int h dt)$, we arrive to the SE for the spinor U.

The SE can be reduced to a set of two independent one-dimensional Schrödinger equations in with generally complex potentials. Consider equations (8.5) for the function $v_s(t)$, $s = 1, 2$. By squaring this set and introducing functions $\psi_s(t)$,

$$v_s = \sqrt{A_s}\psi_s, \quad A_s = F_1 + (-1)^s iF_2\,, \tag{8.13}$$

we obtain an independent linear second-order differential equation for the latter functions,

$$\begin{aligned}
&\ddot{\psi}_s - V_s\psi_s = 0, \quad s = 1, 2\,, \\
&V_s = \frac{3}{4}\left(\frac{\dot{A}_s}{A_s}\right)^2 - \frac{1}{2}\frac{\ddot{A}_s}{A_s} - A_1A_2 - F_3^2 - i(-1)^s\left(F_3\frac{\dot{A}_s}{A_s} - \dot{F}_3\right)\,.
\end{aligned} \tag{8.14}$$

Each of equations (8.14) is a one-dimensional Schrödinger equation with the complex potential V_s.

8.2.2 Dirac-like equation

Consider the SE with the external field $\mathbf{F} = (F_1, 0, F_3,)$. Let us represent the complex spinor V in this SE via two real spinors U and W, i.e. $V = U + iW$. The SE for V implies the following equations for the real spinors U and W:

$$\dot{U} = (\mathbf{G}\boldsymbol{\sigma})\,U + (\mathbf{K}\boldsymbol{\sigma})\,W, \quad \dot{W} = (\mathbf{G}\boldsymbol{\sigma})\,W - (\mathbf{K}\boldsymbol{\sigma})\,U\,, \tag{8.15}$$

where $\mathbf{G}$ and $\mathbf{K}$ are given by equations (8.3). In turn, the latter set can be written as

$$\dot{\Psi} = [(\boldsymbol{\Sigma}\mathbf{G}) + (\boldsymbol{\gamma}\mathbf{K})]\,\Psi, \quad \Psi = \begin{pmatrix} U \\ W \end{pmatrix}\,, \tag{8.16}$$

where $\boldsymbol{\Sigma}$ and $\boldsymbol{\gamma}$ are Dirac gamma-matrices in the standard representation, see Section A.2.1.

8.2.3 Rigid rotator equation

If V is a solution of the SE, then the vectors (A.18) obey the following equations:

$$\dot{\mathbf{L}}^{v,v} = i(\mathbf{F}^* - \mathbf{F})(V, V) + [(\mathbf{F} + \mathbf{F}^*) \times \mathbf{L}^{v,v}] ,$$
$$\dot{\mathbf{L}}^{\bar{v},v} = 2\left[\mathbf{F} \times \mathbf{L}^{\bar{v},v}\right], \quad \dot{\mathbf{L}}^{v,\bar{v}} = 2\left[\mathbf{F}^* \times \mathbf{L}^{v,\bar{v}}\right] . \tag{8.17}$$

At the same time, the following relations take place:

$$\mathbf{L}^{v,\dot{v}} = -i(V, V)\,\mathbf{F} + [\mathbf{F} \times \mathbf{L}^{v,v}], \quad \mathbf{L}^{\bar{v},\dot{v}} = \mathbf{L}^{\dot{\bar{v}},v} = \left[\mathbf{F} \times \mathbf{L}^{\bar{v},v}\right] . \tag{8.18}$$

In addition, the vectors (A.26) obey the equations:

$$\dot{\mathbf{e}}_1 = 2\mathbf{e}_2(\mathbf{Kn}) - 2\mathbf{n}[(\mathbf{Kn}) + (\mathbf{Ge}_1)] ,$$
$$\dot{\mathbf{e}}_2 = 2\mathbf{n}[(\mathbf{Ke}_1) - (\mathbf{Ge}_2)] - 2\mathbf{e}_1(\mathbf{Kn}) ,$$
$$\dot{\mathbf{n}} = 2\mathbf{e}_1[(\mathbf{Ke}_2) + (\mathbf{Ge}_1)] + 2\mathbf{e}_2[(\mathbf{Ge}_2) - (\mathbf{Ke}_1)] , \tag{8.19}$$

with $\mathbf{K}$ and $\mathbf{G}$ given by (8.3). Supposing that V obeys the SE, we can find the equations of motion for the parameters N, α, θ, and φ from the representation (A.29). Taking into account equations (A.30) and (A.16), we have

$$2\dot{V} = \left(2N^{-1}\dot{N} + i\dot{\alpha} - i\dot{\varphi}\cos\theta\right)V + \left(\dot{\theta} + i\dot{\varphi}\sin\theta\right)\exp(i\alpha)\,\bar{V} . \tag{8.20}$$

Then, with the help of (8.20), (A.36), and (8.3), we finally obtain

$$\dot{\theta} = 2(\mathbf{e}_\varphi\mathbf{K}) + 2(\mathbf{e}_\theta\mathbf{G}), \quad \dot{\varphi}\sin\theta = 2(\mathbf{e}_\theta\mathbf{G}) - 2(\mathbf{e}_\varphi\mathbf{K}) , \tag{8.21}$$
$$\dot{\alpha} = \dot{\varphi}\cos\theta - 2(\mathbf{Kn}), \quad \dot{N} = N(\mathbf{Gn}) . \tag{8.22}$$

The set (8.21) is autonomous (since it does not contain the functions N, α) and can be written in the compact form

$$\dot{\mathbf{n}} = 2[\mathbf{G} - \mathbf{n}(\mathbf{Gn})] + 2[\mathbf{K} \times \mathbf{n}] . \tag{8.23}$$

Thus, the time evolution of the vector $\mathbf{n}$ is determined by the external field only. If the set (8.23) can be integrated to obtain $\theta(t)$ and $\varphi(t)$, then we can find from (8.22)

$$\alpha = \int [\dot{\varphi}\cos\theta - 2(\mathbf{Kn})]\,dt, \quad N = \exp\int (\mathbf{Gn})dt . \tag{8.24}$$

Equation (8.23) for $\mathbf{G} = 0$ is the well-known *rigid rotator equation*. It appears in the gyroscope theory, in the theory of precession of a classical gyromagnet in a magnetic field, in the theory of electromagnetic resonance (see [142]), and so on. For $\mathbf{G} \neq 0$, this equation describes a possible damping of the system.

8.3 Some properties of the spin equation

8.3.1 The inverse problem

The inverse problem for the SE can be formulated as follows: provided that a solution V of the SE is known, is it possible to recover the external field $\mathbf{F}$ using this solution? The answer to this question can be found in the general case.

A spinor V (a solution of the SE) gives rise to a triplet of three-vectors $\mathbf{L}^{v,v}$, $\mathbf{L}^{v,\bar{v}}$, and $\mathbf{L}^{\bar{v},v}$ constructed according to equation (A.18). Let us decompose the external field over these vectors:

$$\mathbf{F} = c_1\mathbf{L}^{v,v} + c_2\mathbf{L}^{v,\bar{v}} + c\mathbf{L}^{\bar{v},v}\,, \tag{8.25}$$

where c_1, c_2 and c are some time-dependent coefficients. Substituting this expression into the SE and using formula (A.36), with allowance for (A.16), we find

$$i\dot{V} = (\boldsymbol{\sigma}\mathbf{F})\,V = (V, V)\left(c_1 V + 2c_2\bar{V}\right)\,. \tag{8.26}$$

Multiplying this relation from the left by V^+ and $\bar{V}^+$, we obtain

$$c_1 = \frac{i\,(V, \dot{V})}{(V, V)^2}\,, \quad c_2 = \frac{i\,(\bar{V}, \dot{V})}{2\,(V, V)^2}\,. \tag{8.27}$$

Substituting (8.27) into (8.25), we finally have

$$\mathbf{F} = \frac{i}{2\,(V, V)^2}\left[2\,(V, \dot{V})\,\mathbf{L}^{v,v} + (\bar{V}, \dot{V})\,\mathbf{L}^{v,\bar{v}}\right] + c\mathbf{L}^{\bar{v},v}\,. \tag{8.28}$$

Here, the complex function $c(t)$ remains arbitrary. Thus, there exists an infinite number of external fields $\mathbf{F}$ which admit one and the same solution of the SE, and the corresponding functional arbitrariness is completely described.

One can write (8.28) in a different form:

$$\mathbf{F} = \frac{i}{2\,(V, V)^2}\left\{(V, V)\left(\mathbf{L}^{v,\dot{v}} - \mathbf{L}^{\dot{v},v}\right) + \left[(V, \dot{V}) + (\dot{V}, V)\right]\mathbf{L}^{v,v}\right\} + b\mathbf{L}^{\bar{v},v}\,. \tag{8.29}$$

Here, $b(t)$ is also an arbitrary complex function. Expression (8.29) can be easily reduced to (8.28) with allowance for (A.23).

The functional arbitrariness in the solution of the inverse problem is related to the fact that the spinor V is given by two complex functions, whereas the external field $\mathbf{F}$ is defined by three complex functions.

It should be noted that we have also demonstrated that any complex spinor with an arbitrary time-dependence (provided that this spinor is differentiable) is a solution of a certain family of the SE.

Taking into account the explicit form (A.29) of the spinor V and using formulas (A.27) and (8.20), one easily deduces from (8.28) that

$$\mathbf{F} = \frac{1}{2}\left[(\dot{\varphi}\cos\theta - \dot{\alpha})\,\mathbf{n} - \dot{\varphi}\sin\theta\mathbf{e}_\theta + \dot{\theta}\mathbf{e}_\varphi\right] + i\frac{\dot{N}}{N}\mathbf{n} + a\left(\mathbf{e}_\theta + i\mathbf{e}_\varphi\right)\,, \tag{8.30}$$

where $a(t)$ is an arbitrary complex function.

8.3.2 General solution

The general solution Y_{gen} of the SE can always be written as

$$Y_{\text{gen}} = aV + bU\,, \tag{8.31}$$

where a and b are arbitrary complex constants, while $V(t)$ and $U(t)$ are two linearly-independent particular solutions of the SE. In fact, one needs to know only one particular solution V, since the other solution U can be constructed from V in quadratures. Indeed, according to (A.16), one can always present U in the form

$$U = \alpha V + \beta \bar{V}\,, \tag{8.32}$$

where $\alpha(t)$ and $\beta(t)$ are complex functions of time. Substituting (8.32) into the SE, and taking into account (8.7) and (8.3), we find

$$\dot{\alpha} v + \dot{\beta}\bar{V} = 2\beta(\sigma\mathbf{G})\bar{V}\,. \tag{8.33}$$

Hence, multiplying this relation by V^+ and $\bar{V}^+$, we obtain, with allowance for (A.18), (A.19), (A.20), (A.21), (A.24), and (A.25), that

$$(V,V)\,\dot{\beta} = -2\beta(\mathbf{L}^{v,v}\mathbf{G}), \quad \dot{\alpha} = 2\beta\,(V,V)^{-1}\,(\mathbf{L}^{v,\bar{v}}\mathbf{G})\,. \tag{8.34}$$

Taking into account (A.27) and (8.22), one can rewrite the first of these equations in the form $N\dot{\beta} = -2\dot{N}\beta$. The latter equation can be easily integrated:

$$\beta = \beta_0 N^{-2} = \beta_0\,(V,V)^{-1}\,. \tag{8.35}$$

Here β_0 is an arbitrary complex constant. Then, the second equation in (8.34) implies

$$\dot{\alpha} = 2\beta_0\,(V,V)^{-2}\,(\mathbf{L}^{v,\bar{v}}\mathbf{G})\,. \tag{8.36}$$

Hence, α can be found by an integration:

$$\alpha = \alpha_0 + 2\beta_0 \int (V,V)^{-2}\,(\mathbf{L}^{v,\bar{v}}\mathbf{G})dt\,, \tag{8.37}$$

where α_0 is an arbitrary complex constant. Thus, the general solution $Y_{\text{gen}}(t)$ of the SE, with the known particular solution V of this equation, has the form

$$Y_{\text{gen}} = \left[\alpha_0 + 2\beta_0 \int (V,V)^{-2}\,(\mathbf{L}^{v,\bar{v}}\mathbf{G})dt\right] V + \beta_0\,(V,V)^{-1}\,\bar{V}\,, \tag{8.38}$$

where α_0, β_0 are arbitrary complex constants.

8.3.3 Stationary solutions

Consider the SE with a constant external filed, $\mathbf{F} = \text{const}$. In this case, we can search for stationary solutions of the form $V(t) = \exp(-i\lambda t)V$, where V is a time-independent spinor subject to the equation[1] $(\boldsymbol{\sigma}\mathbf{F})V = \lambda V$. In particular, for $\mathbf{F}^2 \neq 0$, we have two independent solutions $V_\zeta, \lambda_\zeta, \zeta = \pm 1$,

$$V_1 = N_1 \begin{pmatrix} F_3 + \sqrt{\mathbf{F}^2} \\ iF_2 + F_1 \end{pmatrix}, \quad \lambda_1 = \sqrt{\mathbf{F}^2}\,,$$
$$V_{-1} = N_{-1} \begin{pmatrix} iF_2 - F_1 \\ F_3 + \sqrt{\mathbf{F}^2} \end{pmatrix}, \quad \lambda_{-1} = -\sqrt{\mathbf{F}^2}\,, \tag{8.39}$$

where N_ζ are normalization factors.

8.3.4 Reduction of the external field

Suppose that $V'(t)$ is a solution of the SE with an external field $\mathbf{F}' = (F_1', F_2', F_3')$. Let us perform a transformation

$$V = \hat{T}V', \quad \hat{T}(t) = \exp i\alpha\,(\boldsymbol{\sigma}\mathbf{l})\ . \tag{8.40}$$

Here, $\mathbf{l}$ is an arbitrary constant complex vector, and $\alpha(t)$ is an arbitrary complex function.

If $\mathbf{l}^2 \neq 0$, then, without loss of generality, we can set $\mathbf{l}^2 = 1$, so that the matrix (8.40) reads

$$\hat{T} = \cos\alpha + i\,(\boldsymbol{\sigma}\mathbf{l})\sin\alpha\,.$$

One can easily verify that the spinor $V(t)$ also obeys the SE with the external field $\mathbf{F}$ of the form

$$\mathbf{F} = \left[\mathbf{F}' - \mathbf{l}\left(\mathbf{F}'\mathbf{l}\right)\right]\cos 2\alpha + \left[\mathbf{F}' \times \mathbf{l}\right]\sin 2\alpha + \mathbf{l}\left[(\mathbf{F}'\mathbf{l}) - \dot{\alpha}\right]\ . \tag{8.41}$$

Since the transformation matrix $\hat{T}$ is invertible the SE with the external field $\mathbf{F}'$ and the one with the external field $\mathbf{F}$ are equivalent. If we subject, for example, the function α to the relation $\dot{\alpha} = (\mathbf{F}'\mathbf{l})$ then the projection of $\mathbf{F}$ onto the direction $\mathbf{l}$ becomes zero, $(\mathbf{F}\mathbf{l}) = 0$, i.e. the reduced external field $\mathbf{F}$ has only two nonzero (complex) components in the plane that is orthogonal to $\mathbf{l}$.

One can also consider complex constant $\mathbf{l}$ with $\mathbf{l}^2 = 0$. Then $\hat{T} = 1 + i\alpha(\boldsymbol{\sigma}\mathbf{l})$, and

$$\mathbf{F} = \mathbf{F}' + 2\alpha\left[\mathbf{F}' \times \mathbf{l}\right] + \mathbf{l}\left[2\alpha^2\left(\mathbf{F}'\mathbf{l}\right) - \dot{\alpha}\right]\ . \tag{8.42}$$

1 This equation is analyzed in Section A.1.3

By an appropriate choice of a complex $\mathbf{l}$, one can always reduce any component of both $\mathbf{K} = \operatorname{Re}\mathbf{F}$ and $\mathbf{G} = \operatorname{Im}\mathbf{F}$ to zero. However, in this case we cannot imagine $\mathbf{F}$ as a vector in a fixed plane, in contrast to the case of a real $\mathbf{l}$.

Let us choose the vector $\mathbf{l}$ to be a unit vector in the z-direction, $\mathbf{l} = (0, 0, 1)$, and α to be a solution of the equation $\dot{\alpha} = F_3'$. Then the reduced external field $\mathbf{F}$ takes the form

$$\mathbf{F} = (F_1, F_2, 0) , \tag{8.43}$$

where

$$F_1 = F_1' \cos 2\alpha + F_2' \sin 2\alpha, \quad F_2 = F_2' \cos 2\alpha - F_1' \sin 2\alpha ,$$
$$F_1' = F_1 \cos 2\alpha - F_2 \sin 2\alpha, \quad F_2' = F_2 \cos 2\alpha + F_1 \sin 2\alpha .$$

Choosing $\mathbf{l} = (0, 0, 1)$ and selecting α to be a solution of the equations

$$F_1' = F_1 \cos 2\alpha, \quad F_2' = F_1 \sin 2\alpha, \quad F_3' = F_3 + \dot{\alpha} ,$$

we obtain

$$\begin{aligned} &F_1 = F_1' \cos 2\alpha + F_2' \sin 2\alpha, \quad F_3 = F_3' - \dot{\alpha} , \\ &F_2 = F_2' \cos 2\alpha - F_1' \sin 2\alpha = 0 , \end{aligned} \tag{8.44}$$

so that the reduced external field $\mathbf{F}$ takes the form

$$\mathbf{F} = (F_1, 0, F_3) . \tag{8.45}$$

In addition, one can see that if V is a solution of the SE with the external field (8.45) then:

(i) $U = (2)^{-1/2}(1 + i\sigma_1)V$ is a solution of the SE with the external field

$$\mathbf{F} = (F_1, F_3, 0) ; \tag{8.46}$$

(ii) $U = \sigma_1 V$ is a solution of the SE with the external field

$$\mathbf{F} = (F_1, 0, -F_3) ; \tag{8.47}$$

(iii) $U = \sigma_3 V$ is a solution of the SE with the external field

$$\mathbf{F} = (-F_1, 0, F_3) ; \tag{8.48}$$

(iv) $U = \sigma_2 V$ is a solution of the SE with the external field

$$\mathbf{F} = (-F_1, 0, -F_3) ; \tag{8.49}$$

(v) $U = (2)^{-1/2}(\sigma_1 + \sigma_3)V$ is a solution of the SE with the external field

$$\mathbf{F} = (F_3, 0, F_1) . \tag{8.50}$$

8.3.5 Transformation matrix

Suppose that we know a solution V_1 of the SE with an external field $\mathbf{F}_1$ and wish to find such a nonsingular time-dependent matrix $\hat{T}^{21}$ (in what follows, it is called the transformation matrix) that a spinor V_2,

$$V_2(t) = \hat{T}^{21}(t)V_1(t)\,, \tag{8.51}$$

is a solution of the SE with an external field $\mathbf{F}_2$. It is easy to obtain an equation for such a transformation matrix:

$$i\frac{d}{dt}\hat{T}^{21} = (\boldsymbol{\sigma}\mathbf{F}_2)\,\hat{T}^{21} - \hat{T}^{21}\,(\boldsymbol{\sigma}\mathbf{F}_1)\;. \tag{8.52}$$

Like any 2×2 matrix, the matrix $\hat{T}^{21}$ can be written in the form

$$\hat{T}^{21} = a_0 - i(\boldsymbol{\sigma}\mathbf{a}), \quad \mathbf{a} = (a_1, a_2, a_3)\;, \tag{8.53}$$

where $a_s(t), s = 0, 1, 2, 3$ are some complex functions of t. Substituting (8.53) into (8.52), and using elementary properties of the Pauli matrices, one obtains the following set of equations for the functions a_s:

$$\begin{aligned} &\dot{a}_0 + \mathbf{a}\mathbf{F}_{21} = 0, \quad \mathbf{F}_{21} = \mathbf{F}_2 - \mathbf{F}_1\,, \\ &\dot{\mathbf{a}} + 2\,[\mathbf{a}\times\mathbf{F}_1] + [\mathbf{a}\times\mathbf{F}_{21}] - a_0\mathbf{F}_{21} = 0\,. \end{aligned} \tag{8.54}$$

It is easy to find that $\Delta = \det\hat{T}^{21} = a_0^2 + \mathbf{a}^2$ is an integral of motion. Since the matrix $\hat{T}^{21}$ is determined by equation (8.52) only up to a constant complex multiplier, we choose, without loss of generality,

$$\Delta = a_0^2 + \mathbf{a}^2 = 1\,. \tag{8.55}$$

For the inverse matrix $(\hat{T}^{21})^{-1}$, we obtain

$$\left(\hat{T}^{21}\right)^{-1} = \Delta^{-1}\left[a_0 + i(\boldsymbol{\sigma}\mathbf{a})\right] = a_0 + i(\boldsymbol{\sigma}\mathbf{a})\,. \tag{8.56}$$

Given $\mathbf{F}_1$ and $\mathbf{F}_2$, equations (8.54) represent a linear homogenous (complex) set of four ordinary differential equations of first order. Solving this set is completely analogous to solving the SE with the external field $\mathbf{F}_2$, so we do not achieve any simplification. However, assuming that the external field $\mathbf{F}_1$ and the matrix $\hat{T}^{21}$ are known, we can obtain the external field $\mathbf{F}_2$ from (8.54). It turns out that this problem can be easily solved. To this end, we need to consider two cases.

(1) Let $a_0 \neq 0$. We introduce a complex vector $\mathbf{q} = \mathbf{a}/\mathbf{a}_0$ such that $\mathbf{q}^2 \neq -1$. Then (8.55) implies $a_0 = (1 + \mathbf{q}^2)^{-1/2}$. From (8.54), one obtains the equation

$$\dot{\mathbf{q}} - \mathbf{q}\,(\mathbf{q}\mathbf{F}_{21}) + [\mathbf{q}\times\mathbf{F}_{21}] + 2\,[\mathbf{q}\times\mathbf{F}_1] - \mathbf{F}_{21} = 0\,. \tag{8.57}$$

This equation allows one to find a unique representation for the external field $\mathbf{F}_2$,

$$\mathbf{F}_2 = \frac{\dot{\mathbf{q}} + [\mathbf{q}\times\dot{\mathbf{q}}] + 2\,[\mathbf{q}\times\mathbf{F}_1] + 2\mathbf{q}\,(\mathbf{q}\mathbf{F}_1) + (1 - \mathbf{q}^2)\mathbf{F}_1}{1 + \mathbf{q}^2}\,. \tag{8.58}$$

In this case, the transformation matrix reads

$$\hat{T}^{21} = \frac{1 - i(\boldsymbol{\sigma}\mathbf{q})}{\sqrt{1 + \mathbf{q}^2}}, \quad \left(\hat{T}^{21}\right)^{-1} = \frac{1 + i(\boldsymbol{\sigma}\mathbf{q})}{\sqrt{1 + \mathbf{q}^2}}. \tag{8.59}$$

(2) Let $a_0 = 0$. We introduce a vector $\mathbf{q} = \mathbf{a}$. In this case, the condition (8.55) implies $\mathbf{q}^2 = 1$, and we obtain from (8.54)

$$\dot{\mathbf{q}} + [\mathbf{q} \times \mathbf{F}_{21}] + 2[\mathbf{q} \times \mathbf{F}_1] = 0, \quad (\mathbf{q}\mathbf{F}_{21}) = 0. \tag{8.60}$$

From (8.60), we uniquely recover $\mathbf{F}_2$ in the form

$$\mathbf{F}_2 = [\mathbf{q} \times \dot{\mathbf{q}}] + 2\mathbf{q}(\mathbf{q}\mathbf{F}_1) - \mathbf{F}_1, \quad \mathbf{q}^2 = 1. \tag{8.61}$$

The transformation matrix now reads

$$\hat{T}^{21} = \left(R^{21}\right)^{-1} = (\boldsymbol{\sigma}\mathbf{q}). \tag{8.62}$$

Consequently, given an exact solution V_0 that corresponds to the external field $\mathbf{F}_1$, we can construct a family of external fields $\mathbf{F}_2$ and the corresponding exact solutions (8.51), parametrized by an arbitrary complex time-dependent vector $\mathbf{q}$.

In the particular case of a self-adjoint SE, the above statement remains valid if one assumes $\mathbf{q}$ to be a real vector. In this case, the transformation matrix is unitary.

8.3.6 Evolution operator

Let us choose $\mathbf{F}_1 = 0$ and denote $\mathbf{F}_2 = \mathbf{F}$, $\hat{T}^{21} = \hat{T}$ in the above consideration. Then one can choose V_1 as an arbitrary constant spinor, $V_1 = V_0 = \text{const}$. For $\mathbf{F}_1 = 0$, one deduces from (8.52) that the transformation matrix $\hat{T}$ obeys the equation

$$i\frac{d\hat{T}}{dt} = (\boldsymbol{\sigma}\mathbf{F})\,\hat{T}. \tag{8.63}$$

If the transformation matrix $\hat{T}$ is known, the evolution operator $\hat{R}$ can be constructed as the following solution of equation (8.63) with the initial condition $\hat{R}(0) = I$:

$$\hat{R}(t) = \hat{T}(t)\hat{T}^{-1}(0). \tag{8.64}$$

Using the above expressions for the transformation matrix, we can construct the evolution operator for the SE with any external field $\mathbf{F}$ according to (8.64). One has to consider separately two cases.

(1) Let us chose an arbitrary complex time-dependent vector $\mathbf{q}(t)$ ($\mathbf{q}(0) = \mathbf{q}_0$) such that $\mathbf{q}^2 \neq -1$. Then the SE with the external field

$$\mathbf{F} = \frac{\dot{\mathbf{q}} + [\mathbf{q} \times \dot{\mathbf{q}}]}{1 + \mathbf{q}^2} \tag{8.65}$$

has the evolution operator of the form

$$\hat{R} = \frac{[1 - i(\boldsymbol{\sigma}\mathbf{q})]\,[1 + i(\boldsymbol{\sigma}\mathbf{q}_0)]}{\sqrt{(1 + \mathbf{q}^2)\,(1 + \mathbf{q}_0^2)}} = \frac{1 + (\mathbf{q}\mathbf{q})_0 - i(\boldsymbol{\sigma}\mathbf{p})}{\sqrt{(1 + \mathbf{q}^2)\,(1 + \mathbf{q}_0^2)}}\,, \tag{8.66}$$

where $\mathbf{p} = \mathbf{q} - \mathbf{q}_0 + [\mathbf{q}_0 \times \mathbf{q}]$.

(2) Let us select an arbitrary complex unit time-dependent vector $\mathbf{q}(t)$ ($\mathbf{q}(0) = \mathbf{q}_0$), $\mathbf{q}^2 = 1$. Then the SE with the external field $\mathbf{F} = [\mathbf{q} \times \dot{\mathbf{q}}]$, has the evolution operator of the form

$$\hat{R} = (\boldsymbol{\sigma}\mathbf{q})\,(\boldsymbol{\sigma}\mathbf{q}_0) = (\mathbf{q}\mathbf{q}_0) + i(\boldsymbol{\sigma}\,[\mathbf{q} \times \mathbf{q}_0])\,. \tag{8.67}$$

In the case of a self-adjoint SE (real external fields), $\mathbf{q}$ is selected as a real vector, and the operator $\hat{R}$ is unitary.

8.4 Self-adjoint spin equation

8.4.1 General solution and inverse problem

We shall refer to the SE as *self-adjoint SE* if the external field $\mathbf{F}$ is real. In this case, according to (8.3), we have

$$\mathbf{F} = \operatorname{Re}\mathbf{F} = \mathbf{K}, \quad \operatorname{Im}\mathbf{F} = \mathbf{G} = 0\,, \tag{8.68}$$

and the SE has the form of a Schrödinger equation,

$$i\dot{V} = HV, \quad H = (\boldsymbol{\sigma}\mathbf{F}) = H^+\,, \tag{8.69}$$

with a Hermitian Hamiltonian H.

Below, we list some properties of a self-adjoint SE, which, generally speaking, do not take place for a generic complex external field.

The general solution Y_{gen} of a self-adjoint SE has the form

$$Y_{\text{gen}} = aV + b\bar{V} \tag{8.70}$$

where $V(t)$ is any nonzero particular solution of the SE, and a, b are arbitrary complex constants. This fact follows from (8.34).

For any solution V of a self-adjoint SE, the quantity $N^2 = (V, V)$ is conserved in time, which is implied by (8.22) in the case $\mathbf{G} = 0$. However, the reverse is not true. The fact that $N^2 = \text{const}$ does not imply that V is a solution of a self-adjoint SE, since one can indicate, according to (8.28), a family of complex external fields in the SE that admit solutions with $N^2 = \text{const}$.

For an arbitrary nonzero differentiable spinor V subject to the condition $(V, V) = \text{const}$, there exists only one self-adjoint SE (only one real external field), whose particular solution is given by this spinor, and whose general solution has the form (8.70).

Indeed, it follows from (8.29) that in this case a real external field is recovered by the spinor V in a unique manner:

$$\mathbf{F} = i\,[2\,(V,V)]^{-1}\left(\mathbf{L}^{v,\dot{v}} - \mathbf{L}^{\dot{v},v}\right). \tag{8.71}$$

It can be easily verified that the same expression for $\mathbf{F}$ arises in the case when V is replaced by Y_{gen} from (8.70), which confirms the uniqueness of the external field $\mathbf{F}$. Now, presenting V in the form (A.29), and setting $N = \text{const}$, one obtains a decomposition of $\mathbf{F}$ in the basis vectors of a spherical coordinate system (A.28):

$$\mathbf{F} = \frac{1}{2}\left[(\dot{\varphi}\cos\theta - \dot{\alpha})\,\mathbf{n} - \dot{\varphi}\sin\theta\mathbf{e}_{\theta} + \dot{\theta}\mathbf{e}_{\varphi}\right]. \tag{8.72}$$

Hence, one can find the Cartesian components of the external field and calculate its square:

$$\mathbf{F} = \frac{1}{2}(-\dot{\theta}\sin\varphi - \dot{\alpha}\sin\theta\cos\varphi, \dot{\theta}\cos\varphi - \dot{\alpha}\sin\theta\sin\varphi, \dot{\varphi}-\dot{\alpha}\cos\theta)\,,$$
$$F^2 = \mathbf{F}^2 = \frac{1}{4}\left(\dot{\theta}^2 + \dot{\varphi}^2 + \dot{\alpha}^2 - 2\dot{\alpha}\dot{\varphi}\cos\theta\right). \tag{8.73}$$

The possibility of an unambiguous recovery of the real external field $\mathbf{F}$ by a given arbitrary spinor $V(t)$ with a constant norm also signifies the possibility of generating an exactly solvable self-adjoint SE.

For equations that are associated with a self-adjoint SE, one can state some additional properties. For instance, the evolution equations (8.17) for the linearly-independent vectors (A.22) become coincident, so that the vectors (A.22) have to be distinguished by an appropriate choice of initial conditions.

8.4.2 Hamiltonian and Lagrangian forms of self-adjoint spin equation

Consider the set of equations (8.21) for real external fields. In this case, the set can be written as

$$\dot{\theta} = 2\left(F_2\cos\varphi - F_1\sin\varphi\right), \quad \dot{\varphi}\sin\theta = 2F_3\sin\theta - 2\left(F_1\cos\varphi + F_2\sin\varphi\right)\cos\theta\,. \tag{8.74}$$

Without loss of generality, one can always choose

$$F_1 = g(t)\cos 2\alpha(t), \quad F_2 = g(t)\sin 2\alpha(t)\,, \tag{8.75}$$

where $g(t)$ and $\alpha(t)$ are some real functions of time. Let us replace $\varphi(t)$ in equation (8.74) by a new function $\Phi(t)$, introduced as

$$\varphi(t) = \Phi(t) + 2\alpha(t)\,. \tag{8.76}$$

It is easy to see that the set (8.74) transforms to

$$\dot{\theta} = -2g\sin\Phi, \quad \dot{\Phi}\sin\theta = 2f\sin\theta - 2g\cos\Phi\cos\theta\,, \tag{8.77}$$

where $f = F_3(t) - \dot{\alpha}(t)$. Notice that the replacement (8.76) is equivalent to the transformation (8.40), if one chooses $\mathbf{e} = (0, 0, 1)$ and selects $\alpha(t)$ such that the external field takes the form (8.45).

If one introduces the coordinate q, the conjugate momentum p, and the Hamiltonian H as follows:

$$q = \cos\theta, \quad p = -\Phi, \quad H = 2g\sqrt{1-q^2}\cos p + 2qf\,, \tag{8.78}$$

then set (8.77) takes the form of one-dimensional classical Hamilton equations [49, 142],

$$\dot{q} = \frac{\partial H}{\partial p}, \quad \dot{p} = -\frac{\partial H}{\partial q}\,. \tag{8.79}$$

Making canonical transformations, we can obtain different forms of the Hamilton equations that are associated with the self-adjoint SE.

It is straightforward to check that (8.77) are the Euler–Lagrange equations for the Lagrange function

$$\mathcal{L} = \left[(1-\gamma)\,\dot{\theta}\Phi - 2g\cos\Phi\right]\sin\theta + \left[\gamma\dot{\Phi} - 2f\right]\cos\theta\,, \tag{8.80}$$

where γ is an arbitrary real number.

Finally, the set (8.77) leads to a second-order equation for the function $\theta(t)$:

$$\ddot{\theta} - \frac{\dot{g}}{g}\dot{\theta} + 2f\sqrt{4g^2 - \dot{\theta}^2} - \left(4g^2 - \dot{\theta}^2\right)\frac{\cos\theta}{\sin\theta} = 0\,. \tag{8.81}$$

This equation is also the Euler–Lagrange equation for the Lagrange function

$$\mathcal{L} = \dot{\theta}\arcsin\left(\dot{\theta}/2g\right)\sin\theta + \sqrt{4g^2 - \dot{\theta}^2}\sin\theta + 2f\cos\theta\,. \tag{8.82}$$

8.5 Exact solutions of spin equation

We start this section with the following three remarks.

(1) Let $V(t)$ be a solution of the SE with a given external field $\mathbf{F}$. In this equation, we make the following change of the variable $t = T(t')$, where t' is the new real variable (T is a real invertible function). Then the SE takes the form

$$i\frac{dV'(t)}{dt} = (\boldsymbol{\sigma}\mathbf{F}'(t))V'(t)\,, \tag{8.83}$$

where

$$\mathbf{F}'(t) = \mathbf{F}\left(T(t)\right)\dot{T}, \quad V'(t) = V\left(T(t)\right)\,. \tag{8.84}$$

Consequently, if one knows a solution of the SE with an external field $\mathbf{F}$, then one knows a solution of the SE with external fields $\mathbf{F}'$, parametrized by an arbitrary function T. In this sense, all solutions are divided into equivalence classes. Below, we are going to list only those solutions of the SE that belong to different classes.

(2) We have demonstrated that the SE with an arbitrary external field can be reduced to an equivalent SE with the external field (8.45) which has only two nonzero components, $\mathbf{F} = (F_1, 0, F_3)$. Below, we are going to list only solutions for such external fields.

(3) Let the components F_1 and F_3 of the external field be linearly dependent. Then, without loss of the generality, we can write

$$F_1 = q \sin \lambda, \quad F_3 = q \cos \lambda\,, \tag{8.85}$$

where $q(t)$ is an arbitrary function of time, and λ is a complex constant. Let us define the function $\omega(t)$ by the relations

$$\dot{\omega}(t) = q(t), \quad \omega(0) = 0\,. \tag{8.86}$$

Then, the evolution operator for the SE with such an external field has the form

$$\hat{R} = \cos \omega(t) - i \left(\sigma_1 \sin \lambda + \sigma_3 \cos \lambda\right) \sin \omega(t)\,. \tag{8.87}$$

Especially interesting are solutions of the SE that can be written via the known special functions.

Below, we consider external fields with such nonzero components F_1 and F_3 that obey the following properties: if solutions of the SE are known for such external fields, then one can construct solutions for the external fields

$$\mathbf{F} = (\alpha F_1(\varphi), 0, \beta F_3(\varphi)), \quad \varphi = \omega t + \varphi_0\,, \tag{8.88}$$

where α, β, and φ_0 are arbitrary real constants. We have succeeded in finding 26 pairs of linearly independent functions F_1 and F_3 that conform to this condition. Below, we list such pairs and present the spinor u, which is the corresponding exact solution of the SE.

(1) $F_1 = at, \quad F_3 = bt + c/t$:

$$u = \begin{pmatrix} at^{\gamma+2} e^{-z/2} \Phi(\alpha + 1, \gamma + 2; z) \\ 2(i - c)\, t^{\gamma} e^{-z/2} \Phi(\alpha, \gamma; z) \end{pmatrix},$$

$$z = it^2 \sqrt{a^2 + b^2}, \quad \alpha = \frac{\gamma}{2}\left(1 + \frac{b}{\sqrt{a^2 + b^2}}\right), \quad \gamma = ic\,.$$

(2) $F_1 = a/t, \quad F_3 = b/t + ct$:

$$u = \begin{pmatrix} -at^{\gamma-1} e^{-z/2} \Phi(\alpha, \gamma; z) \\ \left(\sqrt{a^2 + b^2} + b\right) t^{\gamma-1} e^{-z/2} \Phi(\alpha + 1, \gamma; z) \end{pmatrix},$$

$$z = ict^2, \quad 2\alpha = i\left(\sqrt{a^2 + b^2} + b\right), \quad \gamma = 1 + i\sqrt{a^2 + b^2}\,.$$

(3) $F_1 = a/t, \quad F_3 = b/t + c$:

$$u = \begin{pmatrix} -at^{(\gamma-1)/2}e^{-z/2}\Phi(\alpha, \gamma; z) \\ -iat^{(\gamma-1)/2}e^{-z/2}\Phi(1+\alpha, \gamma; z) \end{pmatrix},$$

$$z = 2ict, \quad \alpha = i\left(\sqrt{a^2+b^2}+b\right), \quad \gamma = 1 + 2i\sqrt{a^2+b^2}\,.$$

(4) $F_1 = a/\sin 2\varphi, \quad F_3 = (b\cos 2\varphi + c)/\sin 2\varphi$:

$$u = \begin{pmatrix} -az^{\mu}(1-z)^{\nu}F(\alpha+1, \beta; \gamma; z) \\ (-4i\omega\mu + b + c)z^{\mu}(1-z)^{\nu}F(\alpha, \beta+1; \gamma; z) \end{pmatrix},$$

$$z = \sin^2\varphi, \quad \mu = \frac{i}{4\omega}\sqrt{a^2+(b+c)^2}, \quad \nu = \frac{i}{4\omega}\sqrt{a^2+(b-c)^2}\,,$$

$$\alpha = \mu + \nu - ib/2\omega, \quad \beta = \mu + \nu + ib/2\omega, \quad \gamma = 1 + 2\mu\,.$$

(5) $F_1 = a\tan\varphi, \quad F_3 = b\tan\varphi + c\cot\varphi$:

$$u = \begin{pmatrix} 2(c+i\omega)z^{\mu}(1-z)^{\nu}F(\alpha, \beta; 2\mu; z) \\ az^{\mu+1}(1-z)^{\nu}F(\alpha+1, \beta+1; 2\mu+2; z) \end{pmatrix},$$

$$z = \sin^2\varphi, \quad \mu = -\frac{ic}{2\omega}, \quad \nu = \frac{i}{2\omega}\sqrt{a^2+b^2}\,,$$

$$\lambda = \frac{i}{2\omega}\sqrt{a^2+(b-c)^2}, \quad \alpha = \nu + \mu + \lambda, \quad \beta = \nu + \mu - \lambda\,.$$

(6) $F_1 = a/\sin\varphi, \quad F_3 = b\tan\varphi + c\cot\varphi$:

$$u = \begin{pmatrix} -az^{\mu}(1-z)^{\nu+1/2}F(\alpha+1, \beta; 2\mu+1; z) \\ \left(\sqrt{a^2+c^2}+c\right)z^{\mu}(1-z)^{\nu}F(\alpha, \beta; 2\mu+1; z) \end{pmatrix},$$

$$\mu = \frac{i}{2\omega}\sqrt{a^2+c^2}, \quad \nu = -\frac{ib}{2\omega}, \quad z = \sin^2\varphi\,,$$

$$\alpha = \mu - \frac{ic}{2\omega}, \quad \beta = \frac{1}{2} + \mu + 2\nu + \frac{ic}{2\omega}\,.$$

(7) $F_1 = a/\cos\varphi, \quad F_3 = b\tan\varphi + c$:

$$u = \begin{pmatrix} (\omega + 2c - 2ib)z^{\mu}(1-z)^{\nu}F(\alpha, \beta; \gamma; z) \\ 2iaz^{\mu+1/2}(1-z)^{\nu}F(\alpha, \beta+1; \gamma+1; z) \end{pmatrix},$$

$$z = -e^{-2i\varphi}, \quad \mu = \frac{c-ib}{2\omega}, \quad \nu = \frac{i}{\omega}\sqrt{a^2+b^2}\,,$$

$$\alpha = \frac{1}{2} + \frac{c}{\omega} + \nu, \quad \beta = \nu - \frac{ib}{\omega}, \quad \gamma = \frac{1}{2} + 2\mu\,.$$

(8) $F_1 = a/\sinh\varphi, \quad F_3 = b\tanh\varphi + c\coth\varphi$:

$$u = \begin{pmatrix} -az^{\mu}(1-z)^{\nu} F(\alpha,\beta;\gamma;z) \\ (-2i\omega\mu a + c)\, z^{\mu}(1-z)^{\nu+1/2} F(\alpha,\beta+1;\gamma;z) \end{pmatrix},$$

$$z = \tanh^2\varphi, \quad \mu = \frac{i}{2\omega}\sqrt{a^2+c^2}, \quad \nu = \frac{i(b+c)}{2\omega},$$

$$\alpha = \frac{1}{2} + \frac{ib}{\omega} + \beta, \quad \beta = \mu + \frac{ic}{2\omega}, \quad \gamma = 2\mu + 1\,.$$

(9) $F_1 = a/\cosh\varphi, \quad F_3 = b\tanh\varphi + c\coth\varphi$:

$$u = \begin{pmatrix} (2c+i\omega)z^{\mu}(1-z)^{\nu} F(\alpha,\beta;\gamma;z) \\ az^{\mu+1/2}(1-z)^{\nu+1/2} F(\alpha+1,\beta+1;\gamma+1;z) \end{pmatrix},$$

$$z = \tanh^2\varphi, \quad \mu = -\frac{ic}{2\omega}, \quad \nu = \frac{i(b+c)}{2\omega}, \quad \lambda = \frac{1}{2\omega}\sqrt{a^2-b^2}\,,$$

$$\alpha = \frac{ib}{2\omega} + \lambda, \quad \beta = \frac{ib}{2\omega} - \lambda, \quad \gamma = \frac{1}{2} - \frac{ic}{\omega}\,.$$

(10) $F_1 = a/\sinh 2\varphi, \quad F_3 = (b\cosh 2\varphi + c)/\sinh 2\varphi$:

$$u = \begin{pmatrix} -az^{\mu}(1-z)^{\nu} F(\alpha,\beta;\gamma;z) \\ (-4i\omega\mu + b + c)\, z^{\mu}(1-z)^{\nu+1} F(\alpha+1,\beta+1;\gamma;z) \end{pmatrix},$$

$$z = \tanh^2\varphi, \quad \mu = \frac{i}{4\omega}\sqrt{a^2+(b+c)^2}, \quad \lambda = \frac{i}{4\omega}\sqrt{a^2+(b-c)^2}\,,$$

$$\nu = \frac{ib}{2\omega}, \quad \alpha = \mu+\nu+\lambda, \quad \beta = \mu+\nu-\lambda \quad \gamma = 1+2\mu\,.$$

(11) $F_1 = a/\cosh\varphi, \quad F_3 = (b\sinh\varphi + c)/\cosh\varphi$:

$$u = \begin{pmatrix} az^{\mu}(1-z)^{\nu} F(\alpha,\beta;\gamma;z) \\ (2\omega\mu - c + ib)\, z^{\mu}(1-z)^{\nu+1} F(\alpha+1,\beta+1;\gamma;z) \end{pmatrix},$$

$$z = \left(\frac{e^{\varphi}+i}{e^{\varphi}-i}\right)^2, \quad \mu = \frac{1}{2\omega}\sqrt{a^2+(c-ib)^2}, \quad \alpha = \mu+\nu+\lambda\,,$$

$$\lambda = \frac{1}{2\omega}\sqrt{a^2+(c+ib)^2}, \quad \nu = \frac{ib}{\omega}, \quad \beta = \mu+\nu-\lambda, \quad \gamma = 1+2\mu\,.$$

(12) $F_1 = a\tanh\varphi, \quad F_3 = b\tanh\varphi + c\coth\varphi$:

$$u = \begin{pmatrix} 2(c+i\omega)z^{\mu}(1-z)^{\nu} F(\alpha,\beta;\gamma;z) \\ az^{\mu+1}(1-z)^{\nu} F(\alpha+1,\beta+1;\gamma+2;z) \end{pmatrix},$$

$$z = \tanh^2\varphi, \quad \mu = -\frac{ic}{2\omega}, \quad \nu = \frac{i}{2\omega}\sqrt{a^2+(b+c)^2}\,,$$

$$\lambda = \frac{i}{2\omega}\sqrt{a^2+b^2}, \quad \alpha = \mu+\nu+\lambda, \quad \beta = \mu+\nu-\lambda, \quad \gamma = 2\mu\,.$$

(13) $F_1 = a\coth\varphi, \quad F_3 = b\tanh\varphi + c\coth\varphi$:

$$u = \begin{pmatrix} -az^{\mu}(1-z)^{\nu} F(\alpha+1,\beta;\gamma;z) \\ (2\omega\mu + c) z^{\mu}(1-z)^{\nu} F(\alpha,\beta+1;\gamma;z) \end{pmatrix},$$

$$z = \tanh^2\varphi, \quad \mu = \frac{i}{2\omega}\sqrt{a^2+c^2}, \quad \nu = \frac{i}{2\omega}\sqrt{a^2+(b+c)^2},$$

$$\alpha = \mu + \nu + \frac{ib}{2\omega}, \quad \beta = \mu + \nu - \frac{ib}{2\omega}, \quad \gamma = 1 + 2\mu.$$

(14) $F_1 = a/\cosh\varphi, \quad F_3 = b\tanh\varphi + c$:

$$u = \begin{pmatrix} (2b + 2c - i\omega) z^{\mu}(1-z)^{\nu} F(\alpha,\beta;\gamma;z) \\ 2az^{\mu+1/2}(1-z)^{\nu+1/2} F(\alpha+1,\beta+1,\gamma+1;z) \end{pmatrix},$$

$$z = \frac{1}{2}(1 - \tanh\varphi), \quad \alpha = \mu + \nu + \lambda, \quad \beta = \mu + \nu - \lambda,$$

$$\mu = \frac{i(b+c)}{2\omega}, \quad \nu = \frac{i(b-c)}{2\omega}, \quad \gamma = 1/2 + 2\mu, \quad \lambda = \frac{1}{\omega}\sqrt{a^2 - b^2}.$$

(15) $F_1 = a/\sinh\varphi, \quad F_3 = b\coth\varphi + c$:

$$u = \begin{pmatrix} -az^{\mu}(1-z)^{\nu} F(\alpha,\beta;\gamma;z) \\ (-i\omega\mu + b) z^{\mu}(1-z)^{\nu+1/2} F(\alpha,\beta+1;\gamma;z) \end{pmatrix},$$

$$z = 1 - e^{-2\varphi}, \quad \mu = \frac{i}{\omega}\sqrt{a^2+b^2}, \quad \nu = \frac{i(b+c)}{2\omega},$$

$$\alpha = \frac{1}{2} + \mu + \frac{ic}{\omega}, \quad \beta = \mu + \frac{ib}{\omega}, \quad \gamma = 1 + 2\mu.$$

(16) $F_1 = a, \quad F_3 = bt + c$:

$$u = \begin{pmatrix} 2\sqrt{b} D_{\mu}(z) \\ (1+i) a D_{\mu-1}(z) \end{pmatrix},$$

$$z = \frac{1+i}{\sqrt{b}}(bt + c), \quad \mu = -\frac{ia^2}{2b}.$$

(17) $F_1 = a, \quad F_3 = b/t + c$:

$$u = \begin{pmatrix} (1 - 2ib) t^{\gamma} e^{-z/2} \Phi(\alpha, 2\gamma; z) \\ -iat^{\gamma+1} e^{-z/2} \Phi(\alpha+1, 2\gamma+2, z) \end{pmatrix},$$

$$z = 2it\sqrt{a^2+c^2}, \quad \gamma = -ib, \quad \alpha = \gamma\left(1 - \frac{c}{\sqrt{a^2+c^2}}\right).$$

(18) $F_1 = a, \quad F_3 = b/t + ct$:

$$u = \begin{pmatrix} (2b + i) t^{\gamma-1/2} e^{-z/2} \Phi(\alpha,\gamma;z) \\ at^{\gamma+1/2} e^{-z/2} \Phi(\alpha+1,\gamma+1;z) \end{pmatrix},$$

$$z = ict^2, \quad \alpha = \frac{ia^2}{4c}, \quad \gamma = \frac{1}{2} - ib.$$

(19) $F_1 = a, \quad F_3 = (b\cos 2\varphi + c)/\sin 2\varphi$:

$$u = \begin{pmatrix} (b + c + i\omega)z^{\mu}(1 - z)^{\nu} F(\alpha, \beta; \gamma; z) \\ az^{\mu+1/2}(1 - z)^{\nu+1/2} F(\alpha + 1, \beta + 1; \gamma + 1; z) \end{pmatrix},$$

$$z = \sin^2\varphi, \quad \mu = -\frac{i}{4\omega}(b + c), \quad \nu = \frac{i}{4\omega}(c - b), \quad \gamma = \frac{1}{2} + 2\mu,$$

$$\alpha = \frac{1}{2\omega}\left(\sqrt{a^2 - b^2} - ib\right), \quad \beta = -\frac{1}{2\omega}\left(\sqrt{a^2 - b^2} + ib\right).$$

(20) $F_1 = a, \quad F_3 = b\tan\varphi + c\cot\varphi$:

$$u = \begin{pmatrix} (2c + i\omega)z^{\mu}(1 - z)^{\nu} F(\alpha, \beta; \gamma; z) \\ az^{\mu+1/2}(1 - z)^{\nu+1/2} F(\alpha + 1, \beta + 1; \gamma + 1; z) \end{pmatrix},$$

$$z = \sin^2\varphi, \quad \mu = -\frac{ic}{2\omega}, \quad \nu = \frac{ib}{2\omega}, \quad \lambda = \frac{1}{2\omega}\sqrt{a^2 - (b - c)^2},$$

$$\alpha = \mu + \nu + \lambda, \quad \beta = \mu + \nu - \lambda, \quad \gamma = \frac{1}{2} + 2\mu.$$

(21) $F_1 = a, \quad F_3 = b\tan\varphi + c$:

$$u = \begin{pmatrix} az^{\mu}(1 - z)^{\nu} F(\alpha, \beta; \gamma; z) \\ (2\omega\mu - c + ib)\, z^{\mu}(1 - z)^{\nu+1} F(\alpha + 1, \beta + 1; \gamma; z) \end{pmatrix},$$

$$z = -e^{-2i\varphi}, \quad \mu = \frac{1}{2\omega}\sqrt{a^2 + (c - ib)^2}, \quad \alpha = \mu + \nu + \lambda,$$

$$\nu = \frac{ib}{\omega}, \quad \beta = \mu + \nu - \lambda, \quad \gamma = 1 + 2\mu, \quad \lambda = \frac{1}{2\omega}\sqrt{a^2 + (c + ib)^2}.$$

(22) $F_1 = a, \quad F_3 = b\tanh\varphi + c\coth\varphi$:

$$u = \begin{pmatrix} (2c + i\omega)z^{\mu}(1 - z)^{\nu} F(\alpha, \beta; \gamma; z) \\ az^{\mu+1/2}(1 - z)^{\nu} F(\alpha, \beta + 1; \gamma + 1; z) \end{pmatrix},$$

$$z = \tanh^2\varphi, \quad \mu = -\frac{ic}{2\omega}, \quad \nu = \frac{i}{2\omega}\sqrt{a^2 + (b + c)^2},$$

$$\gamma = \frac{1}{2} + 2\mu, \quad \alpha = \gamma + \nu + \frac{i}{2\omega}(b + c), \quad \beta = \nu - \frac{i}{2\omega}(b + c).$$

(23) $F_1 = a, \quad F_3 = (b\cosh 2\varphi + c)/\sinh 2\varphi$:

$$u = \begin{pmatrix} (b + c + i\omega)z^{\mu}(1 - z)^{\nu} F(\alpha, \beta; \gamma; z) \\ az^{\mu+1/2}(1 - z)^{\nu} F(\alpha, \beta + 1; \gamma + 1; z) \end{pmatrix},$$

$$z = \tanh^2\varphi, \quad \mu = -\frac{i(b + c)}{4\omega}, \quad \nu = \frac{i}{2\omega}\sqrt{a^2 + b^2},$$

$$\alpha = \frac{1}{2} + \nu - \frac{ic}{2\omega}, \quad \beta = \nu - \frac{ib}{2\omega}, \quad \gamma = \frac{1}{2} + 2\mu.$$

(24) $F_1 = a, \quad F_3 = (b \sinh\varphi + c)/\cosh\varphi$:

$$u = \begin{pmatrix} (2b + 2ic + i\omega) z^{\mu} (1-z)^{\nu} F(\alpha, \beta; \gamma; z) \\ 2az^{\mu+1/2} (1-z)^{\nu} F(\alpha, \beta+1; \gamma+1; z) \end{pmatrix},$$

$$z = \left(\frac{e^{\varphi} + i}{e^{\varphi} - i}\right)^2, \quad \mu = \frac{c - ib}{2\omega}, \quad \nu = \frac{i}{\omega}\sqrt{a^2 + b^2},$$

$$\alpha = \frac{1}{2} + \nu + \frac{c}{\omega}, \quad \beta = \nu - \frac{ib}{\omega}, \quad \gamma = \frac{1}{2} + 2\mu.$$

(25) $F_1 = a, \quad F_3 = b \tanh\varphi + c$:

$$u = \begin{pmatrix} az^{\mu} (1-z)^{\nu} F(\alpha+1, \beta; \gamma; z) \\ -(i2\omega\mu + b + c) z^{\mu} (1-z)^{\nu} F(\alpha, \beta+1; \gamma; z) \end{pmatrix},$$

$$z = \frac{1}{2}(1 - \tanh\varphi), \quad \alpha = \mu + \nu + \frac{ib}{\omega}, \quad \nu = \frac{i}{2\omega}\sqrt{a^2 + (b-c)^2},$$

$$\beta = \mu + \nu - \frac{ib}{\omega}, \quad \gamma = 1 + 2\mu, \quad \mu = \frac{i}{2\omega}\sqrt{a^2 + (b+c)^2}.$$

(26) $F_1 = a, \quad F_3 = b \coth\varphi + c$:

$$u = \begin{pmatrix} 2(2b + i\omega) z^{\mu} (1-z)^{\nu} F(\alpha, \beta; \gamma; z) \\ az^{\mu+1} (1-z)^{\nu} F(\alpha+1, \beta+1; \gamma+2; z) \end{pmatrix},$$

$$z = 1 - e^{-2\varphi}, \mu = -\frac{ib}{\omega}, \quad \nu = \frac{i}{2\omega}\sqrt{a^2 + (b+c)^2}, \quad \alpha = \nu - \frac{ib}{\omega} + \lambda,$$

$$\beta = \nu - \frac{ib}{\omega} - \lambda, \quad \gamma = -\frac{2ib}{\omega}, \quad \lambda = \frac{i}{2\omega}\sqrt{a^2 + (b-c)^2}.$$

8.6 Darboux transformation for spin equation

Consider the SE with the external field

$$\mathbf{F}_{\varepsilon} = (F_1, 0, F_3), \quad F_1 = \varepsilon = \text{const}, \quad F_3 = F_3(t). \tag{8.89}$$

We consider here the *Darboux transformation* for the SE with the potentials (8.89). Such a transformation allows one to generate new exact solution from the known ones [115].

The SE with the external field (8.89) can be written as the eigenvalue problem for the spinor V_{ε},

$$\hat{h}V_{\varepsilon} = \varepsilon V_{\varepsilon}, \quad \hat{h} = i\sigma_1 \frac{d}{dt} + \Lambda, \quad \Lambda = i\sigma_2 F_3. \tag{8.90}$$

The idea of the Darboux transformation in this case can be formulated as follows: Suppose that the spinor V_{ε} is known for a given function F_3 with any complex ε. If an operator $\hat{L}$ (which is called the intertwining operator) that obeys the equation

$$\hat{L}\hat{h}' = \hat{h}'\hat{L}, \hat{h}' = i\sigma_1 \frac{d}{dt} + \Lambda', \quad \Lambda' = i\sigma_2 F_3' \tag{8.91}$$

with a given function $F_3'(t)$ is known, then the equation $\hat{h}'V_\varepsilon' = \varepsilon V_\varepsilon'$ has the solution $V_\varepsilon' = \hat{L}V_\varepsilon$. If the intertwining operator $\hat{L}$ is chosen as

$$\hat{L} = \frac{d}{dt} + A\,, \tag{8.92}$$

where $A(t)$ is a time-dependent $n \times n$ matrix, then the transformation from V_ε to V_ε' is called the Darboux transformation [245]. There exists a general method of constructing the intertwining operators $\hat{L}$ (see, for example, [251] and references therein) for the given eigenvalue problem (8.90). However, for our purposes the direct application of the general method is not useful. The point is that by applying this method one can violate the specific structure of the initial matrix Λ, so that the final matrix Λ' will not have the specific structure (8.90) with a real function F_3, whereas we wish to maintain the structure (8.91) of the matrix Λ', i.e. the structure (8.89) of the external field. Thus, the peculiarity of our problem is that the matrices Λ and Λ' must have the same block structure and the Darboux transformation must respect these restrictions. The existence of such transformations is a nontrivial fact, which is demonstrated below.

The intertwining relation (8.91) with the operator $\hat{L}$ in the form (8.92) and with the matrix Λ' in the form (8.91) leads to the following relations:

$$\begin{aligned} &\sigma_1 A - A\sigma_1 + \sigma_2\left(F_3' - F_3\right) = 0\,,\\ &\sigma_1 \dot{A} + \sigma_2 A F_3' - \sigma_2 \dot{F}_3 - A\sigma_2 F_3 = 0\,. \end{aligned} \tag{8.93}$$

Let us choose

$$A = \alpha + i\left(F_3 - \beta\right)\sigma_3\,, \tag{8.94}$$

where $\alpha(t)$ and $\beta(t)$ are certain functions. Then for the function F_3' we obtain

$$F_3' = 2\beta - F_3 \tag{8.95}$$

and for the functions α and β the equations

$$\dot{\alpha} - 2\beta\left(F_3 - \beta\right) = 0, \quad \dot{\beta} + 2\alpha\left(F_3 - \beta\right) = 0\,. \tag{8.96}$$

It is easy to see that there exists a first integral of equation (8.96):

$$\alpha^2 + \beta^2 = R^2, \quad R = \text{const}\,, \tag{8.97}$$

where R is a complex constant in the general case. Note that (8.97) is satisfied if we choose

$$\alpha = R\cos\mu, \quad \beta = R\sin\mu\,, \tag{8.98}$$

with $\mu(t)$ being a real function. Substituting (8.98) into (8.96), we obtain a nonlinear differential equation:

$$\dot{\mu} = 2\left(R\sin\mu - F_3\right) \tag{8.99}$$

for the function μ. The time derivative in (8.92) can be taken from equation (8.90). Then we obtain, with allowance for (8.94) and (8.95),

$$V'_\varepsilon = \left[\alpha - i\left(\varepsilon\sigma_1 + \beta\sigma_3\right)\right] V_\varepsilon \,. \tag{8.100}$$

Thus, we can see that for the SE with the external field (8.89) there exists a Darboux transformation that does not violate the structure of the external field. To complete the construction, one has to represent solutions of equations (8.96) or (8.99) in terms of the initial solutions.

Let us fix $\varepsilon = \varepsilon_0$ and construct the vector $\mathbf{L} = (\bar{V}_{\varepsilon_0}, \boldsymbol{\sigma} V_{\varepsilon_0})$, see equations (A.18) and (A.19). According to equation (8.17), this vector satisfies the equation

$$\dot{\mathbf{L}} = 2\left[\mathbf{F}_{\varepsilon_0} \times \mathbf{L}\right], \quad \mathbf{F}_{\varepsilon_0} = \left(\varepsilon_0, 0, F_3\right) \,. \tag{8.101}$$

In addition, equations (A.20), (A.21), (A.24), (A.25) and (A.14) imply that $\mathbf{L}^2 = 0$. We construct solutions α and β of the set (8.96), via the Cartesian components L_i, $i = 1, 2, 3$ of $\mathbf{L}$ as follows:

$$\alpha = -\varepsilon_0 \frac{L_2}{L_3}, \quad \beta = -\varepsilon_0 \frac{L_1}{L_3} \,. \tag{8.102}$$

Equation $\mathbf{L}^2 = 0$ implies $\alpha^2 + \beta^2 = -\varepsilon_0^2$. Thus, we have expressed solutions of the set (8.96) via solutions of the initial equations (8.90) at $\varepsilon_0 = iR$. Substituting (8.102) into (8.100), one can find the final form of the Darboux transformation:

$$V'_\varepsilon = N\sigma_2 \left[(\boldsymbol{\sigma}\mathbf{L})\, L_3^{-1} + \frac{\varepsilon}{\varepsilon_0}\sigma_3\right] V_\varepsilon \,, \tag{8.103}$$

where N is an arbitrary complex number. Using equations (A.36) and (A.34), we transform (8.103) to a different form:

$$V'_\varepsilon = N\left[2\left(\bar{V}_\varepsilon, V_\varepsilon\right) L_3^{-1}\sigma_2 V_{\varepsilon_0} + i\frac{\varepsilon}{\varepsilon_0}\sigma_1 V_\varepsilon\right] \,. \tag{8.104}$$

For the constructed Darboux transformation, one can check the following properties:

If F_3 is real, then the function F'_3 is also real if one chooses the functions α and β to be real. This choice is always possible, according to equations (8.96).

If F_3 is imaginary, then the function F'_3 is also imaginary in case one chooses α to be real and β to be imaginary. This choice is always possible, according to equations (8.96).

Consider, finally, a simple example of Darboux transformation. Let $F_3 = f$ be a constant. Then the general solution of the SE can be obtained from (8.85), (8.86) and (8.87). It has the form

$$V_\varepsilon(t) = \begin{pmatrix} i\left(f - \omega\right) p \exp\left(i\omega t\right) - \varepsilon q \exp\left(-i\omega t\right) \\ i\varepsilon p \exp\left(i\omega t\right) + \left(f - \omega\right) q \exp\left(-i\omega t\right) \end{pmatrix}, \tag{8.105}$$

where $\omega^2 = f^2 + \varepsilon^2$ and p and q are arbitrary complex constants. The functions α and β can be easily found:

$$\alpha = -\frac{\dot{Q}}{2(Q-f)}, \quad \beta = f + \frac{f^2 - R^2}{Q-f},$$
$$Q = R\cosh\varphi, \quad \varphi = 2(\omega_0 t + \varphi_0), \quad \omega_0^2 = R^2 - f^2, \tag{8.106}$$

where R and φ_0 are arbitrary complex constants. Then the function F_3' can be extracted from (8.95):

$$F_3' = f + 2\frac{f^2 - R^2}{Q-f}. \tag{8.107}$$

If f is real, then F_3' is also real in case we choose a real R and a real φ_0 for $R^2 > f_0^2$. For $R^2 < f^2$, replacing φ_0 by $i\varphi_0$, we also obtain a real F_3', which is determined by equation (8.107) with $Q = R\cos\varphi$ and $\omega_0 = \sqrt{|R^2 - f^2|}$. If f is imaginary, then F_3 is also imaginary in case we choose an imaginary R.

New solutions V_ε' of the SE can be easily constructed according to formula (8.100). We do not present here their explicit form, which is quite cumbersome. Note that for any $f \neq 0$ such solutions do not coincide with the solutions presented in the previous section.

9 One-dimensional Schrödinger equation and its solutions

The equation of the form

$$\check{H}\psi(x) = E\psi(x), \quad \check{H} = -d_x^2 + V(x), \quad x \in (a, b) \ , \tag{9.1}$$

we call the stationary *one-dimensional Schrödinger equation*. Here $\check{H}$ is the so-called *Schrödinger differential operation*, the constant E is the particle energy, $V(x)$ the potential, and $\psi(x)$ the particle wave function; the latter depends on the independent variable $x \in (a, b)$. The domain (a, b) where x lies is specified for each problem separately. The Schrödinger differential operation becomes a self-adjoint operator $\hat{H}$ after an appropriate definition domain in the Hilbert space $L^2(a, b)$ is chosen. In general there exist families of different self-adjoint operators associated with the same Schrödinger differential operation $\check{H}$, see [177].

It is known that an infinite number of potentials $V(x)$ admit exact solutions of equation (9.1) in terms of elementary or special functions. However, in this Chapter we consider a restricted class of exactly solvable cases that are relevant to the aim of the present book. In this respect, we stress that the one-dimensional Schrödinger equation (9.1) arises in this book as a result of separating variables in the course of solving relativistic wave equations in 3+1, or 2+1, dimensions. Thus, usually, it turns out that the potential $V(x)$ resulting from these manipulations can depend on the separation constants (being integrals of motion). This is why we consider only such cases when equation (9.1) can be exactly solved for all admissible values of these parameters. To be more precise, we consider here only such solutions where the potential

$$\widetilde{V}(x) = AV(\varphi), \quad \varphi = cx + x_0 \ , \tag{9.2}$$

with arbitrary real constants A, c and x_0, can admit exact solutions simultaneously with $V(x)$. We are also interested to have neither just one, nor several, solutions of equation (9.1) for a given potential and some values of the energy, but, in a sense, a complete set of solutions[1]. Potentials that admit solving equation (9.1) in this sense will be called *exactly solvable potentials (ESP)*. Two ESP are called different if they are linearly independent and cannot be transformed into each other by real changes of the variable x that do not violate the equation form.

There exist eleven known examples of different (in the above-mentioned sense) ESP. The corresponding exact solutions are considered below.

---

1 We do not describe here self-adjoint Schrödinger Hamiltonians associated with the inital differential operation $\check{H}$. That is why solutions of spectral problems given in each item correspond to a possible self-adjoint Hamiltonian $\hat{H}$. Complete construction of all possible self-adjoint $\hat{H}$ and solutions of the corresponding spectral problems can be found in [177].

The qualitative behavior of ESP is illustrated by a number of figures. We stress that the figures are intended to show only the general shapes of the potentials, but do not display any scales, rates of growth or decrease of the potentials, nor the exact location of the coordinate axes or of the turning points of the potentials.

By the letters $V_1, V_2, \alpha, \beta, \gamma, \mu, \nu, \lambda, c, r_1, r_2$, and n we denote various constants. In particular, the constant c is assumed to be always positive ($c > 0$), n is a non-negative integer ($n \in \mathbb{Z}_+$), and the constants r_1 and r_2 characterize the densities of particle fluxes for $x \to \pm\infty$. When extracting square roots, we choose $\operatorname{Re}\sqrt{a} > 0$; if $\operatorname{Re}\sqrt{a} = 0$, the sign of the root is not important, although fixed.

Here we use designations from Section 1.2 that are adopted for the higher transcendental functions (these notations are in agreement with those used in the reference book [191]).

9.1 ESP $I : V(x) = cx$

Here $x \in \mathbb{R}$. It is sufficient to consider only the case when the constant c is positive. The case with negative constant c can be reduced to the previous one by the transformation $x \to -x$. The Schrödinger equation is solvable for any real E.

Introducing the variable y as

$$y^2 = \frac{4c}{9}\left(x - \frac{E}{c}\right)^3 \tag{9.3}$$

and making the replacement of the function $\psi(x) = y^{1/3} f(y)$ in equation (9.1), we obtain the following equation for $f(y)$:

$$f'' + \frac{1}{y} f' - \left(1 + \frac{1}{9y^2}\right) f = 0\,. \tag{9.4}$$

The general solution of this equation has the form

$$f(y) = c_1 I_{1/3}(y) + c_2 K_{1/3}(y)\,. \tag{9.5}$$

Introducing the variable $z = c^{1/3}(x - E/c)$, we obtain for the function $\varphi(z) = \psi(x)$ the equation

$$\varphi''(z) - z\varphi(z) = 0\,. \tag{9.6}$$

One of the solutions to this equation regular near $z = 0$ is the Airy function $\Phi(z)$. The Airy function is an entire function of z, decreasing as $z \to \infty$. Note that the Airy function and its derivatives can be expressed in terms of the MacDonald functions as follows:

$$\Phi(z) = \sqrt{\frac{z}{3\pi}} K_{1/3}\left(\frac{2}{3} z^{3/2}\right), \quad \Phi'(z) = -\frac{z}{\sqrt{3}\pi} K_{2/3}\left(\frac{2}{3} z^{3/2}\right). \tag{9.7}$$

9.2 ESP II : $V(x) = V_1x^2 + V_2x$

Here $x \in \mathbb{R}$. It is understood that $V_1 \neq 0$ (see Section 9.1 otherwise).

Define the variable z as

$$z = \mu\,(x + V_2/2V_1)\,, \quad \mu^4 = 4V_1\,. \tag{9.8}$$

The general solution of equation (9.1) with the potentials under consideration can be for any (complex) value of E written in terms of the parabolic cylinder functions:

$$\psi(x) = r_1 D_p(z) + r_2 D_p(-z), \quad p = \frac{\mu^2}{4V_1}\left(E + \frac{V_2^2}{4V_1}\right) - \frac{1}{2}\,. \tag{9.9}$$

A discrete spectrum of energy is possible only if $V_1 > 0$. In this case, we have

$$E_n = \sqrt{V_1}(2n+1) - \frac{V_2^2}{4V_1}, \quad n \in \mathbb{Z}_+$$
$$\psi_n(x) = U_n(z) = \left(2^n n!\sqrt{\pi}\right)^{-1/2} e^{-z^2/2} H_n(z)\,, \tag{9.10}$$

where H_n are Hermite polynomials.

9.3 ESP III : $V(x) = -V_1/x + V_2/x^2$

If $V_2 \geq 0$, it is sufficient to consider the half-axis $x > 0$ (or the half-axis $x < 0$). In this case, the shape of the potential is shown in Figure 9.1 for $V_1 > 0$ and in Figure 9.2 for $V_1 < 0$. If $V_2 < 0$, one should consider the entire axis $x \in \mathbb{R}$. The corresponding potential is displayed in Figure 9.3 ($V_1 > 0$) and Figure 9.4 ($V_1 < 0$).

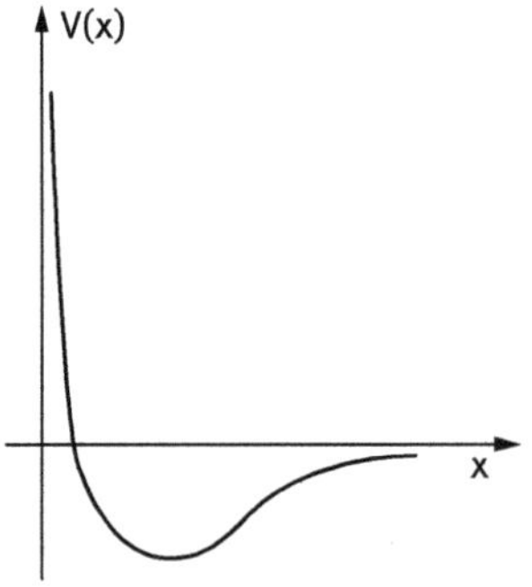

Fig. 9.1. ESP III with $V_1 > 0, V_2 \geq 0$. This plot also describes the shape of the ESP IX with $0 < 2V_1 < -V_2$, and the ESP X with $V_2 < 0, V_1 + V_2 > 0$.

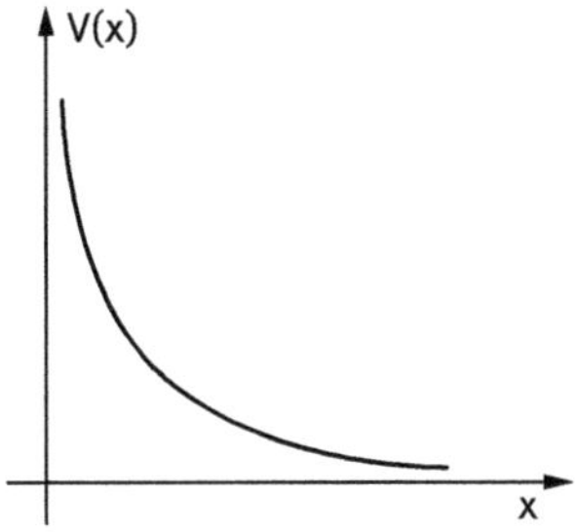

Fig. 9.2. ESP III with $V_1 < 0, V_2 \geq 0$. This plot also describes the ESP IX $V_1 > 0, 2V_1 > -V_2$, and the ESP X with $V_2 > 0, V_1 + V_2 > 0$.

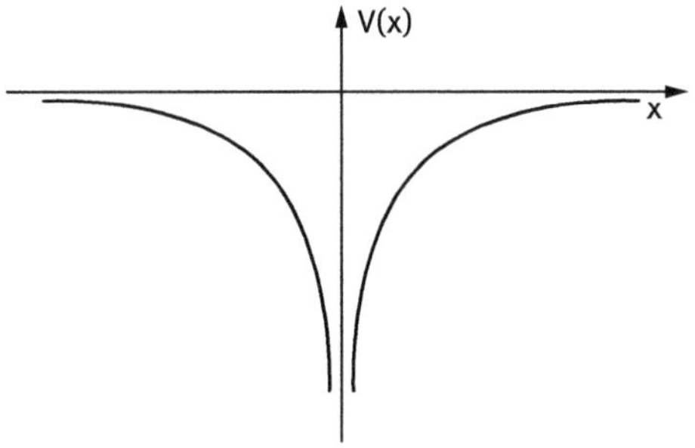

Fig. 9.3. ESP III with $V_1 > 0, V_2 < 0$. This plot also describes the ESP IX for $x > 0$ with $V_1 \leq 0$, $2V_1 \leq -V_2$, and the ESP X with $V_2 < 0, V_1 + V_2 < 0$.

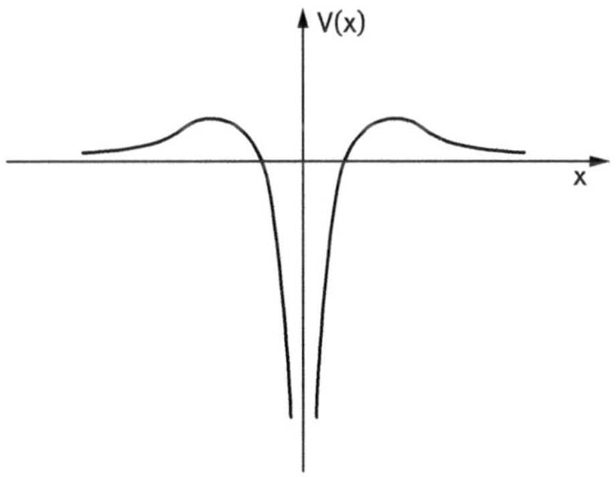

Fig. 9.4. ESP III with $V_1 < 0, V_2 < 0$. This plot also describes the ESP IX for $x > 0$ with $-V_2 < 2V_1 < 0$, and the ESP X with $V_2 > 0, V_1 + V_2 < 0$.

Let us define the constants

$$\nu^2 = -4E, \quad \mu^2 = 1/4 + V_2, \quad \alpha = 1/2 + \mu - V_1/\nu\,, \tag{9.11}$$

and the variable $z = \nu x$. Then the general solution of equation (9.1) with the potentials under consideration has the form

$$\psi(x) = \sqrt{z}e^{-z/2}\left[r_1\varphi(\mu, z) + r_2\varphi(-\mu, z)\right]\,,$$
$$\varphi(\mu, z) = z^{\mu}\Phi\left(\alpha, 1 + 2\mu; z\right)\,. \tag{9.12}$$

A discrete spectrum exists only if $V_2 \geq 0$. In this case, one should consider the half-axis $x > 0$ for $V_1 > 0$ and the half-axis $x < 0$ for $V_1 > 0$. The discrete energy levels are given by the expression

$$E_n = -V_1^2(2n + 1 + 2\mu)^{-2}, \quad n \in \mathbb{Z}_+\,, \tag{9.13}$$

and the corresponding wave functions have the form

$$\psi_n(x) = z^{1/2+\mu}e^{-z/2}L_n^{2\mu}(z)\,. \tag{9.14}$$

For $V_2 \geq 0$, the wave functions that vanish as $x \to 0$ correspond to $r_2 = 0$ and $r_1 \neq 0$ in (9.12).

For $V_2 < 0$, it is more convenient to express solutions as linear combinations of the Whittaker functions

$$\psi(x) = r_1 W_{\lambda,\mu}(z) + r_2 W_{-\lambda,\mu}(-z), \quad \lambda = V_1/\nu\,. \tag{9.15}$$

In particular, when $V_1 = 0$, the solutions can be written in terms of the Bessel functions

$$\psi(x) = \sqrt{x}\left[r_1 J_{\mu}(x\sqrt{E}) + r_2 J_{-\mu}(x\sqrt{E})\right]\,. \tag{9.16}$$

For $V_2 < 0$ and $E < 0$, the particle collapses to the origin $x = 0$, and there always exist solutions decreasing as $|x| \to \infty$, see [232]. These solutions are particular cases of (9.15) and have the form

$$\psi(x) = W_{\epsilon\lambda,\mu}\left(|z|\right), \quad \epsilon = \operatorname{sgn} x\,. \tag{9.17}$$

In the particular case of $V_1 = 0$, we have

$$\psi(x) = \sqrt{x}K_{\mu}\left(|x|\sqrt{|E|}\right)\,. \tag{9.18}$$

9.4 ESP $IV: V(x) = V_1/x^2 + V_2x^2$

Let us define the variable z, and constants μ, ν, α and γ as

$$z = \nu x^2; \quad \mu^2 = \frac{1 + 4V_1}{16}, \quad \nu^2 = V_2; \quad \gamma = 1 + 2\mu, \quad \alpha = \frac{1}{2} + \mu - \frac{E}{4\nu}. \tag{9.19}$$

Then the general solution of equation (9.1) with the potentials under consideration is

$$\psi(x) = z^{1/4} e^{-z/2} \left[r_1 \varphi(\mu, z) + r_2 \varphi(-\mu, z) \right], \quad \varphi(\mu, z) = z^{\mu} \Phi(\alpha, \gamma; z), \tag{9.20}$$

where $\Phi(\alpha, \gamma; z)$ is the confluent hypergeometric function.

For $V_1 > 0$ and $V_2 > 0$, one should consider only the half-axis $x > 0$ (or the half-axis $x < 0$). In this case, we have the well potential drawn in Figure 9.5. For these values of the parameters, there exists a discrete energy spectrum. This spectrum and the corresponding wave function have the form

$$E_n = 2\sqrt{V_2}\left(2n + 2\mu + 1\right), \quad \psi_n(x) = z^{1/4+\mu} e^{-z/2} L_n^{2\mu}(z), \quad n \in \mathbb{Z}_+ . \tag{9.21}$$

For $V_1 > 0$ and $V_2 < 0$, the shape of the potential is shown (for the half-axis $x > 0$) in Figure 9.6. Solutions that vanish at $x = 0$ exist for every energy E and can be obtained from (9.20) for $r_1 \neq 0$, $r_2 = 0$.

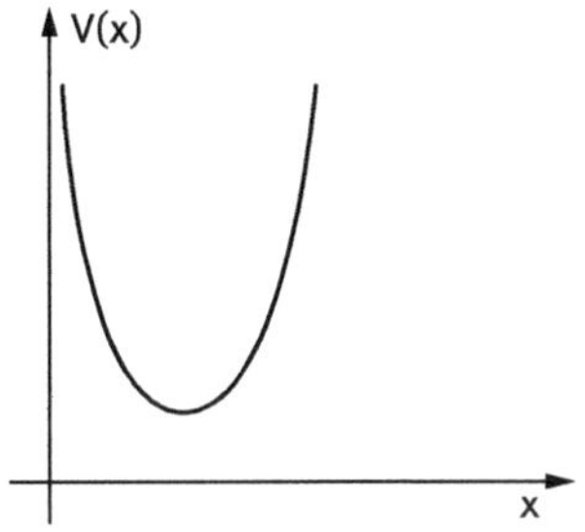

Fig. 9.5. ESP IV with $V_1 > 0, V_2 > 0$.

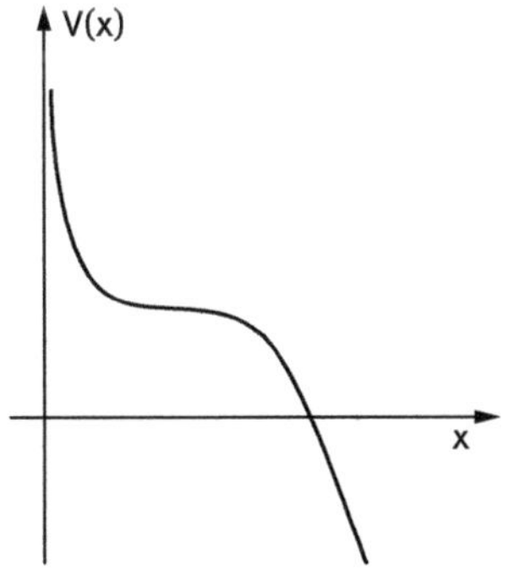

Fig. 9.6. ESP IV with $V_1 > 0, V_2 < 0$.

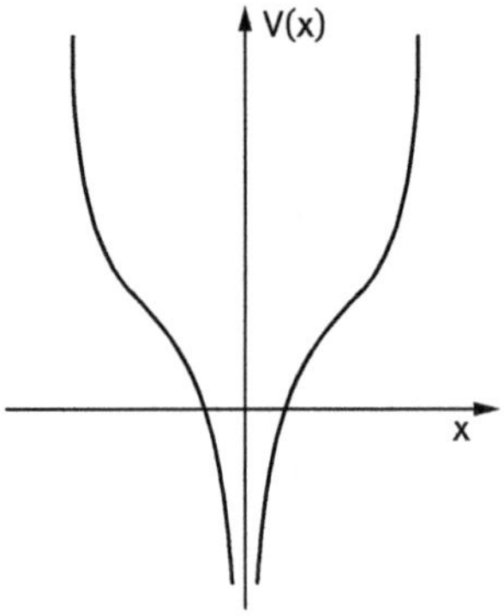

Fig. 9.7. ESP IV with $V_1 < 0, V_2 > 0$.

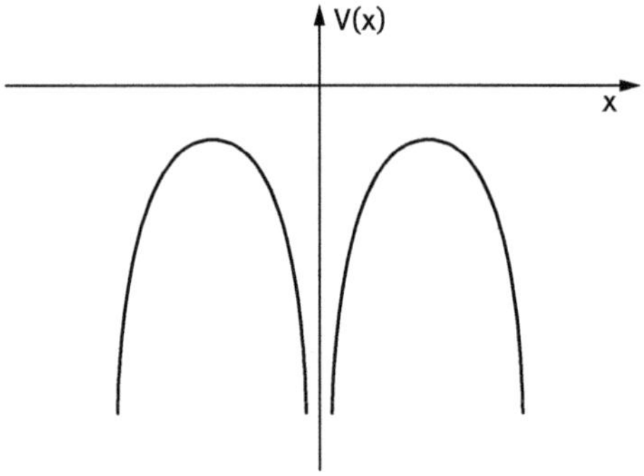

Fig. 9.8. ESP IV 9.4 with $V_1 < 0, V_2 < 0$.

For $V_1 < 0$ and $V_2 > 0$, one should consider the domain $-\infty < x < \infty$. The shape of the potential in this case is shown in Figure 9.7. The wave functions which decrease as $|x| \to \infty$ have the form

$$\psi = z^{-1/4} W_{\lambda,\mu}(z), \quad \lambda = \frac{E}{4\sqrt{V_2}} . \tag{9.22}$$

In this case, the particle collapses to the point $z = 0$.

Finally, at $V_1 < 0$ and $V_2 < 0$, the potential is displayed in Figure 9.8. One should consider the domain $-\infty < x < \infty$, where the solutions are determined by the general expression (9.20).

9.5 ESP V : $V(x) = V_1 e^{-2cx} + V_2 e^{-cx}$

Here $-\infty < x < \infty$. For $V_1 > 0$ and $V_2 \geq 0$, the potential is represented in Figure 9.9, whereas the case $V_1 > 0$ and $V_2 < 0$ corresponds to Figure 9.10, and the case $V_1 < 0$ and $V_2 > 0$ to Figure 9.11. Finally, the case $V_1 < 0$ and $V_2 \leq 0$ is shown in Figure 9.12.

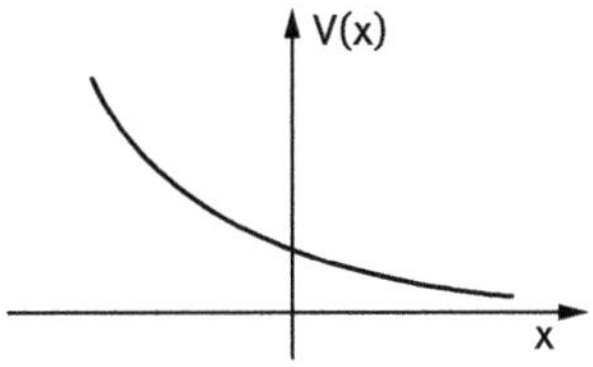

Fig. 9.9. ESP V with $V_1 > 0, V_2 \geq 0$.

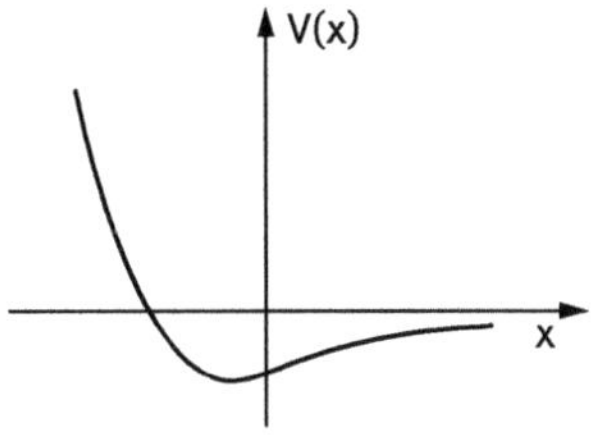

Fig. 9.10. ESP V with $V_1 > 0, V_2 < 0$.

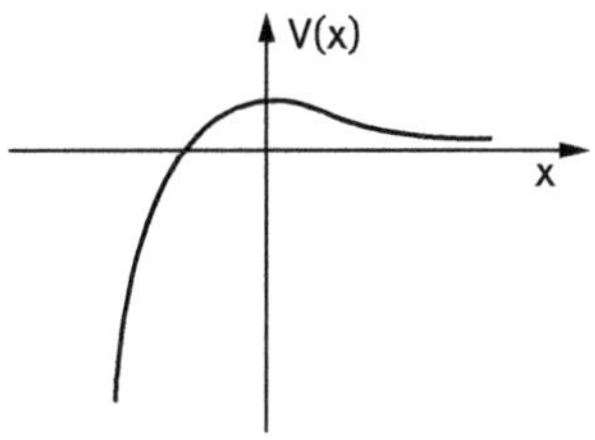

Fig. 9.11. ESP V with $V_1 < 0, V_2 > 0$.

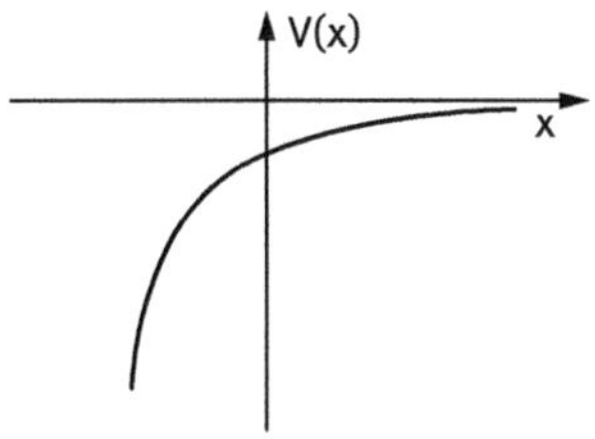

Fig. 9.12. ESP V with $V_1 < 0, V_2 \leq 0$.

Define the variable z and constants μ, α, γ as follows:

$$z = \frac{2\sqrt{V_1}}{c} e^{-cx}; \quad \mu = \frac{\sqrt{-E}}{c}; \quad \alpha = \frac{1}{2} + \mu + \frac{V_2}{2c\sqrt{V_1}}, \quad \gamma = 1 + 2\mu \,. \tag{9.23}$$

The general solution of equation (9.1) with the potentials under consideration has the form

$$\psi(x) = e^{-z/2}\left[r_1\varphi(\mu, z) + r_2\varphi(-\mu, z)\right], \quad \varphi(\mu, z) = e^{-\mu cx}\Phi\left(\alpha, \gamma; z\right) . \tag{9.24}$$

A discrete energy spectrum exists for $V_1 > 0, V_2 < 0$ and $E < 0$. The discrete energy levels and the corresponding wave functions are given by the expressions

$$E_n = -\frac{1}{4V_1}\left[|V_2| - c\sqrt{V_1}(2n+1)\right]^2 ,$$
$$\psi_n(x) = z^{\mu} e^{-z/2} L_n^{2\mu}(z), \quad n \in \mathbb{Z}_+ \,. \tag{9.25}$$

The number of the discrete levels is finite and determined by the condition

$$|V_2| > c\sqrt{V_1}(2n+1) \,. \tag{9.26}$$

In order that at least one level might exist, it is necessary that $|V_2| > c\sqrt{V_1}$.

The solutions which decrease at $x \to -\infty$, for $V_1 > 0$ and $E > 0$, have the form

$$\psi(x) = z^{-1/2} W_{\lambda, i\mu}(z), \quad \lambda = -\frac{V_2}{2c\sqrt{V_1}} \,. \tag{9.27}$$

In particular, for $V_2 = 0$ we obtain

$$\psi(x) = K_{i\mu}\left(z/2\right) , \tag{9.28}$$

where $K_\mu(x)$ is the MacDonald function.

9.6 ESP *VI* : $V(x) = V_1/\sin^2 cx + V_2/\cos^2 cx$

If $V_1 > 0$ and $V_2 > 0$, we shall consider only the domain $0 < 2cx < \pi$ referring to $V(x)$ as equal to ∞ for x outside this domain. The corresponding potential shape is given in Figure 9.13. If $V_1 < 0$ and $V_2 > 0$, we consider the domain $-\pi < 2cx < \pi$, with $V(x) = \infty$ outside. If $V_1 > 0$ and $V_2 < 0$, the potential is drawn in Figure 9.14 (the change of variable $cx \to -cx + \pi/2$ reduces the case of $V_1 > 0$ and $V_2 < 0$ to that of $V_1 < 0$ and $V_2 > 0$). Finally, if $V_1 < 0$ and $V_2 < 0$, we consider the entire axis $-\infty < x < \infty$ with an infinite number of potential hills inside (see Figure 9.15).

We define the constants

$$\mu^2 = \frac{c^2 + 4V_1}{4c^2}, \quad \nu^2 = \frac{c^2 + 4V_2}{4c^2}, \quad \lambda^2 = \frac{E}{c^2};$$
$$2\alpha = 1 + \mu + \nu + \lambda, \quad 2\beta = 1 + \mu + \nu - \lambda, \quad \gamma = 1 + \mu \,. \tag{9.29}$$

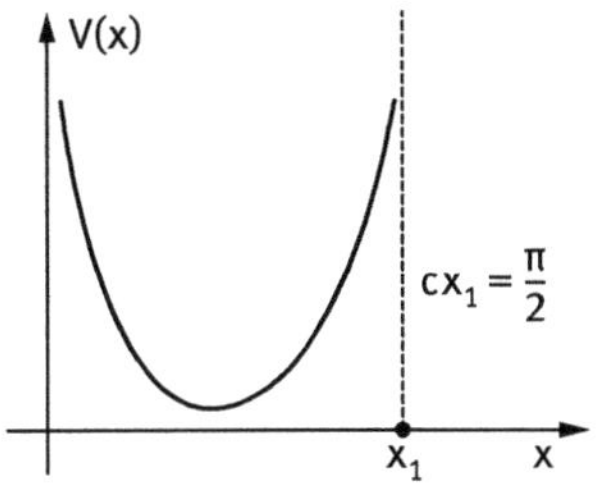

Fig. 9.13. ESP VI with $V_1 > 0, V_2 > 0$. This plot also describes the ESP VII with $V_1 > 0, V_2 > 0$ and the origin shifted by $\pi/2c$ to the right.

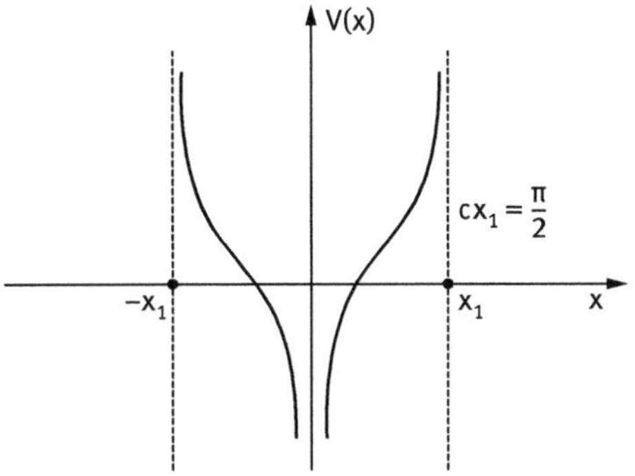

Fig. 9.14. ESP VI with $V_1 < 0, V_2 > 0$.

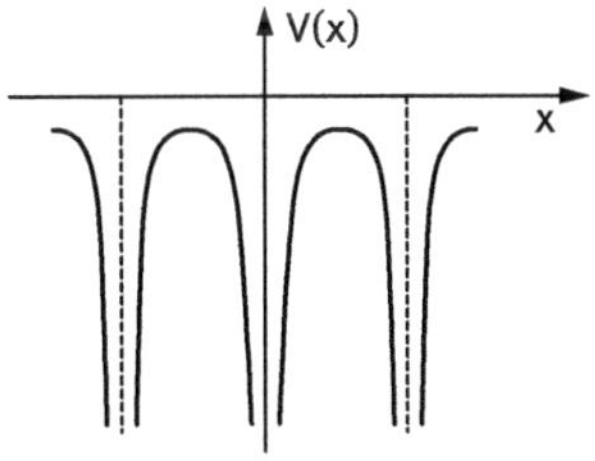

Fig. 9.15. ESP VI with $V_1 < 0, V_2 < 0$. This plot also describes the ESP VII with $V_1 < 0$.

The general solution of equation (9.1) with the potentials under consideration has the form

$$\psi(x) = \cos^{1/2+\nu} cx \left[r_1\varphi(\mu, x) + r_2\varphi(-\mu, x)\right] ,$$
$$\varphi(\mu, x) = \sin^{1/2+\mu} cx F(\alpha, \beta; \gamma; \sin^2 cx) . \tag{9.30}$$

For $V_1 > 0$ and $V_2 > 0$, there exists a discrete spectrum and the corresponding wave functions are bounded,

$$E_n = c^2 (2n + 1 + \mu + \nu)^2 , \quad n \in \mathbb{Z}_+ ,$$
$$\psi_n(x) = P_n^{(\mu,\nu)}(\cos 2cx) \sin^{1/2+\mu} cx \cos^{1/2+\nu} cx, \quad 0 < 2cx < \pi . \tag{9.31}$$

9.7 ESP VII : $V(x) = V_1 \tan^2 cx + V_2 \tan cx$

Here $V_2 \geq 0$. If $V_1 > 0$, one should consider only the domain $-\pi < 2cx < \pi$, referring to $V(x)$ as equal to infinity for $2cx$ outside this domain. The shape of the potential is approximately the same as that drawn in Figure 9.13. For $V_1 = 0$, the shape of the potential is shown in Figure 9.16. For $V_1 < 0$, one has a set of potential barriers ($-\infty < x < \infty$) analogous to those displayed in Figure 9.15.

Define the variable z and the constants $\mu, \nu, \lambda, \alpha, \beta, \gamma$ as

$$z = -e^{-2icx}; \quad \mu^2 = \frac{V_1 + E + iV_2}{4c^2}, \quad \nu^2 = \frac{c^2 + 4V_1}{4c^2}, \quad \lambda^2 = \frac{V_1 + E - iV_2}{4c^2} ;$$
$$\alpha = 1/2 + \mu + \nu + \lambda, \quad \beta = 1/2 + \mu + \nu - \lambda, \quad \gamma = 1 + 2\mu . \tag{9.32}$$

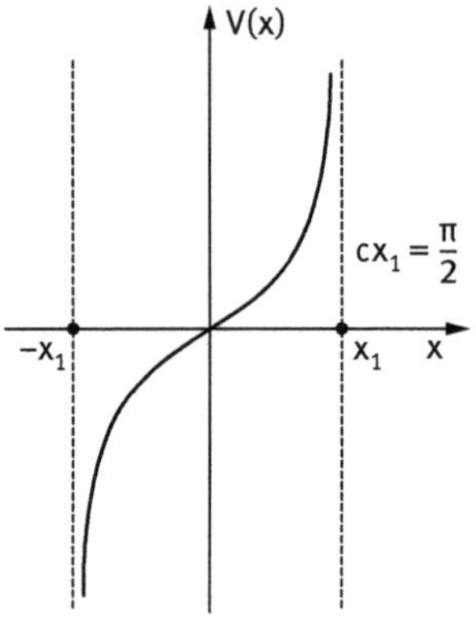

Fig. 9.16. ESP VII with $V_2 = 0$.

The general solution of equation (9.1) with the potentials under consideration has the form

$$\psi(x) = (1-z)^{1/2+\nu}\left[r_1\varphi(\mu,x) + r_2\varphi(-\mu,x)\right],$$
$$\varphi(\mu,x) = z^{\mu}F(\alpha,\beta;\gamma;z)\,. \tag{9.33}$$

If $V_1 > 0$, the energy levels are discrete,

$$E_n = \frac{1}{4}\left[\sqrt{c^2+4V_1} + c(2n+1)\right]^2 - \frac{V_2^2}{\left[\sqrt{c^2+4V_1} + c(2n+1)\right]^2} - V_1,$$
$$\psi_n(x) = (1-i\tan cx)^{\mu}\,(1+i\tan cx)^{\mu^*}\,P_n^{(2\mu,2\mu^*)}(i\tan cx)\,,$$
$$\mu = -\frac{1}{4c}\left[\sqrt{c^2+4V_1} + c(2n+1) + \frac{2iV_2}{\sqrt{c^2+4V_1} + c(2n+1)}\right]. \tag{9.34}$$

Here $*$ stands for complex conjugation.

9.8 ESP VIII : $V(x) = V_1 \tanh^2 cx + V_2 \tanh cx$

Here $V_2 \ge 0$ and $-\infty < x < \infty$. The case $2V_1 \ge V_2$ corresponds to the potential shown in Figure 9.17. If $2|V_1| < V_2$, we have the step potential, as the one drawn in Figure 9.18. For $V_1 < 0$ and $2|V_1| \ge V_2$, the potential is displayed in Figure 9.19.

Define the variable z and constants $\mu, \nu, \lambda, \alpha, \beta, \gamma$ by the relations

$$z = \frac{1}{2}(1-\tanh cx)\,;\quad \mu^2 = \frac{V_1+V_2-E}{4c^2},\quad \nu^2 = \frac{V_1-V_2-E}{4c^2},\quad \lambda^2 = \frac{c^2+4V_1}{4c^2}\,;$$
$$\alpha = 1/2+\mu+\nu+\lambda,\quad \beta = 1/2+\mu+\nu-\lambda,\quad \gamma = 1+2\mu\,. \tag{9.35}$$

Then the general solution of equation (9.1) for the potential under consideration can be written as

$$\psi(x) = (1-z)^{\nu}\left[r_1\varphi(\mu,x) + r_2\varphi(-\mu,x)\right],\quad \varphi(\mu,x) = z^{\mu}F(\alpha,\beta;\gamma;z)\,. \tag{9.36}$$

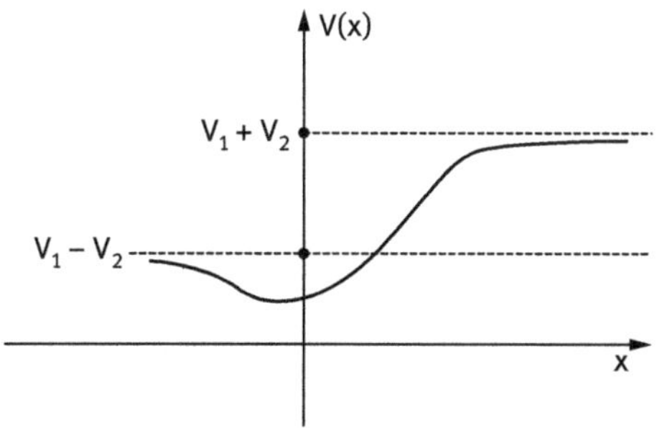

Fig. 9.17. ESP VIII with $2V_1 > V_2 \ge 0$.

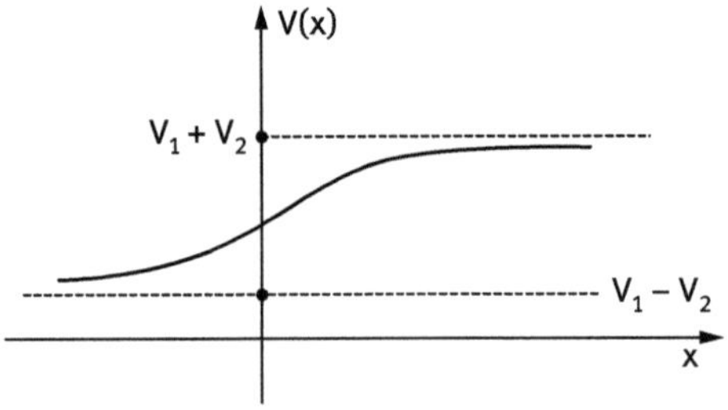

Fig. 9.18. ESP VIII with $V_2 > 2|V_1|$.

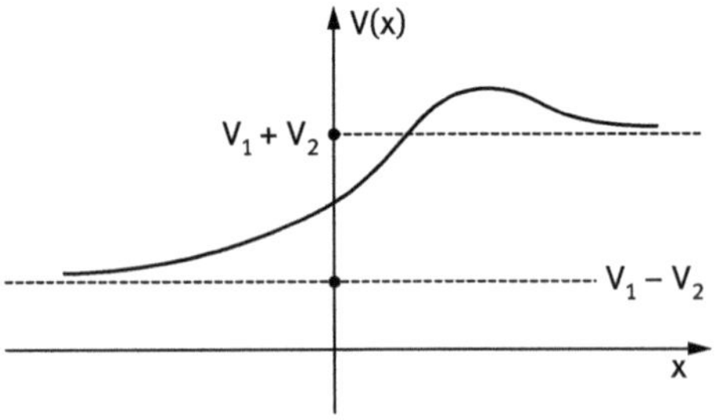

Fig. 9.19. ESP VIII with $V_1 < 0, 0 \leq V_2 < 2|V_1|$.

Bound states and the discrete energy spectrum can exist only provided that the well potential exists, which is the case if $2V_1 \geq V_2$. In such a case the energy levels and the corresponding wave functions have the form

$$\begin{aligned} E_n &= V_1 - V_2 - c^2 \frac{\left[(\lambda - n - 1/2)^2 - V_2/2c^2\right]^2}{(\lambda - n - 1/2)^2} , \\ 2\mu &= \lambda - n - 1/2 + \left(V_2/2c^2\right)(\lambda - n - 1/2)^{-1} , \\ 2\nu &= \lambda - n - 1/2 - \left(V_2/2c^2\right)(\lambda - n - 1/2)^{-1} , \\ \psi_n(x) &= (1 - \tanh cx)^{\mu} (1 + \tanh cx)^{\nu} P_n^{(2\mu,2\nu)}(\tanh cx) . \end{aligned} \tag{9.37}$$

The number of levels is finite. Possible values of n in the expression (9.37) are determined by the inequality

$$\sqrt{c^2 + 4V_1} - \sqrt{2V_2} > c(2n + 1) . \tag{9.38}$$

In order that at least one level might exist, it is necessary that the inequality $2V_1 > V_2 + c\sqrt{2V_2}$ be fulfilled.

For $V_1 - V_2 < E < V_1 + V_2$, the wave function should decrease when $x \to \infty$, which is provided by the condition $r_2 = 0$.

9.9 ESP $IX : V(x) = V_1 \coth^2 cx + V_2 \coth cx$

Here only the half-axis $x > 0$ is considered, and $V(x)$ is set equal to ∞ for $x < 0$. If $0 < 2V_1 < -V_2$, the potential has the form shown in Figure 9.1, whereas for $V_1 > 0$, $2V_1 > -V_2$ this is as shown in Figure 9.2. If $V_1 \leq 0$, the form of the potential can be illustrated by the right hand side either of Figure 9.3, $(2V_1 \leq -V_2)$, or of Figure 9.4, $(-V_2 < 2V_1 < 0)$.

Define the variable z and the constants $\mu, \nu, \lambda, \alpha, \beta, \gamma$ as

$$z = 1 - e^{-2cx}; \quad \mu^2 = \frac{c^2 + 4V_1}{4c^2}, \quad \nu^2 = \frac{V_1 + V_2 - E}{4c^2}, \quad \lambda^2 = \frac{V_1 - V_2 - E}{4c^2};$$
$$\alpha = 1/2 + \mu + \nu + \lambda, \quad \beta = 1/2 + \mu + \nu - \lambda, \quad \gamma = 1 + 2\mu. \tag{9.39}$$

The general solution of equation (9.1) with the potentials under consideration has the form

$$\psi(x) = (1 - z)^{\nu} \left[r_1 \varphi(\mu, x) + r_2 \varphi(-\mu, x) \right],$$
$$\varphi(\mu, x) = z^{1/2+\mu} F(\alpha, \beta; \gamma; z). \tag{9.40}$$

Bound states can exist only if $0 \leq 2V_1 < -V_2$. In such a case

$$E_n = V_1 + V_2 - c^2 \left[\mu + n + 1/2 + V_2 c^{-2} (2\mu + 2n + 1)^{-1} \right]^2,$$
$$\psi_n(x) = z^{1/2+\mu} (1 - z)^{\nu} P_n^{(2\nu, 2\mu)}(2z - 1), \quad n \in \mathbb{Z}_+,$$
$$\nu = -\left[V_2/2c^2 + (\mu + n + 1/2)^2 \right] (2\mu + 2n + 1)^{-1}. \tag{9.41}$$

The number of discrete levels is finite, the quantum number n being restricted by the inequality

$$(2n + 1)c < \sqrt{-2V_2} - \sqrt{c^2 + 4V_1}, \tag{9.42}$$

which implies that the condition

$$-2V_2 > \left(c + \sqrt{c^2 + 4V_1} \right)^2 \tag{9.43}$$

must be satisfied in order that there might exist at least one energy level.

Among the continuum states, the wave functions that vanish at $x = 0$ correspond to the choice $r_2 = -r_1$ in (9.40). For $V_1 < 0$, the states with $E < V_1 + V_2$ also belong to a continuum, with the particle collapsing to the point $x = 0$. The wave functions of such states that decrease as $x \to \infty$ are expressed by (9.40) for $r_2 = 0$.

9.10 ESP $X : V(x) = (V_1 + V_2 \cosh 2x)/(\sinh^2 2x)$

We consider here the case $V_2 \neq 0$ only, since otherwise the potential is reduced to a particular case of the one considered in Section 9.9. Characteristic shapes of the potentials for various domains of the parameters V_1 and V_2 can be seen in Figures 9.1–9.4 following the correspondence rules

(a) $V_1 + V_2 > 0, \quad V_2 < 0; \quad$ (Figure 9.1) $\quad x > 0\,,$
(b) $V_1 + V_2 > 0, \quad V_2 > 0; \quad$ (Figure 9.2) $\quad x > 0\,,$
(c) $V_1 + V_2 < 0, \quad V_2 < 0; \quad$ (Figure 9.3) $\quad -\infty < x < \infty\,,$
(d) $V_1 + V_2 < 0, \quad V_2 > 0; \quad$ (Figure 9.4) $\quad -\infty < x < \infty\,.$

Let us introduce the new variable z and constants $\mu, \nu, \lambda, \alpha, \beta\gamma$ as

$$z = \tanh^2 cx; \quad \mu^2 = \frac{V_1 + V_2 + c^2}{16c^2}, \quad \nu^2 = -\frac{E}{4c^2}, \quad \lambda^2 = \frac{V_1 - V_2 + c^2}{16c^2}\,;$$
$$\alpha = 1/2 + \mu + \nu + \lambda, \quad \beta = 1/2 + \mu + \nu - \lambda, \quad \gamma = 1 + 2\mu\,. \tag{9.44}$$

The general solution of equation (9.1) with the potentials under consideration has the form

$$\psi(x) = z^{1/4}(1-z)^{\nu}\left[r_1\varphi(\mu, x) + r_2\varphi(-\mu, x)\right]\,,$$
$$\varphi(\mu, x) = z^{\mu}F(\alpha, \beta; \gamma; z)\,. \tag{9.45}$$

Bound states and the discrete energy spectrum exist in the case $V_1 + V_2 > 0$, $V_2 < 0$ and $E < 0$,

$$E_n = -4c^2(\lambda - \mu - n - 1/2)^2, \quad \nu = \lambda - \mu - n - 1/2\,,$$
$$\psi_n(x) = z^{1/4+\mu}(1-z)^{\nu}P_n^{(2\mu, 2\nu)}(1-2z)\,. \tag{9.46}$$

The number of levels is finite and determined from the inequality $n \le \lambda - \mu - 1/2$. For at least one level to exist it is necessary that

$$V_1 \ge |V_2| > 2c\sqrt{V_1} > 4c^2\,. \tag{9.47}$$

In the other cases, the spectrum is continuous. In cases (c) and (d), continuum states occur also at $E < 0$. The wave function decreasing as $|x| \to \infty$ corresponds to $r_2 = 0$ in (9.45). In these cases, the particle collapses to the origin $x = 0$.

9.11 ESP XI : $V(x) = (V_1 + V_2 \sinh cx)/(\cosh^2 cx)$

We consider here $V_2 > 0$ and $-\infty < x < \infty$. The case with a negative V_2 can be reduced to the previous one by the transformation $x \to -x$. If $V_2 = 0$, we have a particular case of the potential considered in Section 9.7. The character of the potential is shown by Figure 9.20.

We define the variable z and the constants $\mu, \nu, \lambda, \alpha, \beta, \gamma$ as

$$z = \left(\frac{e^{cx} + i}{e^{cx} - i}\right)^2; \quad \mu^2 = \frac{c^2 - 4V_1 + 4iV_2}{16c^2}, \quad \nu^2 = -\frac{E}{c^2}, \quad \lambda^2 = \frac{c^2 - 4V_1 - 4iV_2}{16c^2}\,;$$
$$\alpha = 1/2 + \mu + \nu + \lambda, \quad \beta = 1/2 + \mu + \nu - \lambda, \quad \gamma = 1 + 2\mu\,. \tag{9.48}$$

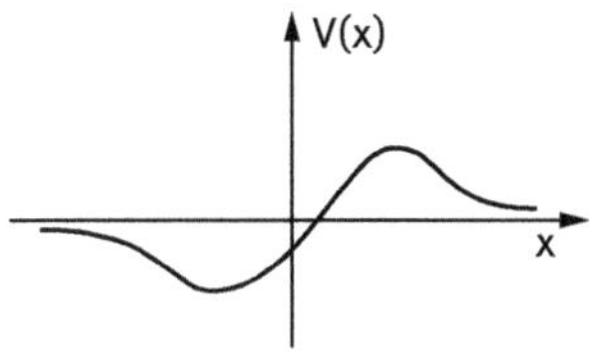

Fig. 9.20. ESP XI.

The general solution of equation (9.1) with the potentials under consideration has the form

$$\begin{aligned} \psi(x) &= z^{1/4}(1-z)^{\nu}\left[r_1\varphi(\mu,x) + r_2\varphi(-\mu,x)\right] , \\ \varphi(\mu,x) &= z^{\mu}F(\alpha,\beta;\gamma;z) . \end{aligned} \tag{9.49}$$

For $E < 0$, a finite number of bound states is possible,

$$\begin{aligned} E_n &= -c^2(a-1/2-n)^2 , \quad n \le a - 1/2 , \\ a &= \frac{1}{\sqrt{8c^2}}\left\{c^2 - 4V_1 + \left[\left(c^2-4V_1\right)^2 + 16V_2^2\right]^{1/2}\right\}^{1/2} , \\ \psi_n(x) &= (1 - i\sinh cx)^{1/4+\mu}(1 + i\sinh cx)^{1/4+\lambda} P_n^{(2\lambda,2\mu)}(i\sinh cx) . \end{aligned} \tag{9.50}$$

The existence of at least one level is possible only if the inequality $V_2^2 > c^2V_1$ is fulfilled.

We do not know other exact solvable potentials $V(x)$ for which solutions of the one-dimensional Schrödinger equation are studied well enough. As an example of solutions that are not studied sufficiently well, we can mention the potentials of the form

$$\begin{aligned} V(x) &= V_1\sin^2 cx + V_2\cos cx + V_3\sin^{-2} cx , \\ V(x) &= V_1\sinh^2 cx + V_2\cosh cx + V_3\sinh^{-2} cx . \end{aligned}$$

For such potentials, one might obtain solutions of equation (9.1) expressed in terms of the Coulomb spheroidal functions [223]. For the particular case $V_2 = 0$, the solutions are expressed in terms of spherical functions or (if also $V_3 = 0$) of the Mathieu functions. These functions, however, do not seem sufficiently well studied. As a consequence, for bound states there exist neither explicit answers for the energy spectrum, integral expressions in terms of elementary functions, nor manifest expansions in terms of other special functions.

One example is known of a potential that, although given implicitly, admits an exact solution. This example is presented, apparently for the first time, in Ref. [228].

10 Coherent states

10.1 Introduction

Coherent states (*CS*) play an important role in modern quantum theory due to their fundamental theoretical importance and wide range of applications, for example in semiclassical descriptions of quantum systems, in quantization theory, in condensed matter physics, in radiation theory, in quantum computations and so on, see, for example Refs. [164, 216, 252]. The first example of CS was given by Schrödinger for the harmonic oscillator [289]. Especially useful CS are in the theory of the electromagnetic field. Such states were introduced and systematically studied in the works by Schwinger [293], Rashevsky [272], and Glauber [184]. Perelomov and Gilmore [169, 263] pointed out the group-theoretical aspect of CS. This enabled them to generalize the construction of CS. Using Perelomov's method, CS for $SU(N)$ and $SU(N, 1)$ groups were constructed in Refs. [178, 179]. Their application to semiclassical descriptions of the quantum rotator was considered in [180]. As they are usually labeled by phase space variables, CS allow one to construct classical symbols of operators, thus they obtain a special status of a quantizer *à la* Berezin–Klauder [87, 89, 90, 164, 217, 218]. Developing the Glauber and Malkin–Man'ko initial approach [110, 239, 240, 242], Dodonov and Man'ko had constructed CS for arbitrary nonrelativistic systems with quadratic Hamiltonians [125]. Some nontrivial generalizations of the Glauber approach are developed by Klauder and Gazeau (see [165]). For constructing CS of relativistic particles described by the K–G or Dirac equation, new methods were required. For specific cases, the problem was solved in Refs. [34, 35, 63, 122, 123], where two methods were proposed. Namely, in Refs. [34, 35, 63] the light-cone formulation was used for this purpose, whereas Ref. [122, 123] dealt with the proper-time method.

It should be noted that a universally accepted definition of CS for an arbitrary physical system is still lacking. Nevertheless, constructing CS one always tries to maintain basic properties of already known CS for simple system. In particular, CS have to form a complete system, they have to minimize uncertainty relations for some physical quantities (e.g. coordinates and momenta) at a fixed time instant and mean values of some physical quantities, calculated with respect to time-dependent CS and have to move along the corresponding classical trajectories. It is also desirable for time-dependent CS to maintain their form under the time evolution, such that this evolution affects only their parameters. Besides, CS have to be labeled by parameters that have a direct classical analog, say, by points in a phase space.

10.2 Coherent states of the Heisenberg–Weyl group

10.2.1 HW algebra and HW group

Here we consider Perelomov CS of *the Heisenberg–Weyl (HW) group* and their relation to Glauber CS.

Remember that the HW algebra w_1 is a real three-dimensional Lie algebra with generators $\mathbf{e}_i$, $i = 1, 2, 3$ that satisfy the following commutation relations:

$$[\mathbf{e}_1, \mathbf{e}_2] = \mathbf{e}_3, \quad [\mathbf{e}_3, \mathbf{e}_\lambda] = 0, \quad \lambda = 1, 2 , \tag{10.1}$$

where $[\mathbf{a}, \mathbf{b}]$ is a Lie product, or simple commutator, of two elements $\mathbf{a}$ and $\mathbf{b}$ of the algebra. A general element $\mathbf{a}$ of a Lie algebra is a linear combination of the generators, $\mathbf{a} = \sum_i a^i \mathbf{e}_i \in w_1$. The commutator of two elements $\mathbf{a}$ and $\mathbf{b}$ has the following form:

$$[\mathbf{a}, \mathbf{b}] = (a^1 b^2 - a^2 b^1)\mathbf{e}_3 . \tag{10.2}$$

One can construct a simple realization of the HW algebra in the three-dimensional vector space. As generators we take three orthogonal unit vectors $\mathbf{i}, \mathbf{j}, \mathbf{k}$ and define the commutator as $[\mathbf{a}, \mathbf{b}] = ([\mathbf{a} \times \mathbf{b}]\mathbf{k})\mathbf{k}$. Then $[\mathbf{i}, \mathbf{j}] = \mathbf{k}$, $[\mathbf{k}, \mathbf{i}] = [\mathbf{k}, \mathbf{j}] = 0$.

Let us consider commutation relations for the basic quantum mechanics operators,

$$[\hat{x}, \hat{p}] = i\hbar I , \tag{10.3}$$

where I is an identical operator. We can interpret them as a realization of the HW algebra in the Hilbert space. Indeed, we can take for generators of the HW algebra the following three operators:

$$\mathbf{e}_1 = \frac{i}{\sqrt{\hbar}}\hat{p}, \quad \mathbf{e}_2 = \frac{i}{\sqrt{\hbar}}\hat{x}, \quad \mathbf{e}_3 = iI . \tag{10.4}$$

If we select the ordinary commutator as the Lie product, then one can see that generators (10.4) obey just commutation relations (10.1). A generalization to the n-dimensional case is obvious and leads to a real $(2n + 1)$-dimensional HW algebra with generators

$$\mathbf{e}_1^a = \frac{i}{\sqrt{\hbar}}\hat{p}_a, \quad \mathbf{e}_2^a = \frac{i}{\sqrt{\hbar}}\hat{x}^a, \quad \mathbf{e}_3 = iI , \tag{10.5}$$

obeying the commutation relation

$$[\mathbf{e}_1^a, \mathbf{e}_2^b] = \delta_{ab}\mathbf{e}_3, \quad [\mathbf{e}_3, \mathbf{e}_\lambda^a] = 0, \quad \lambda = 1, 2, \quad a = 1, \dots, n . \tag{10.6}$$

Sometimes a result of a quantization of classical mechanics with the flat phase space is formulated in terms of commutation relations for mutually conjugated creation and annihilation operators $\hat{c}^+$ and $\hat{c}$,

$$[\hat{c}, \hat{c}^+] = I , \tag{10.7}$$

which can be realized in a Fock space [292]. If we chose

$$\mathbf{e}_1' = \hat{c}, \quad \mathbf{e}_2' = \hat{c}^+, \quad \mathbf{e}_3' = I\,, \tag{10.8}$$

then equations (10.7) appear to be commutation relations for the HW algebra with the generators $\mathbf{e}'$.

Thus, the commutation relations of the HW algebra arise as a result of quantization of classical mechanics with the flat phase space. Generators of the HW algebra can be connected both with operators of phase variables or with operators of creation and annihilation. That is not by chance, because each set of the former operators creates a set of the latter operators and vice versa,

$$\hat{c} = \frac{\hat{x} + i\hat{p}}{\sqrt{2\hbar}}, \quad \hat{c}^+ = \frac{\hat{x} - i\hat{p}}{\sqrt{2\hbar}},$$
$$\hat{x} = \sqrt{\frac{\hbar}{2}}(\hat{c} + \hat{c}^+), \quad i\hat{p} = \sqrt{\frac{\hbar}{2}}(\hat{c} - \hat{c}^+)\,. \tag{10.9}$$

From the point of view of the HW algebra a transition from the canonical operators to the operators of creation and annihilation means a transition from one set of generators to another one by a linear transformation. For example, the generators (10.4) are connected with ones (10.8) in the following way:

$$\mathbf{e}_1 = \frac{1}{\sqrt{2}}(\mathbf{e}_1' - \mathbf{e}_2'), \quad \mathbf{e}_2 = \frac{i}{\sqrt{2}}(\mathbf{e}_1' + \mathbf{e}_2'), \quad \mathbf{e}_3 = i\mathbf{e}_3'\,.$$

Consider now the HW group W_1, which, for example, can be constructed on the base of the HW algebra realized in a Hilbert space. We construct the HW group, considering operators of the form

$$D(a) = \exp\{\mathbf{a}\} \in W_1, \quad \mathbf{a} = \sum_{i=1}^{3} a^i \mathbf{e}_i \in w_1, \quad a = (a^i)\,. \tag{10.10}$$

The unit and the inverse elements are: $D(0) = I$, $D^{-1}(a) = D(-a)$. The multiplication low of the group can be calculated using the Baker–Hausdorff formula

$$\exp(\hat{b} + \hat{a}) = \exp(\hat{b})\exp(\hat{a})\exp\left\{-\frac{1}{2}[\hat{b}, \hat{a}]\right\} \tag{10.11}$$

and the Lie commutator (10.2),

$$\begin{aligned} D(a)D(b) &= D(c) = \exp\{\mathbf{c}\} = \exp\{\mathbf{a}\}\exp\{\mathbf{b}\} = \exp\{\mathbf{a} + \mathbf{b}\}\exp\left\{\frac{1}{2}[\mathbf{a}, \mathbf{b}]\right\} \\ &= \exp\left\{(a^1 + b^1)\mathbf{e}_1 + (a^2 + b^2)\mathbf{e}_2 + [a^3 + b^3 + \frac{1}{2}(a^1 b^2 - a^2 b^1)]\mathbf{e}_3\right\}, \end{aligned}$$

so that

$$c^1 = a^1 + b^1, \quad c^2 = a^2 + b^2, \quad c^3 = a^3 + b^3 + \frac{1}{2}(a^1 b^2 - a^2 b^1)\,.$$

10.2.2 CS of the HW group and Glauber CS

Let us consider the HW algebra w_1 with the generators given by (10.8). The corresponding W_1 group consists of the operators (10.10) acting in a Hilbert space $\mathcal{R}$, which is, in fact, the Fock space. However, we restrict ourselves only by unitary operators of the form (10.10), which means that we deal with a subgroup and respectively with a subalgebra with anti-Hermitian elements w,

$$w = irI + z\hat{c}^+ - \bar{z}\hat{c} \in w_1, \quad w^+ = -w\,, \tag{10.12}$$

where r is a real and z is a complex number. Then

$$D(w) = D(r, z) = \exp\{irI\}\mathcal{D}(z), \quad \mathcal{D}(z) = \exp\{z\hat{c}^+ - \bar{z}\hat{c}\}\,. \tag{10.13}$$

In fact we deal with a unitary representation of the group.

As before, the multiplication low of the subgroup can be calculated using the Baker–Hausdorff formula (10.11),

$$D(r_1, z_1)D(r_2, z_2) = D(r_1 + r_2 + \operatorname{Im}(z_1\bar{z}_2), z_1 + z_2)\,. \tag{10.14}$$

Operators $\mathcal{D}(z)$, which are called displacement operators, satisfy the following properties:

$$\begin{aligned}
&\mathcal{D}^+(z) = \mathcal{D}^{-1}(z) = \mathcal{D}(-z)\,,\\
&\mathcal{D}(z) = \exp\left\{-\frac{|z|^2}{2}\right\}\exp\{z\hat{c}^+\}\exp\{-z\hat{c}\},\\
&\mathcal{D}^{-1}(z)\hat{c}\mathcal{D}(z) = \hat{c} + z, \quad \mathcal{D}^{-1}(z)\hat{c}^+\mathcal{D}(z) = \hat{c}^+ + \bar{z},\\
&\mathcal{D}(z_1)\mathcal{D}(z_2) = \exp\{2i \operatorname{Im} z_1\bar{z}_2\}\mathcal{D}(z_2)\mathcal{D}(z_1) = \exp\{i \operatorname{Im} z_1\bar{z}_2\}\mathcal{D}(z_1 + z_2)\,.
\end{aligned} \tag{10.15}$$

Following Perelomov [263], we construct the orbit of an element $|\psi_0\rangle \in \mathcal{R}$. To this end, we act by operators (10.13) on this element,

$$D(r, z)|\psi_0\rangle = |r, z\rangle = \exp\{ir\}\exp\{z\hat{c}^+ - \bar{z}\hat{c}\}|\psi_0\rangle\,. \tag{10.16}$$

It is clear that all the vectors of the orbit (10.16) with different r represent the same physical state, such that we can consider r as a gauge parameter. Fixing the gauge, let's say by setting $r = 0$, we obtain a set of vectors

$$|z\rangle = D(0, z)|\psi_0\rangle = \mathcal{D}(z)|\psi_0\rangle\,, \tag{10.17}$$

which are Perelomov CS of the group W_1. Properties of these CS depend on the choice of the vector $|\psi_0\rangle$. There exists the physically most interesting choice $|\psi_0\rangle = |0\rangle$, where $|0\rangle$ is the vacuum vector, $\hat{c}|0\rangle = 0$, $\langle 0|0\rangle = 1$. In this case, we obtain

$$|z\rangle = \mathcal{D}(z)|0\rangle = \exp\{z\hat{c}^+ - \bar{z}\hat{c}\}|0\rangle\,. \tag{10.18}$$

The CS of the form (10.18) are, in fact Glauber CS, see [184, 216].

Using properties of the displacement operator, we can represent Glauber CS in the following form:

$$|z\rangle = \exp\left\{-\frac{|z|^2}{2}\right\} \exp\{z\hat{c}^+\}|0\rangle = \exp\left\{-\frac{|z|^2}{2}\right\} \sum_{n=0}^{\infty} \frac{z^n}{\sqrt{n!}}|n\rangle\,, \tag{10.19}$$

where $|n\rangle$ are n-particle states,

$$|n\rangle = \frac{(\hat{c}^+)^n}{\sqrt{n!}}|0\rangle, \quad \langle n|m\rangle = \delta_{nm}\,. \tag{10.20}$$

Using equations (10.15) and (10.18), we obtain

$$\hat{c}|z\rangle = \hat{c}\mathcal{D}(z)|0\rangle = \mathcal{D}(z)\mathcal{D}^{-1}(z)\hat{c}\mathcal{D}(z)|0\rangle = \mathcal{D}(z)\,(\hat{c} + z)\,|0\rangle = z|z\rangle\,,$$

which means that Glauber CS are eigenvectors of the annihilation operator. In such a way Glauber defined his CS.

In the general case of d-dimensional system we have

$$|z|^2 \to \sum_{k=1}^{d} |z_k|^2, \quad n \to \{n_1\,, \ldots n_d\}\,, \quad |n\rangle = \prod_{k=1}^{d} \frac{(\hat{c}_k^+)^{n_k}}{\sqrt{n_k!}}|0\rangle,$$
$$\sum_{n=0}^{\infty} \frac{z^n}{\sqrt{n!}}|n\rangle \to \sum_{\{n\}} \left(\prod_{k=1}^{d} \frac{z_k^{n_k}}{\sqrt{n_k!}}\right)|n\rangle\,. \tag{10.21}$$

By the help of equation (10.19) we obtain Glauber CS in the occupation number representation:

$$\langle n|z\rangle = \exp\left\{-\frac{|z|^2}{2}\right\} \frac{z^n}{\sqrt{n!}}\,. \tag{10.22}$$

Calculating derivatives in z and $\bar{z}$ we obtain the action of the creation operator on the Glauber CS,

$$\frac{\partial}{\partial z}|z\rangle = \left(\hat{c}^+ - \frac{\bar{z}}{2}\right)|z\rangle \Rightarrow \hat{c}^+|z\rangle = \left(\frac{\bar{z}}{2} + \frac{\partial}{\partial z}\right)|z\rangle,$$
$$\frac{\partial}{\partial \bar{z}}\langle z| = \langle z|\left(\hat{c} - \frac{z}{2}\right) \Rightarrow \langle z|\hat{c} = \left(\frac{z}{2} + \frac{\partial}{\partial \bar{z}}\right)\langle z|\,. \tag{10.23}$$

One can demonstrate that the creation operators have no eigenvectors in the Fock space.

The overlapping of the CS can be easily calculated using the properties (10.15),

$$\langle z_1|z_2\rangle = \exp\left\{-\operatorname{Im}(z_1\bar{z}_2) - \frac{1}{2}|z_2 - z_1|^2\right\} = \exp\left\{\bar{z}_1 z_2 - \frac{1}{2}\left(|z_1|^2 + |z_2|^2\right)\right\}\,.$$

Its modulus square reads

$$|\langle z_1|z_2\rangle|^2 = \exp\left\{-|z_2 - z_1|^2\right\}\,. \tag{10.24}$$

The CS with different z are not orthogonal. But the set of CS is complete

$$\int |z\rangle\langle z|d\mu(z) = 1, \quad d\mu(z) = \frac{1}{\pi}d^2z = \frac{1}{\pi}d\,(\mathrm{Re}\,z)\,d\,(\mathrm{Im}\,z)\ . \tag{10.25}$$

The proof of this relation is reduced to a calculation of the integral

$$I = \int (\bar{z})^n z^m \exp\left\{-|z|^2\right\} d\mu(z) = \delta_{nm} n!\ ,$$

which can be done in the variables ρ and ϕ where $z = \rho\exp(i\phi)$. Relation (10.25) can be represented as

$$\begin{aligned}\int |z\rangle\langle z|d\mu(z) &= \int |z\rangle\langle z|z'\rangle\langle z'|d\mu(z)d\mu(z') \\ &= \int \exp\left\{\bar{z}z' - \frac{1}{2}\left(|z|^2 + |z'^2|\right)\right\} |z\rangle\langle z'|d\mu(z)d\mu(z') = 1\ .\end{aligned}$$

The physical sense of the parameter z is revealed if we calculate the mean value $\bar{n}$ of the particle number operator $\hat{n} = \hat{c}^+\hat{c}$,

$$\bar{n} = \langle z|\hat{n}|z\rangle = |z|^2 \Longrightarrow z = \sqrt{\bar{n}}\exp\{i\phi\}\ . \tag{10.26}$$

The probability to find n particles in the CS is $P_n = |\langle n|z\rangle|^2$. Using equations (10.22) and (10.26), we obtain the Poisson distribution

$$P_n = \frac{\bar{n}^n}{n!}\exp(-\bar{n})\ . \tag{10.27}$$

The standard deviation of the particle number reads $\Delta n = \sqrt{\bar{n}}$. Indeed,

$$\Delta n = \sqrt{\langle z|\hat{n}^2|z\rangle - \langle z|\hat{n}|z\rangle^2} = \sqrt{\overline{n^2} - \bar{n}^2} = \sqrt{|z|^2\left(|z|^2+1\right) - |z|^4} = \sqrt{\bar{n}}\ .$$

10.2.3 Heisenberg uncertainty relation and CS

Considering minimization of the *Heisenberg uncertainty relation* one can arrive at the Glauber definition of the CS described in the previous Section. Indeed, let us consider a normalized state $|\psi\rangle$. Mean values of any operator in this state will be denoted as $\langle\hat{A}\rangle = \langle\psi|\hat{A}|\psi\rangle$. Taking into account the commutation relation for the coordinate and momentum operators, $[\hat{x},\hat{p}] = i\hbar$, we can write

$$|\langle[\hat{x},\hat{p}]\rangle| = |\langle[\Delta\hat{x},\Delta\hat{p}]\rangle| = \left|\langle\Delta\hat{x}\Delta\hat{p}\rangle - \overline{\langle\Delta\hat{x}\Delta\hat{p}\rangle}\right| = 2\left|\mathrm{Im}\langle\Delta\hat{x}\Delta\hat{p}\rangle\right| = \hbar\ ,$$

where $\Delta\hat{x} = \hat{x} - \langle\hat{x}\rangle$ and $\Delta\hat{p} = \hat{p} - \langle\hat{p}\rangle$. Then using the obvious inequality $|\,\mathrm{Im}\langle\Delta\hat{x}\Delta\hat{p}\rangle| \le |\langle\Delta\hat{x}\Delta\hat{p}\rangle|$ and the inequality $|\,\mathrm{Im}\langle\Delta\hat{x}\Delta\hat{p}\rangle| \le \sqrt{\langle\Delta\hat{x}^2\rangle\langle\Delta\hat{p}^2\rangle}$, which is a consequence of the the Schwartz inequality $|\langle\psi|\phi\rangle| \leq \sqrt{\langle\psi|\psi\rangle\langle\phi|\phi\rangle}$, we obtain the Heisenberg inequality

$$\Delta x \Delta p \geq \frac{\hbar}{2}\,, \tag{10.28}$$

for the *standard deviations* Δx and Δp,

$$\Delta x = \sqrt{\langle \Delta \hat{x}^2 \rangle} = \sqrt{\langle \hat{x}^2 \rangle - \langle \hat{x} \rangle^2}, \Delta p = \sqrt{\langle \Delta \hat{p}^2 \rangle} = \sqrt{\langle \hat{p}^2 \rangle - \langle \hat{p} \rangle^2}.$$

The Heisenberg inequality turns out to equality when both inequalities that were used in deriving (10.28) are equalities. The Schwartz inequality turns out to equality when

$$\Delta \hat{x}|\psi\rangle = \kappa \Delta \hat{p}|\psi\rangle\,, \tag{10.29}$$

which means that vectors $\Delta \hat{x}|\psi\rangle$ and $\Delta \hat{p}|\psi\rangle$ are parallel. The inequality $|\operatorname{Im}\langle\Delta\hat{x}\Delta\hat{p}\rangle| \leq |\langle\Delta\hat{x}\Delta\hat{p}\rangle|$ turns out to equality when $\operatorname{Re}\langle\Delta\hat{x}\Delta\hat{p}\rangle = 0$. Combining the both latter equalities, we obtain

$$\bar{\kappa}\langle\Delta\hat{p}^2\rangle + \kappa\langle\Delta\hat{p}^2\rangle = \langle\Delta\hat{p}^2\rangle\,(\bar{\kappa} + \kappa) = 0\,,$$

which results in $\operatorname{Re}\kappa = 0$, or $\kappa = -ir$, $\operatorname{Im} r = 0$. Taking this into account and applying the operator $\Delta\hat{x}$ to both sides of equation (10.29), we derive that $r > 0$. Indeed,

$$\Delta\hat{x}^2|\psi\rangle = -ir\Delta\hat{x}\Delta\hat{p}|\psi\rangle = r\hbar|\psi\rangle - r^2\Delta\hat{p}^2|\psi\rangle \Longrightarrow \langle\Delta\hat{x}^2\rangle + r^2\langle\Delta\hat{p}^2\rangle = r\hbar \Longrightarrow r > 0\,.$$

Then, remembering definitions of the operators $\Delta\hat{x}$ and $\Delta\hat{p}$, we represent equation (10.29) as follows:

$$(\hat{x} + ir\hat{p})\,|\psi\rangle = (\langle\hat{x}\rangle + ir\langle\hat{p}\rangle)\,|\psi\rangle = \langle\psi|\hat{x} + ir\hat{p}|\psi\rangle|\psi\rangle, \quad r > 0\,. \tag{10.30}$$

Thus, any state $|\psi\rangle$ that satisfies equation (10.30) "minimizes" the Heisenberg inequality i.e. the latter is reduced to the equality $\Delta x \Delta p = \hbar/2$.

To study solutions of equation (10.30) we introduce creation and annihilation operators $\hat{A}_r$ and $\hat{A}_r^+$,

$$\hat{A}_r = \frac{\hat{x} + ir\hat{p}}{\sqrt{2\hbar r}}, \quad \hat{A}_r^+ = \frac{\hat{x} - ir\hat{p}}{\sqrt{2\hbar r}}, \quad \left[\hat{A}_r, \hat{A}_r^+\right] = 1 \tag{10.31}$$

and in their terms we rewrite equation (10.30) as follows:

$$\hat{A}_r|\psi\rangle = \langle\psi|\hat{A}_r|\psi\rangle|\psi\rangle\,. \tag{10.32}$$

Now is is obvious that for any fixed $r > 0$ this equation has a set of solutions $|\psi\rangle_r = |z\rangle_r$ which are Glauber CS. Namely, they satisfy the equation

$$\hat{A}_r|z\rangle_r = z|z\rangle_r\,. \tag{10.33}$$

Obviously, if the set of states $|z\rangle_r$ satisfies equation (10.33) then this set satisfies equation (10.32) as well.

Thus, any set of Glauber CS $|z\rangle_r$ minimizes the Heisenberg inequality for the standard deviations Δx and Δp. Changing the parameter r one can control the ratio between the standard deviations Δx and Δp. Because the operators $\hat{A}_r$ and $\hat{A}_r^+$ satisfy canonical commutation relations (10.31) for any admissible r, this means that two sets of such operators with different r are related by a linear canonical transformation, see Section 7.8.

10.2.4 Schrödinger–Glauber CS of a harmonic oscillator

Here we consider a harmonic oscillator and its *Schrödinger–Glauber CS*. The quantum Hamiltonian for this system reads

$$\hat{H} = \frac{\hat{p}^2}{2m} + \frac{m\omega_0^2}{2}x^2, \quad \hat{p} = -i\hbar\frac{d}{dx}. \tag{10.34}$$

Introducing dimensionless coordinate $\xi = x\sqrt{m\omega/\hbar}$, and related to these coordinate creation and annihilation operators $\hat{c}^+$ and $\hat{c}$, ($[\hat{c}, \hat{c}^+] = 1$),

$$\hat{c} = \frac{1}{\sqrt{2}}\left(\xi + \frac{d}{d\xi}\right), \quad \hat{c}^+ = \frac{1}{\sqrt{2}}\left(\xi - \frac{d}{d\xi}\right), \tag{10.35}$$

we reduce this operator to the following form:

$$\hat{H} = \frac{\hbar\omega}{2}\left(\xi^2 - \frac{d^2}{d\xi^2}\right) = \hbar\omega\left(\hat{n} + \frac{1}{2}\right), \tag{10.36}$$

where $\hat{n} = \hat{c}^+\hat{c}$ is the particle number operator. The eigenvectors of this operator

$$\hat{n}|n\rangle = n|n\rangle, \quad |n\rangle = \frac{1}{\sqrt{n!}}\left(\hat{c}^+\right)^n|0\rangle, \hat{c}|0\rangle = 0, \quad n \in \mathbb{Z}_+$$

are stationary states of the quantum oscillator,

$$\hat{H}|n\rangle = E_n|n\rangle, \quad E_n = \hbar\omega\left(n + \frac{1}{2}\right), \quad n \in \mathbb{Z}_+ .$$

Instantaneous CS in the Glauber definition $\hat{c}|z\rangle = z|z\rangle$ are given by equation (10.19). For the first time they were derived by Schrödinger in the coordinate representation, see [289]. In this representation, we consider state vectors as functions $\psi(\xi) = \langle\xi|\psi\rangle$ belonging to the Hilbert space $L^2(\mathbb{R})$ with the scalar product

$$\langle\psi|\varphi\rangle = \int_{-\infty}^{\infty} \bar{\psi}(\xi)\varphi(\xi)d\xi .$$

In this representation the equation for the coherent states $\hat{c}\Psi_z(\xi) = z\Psi_z(\xi)$ with account taken of (10.35) is an ordinary first-order differential equation

$$\frac{d}{d\xi}\Psi_z(\xi) = \left(\sqrt{2}z - \xi\right)\Psi_z(\xi), \tag{10.37}$$

which has the following normalized solutions:

$$\Psi_z(\xi) = \frac{1}{\sqrt[4]{\pi}} \exp\left\{-\frac{1}{2}\left(\xi - \sqrt{2}z\right)^2 + \frac{(z-\bar{z})^2}{4}\right\} . \tag{10.38}$$

One can express the initial operators $\hat{x}$ and $\hat{p}$ via operators $\hat{c}$ and $\hat{c}^+$,

$$\hat{x} = \sqrt{\frac{\hbar}{2m\omega}}\left(\hat{c} + \hat{c}^+\right), \quad \hat{p} = -i\sqrt{\frac{m\hbar\omega}{2}}\left(\hat{c} - \hat{c}^+\right),$$
$$\hat{c} = \sqrt{\frac{m\omega}{2\hbar}}\left(\hat{x} + i\frac{\hat{p}}{m\omega}\right), \quad \hat{c}^+ = \sqrt{\frac{m\omega}{2\hbar}}\left(\hat{x} - i\frac{\hat{p}}{m\omega}\right).$$

With the help of these relations, one easily obtains for the mean values in the CS,

$$\langle x\rangle = \sqrt{\frac{2\hbar}{m\omega}}\,\mathrm{Re}\,z, \quad \langle p\rangle = \sqrt{2m\hbar\omega}\,\mathrm{Im}\,z;\; z = \sqrt{\frac{m\omega}{2\hbar}}\left(\langle x\rangle + i\frac{\langle p\rangle}{m\omega}\right).$$

Applying the evolution operator to the instantaneous CS, we obtain their time evolution:

$$\Psi_z(\xi, t) = \exp\left\{-\frac{i}{\hbar}\hat{H}t\right\}\Psi_z(\xi) = e^{-\frac{i}{2}\omega t}\Psi_{z(t)}(\xi), \quad z(t) = ze^{-i\omega t} . \tag{10.39}$$

We see that in the case under consideration, the time evolution does not destroy the form of the CS and is reduced only to a time evolution of the CS parameter z.

Calculating mean values of the coordinate and momentum in the time-dependent CS (10.39), we obtain:

$$\langle x\rangle = \sqrt{\frac{2\hbar}{m\omega}}\,\mathrm{Re}\,z(t) = x_0\cos\omega t + \frac{p_0}{m\omega}\sin\omega t, \quad x_0 = \langle x\rangle|_{t=0}, p_0 = \langle p\rangle|_{t=0},$$
$$\langle p\rangle = \sqrt{2m\hbar\omega}\,\mathrm{Im}\,z(t) = p_0\cos\omega t - m\omega x_0\sin\omega t .$$

Then the mean values of the operators $\hat{x}$ and $\hat{p}$ in the time-dependent CS follow the corresponding classical trajectories.

10.3 Coherent states for systems with quadratic Hamiltonians

As was already mentioned, using the method of integrals of motion [242], Dodonov and Man'ko have constructed different kinds of CS, generalized and squeezed, for systems with nonstationary quadratic Hamiltonians [125]. Below, we, in the same manner, construct some kind of generalized (squeezed) one-dimensional CS and represent some simple examples.

10.3.1 Basic equations

Consider quantum motion of a one-dimensional system with the generalized coordinate x on the whole real axis, $x \in \mathbb{R} = (-\infty, \infty)$, supposing that the corresponding quantum Hamiltonian $\hat{H}_x$ is given by a quadratic form of the operator x and the momentum operator $\hat{p}_x = -i\hbar\partial_x$,

$$\hat{H}_x = r_1\hat{p}_x^2 + r_2x^2 + r_3(x\hat{p}_x + \hat{p}_x x) + r_4x + r_5\hat{p}_x + r_6 , \tag{10.40}$$

where $r_s = r_s(t)$, $s = 1, \ldots, 6$ are some given functions of the time t. We suppose that these functions are real and both $\hat{H}_x$ and $\hat{p}_x$ are self-adjoint on their natural domains D_{H_x} and D_{p_x} respectively, see for example [177].

Quantum states of the system under consideration are described by a wave function $\Psi(x, t)$ which satisfies the Schrödinger equation

$$i\hbar\partial_t\Psi(x, t) = \hat{H}_x\Psi(x, t) . \tag{10.41}$$

In what follows, we restrict ourselves by a physically reasonable case $r_1(t) > 0$. In this case, we introduce dimensionless variables, a coordinate q and a time τ as follows:

$$q = xl^{-1}, \quad \tau = \int_0^t \frac{ds}{T(s)} = \frac{2\hbar}{l^2}\int_0^t r_1(s)ds, \quad T(t) = \frac{l^2}{2\hbar r_1(t)} , \tag{10.42}$$

where l is an arbitrary constant of the dimension of the length. The new momentum operator $\hat{p}$ and the new wave function $\psi(q, \tau)$ read

$$\hat{p} = \frac{l}{\hbar}\hat{p}_x = -i\partial_q, \psi(q, \tau) = \sqrt{l}\Psi\left(lq, \frac{ml^2}{\hbar}\tau\right) , labelCS2.2 \tag{10.43}$$

so that $|\Psi(x, t)|^2dx = |\psi(q, \tau)|^2dq$.

In the new variables, equation (10.41) takes the form

$$\hat{S}\psi(q, \tau) = 0, \quad \hat{S} = i\partial_\tau - \hat{H} , \tag{10.44}$$

where the new Hamiltonian reads

$$\hat{H} = \frac{\hat{p}^2}{2} + \alpha\hat{q}^2 + \beta(\hat{q}\hat{p} + \hat{p}\hat{q}) + \mu\hat{q} + \nu\hat{p} + \varepsilon . \tag{10.45}$$

Here $\alpha = \alpha(\tau)$, $\beta = \beta(\tau)$, $\varrho = \varrho(\tau)$, $\nu = \nu(\tau)$ and $\varepsilon = \varepsilon(\tau)$ are dimensionless real functions on the new time τ. They are given by the following relations:

$$\begin{aligned} &\alpha(\tau) = \frac{l^4}{2\hbar^2}\frac{r_2(t)}{r_1(t)}, \quad \beta(\tau) = \frac{l^2}{2\hbar}\frac{r_3(t)}{r_1(t)}, \\ &\varrho(\tau) = \frac{l^3}{2\hbar^2}\frac{r_4(t)}{r_1(t)}, \quad \nu(\tau) = \frac{l}{2\hbar}\frac{r_5(t)}{r_1(t)}, \quad \varepsilon(\tau) = \frac{l^2}{2\hbar^2}\frac{r_6(t)}{r_1(t)} , \end{aligned} \tag{10.46}$$

where t has to be expressed via τ by the help of equations (10.42). We call the operator $\hat{S}$ the equation operator.

10.3.2 Integrals of motion linear in canonical operators $\hat{q}$ and $\hat{p}$

Let us construct an integral of motion $\hat{A}(\tau)$ linear in $\hat{q}$ and $\hat{p}$. The general form of such an integral of motion reads

$$\hat{A}(\tau) = f(\tau)\hat{q} + ig(\tau)\,\hat{p} + \varphi(\tau)\,, \tag{10.47}$$

where $f(\tau)$, $g(\tau)$ and $\varphi(\tau)$ are some complex functions on the time τ. The operator $\hat{A}(\tau)$ is an integral of motion if it commutes with equation operator (10.44),

$$\left[\hat{S}, \hat{A}(\tau)\right] = 0\,. \tag{10.48}$$

In the case where the Hamiltonian is self-adjoint, the adjoint operator $\hat{A}^{\dagger}(\tau)$ is also an integral of motion,

$$\left[\hat{S}, \hat{A}^{\dagger}(\tau)\right] = 0\,. \tag{10.49}$$

The commutator $[\hat{A}(\tau), \hat{A}^{\dagger}(\tau)]$ reads

$$\left[\hat{A}(\tau), \hat{A}^{\dagger}(\tau)\right] = \delta = 2\,\mathrm{Re}\,g(\tau)f^{*}(\tau)\,. \tag{10.50}$$

Substituting representation (10.47) into equations (10.48), we obtain the following equations for the functions $f(\tau)$, $g(\tau)$, and $\varphi(\tau)$:

$$\begin{aligned} \dot{f}(\tau) + 2\beta(\tau)f(\tau) - 2i\alpha(\tau)g(\tau) &= 0\,,\\ \dot{g}(\tau) - if(\tau) - 2\beta(\tau)\,g(\tau) &= 0\,,\\ \dot{\varphi}(\tau) + \nu(\tau)f(\tau) - i\varrho(\tau)g(\tau) &= 0\,. \end{aligned} \tag{10.51}$$

We see that it is enough to find the functions $f(\tau)$ and $g(\tau)$, then the function $\varphi(\tau)$ can be found by a simple integration. Without loss of generality we can set $\varphi(0) = 0$.

Equations (10.51) imply that δ is a real integral of motion, $\delta = \mathrm{const}$. In what follows we suppose that $\delta = 1$, which means

$$\mathrm{Re}\,f^{*}(\tau)g(\tau) = 1/2\,. \tag{10.52}$$

Any nontrivial solution of two first equations (10.51) consists of both nonzero functions $f(\tau)$ and $g(\tau)$. That is why we can chose arbitrary integration constants in these equations as

$$g(0) = c_2 = |c_2|\,e^{i\mu_2}, \quad f(0) = c_1 = |c_1|\,e^{i\mu_1}, \quad |c_2| \neq 0, \quad |c_1| \neq 0\,. \tag{10.53}$$

In terms of the latter constants, condition (10.52) implies

$$2\,\mathrm{Re}\,(c_1^{*}c_2)\,|c_2|\,|c_1|\cos(\mu_2 - \mu_1) = 1/2\,. \tag{10.54}$$

Under the choice $\delta = 1$, the operators $\hat{A}(\tau)$ and $\hat{A}^{\dagger}(\tau)$ are annihilation and creation operators,

$$\left[\hat{A}(\tau), \hat{A}^{\dagger}(\tau)\right] = 1\,. \tag{10.55}$$

It follows from equation (10.47) and (10.52) that

$$\hat{q} = g^*(\tau)\left[\hat{A}(\tau) - \varphi(\tau)\right] + g(\tau)\left[\hat{A}^\dagger(\tau) - \varphi^*(\tau)\right],$$
$$i\hat{p} = f^*(\tau)\left[\hat{A}(\tau) - \varphi(\tau)\right] - f(\tau)\left[\hat{A}^\dagger(\tau) - \varphi^*(\tau)\right]. \tag{10.56}$$

It should be noted that the two first equations (10.51) can be identified with the spin equation (8.1), with

$$V = \begin{pmatrix} f(\tau) \\ g(\tau) \end{pmatrix}, \quad \mathrm{F}(\tau) = \left(-\frac{1+2\alpha}{2}, i\frac{1-2\alpha}{2}, -2i\beta\right). \tag{10.57}$$

At the same time, the two first equations (10.51) can be reduced to a one second-order differential equation for the function $g(\tau)$, which has the form of the oscillator equation with a time-dependent frequency $\omega^2(\tau)$,

$$\ddot{g}(\tau) + \omega^2(\tau)g(\tau) = 0, \; \omega^2(\tau) = 2\alpha(\tau) - 4\beta^2(\tau) - 2\dot{\beta}(\tau). \tag{10.58}$$

The function $f(\tau)$ can be found then via the function $g(\tau)$ as

$$f(\tau) = 2i\beta(\tau)g(\tau) - i\dot{g}(\tau). \tag{10.59}$$

10.3.3 Time dependent generalized CS

Let us consider eigenvectors $|z, \tau\rangle$ of the annihilation operator $\hat{A}(\tau)$ corresponding to the eigenvalue z

$$\hat{A}(\tau)|z, \tau\rangle = z|z, \tau\rangle. \tag{10.60}$$

In the general case z is a complex number.

It follows from equations (10.56) and (10.60) that

$$q(\tau) \equiv \langle z, \tau|\hat{q}|z, \tau\rangle = g^*(\tau)\left[z - \varphi(\tau)\right] + g(\tau)\left[z^* - \varphi^*(\tau)\right],$$
$$p(\tau) \equiv \langle z, \tau|\hat{p}|z, \tau\rangle = i f(\tau)\left[z^* - \varphi^*(\tau)\right] - i f^*(\tau)\left[z - \varphi(\tau)\right],$$
$$z = f(\tau)q(\tau) + ig(\tau)p(\tau) + \varphi(\tau). \tag{10.61}$$

Using (10.51), one can easily verify that the functions $q(\tau)$ and $p(\tau)$ satisfy Hamilton equations

$$\dot{q}(\tau) = \frac{\partial H}{\partial p}, \quad \dot{p}(\tau) = -\frac{\partial H}{\partial q},$$

where $H = H(q, p)$ is the classical Hamiltonian that corresponds to the quantum Hamiltonian (10.45). Thus, $q(\tau)$ and $p(\tau)$ represent a classical trajectory of the system under consideration. All such trajectories can be parametrized by the initial data, $q_0 = q(0)$, $p_0 = p(0)$. Thus, we have the following correspondence:

$$z = f(0)q_0 + ig(0)p_0 + \varphi(0). \tag{10.62}$$

Being written in the q-representation, equation (10.60) reads

$$\left[f(\tau)q + g(\tau)\partial_q + \varphi(\tau)\right]\langle q|z,\tau\rangle = z\langle q|z,\tau\rangle\ . \tag{10.63}$$

The general solution of this equation has the form

$$\langle q|z,\tau\rangle = \Phi_z^{c_1c_2}(q,\tau) = \exp\left[-\frac{f(\tau)}{g(\tau)}\frac{q^2}{2} + \frac{z-\varphi(\tau)}{g(\tau)}q + \chi(\tau)\right], \tag{10.64}$$

where $\chi(\tau)$ is an arbitrary function on τ.

One can see that the functions $\Phi_z(q,\tau)$ can be written in terms of the mean values $q(\tau)$ and $p(\tau)$,

$$\Phi_z^{c_1c_2}(q,\tau) = \exp\left\{ip(\tau)q - \frac{f(\tau)}{2g(\tau)}\left[q - q(\tau)\right]^2 + \tilde{\chi}(\tau)\right\}, \tag{10.65}$$

where $\tilde{\chi}(\tau)$ is again an arbitrary function on τ.

The functions Φ_z satisfy the following equation:

$$\hat{S}\Phi_z^{c_1c_2}(q,\tau) = \lambda(\tau)\,\Phi_z^{c_1c_2}(q,\tau)\ , \tag{10.66}$$

where

$$\lambda(\tau) = i\partial_\tau\tilde{\chi}(\tau) + \alpha q^2(\tau) - \frac{1}{2}\left[p^2(\tau) + f/g\right] - \nu p(\tau) + i\beta - \varepsilon\ . \tag{10.67}$$

If we wish that the functions (10.65) satisfy Schrödinger equation (10.44), we have to fix $\tilde{\phi}(\tau)$ from the condition $\lambda(\tau) = 0$. Thus, we obtain for the function $\tilde{\chi}(\tau)$ the following result:

$$\begin{aligned}&\tilde{\chi}(\tau) = \phi(\tau) + \ln N,\\ &\phi(\tau) = \int_0^\tau \left\{i\alpha q^2(\tau) - \frac{i}{2}\left[p^2(\tau) + f/g\right] - i\nu p(\tau) - \beta - i\varepsilon\right\} d\tau\ ,\end{aligned} \tag{10.68}$$

were $\phi(0) = 0$ and N is a normalization constant, which we suppose to be real.

The density probability generated by function (10.65) reads

$$\rho_z^{c_1c_2}(q,\tau) = \left|\Phi_z^{c_1c_2}(q,\tau)\right|^2 = N^2\exp\left\{-\frac{\left[q - q(\tau)\right]^2}{2\left|g(\tau)\right|^2} + 2\,\mathrm{Re}\,\phi(\tau)\right\}. \tag{10.69}$$

Considering the normalization integral, we find the constant N,

$$\int_{-\infty}^{\infty}\rho_z^{c_1c_2}(q,\tau)\,dq = 1 \Rightarrow N = \frac{\exp\left(-\mathrm{Re}\,\phi(\tau)\right)}{\sqrt{\sqrt{2\pi}\left|g(\tau)\right|}}\ . \tag{10.70}$$

Thus, normalized solutions of the Schrödinger equation that are at the same time eigenfunctions of the annihilation operator $\hat{A}(\tau)$ have the form

$$\Phi_z^{c_1c_2}(q,\tau) = \frac{1}{\sqrt{\sqrt{2\pi}\,|g(\tau)|}} \exp\left\{ip(\tau)\,q - \frac{f(\tau)}{g(\tau)}\frac{[q-q(\tau)]^2}{2} + i\,\mathrm{Im}\,\phi(\tau)\right\} \tag{10.71}$$

and the corresponding probability density reads

$$\rho_z^{c_1c_2}(q,\tau) = \frac{1}{\sqrt{2\pi}\,|g(\tau)|} \exp\left\{-\frac{[q-q(\tau)]^2}{2\,|g(\tau)|^2}\right\} . \tag{10.72}$$

In what follows we call the solutions (10.71) the time-dependent generalized CS.

10.3.4 Standard deviations and uncertainty relations

Using equations (10.56) and (10.60) we can calculate *standard deviations* $\sigma_q(\tau)$, σ_p, and the quantity $\sigma_{qp}(\tau)$, in the generalized CS,

$$\begin{aligned}
\sigma_q(\tau) &= \sqrt{\langle(\hat{q}-\langle q\rangle)^2\rangle} = \sqrt{\langle q^2\rangle - \langle q\rangle^2} = |g(\tau)|\,, \\
\sigma_p(\tau) &= \sqrt{\langle(\hat{p}-\langle p\rangle)^2\rangle} = \sqrt{\langle \hat{p}^2\rangle - \langle p\rangle^2} = |f(\tau)|\,, \\
\sigma_{qp}(\tau) &= \frac{1}{2}\,\langle(\hat{q}-\langle q\rangle)(\hat{p}-\langle p\rangle) + (\hat{p}-\langle p\rangle)(\hat{q}-\langle q\rangle)\rangle \\
&= i\left[1/2 - g(\tau)f^*(\tau)\right] .
\end{aligned} \tag{10.73}$$

One can easily see that the generalized CS (10.71) minimize the Robertson–Schrödinger uncertainty relation [290],

$$\sigma_q^2(\tau)\sigma_p^2 - \sigma_{qp}^2(\tau) = 1/4 . \tag{10.74}$$

This means that the generalized CS are squeezed states [125].

Let us analyze the Heisenberg uncertainty relation in the generalized CS taking into account restriction (10.52) or (10.54),

$$\sigma_q(\tau)\sigma_p(\tau)\big|_{2\,\mathrm{Re}(c_1^*c_2)} = \frac{1}{2}\sqrt{1+4\left(\mathrm{Im}\,(gf^*)\right)^2} \geq \frac{1}{2} . \tag{10.75}$$

Then using (10.73), we find $\sigma_q(0) = \sigma_q = |c_2|$ and $\sigma_p(0) = \sigma_p = |c_1|$, such that at $\tau = 0$ this relation reads

$$\sigma_q\sigma_p\big|_{2\,\mathrm{Re}(c_1^*c_2)} = \sqrt{\frac{1}{4} + \left[|c_2|\,|c_1|\sin(\mu_2-\mu_1)\right]^2} . \tag{10.76}$$

Taking into account equations (10.53), we see that if $\mu_1 = \mu_2 = \mu$ the left hand side of (10.76) is minimal, such that

$$\sigma_q\sigma_p = 1/2, \sigma_{qp} = 0 . \tag{10.77}$$

In what follows we consider generalized CS with the restriction $\mu_1 = \mu_2$. Namely, such states we call simply CS.

Now restriction (10.52), $2\,\mathrm{Re}(c_1^* c_2) = 1$ takes the form

$$|c_2|\,|c_1| = 1/2 \Longrightarrow c_2^* = c_1^{-1}/2\,. \tag{10.78}$$

One can see that the constant μ does not enter CS (10.71). Thus, we set $\mu = 0$ in what follows. Then

$$c_2 = |c_2| = |g(0)| = \sigma_q, \quad c_1 = |c_1| = |f(0)| = \sigma_p = 1/(2\sigma_q)\,. \tag{10.79}$$

With account taken of equations (10.71), (10.79) and (10.68), we obtain the following expression for the CS:

$$\Phi_z^{\sigma_q}(q,\tau) = \frac{1}{\sqrt{2\sqrt{\pi}\sigma_q(\tau)}} \exp\left\{ip(\tau)q - \frac{f(\tau)}{g(\tau)}\frac{[q-q(\tau)]^2}{2} + i\,\mathrm{Im}\,\phi(\tau)\right\},$$

$$\phi(\tau) = \int_0^\tau \left\{i\alpha q^2(\tau) - \frac{i}{2}\left[p^2(\tau) + f/g\right] - i\nu p(\tau) - \beta - i\varepsilon\right\} d\tau\,. \tag{10.80}$$

In fact, we have a family of CS parametrized by one real parameter – the initial standard deviation $\sigma_q > 0$. Each set of CS in the family has its specific initial standard deviations σ_q. Different CS from a family with a given σ_q have different quantum numbers z, which are in one to one correspondence with trajectory initial data q_0 and p_0. It follows from equation (10.62) that

$$z = \frac{q_0}{2\sigma_q} + i\sigma_q p_0,$$

$$q_0 = 2\sigma_q\,\mathrm{Re}\,z\ \ p_0 = \frac{\mathrm{Im}\,z}{\sigma_q}\,. \tag{10.81}$$

The probability density that corresponds to the CS (10.80) reads

$$\rho_z^{\sigma_q}(q,\tau) = \frac{1}{\sqrt{2\pi}\sigma_q(\tau)} \exp\left\{-\frac{[q-q(\tau)]^2}{2\sigma_q^2(\tau)}\right\}\,. \tag{10.82}$$

10.3.5 Simple examples

CS of a free particle

To obtain CS for a free particle from the general results, we have to set their $\alpha = \beta = \varrho = \nu = \varepsilon = 0$. In particular, the general solution of equations (10.51) has the form

$$f(\tau) = c_1, \quad g(\tau) = c_2 + ic_1\tau, \quad \varphi(\tau) = 0\,, \tag{10.83}$$

where c_1 and c_2 are arbitrary constants. Without loss of the generality we can set $\varphi(\tau) = 0$. The restriction (10.52) takes the form

$$2\,\mathrm{Re}\,(c_1^* c_2) = 1 \Longrightarrow |c_2|\,|c_1| \cos(\mu_2 - \mu_1) = 1/2\,. \tag{10.84}$$

Generalized CS of the free particle have the form

$$\Phi_z^{c_{1,2}}(q,\tau) = \frac{1}{\sqrt{\sqrt{2\pi} g(\tau)}} \exp\left\{ i\left(pq - \frac{1}{2}p^2\tau\right) - \frac{c_1}{g(\tau)} \frac{[q - q(\tau)]^2}{2} \right\} \tag{10.85}$$

and the corresponding probability density reads

$$\rho_z^{c_{1,2}}(q,\tau) = \left|\Phi_z^{c_{1,2}}(q,\tau)\right|^2 = \frac{1}{\sqrt{2\pi}\,|g(\tau)|} \exp\left\{ -\frac{[q - q(\tau)]^2}{2\,|g(\tau)|^2} \right\}, \tag{10.86}$$

where the mean trajectory $q(\tau)$ is

$$q(\tau) = \langle z,\tau\,|\hat{q}|\,z,\tau\rangle = q_0 + p\tau = (c_2^* - ic_1\tau)\,z + (c_2 + ic_1\tau)\,z^*\,,$$
$$p = p(\tau) \equiv \langle z,\tau\,|\hat{p}|\,z,\tau\rangle = i\,(c_1 z^* - c_1^* z)\,.$$

All the general results obtained for standard deviations (10.73) hold with $f(\tau)$ and $\varphi(\tau)$ given by (10.83).

As in the general case, we call the generalized CS of the free particle simply CS if $\mu_1 = \mu_2 = \mu$, which provides the minimization of the Heisenberg uncertainty at the initial time instant, (10.77). We set $\mu = 0$ in what follows. Then CS of a free particle read

$$\Phi_z^{\sigma_q}(q,\tau) = \frac{\exp\left\{ i\left(pq - \frac{p^2}{2}\tau\right) - \frac{[q-q(\tau)]^2}{4(\sigma_q^2 + i\tau/2)} \right\}}{\sqrt{\left(\sigma_q + \frac{i\tau}{2\sigma_q}\right)\sqrt{2\pi}}}\,, \tag{10.87}$$

where σ_q is the standard deviation $\sigma_q(\tau)$ at the initial time instant,

$$\sigma_q = c_2 = |c_2|\,,\quad c_1 = |c_1| = \sigma_p = 1/(2\sigma_q)\,,$$
$$g(\tau) = \left(\sigma_q + \frac{i\tau}{2\sigma_q}\right),\quad \sigma_q(\tau) = |g(\tau)| = \sqrt{\sigma_q^2 + \frac{\tau^2}{4\sigma_q^2}}\,. \tag{10.88}$$

With account taken of equation (10.88), we obtain that for any τ the Heisenberg uncertainty relation takes the form

$$\sigma_q(\tau)\sigma_p = \frac{1}{2}\sqrt{1 + \frac{\tau^2}{4\sigma_q^4}} \geq \frac{1}{2}\,. \tag{10.89}$$

In fact, we have a family of CS parametrized by one real parameter σ_q. Each set of CS in the family has its specific initial standard deviations $\sigma_q > 0$. CS from a family with a given σ_q are labeled by quantum numbers z that are related to the trajectory initial data, see (10.81).

The probability densities that corresponds to the CS are

$$\rho_z^{\sigma_q}(q,\tau) = \frac{1}{\sqrt{\left(\sigma_q^2 + \frac{\tau^2}{4\sigma_q^2}\right)2\pi}} \exp\left\{-\frac{1}{2}\frac{[q-q(\tau)]^2}{\sigma_q^2 + \frac{\tau^2}{4\sigma_q^2}}\right\} . \tag{10.90}$$

One can see that at any time instant τ the probability densities (10.90) are given by Gaussian distributions with standard deviations $\sigma_q(\tau)$. The mean values $\langle q\rangle = q(\tau) = q_0 + p\tau$ are moving along the classical trajectory with the particle velocity p. Moving with the same velocity are the maxima of the probability densities (10.90). It should be noted that similar states were obtained in Ref. [236]. More detailed study of the free particle CS is presented in Ref. [68].

CS of a harmonic oscillator

To obtain *CS of a harmonic oscillator* with a frequency ω from the general results of Section 10.3, we have to set their

$$r_1 = 1/2m, \quad r_2 = \frac{m\omega^2}{2}, \quad \alpha = \frac{\varpi^2}{2}, \quad \varpi^2 = \frac{m^2 l^4}{\hbar^2}\omega^2 . \tag{10.91}$$

Then equations (10.51) take the form

$$\dot f(\tau) - i\varpi^2 g(\tau) = 0, \quad \dot g(\tau) - if(\tau) = 0, \quad \dot\varphi(\tau) = 0 .$$

Their general solution reads

$$\begin{aligned} f(\tau) &= c_1 \cos(\varpi\tau) + i\varpi c_2 \sin(\varpi\tau) , \\ g(\tau) &= c_2 \cos(\varpi\tau) + \frac{ic_1}{\varpi}\sin(\varpi\tau), \ \varphi(\tau) = c_3 , \end{aligned} \tag{10.92}$$

where $f(0) = c_1 = |c_1|e^{i\mu_1}$, $g(0) = c_2 = |c_2|e^{i\mu_2}$ and c_3 are arbitrary constants. Without loss of generality we can set $c_3 = 0$. We chose $\mu_2 = \mu_1 = 0$ to deal with CS of the harmonic oscillator. Then equation (10.78) implies $c_1 = |c_1|$, $c_2 = |c_2|$, $|c_1||c_2| = 1/2$.

The mean trajectories are

$$\begin{aligned} q(\tau) &= 2\sigma_q(\operatorname{Re} z)\cos(\varpi\tau) + \frac{\operatorname{Im} z}{\varpi\sigma_q}\sin(\varpi\tau) , \\ p(\tau) &= \frac{\operatorname{Im} z}{\sigma_q}\cos(\varpi\tau) - \varpi 2\sigma_q(\operatorname{Re} z)\sin(\varpi\tau) , \\ \sigma_q &= \sigma_q(0) = |c_2| . \end{aligned} \tag{10.93}$$

Standard deviations $\sigma_q(\tau), \sigma_p(\tau)$, and the quantity $\sigma_{qp}(\tau)$, are in the case under consideration

$$
\begin{aligned}
&\sigma_q(\tau) = \sigma_q\sqrt{\cos^2(\varpi\tau) + \frac{\sin^2(\varpi\tau)}{4\sigma_q^4\varpi^2}}, \quad \sigma_q = \sigma_q(0) = |c_2| . \\
&\sigma_p(\tau) = \frac{1}{2\sigma_q}\sqrt{\cos^2(\varpi\tau) + 4\varpi^2\sigma_q^4\sin^2(\varpi\tau)}, \\
&\sigma_q(\tau)\sigma_p(\tau) = \frac{1}{2}\sqrt{1 + \frac{\left[\left(1 - 4\varpi^2\sigma_q^4\right)\sin(2\varpi\tau)\right]^2}{8\varpi^2\sigma_q^4}} \geq \frac{1}{2}, \\
&\sigma_q^2(\tau)\sigma_p^2(\tau) - \sigma_{qp}^2(\tau) = \frac{1}{4} .
\end{aligned}
\tag{10.94}
$$

The CS of the harmonic oscillator have the following form:

$$
\begin{aligned}
\Phi_z^{\sigma_q}(q,\tau) &= \frac{1}{\sqrt{\sqrt{2\pi}\left[\sigma_q\cos(\varpi\tau) + \frac{i}{2\varpi\sigma_q}\sin(\varpi\tau)\right]}} \\
&\times \exp\left\{i\left[q - \frac{1}{2}q(\tau)\right]p(\tau) - \frac{\left[\cos(\varpi\tau) + 2i\varpi\sigma_q^2\sin(\varpi\tau)\right]}{\left[2\sigma_q^2\cos(\varpi\tau) + \frac{i}{\varpi}\sin(\varpi\tau)\right]}\frac{[q - q(\tau)]^2}{2}\right\} .
\end{aligned}
\tag{10.95}
$$

In fact, we have again a family of CS parametrized by one real parameter σ_q. Each set of CS in the family has its specific initial standard deviations $\sigma_q > 0$. CS from a family with a given σ_q are labeled by quantum numbers z that are related to the trajectory initial data as

$$
q_0 = 2\sigma_q(\mathrm{Re}\,z), \quad p_0 = \frac{\mathrm{Im}\,z}{\sigma_q} . \tag{10.96}
$$

CS of an inverse oscillator, see [67], can be obtained from results of this section by the replacement $\omega \longrightarrow i\omega$.

It should be noted that the Heisenberg uncertainty relation has the form

$$
\sigma_q(\tau)\sigma_p(\tau) = \frac{1}{2}\theta(\tau) ,
$$

where the function $\theta(\tau)$ is periodically changed from 1 to

$$
\sqrt{1 + \frac{\left(1 - 4\varpi^2\sigma_q^4\right)^2}{8\varpi^2\sigma_q^4}} .
$$

For the CS with

$$
\sigma_q = \frac{1}{\sqrt{2\varpi}} \Longrightarrow \sigma_x = \frac{\hbar}{\sqrt{2m\omega}} \tag{10.97}
$$

the Heisenberg uncertainty relation is minimized at any time instant. Namely such CS are always considered as CS of the harmonic oscillator.

CS of an inverse oscillator, see [67], can be obtained from results of this section by the replacement $\omega \to i\omega$.

A Appendix 1

A.1 Pauli matrices

A.1.1 General properties

Three nonsingular, linearly independent, complex 2×2 matrices $\sigma_k, k = 1, 2, 3$ are designated as *Pauli matrices* if they obey the following relations:

$$\sigma_1 \sigma_2 = i\sigma_3, \quad \sigma_2 \sigma_3 = i\sigma_1, \quad \sigma_3 \sigma_1 = i\sigma_2 \,, \tag{A.1}$$

$$(\sigma_k)^+ = \sigma_k, \quad k = 1, 2, 3 \,. \tag{A.2}$$

If a set $\sigma_k, k = 1, 2, 3$ represents Pauli matrices, then the set $\tilde{\sigma}_k = U\sigma_k U^{-1}$, where U is an arbitrary nonsingular matrix, obeys equations (A.1). If U is unitary, $U^+ = U^{-1}$, then $\tilde{\sigma}_k$ satisfies also equations (A.1) and (A.2) is thus another set of Pauli matrices.

One can see that the matrices (2.59) are Pauli matrices. The representation (2.59) was introduced by Pauli in his work [261], and such a representation is called the *Pauli representation* and is often used in physical applications.

One can demonstrate (see below) that relations (A.1) imply

$$[\sigma_k, \sigma_s]_+ = 2I\delta_{ks} \Longleftrightarrow \sigma_k \sigma_s = -\sigma_s \sigma_k, \quad k \neq s; \quad (\sigma_k)^2 = I, \quad \forall k \,, \tag{A.3}$$

where δ_{ks} is the Kronecker symbol and I is 2×2 unit matrix. It follows from (A.3) that

$$\operatorname{tr} \sigma_k = 0; \ \det \sigma_k = -1, \quad \forall k \,. \tag{A.4}$$

It is convenient to introduce a three-vector $\boldsymbol{\sigma} = (\sigma_1, \sigma_2, \sigma_3)$ with σ_k being its Cartesian coordinates. Then any linear combination of Pauli matrices can be written as $(\boldsymbol{\sigma}\mathbf{a})$, $\mathbf{a} = (a^k)$, and $\det(\boldsymbol{\sigma}\mathbf{a}) = -\mathbf{a}^2$.

The matrices I and σ_k are linearly independent (see below) and form a basis in the space of all complex 2×2 matrices. Then any 2×2 matrix A can be represented as

$$A = Ia^0 + (\boldsymbol{\sigma}\mathbf{a}) \,, \quad \mathbf{a} = (a^k) \,. \tag{A.5}$$

The complex coefficients a^μ, $\mu = 0, k$ are completely determined by A,

$$a^0 = \frac{1}{2} \operatorname{tr} A, \quad a^k = \frac{1}{2} \operatorname{tr} A\sigma_k, \ \forall k \,, \tag{A.6}$$

as it follows from equations (A.1), (A.3), and (A.4).

One obtains from (A.3) and (A.6) that

$$[\boldsymbol{\sigma} \times \boldsymbol{\sigma}] = 2i\boldsymbol{\sigma} \Longleftrightarrow [\sigma_s, \sigma_k] = -2i\epsilon_{skl}\sigma_l \,, \tag{A.7}$$

where summation over repeated indices is supposed and ϵ_{skl} is three-dimensional Levi-Civita symbol with the normalization $\epsilon_{123} = -1$. Then, we have

$$\sigma_s \sigma_k = \delta_{sk} - i\epsilon_{skl}\sigma_l \,. \tag{A.8}$$

The following relations hold as well:

$$\begin{aligned}
&(\boldsymbol{\sigma}\mathbf{a})\boldsymbol{\sigma} = \mathbf{a} + i[\boldsymbol{\sigma} \times \mathbf{a}], \quad \boldsymbol{\sigma}(\boldsymbol{\sigma}\mathbf{a}) = \mathbf{a} - i[\boldsymbol{\sigma} \times \mathbf{a}]\,,\\
&(\boldsymbol{\sigma}\mathbf{a})\boldsymbol{\sigma} + \boldsymbol{\sigma}(\boldsymbol{\sigma}\mathbf{a}) = 2\mathbf{a}, \quad (\boldsymbol{\sigma}\mathbf{a})\boldsymbol{\sigma} - \boldsymbol{\sigma}(\boldsymbol{\sigma}\mathbf{a}) = 2i[\boldsymbol{\sigma} \times \mathbf{a}]\,,\\
&(\boldsymbol{\sigma}\mathbf{a})(\boldsymbol{\sigma}\mathbf{b}) = (\mathbf{ab}) + i(\boldsymbol{\sigma}[\mathbf{a} \times \mathbf{b}]), \quad (\boldsymbol{\sigma}, \boldsymbol{\sigma}(\boldsymbol{\sigma}\mathbf{a})) = -(\boldsymbol{\sigma}\mathbf{a})\,,\\
&(\boldsymbol{\sigma}\mathbf{a})(\boldsymbol{\sigma}\mathbf{b})(\boldsymbol{\sigma}\mathbf{c}) = i(\mathbf{a}[\mathbf{b} \times \mathbf{c}]) + (\boldsymbol{\sigma}\mathbf{p}), \quad \mathbf{p} = \mathbf{a}(\mathbf{bc}) + \mathbf{c}(\mathbf{ab}) - \mathbf{b}(\mathbf{ac})\,,\\
&\exp\left[i\,(\boldsymbol{\sigma}\mathbf{n})\,\theta\right] = \cos\theta + i\,(\boldsymbol{\sigma}\mathbf{n})\sin\theta, \quad \mathbf{n}^2 = 1\,,
\end{aligned} \tag{A.9}$$

where **a**, **b**, and **c** are arbitrary complex vectors.

Let us derive relations (A.3) from (A.1).

It follows from the definition (A.1) that none of σ_k is a multiple of the unit matrix. Indeed, let us suppose that $\sigma_1 = \alpha I$, where α is a complex number. Then it follows from the first equation (A.1) that $\alpha\sigma_2 = i\sigma_3$, which contradicts the supposition that all the sigma matrices are independent.

It follows from the definition (A.1) that $(\sigma_k)^2 = q, \forall k$, where q is a nonsingular matrix. Indeed, multiplying the first equation (A.1) from the right by σ_3 and using then the second equation (A.1), we obtain $-i\sigma_1\sigma_2\sigma_3 = (\sigma_3)^2 = (\sigma_1)^2$. Multiplying the second equation (A.1) from the right by σ_1 and using then the third equation (A.1), we obtain $-i\sigma_2\sigma_3\sigma_1 = (\sigma_1)^2 = (\sigma_2)^2$.

Combining equations (A.1) one can easily obtain

$$\sigma_k\sigma_s\sigma_k = -\sigma_s,\, s \neq k\,. \tag{A.10}$$

Equations (A.1) allow one to find the relations $\Delta_s\Delta_k = -\Delta_l$, where $s \neq k \neq l$, for the quantity $\Delta_s = \det\sigma_s$. Then

$$\Delta_s^{\,2} = \det q = 1 \Longrightarrow \Delta_s = \pm 1\,,$$

with at least one of Δ_s being -1. Thus, q is nonsingular. By squaring both sides of (A.10) with account taken of the obvious relation $\sigma^k q = q\sigma^k$, we obtain

$$\sigma_k\sigma_s\sigma_k\sigma_k\sigma_s\sigma_k = \sigma_s^{\,2} = q = \sigma_k\sigma_s q\sigma_s\sigma_k = q^3\,,$$

which implies $q^2 = I$. Using equation (A.32), we find

$$\det(q + I) = 2 + \operatorname{tr} q, \quad \det(q - I) = 2 - \operatorname{tr} q\,,$$

so that at least one of the matrices $q+I$ or $q-I$ is nonsingular. Taking into account that $q^2 = I$, we have $(q + I)(q - I) = 0$. Therefore, either $q = I$ or $q = -I$. But the choice $q = -I$ leads to a contradiction. Indeed, let us consider the matrices $A_\pm^{(k)} = \sigma_k \pm iI$. According to equation (A.32), we have

$$\det A_\pm^{(k)} = \Delta_k - 1 \pm i\operatorname{tr}\sigma_k\,.$$

Let $\Delta_{k_0} = -1$, such k_0 always exists, as was proven above. Then one of the matrices $A_{\pm}^{(k_0)}$, for any value of $\mathrm{tr}\sigma^k$, is nonsingular. But for any s the relation

$$A_{+}^{(s)} A_{-}^{(s)} = (\sigma_s + iI)(\sigma_s - iI) = q + I = 0$$

holds. This means that one of the matrices $A_{\pm}^{(k_0)}$ is zero, which means that σ_{k_0} is a multiple of the unit matrix. The latter contradicts the initial supposition. Thus, there remains only one possibility that $q = I$, which implies relationships (A.3). It is easily seen that $\Delta_k = \det \sigma_k = -1, \forall k$.

Relationships (A.3) and (A.4) provide the linear independence of the matrices I and σ_k. Indeed, if we set $A = 0$ in equation (A.5), then $a^\mu = 0$ according to (A.6), which proves the statement.

A.1.2 Vectors and spinors associated with Pauli matrices

(I) Let us consider a complex linear space of two-component columns, we call them *spinors* in what follows. An arbitrary spinor has the form

$$v = \begin{pmatrix} v_1 \\ v_2 \end{pmatrix},$$

where complex numbers v_1 and v_2 are its components. The conjugate spinor v^+ has the form $v^+ = (v_1^*, v_2^*)$. The scalar product (u, v) of two spinors u and v is defined as

$$(u, v) = u^+ v = u_1^* v_1 + u_2^* v_2, \quad (u, v)^* = (v, u), \quad v = 0 \Leftrightarrow (v, v) = 0. \tag{A.11}$$

Let us introduce a (singular) matrix vu^+,

$$vu^+ = \begin{pmatrix} v_1 u_1^* & v_1 u_2^* \\ v_2 u_1^* & v_2 u_2^* \end{pmatrix}, \quad \det vu^+ = 0. \tag{A.12}$$

We define the anticonjugate spinor $\overline{v}$ as follows:

$$\overline{v} = \begin{pmatrix} -v_2^* \\ v_1^* \end{pmatrix} = -i\sigma_2 v^*, \tag{A.13}$$

where σ_2 is the Pauli matrix in the representation (2.59). The following relations hold:

$$\overline{(\overline{v})} = -v, \ (\overline{v}, \overline{v}) = (v, v), \ (\overline{v}, v) = (v, \overline{v}) = 0. \tag{A.14}$$

If u and v are two arbitrary spinors then there are relationships:

$$\begin{aligned} &(u, \overline{v}) = -(v, \overline{u}), \ (\overline{u}, v) = -(\overline{v}, u), \ (\overline{u}, \overline{v}) = (v, u), \\ &(u, \overline{v})(\overline{v}, u) = (u, u)(v, v) - (u, v)(v, u) = (\overline{u}, v))(v, \overline{u}) \geq 0. \end{aligned} \tag{A.15}$$

If v is a nonzero spinor, then the set $v, \overline{v}$ forms a basis in the linear space of complex spinors, such that for any spinor u the following decomposition holds:

$$u = (v, v)^{-1}[(v, u)v + (\overline{v}, u)\overline{v}] . \tag{A.16}$$

It follows from (A.12) and (A.16) that

$$\frac{vv^+ + \overline{v}\,\overline{v}^+}{(v, v)} = I . \tag{A.17}$$

(II) Let us introduce a complex vector $\mathbf{L}^{u,v}$ generated by spinors u and v as follows:

$$\mathbf{L}^{u,v} = (u, \sigma v) . \tag{A.18}$$

The vectors (A.18) obey the simple relations

$$(\mathbf{L}^{u,v})^* = \mathbf{L}^{v,u}, \quad \mathbf{L}^{\overline{u},\overline{v}} = -\mathbf{L}^{v,u}, \quad \mathbf{L}^{\overline{u},v} = \mathbf{L}^{\overline{v},u}, \quad \mathbf{L}^{u,\overline{v}} = \mathbf{L}^{v,\overline{u}}.$$

Taking the Pauli matrices in the representation (2.59), we obtain:

$$\mathbf{L}^{u,v} = (u_1^* v_2 + u_2^* v_1, iu_2^* v_1 - iu_1^* v_2, u_1^* v_1 - u_2^* v_2) . \tag{A.19}$$

If u, v, u', and v' are arbitrary spinors, then

$$(\mathbf{L}^{u,v} \cdot \mathbf{L}^{u',v'}) = 2(u, v')(u', v) - (u, v)(u', v') . \tag{A.20}$$

It follows from (A.20) that

$$\begin{aligned} &(\mathbf{L}^{v,v})^2 = (\mathbf{L}^{v,v} \cdot \mathbf{L}^{v,v}) = (v, v)^2, \quad (\mathbf{L}^{v,\overline{v}})^2 = (\mathbf{L}^{\overline{v},v})^2 = 0, \\ &(\mathbf{L}^{v,\overline{v}} \cdot \mathbf{L}^{\overline{v},v}) = 2(v, v)^2, \quad (\mathbf{L}^{v,v} \cdot \mathbf{L}^{v,\overline{v}}) = (\mathbf{L}^{v,v} \cdot \mathbf{L}^{\overline{v},v}) = 0 . \end{aligned} \tag{A.21}$$

Therefore the vectors

$$\mathbf{L}^{v,v}, \quad \mathbf{L}^{v,\overline{v}}, \quad \mathbf{L}^{\overline{v},v} = (\mathbf{L}^{v,\overline{v}})^* \tag{A.22}$$

are linearly independent, provided $(v, v) \neq 0$. Any complex vector $\mathbf{a}$ can be decomposed in these vectors,

$$\mathbf{a} = [2(v, v)^2]^{-1}[2(\mathbf{a} \cdot \mathbf{L}^{v,v})\mathbf{L}^{v,v} + (\mathbf{a} \cdot \mathbf{L}^{\overline{v},v})\mathbf{L}^{v,\overline{v}} + (\mathbf{a} \cdot \mathbf{L}^{v,\overline{v}})\mathbf{L}^{\overline{v},v}] . \tag{A.23}$$

In particular, taking into account (A.20), we get

$$\mathbf{L}^{u,v} = (v, v)^{-1}[(u, v)\mathbf{L}^{v,v} + (u, \overline{v})\mathbf{L}^{\overline{v},v}] = (u, u)^{-1}[(u, v)\mathbf{L}^{u,u} + (\overline{u}, v)\mathbf{L}^{u,\overline{u}}].$$

For vector products of vectors (A.18), we have

$$[\mathbf{L}^{v,u} \times \mathbf{L}^{u,v}] = i(u, u)\mathbf{L}^{v,v} - i(v, v)\mathbf{L}^{u,u}, \quad [\mathbf{L}^{v,\overline{v}} \times \mathbf{L}^{\overline{v},v}] = 2i(v, v)\mathbf{L}^{v,v} , \tag{A.24}$$

$$[\mathbf{L}^{\overline{v},v} \times \mathbf{L}^{v,v}] = i(v, v)\mathbf{L}^{\overline{v},v}, [\mathbf{L}^{v,v} \times \mathbf{L}^{v,\overline{v}}] = i(v, v)\mathbf{L}^{v,\overline{v}} . \tag{A.25}$$

Let us introduce real vectors $\mathbf{e}_k, k = 1, 2, 3$, as

$$\mathbf{e}_1 = [2(v,v)]^{-1}(\mathbf{L}^{v,\overline{v}} + \mathbf{L}^{\overline{v},v}) ,$$
$$\mathbf{e}_2 = i[2(v,v)]^{-1}(\mathbf{L}^{v,\overline{v}} - \mathbf{L}^{\overline{v},v}) , \quad \mathbf{e}_3 = (v,v)^{-1}\mathbf{L}^{v,v} . \tag{A.26}$$

With the help of equations (A.20) and (A.25), we obtain

$$(\mathbf{e}_k, \mathbf{e}_s) = \delta_{k,s} , \quad [\mathbf{e}_1 \times \mathbf{e}_2] = \mathbf{e}_3 , \ [\mathbf{e}_2 \times \mathbf{e}_3] = \mathbf{e}_1 , \ [\mathbf{e}_3 \times \mathbf{e}_1] = \mathbf{e}_2 .$$

The unit mutually orthogonal vectors $\mathbf{e}_k$ form a complete set in the space of three-vectors. Obviously, vectors (A.26) can be expressed via this set,

$$\mathbf{L}^{v,v} = (v,v)\mathbf{e}_3, \ \mathbf{L}^{\overline{v},v} = (v,v)(\mathbf{e}_1 + i\mathbf{e}_2) , \ \mathbf{L}^{v,\overline{v}} = (v,v)(\mathbf{e}_1 - i\mathbf{e}_2) . \tag{A.27}$$

For vectors $\mathbf{e}_k$, we obtain

$$\begin{aligned}
&\mathbf{e}_1 = \mathbf{e}_\theta \cos\alpha - \mathbf{e}_\varphi \sin\alpha, \ \mathbf{e}_2 = \mathbf{e}_\theta \sin\alpha + \mathbf{e}_\varphi \cos\alpha, \ \mathbf{e}_3 = \mathbf{e}_r, \\
&\mathbf{e}_r = (\sin\theta\cos\varphi, \sin\theta\sin\varphi, \cos\theta) , \quad \mathbf{L}^{v,v} = N^2\mathbf{e}_r , \\
&\mathbf{e}_\theta = (\cos\theta\cos\varphi, \cos\theta\sin\varphi, -\sin\theta) , \quad \mathbf{L}^{\overline{v},v} = N^2(\mathbf{e}_\theta + i\mathbf{e}_\varphi)e^{i\alpha}, \\
&\mathbf{e}_\varphi = (-\sin\varphi, \cos\varphi, 0) , \quad \mathbf{L}^{v,\overline{v}} = N^2(\mathbf{e}_\theta - i\mathbf{e}_\varphi)e^{-i\alpha}.
\end{aligned} \tag{A.28}$$

If we treat θ and φ as angular coordinates in a spherical coordinate system r, θ, and φ, then vectors $\mathbf{e}_r, \mathbf{e}_\theta$, and $\mathbf{e}_\varphi$ represent a local right-handed orthonormalized basis of such a coordinate system, and vectors $\mathbf{e}_k$ can be obtained by a rotation on the angle α in turn of the vector $\mathbf{e}_r = \mathbf{e}_3$.

Any complex spinor v can be represented as

$$v = Ne^{i\alpha/2}\begin{pmatrix} e^{-i\varphi/2}\cos\theta/2 \\ e^{i\varphi/2}\sin\theta/2 \end{pmatrix} , \quad (v,v) = N^2 , \tag{A.29}$$

where N, α, φ, and θ are real parameters. Then the anticonjugate spinor $\overline{v}$ reads

$$\overline{v} = Ne^{-i\alpha/2}\begin{pmatrix} -e^{-i\varphi/2}\sin\theta/2 \\ e^{i\varphi/2}\cos\theta/2 \end{pmatrix} , \quad (\overline{v},\overline{v}) = N^2 . \tag{A.30}$$

A.1.3 Eigenvalue problem in space of complex spinors

Any eigenvalue problem in the space of complex spinors is reduced to the following set of algebraic equations:

$$Av = \delta v , \tag{A.31}$$

where A is a given complex 2×2 matrix, δ is an eigenvalue, and v is an eigenvector (a nonzero spinor). The problem has a nontrivial ($v \neq 0$) solution if

$$\det(A - \delta I) = 0 .$$

Because of the relation

$$\det(A - \delta I) = \delta^2 - \delta \mathrm{tr} A + \det A\,, \tag{A.32}$$

we have two possible values for δ,

$$2\delta = \mathrm{tr} A \pm \sqrt{\mathrm{tr}^2 A - 4\det A}\,.$$

Taking (A.5) into account, we rewrite equation (A.31) in the following form

$$(\boldsymbol{\sigma}\mathbf{a})v = \lambda v, \quad \lambda = \delta - \mathrm{tr}\, A/2\,. \tag{A.33}$$

Multiplying equation (A.33) by $(\boldsymbol{\sigma}\mathbf{a})$, we obtain

$$(\boldsymbol{\sigma}\mathbf{a})(\boldsymbol{\sigma}\mathbf{a})v = \lambda(\boldsymbol{\sigma}\mathbf{a})v = \lambda^2 v\,.$$

It follows from (A.9) that $(\boldsymbol{\sigma}\mathbf{a})(\boldsymbol{\sigma}\mathbf{a}) = \mathbf{a}^2$ and, therefore,

$$\lambda = \zeta\sqrt{\mathbf{a}^2}, \quad \zeta = \pm 1\,.$$

If $\mathbf{a}^2 = 0$, then the matrix $(\boldsymbol{\sigma}\mathbf{a})$ is singular and equation (A.33) has a nontrivial solution for $\lambda = 0$ only. Namely,

$$\left.\begin{array}{r}\mathbf{a}^2 = 0\\ \lambda = 0\end{array}\right\} \Longrightarrow v = v_0 = N\begin{pmatrix} ia_2 - a_1 \\ a_3 \end{pmatrix},$$

where N is an arbitrary complex number.

If $\mathbf{a}^2 \neq 0$, then equation (A.33) has two nontrivial solutions $v = v_\zeta, \zeta = \pm 1$. They can be written as

$$\begin{aligned}
&(\boldsymbol{\sigma}\mathbf{a})v_\zeta = \zeta\sqrt{\mathbf{a}^2}\, v_\zeta, \quad \zeta = \pm 1, \quad \mathbf{a}^2 \neq 0\,,\\
&v_1 = N_1\begin{pmatrix} a_3 + \sqrt{\mathbf{a}^2} \\ a_1 + ia_2 \end{pmatrix}, \quad v_{-1} = N_{-1}\begin{pmatrix} ia_2 - a_1 \\ a_3 + \sqrt{\mathbf{a}^2} \end{pmatrix},
\end{aligned}$$

where N_ζ are arbitrary complex numbers.

Let us consider a matrix $A^{u,v}$, associated with two arbitrary spinors u and v,

$$A^{u,v} = (\boldsymbol{\sigma}\mathbf{L}^{u,v}) = 2vu^+ - (u,v)I, \quad \det A^{u,v} = -(u,v)^2, \quad (A^{u,v})^+ = A^{v,u}\,. \tag{A.34}$$

With the help of (A.12), we find the representation for this matrix

$$A^{u,v} = \begin{pmatrix} u_1^* v_1 - u_2^* v_2 & 2u_2^* v_1 \\ 2u_1^* v_2 & u_2^* v_2 - u_1^* v_1 \end{pmatrix}$$

with the properties

$$\begin{aligned}
A^{u,v}v = (u,v)v, \quad & A^{u,v}\bar{u} = -(u,v)\bar{u},\\
u^+ A^{u,v} = (u,v)u^+, \quad & \bar{v}^+ A^{u,v} = -(u,v)\bar{v}^+\,.
\end{aligned} \tag{A.35}$$

As it follows from equation (A.16), the following relation takes place for any vector $\mathbf{p}$:

$$(\boldsymbol{\sigma}\mathbf{p})v = (v,v)^{-1}[(\mathbf{L}^{v,v}\cdot\mathbf{p})v + (\mathbf{L}^{\bar{v},v}\cdot\mathbf{p})\bar{v}]\,. \tag{A.36}$$

Taking all above into account, one can easily verify the following assertions:

(1) For any two arbitrary spinors u and v, there exists a (complex) vector $\mathbf{a}$, such that the eigenvectors in equation (A.33) are:

$$v_1 = v, \quad v_{-1} = u\,.$$

The vector $\mathbf{a}$ has the form $\mathbf{a} = N\mathbf{L}^{\overline{u},v}$, where N is an arbitrary factor. The proof is based on equations (A.34) and (A.35).

(2) If we take spinors u and v to be orthogonal,

$$(u,v) = 0 \Leftrightarrow u = \alpha\overline{v}, \quad \alpha = \text{const}\,, \tag{A.37}$$

then it follows from (A.37) and (A.14) that

$$\mathbf{a} = N_0\mathbf{L}^{v,v}, \quad N_0 = -\alpha^* N\,. \tag{A.38}$$

The vector $\mathbf{L}^{v,v}$ is real. Therefore, the spinors $v_1 = v$ and $v_{-1} = \overline{v}$ are orthogonal if the vector $\mathbf{a}$ in (A.33) is a product of a real vector and a complex number. Then the spinors v_ζ obey the completeness relation (A.17).

A.1.4 Calculations of matrix elements

Let us consider spinors $v_\zeta(\mathbf{l})$, $\zeta = \pm 1$, that are solutions of the eigenvalue problem

$$(\boldsymbol{\sigma}\mathbf{l})v_\zeta(\mathbf{l}) = \zeta v_\zeta(\mathbf{l}), \quad \mathbf{l}^2 = 1\,.$$

Here $\mathbf{l}$ is a real unit vector, we call it the spin orientation vector. According to equations (A.37) and (A.38), we have $v_{-1}(\mathbf{l}) = \overline{(v_1(\mathbf{l}))}$ and

$$\zeta\mathbf{l} = \mathbf{L}^{v_\zeta(\mathbf{l}),\,v_\zeta(\mathbf{l})} = (v_\zeta(\mathbf{l}), \boldsymbol{\sigma}\, v_\zeta(\mathbf{l})), \quad (v_\zeta(\mathbf{l}), v_{\zeta'}(\mathbf{l})) = \delta_{\zeta,\zeta'}\,.$$

Equations (A.34) and (A.20) imply

$$|(v_\zeta(\mathbf{l}_1), v_{\zeta'}(\mathbf{l}_2)|^2 = \frac{1}{2}\left[1 + \zeta\zeta'(\mathbf{l}_1,\mathbf{l}_2)\right], \quad v_\zeta(\mathbf{l})v_\zeta^+(\mathbf{l}) = \frac{1}{2}\left[I + \zeta(\boldsymbol{\sigma}\mathbf{l})\right]\,. \tag{A.39}$$

When calculating matrix elements of physical processes in relativistic quantum mechanics or QED, one usually arrives at the following expression

$$M = v_{\zeta'}^+(\mathbf{l}_2)[\alpha + (\boldsymbol{\sigma}\mathbf{a})]v_\zeta(\mathbf{l}_1)\,. \tag{A.40}$$

Then the modulus squared of such a matrix element reads

$$|M|^2 = v_\zeta^+(\mathbf{l}_1)[\alpha^+ + (\boldsymbol{\sigma}\mathbf{a}^+)]v_{\zeta'}(\mathbf{l}_2)v_{\zeta'}^+(\mathbf{l}_2)[\alpha + (\boldsymbol{\sigma}\mathbf{a})]v_\zeta(\mathbf{l}_1)\,.$$

It follows from equation (A.39) that

$$2|M|^2 = v_\zeta^+(\mathbf{l}_1)[\alpha^+ + (\boldsymbol{\sigma}\mathbf{a}^+)][1 + \zeta'(\boldsymbol{\sigma}\mathbf{l}_2)][\alpha + (\boldsymbol{\sigma}\mathbf{a})]v_\zeta(\mathbf{l}_1)\,.$$

Then with the help of equations (A.9), we find

$$\begin{aligned}2|M|^2 &= v_\zeta^+(\mathbf{l}_1)\left\{\alpha^+\alpha + \alpha(\boldsymbol{\sigma}\mathbf{a}^+) + \alpha^+(\boldsymbol{\sigma}\mathbf{a}^+) + \zeta'\alpha(\mathbf{a}^+\mathbf{l}_2) + \zeta'\alpha^+(\mathbf{a}\mathbf{l}_2)\right.\\ &\quad + i\zeta'\alpha(\boldsymbol{\sigma}[\mathbf{a}^+\times\mathbf{l}_2]) - i\zeta'\alpha^+(\boldsymbol{\sigma}[\mathbf{a}\times\mathbf{l}_2]) + \zeta'\alpha\alpha^+(\boldsymbol{\sigma}\mathbf{l}_2) + (\mathbf{a}^+\mathbf{a}) + i(\boldsymbol{\sigma}[\mathbf{a}^+\times\mathbf{a}])\\ &\quad \left.+ \zeta'(\boldsymbol{\sigma}\mathbf{a}^+)(\mathbf{a}\mathbf{l}_2) + \zeta'(\mathbf{a}^+\mathbf{l}_2)(\boldsymbol{\sigma}\mathbf{a}) - i\zeta'(\mathbf{l}_2[\mathbf{a}^+\times\mathbf{a}]) - \zeta'(\boldsymbol{\sigma}\mathbf{l}_2)(\mathbf{a}^+\mathbf{a})\right\} v_\zeta(\mathbf{l}_1)\\ &= \alpha^+\alpha[1+\zeta\zeta'(\mathbf{l}_1\mathbf{l}_2)] + \alpha\{\zeta(\mathbf{a}^+\mathbf{l}_1) + \zeta'(\mathbf{a}^+\mathbf{l}_2) - i\zeta\zeta'(\mathbf{a}^+[\mathbf{l}_1\times\mathbf{l}_2])\}\\ &\quad + \alpha^+\{\zeta(\mathbf{a}\mathbf{l}_1) + \zeta'(\mathbf{a}\mathbf{l}_2) + i\zeta\zeta'(\mathbf{a}[\mathbf{l}_1\times\mathbf{l}_2])\} + (\mathbf{a}^+\mathbf{a})[1-\zeta\zeta'(\mathbf{l}_1\mathbf{l}_2)]\\ &\quad + i\zeta(\mathbf{l}_1[\mathbf{a}^+\times\mathbf{a}]) - i\zeta'(\mathbf{l}_2[\mathbf{a}^+\times\mathbf{a}]) + \zeta\zeta'(\mathbf{a}^+\mathbf{l}_1)(\mathbf{a}\mathbf{l}_2) + \zeta\zeta'(\mathbf{a}^+\mathbf{l}_2)(\mathbf{a}\mathbf{l}_1)\,. \end{aligned} \tag{A.41}$$

In the particular case $\mathbf{l}_1 = \mathbf{l}_2 = \mathbf{l}$, we have a simplification

$$\begin{aligned}|M|^2 &= \delta_{\zeta,\zeta'}\left[\alpha^+\alpha + \zeta\alpha(\mathbf{a}^+\mathbf{l}) + \zeta\alpha^+(\mathbf{a}\mathbf{l}) + (\mathbf{a}^+\mathbf{l})(\mathbf{a}\mathbf{l})\right]\\ &\quad + \delta_{\zeta,-\zeta'}\left[(\mathbf{a}^+\mathbf{a}) - (\mathbf{a}^+\mathbf{l})(\mathbf{a}\mathbf{l}) + i\zeta(\mathbf{l}[\mathbf{a}^+\times\mathbf{a}])\right]\,.\end{aligned}$$

Let us calculate an average value over initial spinning states in equation (A.41). Thus, we obtain

$$\sum_\zeta |M|^2 = \alpha^+\alpha + (\mathbf{a}^+\mathbf{a}) + \zeta'(\mathbf{l}_2\mathbf{q}),\quad \mathbf{q} = \alpha\mathbf{a}^+ + \alpha^+\mathbf{a} - i[\mathbf{a}^+\times\mathbf{a}]\,. \tag{A.42}$$

After summing over final spinning states in equation (A.41), we obtain

$$\sum_{\zeta'} |M|^2 = \alpha^+\alpha + (\mathbf{a}^+\mathbf{a}) + \zeta(\mathbf{l}_1\mathbf{p}),\quad \mathbf{p} = \alpha\mathbf{a}^+ + \alpha^+\mathbf{a} + i[\mathbf{a}^+\times\mathbf{a}]\,. \tag{A.43}$$

Finally, summing over final spinning states and averaging over initial spinning states in equation (A.41), we obtain

$$\sum_{\zeta\zeta'} |M|^2 = 2\left[\alpha^+\alpha + (\mathbf{a}^+\mathbf{a})\right]\,. \tag{A.44}$$

A.2 Dirac gamma-matrices

A.2.1 General properties

Dirac *gamma-matrices* (γ-matrices) are defined by relations (2.56) and (2.57), which we repeat here

$$\{\gamma^\mu,\gamma^\nu\} = 2\eta^{\mu\nu}\mathbb{I},\quad \eta^{\mu\nu} = \mathrm{diag}(1,-1,-1,-1)\,, \tag{A.45}$$

$$(\gamma^\mu)^+ = \gamma^0\gamma^\mu\gamma^0\,. \tag{A.46}$$

It follows from (A.45) that

$$(\gamma^0)^2 = -(\gamma^k)^2 = \mathbb{I},\quad (\gamma^0)^{-1} = \gamma^0,\quad (\gamma^k)^{-1} = -\gamma^k,\quad \gamma^{\mu+} = (\gamma^\mu)^{-1}\,. \tag{A.47}$$

Relation (A.46) is not a consequence of (A.45), but is consistent with the latter. If equation (A.46) holds then all γ^μ are unitary matrices.

Relations (A.45) and (A.46) do not fix γ^μ uniquely, namely, if a set γ^μ represents Dirac γ-matrices, then the set $\tilde{\gamma}^\mu = U\gamma^\mu U^{-1}$, where U is an arbitrary nonsingular matrix, obeys equations (A.45). If U is unitary, $U^+ = U^{-1}$, then $\tilde{\gamma}^\mu$ obey also (A.46).

In addition, a matrix γ^5 is introduced,

$$\gamma^5 = -i\gamma^0\gamma^1\gamma^2\gamma^3, \quad \left[\gamma^5, \gamma^\mu\right]_+ = 0, \quad (\gamma^5)^2 = \mathbb{I}, \quad \left(\gamma^5\right)^+ = \gamma^5 = (\gamma^5)^{-1}, \tag{A.48}$$

and different types of Dirac γ-matrices are used:

$$\begin{aligned}
&\rho_1 = -\gamma^5, \quad \rho_2 = i\gamma^0\gamma^5 = \gamma^1\gamma^2\gamma^3, \quad \rho_3 = \gamma^0, \quad \rho_k^+ = \rho_k,\\
&[\rho_k, \rho_l]_+ = \delta_{kl}, \quad \rho_1\rho_2 = i\rho_3, \quad \rho_2\rho_3 = i\rho_1, \quad \rho_3\rho_1 = i\rho_2;\\
&\boldsymbol{\alpha} = \gamma^0\boldsymbol{\gamma} = (\alpha_1, \alpha_2, \alpha_3), \quad \boldsymbol{\alpha}^+ = \boldsymbol{\alpha}, \quad [\alpha_k, \alpha_l]_+ = 2\delta_{kl},\\
&\alpha_1\alpha_2 = i\rho_1\alpha_3, \quad \alpha_2\alpha_3 = i\rho_1\alpha_1, \quad \alpha_3\alpha_1 = i\rho_1\alpha_2 \Longrightarrow [\boldsymbol{\alpha}\times\boldsymbol{\alpha}] = 2i\rho_1\boldsymbol{\alpha},\\
&\rho_1\boldsymbol{\alpha} = \boldsymbol{\alpha}\rho_1, \quad \rho_2\boldsymbol{\alpha} = -\boldsymbol{\alpha}\rho_2, \quad \rho_3\boldsymbol{\alpha} = -\boldsymbol{\alpha}\rho_3;\\
&\boldsymbol{\Sigma} = \gamma^0\gamma^5\boldsymbol{\gamma} = \rho_1\boldsymbol{\alpha} = -i\rho_2\boldsymbol{\gamma}, \quad \boldsymbol{\Sigma}^+ = \Sigma, \quad \rho_l\Sigma = \Sigma\rho_l,\\
&\Sigma = (\Sigma_1, \Sigma_2, \Sigma_3), \ \{\Sigma_l, \Sigma_k\} = 2\delta_{kl}, \quad [\Sigma_l, \alpha_k]_+ = 2\rho_1\delta_{kl},\\
&\Sigma_1\Sigma_2 = i\Sigma_3, \quad \Sigma_2\Sigma_3 = i\Sigma_1, \quad \Sigma_3\Sigma_1 = i\Sigma_2 \Longrightarrow [\Sigma\times\Sigma] = 2i\Sigma\,.
\end{aligned}\tag{A.49}$$

The 16 matrices I, Σ_k, ρ_l, and $\rho_l\Sigma_k$ are linearly independent and form a basis in the linear space of all 4×4 matrices. They are all Hermitian, unitary, and their squares are $\mathbb{I}$. The matrices Σ_k, ρ_l, and $\rho_l\Sigma_k$ are traceless.

The spin tensor $\sigma^{\mu\nu}$ and its dual $\tilde{\sigma}^{\mu\nu}$ are defined by the relations

$$\sigma^{\mu\nu} = \frac{i}{2}\left[\gamma^\mu, \gamma^\nu\right], \ \tilde{\sigma}^{\mu\nu} = \frac{1}{2}\varepsilon^{\mu\nu\lambda\delta}\sigma_{\lambda\delta}\,, \tag{A.50}$$

where $\varepsilon^{\mu\nu\lambda\delta}$ is the Levi-Civita symbol in four dimensions normalized as $\varepsilon^{0123} = 1$. The spin tensors obey some useful relations:

$$\begin{aligned}
&\gamma^\mu\gamma^\nu = \eta^{\mu\nu} - i\sigma^{\mu\nu}, \quad \sigma^{\mu\nu}\sigma_{\mu\nu} = 12,\\
&\gamma^\lambda\sigma^{\mu\nu} - \sigma^{\mu\nu}\gamma^\lambda = 2i(\eta^{\mu\lambda}\gamma^\nu - \eta^{\nu\lambda}\gamma^\mu)\,,\\
&i\sigma^{\mu\nu}\gamma_\nu = i\gamma_\nu\sigma^{\nu\mu} = -3\gamma^\mu\,.
\end{aligned}\tag{A.51}$$

We note that these tensors have the following representation:

$$\begin{aligned}
&\sigma^{\mu\nu} = F(i\rho_1\Sigma, -\Sigma) = F(i\boldsymbol{\alpha}, -\Sigma)\,,\\
&\tilde{\sigma}^{\mu\nu} = i\gamma^5\sigma^{\mu\nu} = F(\Sigma, i\rho_1\Sigma) = F(\Sigma, i\boldsymbol{\alpha})\,,
\end{aligned}\tag{A.52}$$

where the matrix $F(\mathbf{a}, \mathbf{b})$ is given by (2.6).

We shall also need the antisymmetric tensor $W^{\mu\nu}$,

$$W^{\mu\nu} = \sigma^{\mu\alpha} F_{\alpha\beta} \eta^{\beta\nu} - \sigma^{\nu\alpha} F_{\alpha\beta} \eta^{\beta\mu} = -F(i\rho_1 [\Sigma \times \mathbf{H}] + [\Sigma \times \mathbf{E}], i\rho_1 [\Sigma \times \mathbf{E}] - [\Sigma \times \mathbf{H}]) . \tag{A.53}$$

The tensor has the properties

$$W^{*\mu\nu} = i\gamma^5 W^{\mu\nu}, \quad W^{\mu\nu} F_{\mu\nu} = W^{\mu\nu} F^*_{\mu\nu} = 0 , \tag{A.54}$$

and can be represented in the form

$$W_{\mu\nu} = \frac{i}{2} \sigma_{\mu\nu} \sigma^{\alpha\beta} F_{\alpha\beta} - iF_{\mu\nu} + \rho_1 F^*_{\mu\nu} . \tag{A.55}$$

Below we shall also exploit the angular momentum tensor $\Lambda^{\mu\nu}$,

$$\Lambda^{\mu\nu} = x^\mu P^\nu - x^\nu P^\mu + \frac{1}{2}\sigma^{\mu\nu} = -F(\digamma, \mathbf{J}) , \tag{A.56}$$

where

$$\digamma = \mathbf{r} P^0 - x^0 \mathbf{P} - \frac{i}{2}\boldsymbol{\alpha}, \boldsymbol{J} = [\mathbf{r} \times \mathbf{P}] + \frac{1}{2}\boldsymbol{\Sigma} . \tag{A.57}$$

Below, we list some representations for γ-matrices in terms of the 2×2 unit matrix I and the Pauli matrices $\boldsymbol{\sigma} = (\sigma_1, \sigma_2, \sigma_3)$, see Section A.1.1. The latter are taken in the Pauli representation (2.59).

(a) *Standard representation*:

$$\gamma = \begin{pmatrix} 0 & \boldsymbol{\sigma} \\ -\boldsymbol{\sigma} & 0 \end{pmatrix}, \quad \boldsymbol{\alpha} = \begin{pmatrix} 0 & \boldsymbol{\sigma} \\ \boldsymbol{\sigma} & 0 \end{pmatrix},$$

$$\rho_1 = \begin{pmatrix} 0 & I \\ I & 0 \end{pmatrix}, \quad \rho_2 = \begin{pmatrix} 0 & -iI \\ iI & 0 \end{pmatrix}, \quad \rho_3 = \begin{pmatrix} I & 0 \\ 0 & -I \end{pmatrix} . \tag{A.58}$$

(b) *Spinor representation*:

$$\gamma = \begin{pmatrix} 0 & -\boldsymbol{\sigma} \\ \boldsymbol{\sigma} & 0 \end{pmatrix}, \quad \boldsymbol{\alpha} = \begin{pmatrix} \boldsymbol{\sigma} & 0 \\ 0 & -\boldsymbol{\sigma} \end{pmatrix},$$

$$\rho_1 = \begin{pmatrix} I & 0 \\ 0 & -I \end{pmatrix}, \quad \rho_2 = \begin{pmatrix} 0 & iI \\ -iI & 0 \end{pmatrix}, \quad \rho_3 = \begin{pmatrix} 0 & I \\ I & 0 \end{pmatrix} . \tag{A.59}$$

There exists a unitary transformation U that relates matrices in the standard representation Γ to ones in the spinor representation $\tilde{\Gamma}$,

$$\tilde{\Gamma} = U\Gamma U^+, \quad U = U^+ = U^{-1} = (\rho_1 + \rho_3)/\sqrt{2} . \tag{A.60}$$

Here the matrix $\rho_1 + \rho_3$ is taken in the standard representation.
The matrices Σ are block-diogonal both in the standard and the spinor representations

$$\Sigma = \begin{pmatrix} \boldsymbol{\sigma} & 0 \\ 0 & \boldsymbol{\sigma} \end{pmatrix} . \tag{A.61}$$

(c) *Majorana representation*:

$$\gamma^1 = i\begin{pmatrix} \sigma_3 & 0 \\ 0 & \sigma_3 \end{pmatrix}, \quad \gamma^2 = \begin{pmatrix} 0 & -\sigma_2 \\ \sigma_2 & 0 \end{pmatrix}, \quad \gamma^3 = -i\begin{pmatrix} \sigma_1 & 0 \\ 0 & \sigma_1 \end{pmatrix},$$
$$\alpha_1 = -\begin{pmatrix} 0 & \sigma_1 \\ \sigma_1 & 0 \end{pmatrix}, \quad \alpha_2 = \begin{pmatrix} 1 & 0 \\ 0 & -1 \end{pmatrix}, \quad \alpha_3 = -\begin{pmatrix} 0 & -\sigma_3 \\ \sigma_3 & 0 \end{pmatrix},$$
$$\rho_1 = \begin{pmatrix} -\sigma_2 & 0 \\ 0 & \sigma_2 \end{pmatrix}, \quad \rho_2 = \begin{pmatrix} 0 & iI \\ -iI & 0 \end{pmatrix}, \quad \rho_3 = \gamma^0 = \begin{pmatrix} 0 & \sigma_2 \\ \sigma_2 & 0 \end{pmatrix},$$
$$\Sigma_1 = i\begin{pmatrix} 0 & \sigma_3 \\ -\sigma_3 & 0 \end{pmatrix}, \quad \Sigma_2 = \begin{pmatrix} \sigma_2 & 0 \\ 0 & \sigma_2 \end{pmatrix}, \quad \Sigma_3 = i\begin{pmatrix} 0 & -\sigma_1 \\ \sigma_1 & 0 \end{pmatrix}. \tag{A.62}$$

Here we have the asterix $*$ denoting the complex conjugation

$$\gamma^* = -\gamma, \quad \boldsymbol{\alpha}^* = \boldsymbol{\alpha}, \quad \rho_k^{\ *} = -\rho_k, \quad \Sigma^* = -\Sigma. \tag{A.63}$$

There exists a unitary transformation U that relates matrices in the standard representation Γ to ones in the Majorana representation $\bar{\Gamma}$,

$$\bar{\Gamma} = U_1 \Gamma U_1^+, \quad U_1 = U_1^+ = U_1^{-1} = (\alpha_2 + \rho_3)/\sqrt{2}. \tag{A.64}$$

Here the matrix $\rho_1 + \rho_3$ is taken in the standard representation.

A.2.2 Gamma-matrix structure of the Lorentz transformation

Here we are going to prove the following formula:

$$\exp\left(-\frac{i}{4}\omega_{\mu\nu}\sigma^{\mu\nu}\right) = M(\omega),$$
$$M(\omega) = \sqrt{\det\cosh\frac{\omega}{2}} \times \left[1 - \frac{i}{2}\left(\tanh\frac{\omega}{2}\right)_{\mu\nu}\sigma^{\mu\nu} + \frac{i}{8}\epsilon^{\alpha\beta\mu\nu}\left(\tanh\frac{\omega}{2}\right)_{\alpha\beta}\left(\tanh\frac{\omega}{2}\right)_{\mu\nu}\gamma^5\right], \tag{A.65}$$

where $\omega_{\mu\nu}$ is an arbitrary antisymmetric tensor. In fact, this formula presents a linear decomposition of a finite Lorentz transformation in the independent γ-matrix structures. We are going to check that the matrix-valued function $M(\lambda\omega)$ of a real parameter λ satisfies the differential equation

$$\frac{d}{d\lambda}M(\lambda\omega) = -\frac{i}{4}\omega_{\mu\nu}\sigma^{\mu\nu}M(\lambda\omega) \tag{A.66}$$

and the initial condition $M(0) = 1$.

Using the identity[1]

$$4\epsilon^{\alpha\beta\mu\nu}(ST)_{\alpha\beta}T_{\mu\nu} = \epsilon^{\alpha\beta\mu\nu}T_{\alpha\beta}T_{\mu\nu}\operatorname{tr} S,$$

1 We remind that, in accordance with our notation, tr$S = S^{\alpha}{}_{\alpha} = \eta^{\alpha\beta}S_{\beta\alpha}$, etc.

where S is an arbitrary tensor and T is antisymmetric, one obtains

$$\frac{d}{d\lambda}M(\lambda\omega) = \left(\det\cosh\frac{\lambda\omega}{2}\right)^{1/2}\left\{\frac{1}{4}\mathrm{tr}\left(\omega\tanh\frac{\lambda\omega}{2}\right)\right.$$
$$-\frac{i}{8}\left[\mathrm{tr}\left(\omega\tanh\frac{\lambda\omega}{2}\right)\tanh\frac{\lambda\omega}{2} + 2\omega\left(\eta - \tanh^2\frac{\lambda\omega}{2}\right)\right]_{\mu\nu}\sigma^{\mu\nu}$$
$$\left.+\frac{i}{8}\epsilon^{\alpha\beta\mu\nu}\omega_{\alpha\beta}\left(\tanh\frac{\lambda\omega}{2}\right)_{\mu\nu}\gamma^5\right\}. \tag{A.67}$$

On the other hand, with the aid of the identities

$$\sigma^{\alpha\beta}\sigma^{\mu\nu} = \eta^{\alpha\mu}\eta^{\beta\nu} - \eta^{\alpha\nu}\eta^{\beta\mu} - i\left(\eta^{\alpha\mu}\sigma^{\beta\nu} + \eta^{\beta\nu}\sigma^{\alpha\mu} - \eta^{\alpha\nu}\sigma^{\beta\mu} - \eta^{\beta\mu}\sigma^{\alpha\nu}\right)$$
$$- i\epsilon^{\alpha\beta\mu\nu}\gamma^5, \quad i\sigma^{\kappa\rho}\gamma^5 = \frac{1}{2}\epsilon^{\kappa\rho\tau\sigma}\sigma_{\tau\sigma},$$

and using the antisymmetry property $\omega_{\mu\nu} = -\omega_{\nu\mu}$ of the matrix ω, one finds

$$-\frac{i}{4}\omega_{\mu\nu}\sigma^{\mu\nu}M(\lambda\omega) = \left(\det\cosh\frac{\lambda\omega}{2}\right)^{1/2}\left[\frac{1}{4}\mathrm{tr}\left(\omega\tanh\frac{\lambda\omega}{2}\right)\right.$$
$$-\frac{i}{64}\epsilon^{\kappa\rho\mu\nu}\left(\tanh\frac{\lambda\omega}{2}\right)_{\kappa\rho}\left(\tanh\frac{\lambda\omega}{2}\right)_{\mu\nu}\epsilon_{\alpha\beta\lambda\tau}\omega^{\alpha\beta}\sigma^{\lambda\tau}$$
$$\left.-\frac{i}{4}\omega_{\alpha\beta}\sigma^{\alpha\beta} + \frac{i}{8}\epsilon^{\alpha\beta\mu\nu}\omega_{\alpha\beta}\left(\tanh\frac{\lambda\omega}{2}\right)_{\mu\nu}\gamma^5\right].$$

Then, using the identity

$$\epsilon_{\alpha_1\alpha_2\alpha_3\alpha_4}\epsilon^{\beta_1\beta_2\beta_3\beta_4} = -\sum_P(-1)^{[P]}\delta^{\beta_1}_{P(\alpha_1)}\delta^{\beta_2}_{P(\alpha_2)}\delta^{\beta_3}_{P(\alpha_3)}\delta^{\beta_4}_{P(\alpha_4)},$$

we get

$$-\frac{i}{4}\omega_{\mu\nu}\sigma^{\mu\nu}M(\lambda\omega) = \left(\det\cosh\frac{\lambda\omega}{2}\right)^{1/2}\left[\frac{1}{4}\mathrm{tr}\left(\omega\tanh\frac{\lambda\omega}{2}\right)\right.$$
$$-\frac{i}{8}\left[\mathrm{tr}\left(\omega\tanh\frac{\lambda\omega}{2}\right)\tanh\frac{\lambda\omega}{2} + 2\omega\left(\eta - \tanh^2\frac{\lambda\omega}{2}\right)\right]_{\mu\nu}\sigma^{\mu\nu}$$
$$\left.+\frac{i}{8}\epsilon^{\alpha\beta\mu\nu}\omega_{\alpha\beta}\left(\tanh\frac{\lambda\omega}{2}\right)_{\mu\nu}\gamma^5\right\}. \tag{A.68}$$

The right hand sides of (A.67) and (A.68) coincide. Therefore $M(\lambda\omega)$ obeys equation (A.66). This completes the proof of formula (A.65).

A disentanglement of more complicated operator functions of the form

$$\underbrace{\gamma^\alpha\cdots\gamma^\beta}_{k}\exp\{\omega_{\mu\nu}\gamma^\mu\gamma^\nu\}, k < D,$$

where the constant matrix ω is antisymmetric and γ^μ, are γ-matrices in D-dimensions was considered in Ref. [185]. For example in $D = 4$ and for $k = 1$, we have

$$\gamma^\alpha \exp\left\{\omega_{\mu\nu}\gamma^\mu\gamma^\nu\right\} = (\eta + \tanh 2\omega)^{\alpha\kappa}\,\gamma_\kappa + \frac{i}{2}\epsilon^{\kappa\mu\nu\lambda}\,(\eta + \tanh 2\omega)^\alpha{}_\kappa\,(\tanh 2\omega)_{\mu\nu}\,\gamma^5\gamma_\lambda\,. \tag{A.69}$$

Another equivalent representation for the left hand side of (A.69) has been derived in $D = 4$ using concrete properties of γ matrices in such dimensions [71],

$$\gamma^\alpha \exp\left\{\omega_{\mu\nu}\gamma^\mu\gamma^\nu\right\} = \left(e^{2\omega}\cos 2\omega^*\right)^\alpha_\kappa\,\gamma^\kappa + i\left(e^{2\omega}\sin 2\omega^*\right)^\alpha_\kappa\,\gamma^5\gamma^\kappa\,. \tag{A.70}$$

B Appendix 2

B.1 Laguerre functions

(1) The *Laguerre functions* are defined by the relation

$$I_{n,m}(x) = \sqrt{\frac{\Gamma(1+n)}{\Gamma(1+m)}} \frac{\exp(-x/2)}{\Gamma(1+n-m)} x^{\frac{n-m}{2}} \Phi(-m, n-m+1; x) \tag{B.1}$$

for any complex x and such n, m, for which the right hand side of (B.1) makes sense. Here $\Phi(\alpha, \gamma; x)$ is the confluent hypergeometric function, [191]. For $\gamma \neq -s$, where $s \in \mathbb{Z}_+$, the latter function can be given by a series, which is convergent for any complex x

$$\Phi(\alpha, \gamma; x) = \sum_{k=0}^{\infty} \frac{(\alpha)_k}{(\gamma)_k} \frac{x^k}{k!} = \frac{\Gamma(\gamma)}{\Gamma(\alpha)} \sum_{k=0}^{\infty} \frac{\Gamma(\alpha+k)}{\Gamma(\gamma+k)} \frac{x^k}{k!}, \tag{B.2}$$

where for any complex α the Pochhammer symbols $(\alpha)_k$ are defined as follows:

$$(\alpha)_k = \alpha(\alpha+1)\cdots(\alpha+k-1) = \frac{\Gamma(\alpha+k)}{\Gamma(\alpha)}. \tag{B.3}$$

(2) If $m \in \mathbb{Z}_+$, then the Laguerre functions are related to the *Laguerre polynomials* by the equation

$$I_{n,m}(x) = \sqrt{\frac{\Gamma(1+m)}{\Gamma(1+n)}} \exp(-x/2) x^{\frac{n-m}{2}} L_m^{n-m}(x), \quad m \in \mathbb{Z}_+. \tag{B.4}$$

We use the standard definition of the latter (see [191]),

$$\begin{aligned} L_n^{\alpha}(x) &= \frac{1}{n!} e^x x^{-\alpha} \frac{d^n}{dx^n} e^{-x} x^{n+\alpha} \\ &= \sum_{k=0}^{n} \binom{n+\alpha}{n-k} \frac{(-x)^k}{k!} = \binom{n+\alpha}{n} \Phi(-n, 1+\alpha; x). \end{aligned} \tag{B.5}$$

Here $n \in \mathbb{Z}_+$, so that the binomial coefficients defined for any complex α, n as

$$\binom{\alpha}{n} = \frac{\Gamma(1+\alpha)}{\Gamma(1+n)\Gamma(1+\alpha-n)}, \tag{B.6}$$

take the form

$$\binom{\alpha}{n} = \frac{\alpha(\alpha-1)\ldots(\alpha-n+1)}{n!}. \tag{B.7}$$

The following properties can be easily proved:

$$\begin{aligned} &\binom{\alpha+1}{k} = \frac{\alpha+1}{\alpha+1-k}\binom{\alpha}{k}, \quad \binom{\alpha}{0} = 1, \quad \binom{\alpha}{1} = \alpha, \ldots, \\ &\binom{\alpha}{k} = \binom{\alpha}{\alpha-k}, \quad \binom{\alpha}{k} = \frac{\sin(\alpha-k)\pi}{\sin\alpha\pi}\binom{k-\alpha-1}{k}. \end{aligned} \tag{B.8}$$

The latter property becomes obvious if one takes into account that

$$\Gamma(x)\Gamma(1-x) = \pi/\sin \pi x \,, \tag{B.9}$$

and

$$\sum_{k=0}^{n} \binom{\alpha}{k} \binom{\beta}{n-k} = \binom{\alpha+\beta}{n} \,. \tag{B.10}$$

(3) Using well-known properties of the confluent hypergeometric function $\Phi(\alpha, \gamma; x)$ (see [191], 9.212; 9.213; 9.216), one can easily get the following relations for the Laguerre function:

$$2\sqrt{x(n+1)}I_{n+1,m}(x) = (n-m+x)I_{n,m}(x) - 2xI'_{n,m}(x) \,, \tag{B.11}$$
$$2\sqrt{x(m+1)}I_{n,m+1}(x) = (n-m-x)I_{n,m}(x) + 2xI'_{n,m}(x) \,, \tag{B.12}$$
$$2\sqrt{xn}I_{n-1,m}(x) = (n-m+x)I_{n,m}(x) + 2xI'_{n,m}(x) \,, \tag{B.13}$$
$$2\sqrt{xm}I_{n,m-1}(x) = (n-m-x)I_{n,m}(x) - 2xI'_{n,m}(x) \,, \tag{B.14}$$
$$2\sqrt{nm}I_{n-1,m-1}(x) = (n+m-x)I_{n,m}(x) - 2xI'_{n,m}(x) \,, \tag{B.15}$$
$$2\sqrt{(n+1)(m+1)}I_{n+1,m+1}(x) = (n+m+2-x)I_{n,m}(x) + 2xI'_{n,m}(x) \,, \tag{B.16}$$

and the differential equation

$$4x^2 I''_{n,m}(x) + 4xI'_{n,m}(x) - [x^2 - 2x(1+n+m) + (n-m)^2]I_{n,m}(x) = 0 \,. \tag{B.17}$$

Its general solution I has the form

$$I = AI_{n,m}(x) + BI_{m,n}(x) \,, \tag{B.18}$$

if $I_{n,m}(x)$ and $I_{m,n}(x)$ are linearly independent. They are dependent only when the condition (B.23) below holds. The formulas (B.11)–(B.16) and the equation (B.17) make sense for any complex n, m, x. One has to be careful when applying equations (B.13)–(B.15) at $n, m = 0$. A straightforward calculation which uses (B.1) and (B.2) gives

$$\lim_{n\to 0} \sqrt{n}I_{n-1,m}(x) = -\frac{\sin m\pi}{\pi} \sqrt{\Gamma(1+m)} x^{-\frac{1+m}{2}} \exp(x/2) \,, \tag{B.19}$$
$$\lim_{m\to 0} \sqrt{m}I_{n,m-1}(x) = 0 \,. \tag{B.20}$$

A combination of the equations (B.11)–(B.14) results in the following relations:

$$2\sqrt{x}I'_{n,m}(x) = \begin{cases} \sqrt{n}I_{n-1,m}(x) - \sqrt{n+1}I_{n+1,m}(x) = \sqrt{m+1}I_{n,m+1}(x) - \sqrt{m}I_{n,m-1}(x) \\ \sqrt{(n+1)(m+1)}I_{n+1,m+1}(x) - \sqrt{nm}I_{n-1,m-1}(x) - I_{n,m}(x) \end{cases} , \tag{B.21}$$

$$\sqrt{x(n+1)}I_{n+1,m}(x) - (n-m+x)\, I_{n,m}(x) + \sqrt{xn}I_{n-1,m}(x) = 0 \,,$$
$$\sqrt{x\,(m+1)}I_{n,m+1}(x) - (n-m-x)\, I_{n,m}(x) + \sqrt{xm}I_{n,m-1}(x) = 0 \,.$$

(4) By using properties of the confluent hypergeometric function (see [191], 9.214), one can get the representation

$$I_{n,m}(x) = \sqrt{\frac{\Gamma(1+n)}{\Gamma(1+m)}\frac{\exp(x/2)}{\Gamma(1+n-m)}}x^{\frac{n-m}{2}}\Phi(1+n, 1+n-m; -x)\,, \tag{B.22}$$

and the relation

$$I_{n,m}(x) = (-1)^{n-m}I_{m,n}(x), \quad n-m \text{ integer}\,. \tag{B.23}$$

(5) Using (B.2) one can derive the following series from (B.1) and (B.22)

$$\begin{aligned} I_{n,m}(x) &= -\frac{\sin m\pi}{\pi}\sqrt{\Gamma(1+n)\Gamma(1+m)}e^{-\frac{x}{2}}x^{\frac{n-m}{2}}\sum_{k=0}^{\infty}\frac{\Gamma(k-m)}{\Gamma(k-m+n+1)}\frac{x^k}{k!} \\ &= \sqrt{\Gamma(1+n)\Gamma(1+m)}e^{-\frac{x}{2}}x^{\frac{n-m}{2}}\sum_{k=0}^{\infty}\frac{(-x)^k}{\Gamma(m-k+1)\Gamma(n-m+k+1)k!}\,, \end{aligned} \tag{B.24}$$

$$I_{n,m}(x) = \frac{\exp(x/2)x^{\frac{n-m}{2}}}{\sqrt{\Gamma(1+n)\Gamma(1+m)}}\sum_{k=0}^{\infty}\frac{\Gamma(n+k+1)}{\Gamma(n-m+k+1)}\frac{(-x)^k}{k!}\,. \tag{B.25}$$

(6) The following asymptotic formula takes place:

$$\Phi(a,c;x) \approx \frac{\Gamma(c)}{\Gamma(a)}e^x x^{a-c}, \quad \operatorname{Re} x \to \infty\,. \tag{B.26}$$

This results in the following asymptotic behavior of $I_{n,m}(x)$ when m is not integer:

$$I_{n,m}(x) = -\frac{\sin m\pi}{\pi}\sqrt{\Gamma(1+n)\Gamma(1+m)}x^{-\frac{n+m+2}{2}}\exp(x/2), \quad \operatorname{Re} x \to \infty\,, \tag{B.27}$$

and

$$I_{n,m}(x) = (-1)^m\frac{x^{\frac{n+m}{2}}exp(-x/2)}{\sqrt{\Gamma(1+n)\Gamma(1+m)}}, \quad \operatorname{Re} x \to \infty\,, \tag{B.28}$$

when m is integer.

(7) Below we analyze asymptotics of $I_{n,m}(x)$ for large values of parameters. It is known (see equation 8.328.2 in Ref. [191]) that

$$\lim_{|z|\to\infty}\frac{\Gamma(z+a)}{\Gamma(z)}e^{-a\ln z} = 1\,, \tag{B.29}$$

which yields the relations

$$\lim_{|z|\to\infty}\frac{\Gamma(z+a)}{\Gamma(z+b)}e^{(b-a)\ln z} = \lim_{|z|\to\infty}\frac{\Gamma(z+a)}{\Gamma(z)}e^{-a\ln z}\frac{\Gamma(z)}{\Gamma(z+b)}e^{b\ln z} = 1\,, \tag{B.30}$$

or an equivalent relation

$$\frac{\Gamma(z+a)}{\Gamma(z+b)} \approx e^{(a-b)\ln z}, \quad |z| \to \infty\,. \tag{B.31}$$

Taking into account (B.31) one can see that

$$\frac{\Gamma(n+k+1)}{\Gamma(n-m+k+1)} \approx n^m, \quad n \to \infty\,. \tag{B.32}$$

Thus, for m, x fixed and in the limit $n \to \infty$, we have

$$I_{n,m}(x) \approx \frac{\exp(x/2) n^m x^{\frac{n-m}{2}}}{\sqrt{\Gamma(1+n)\Gamma(1+m)}} \sum_{k=0}^{\infty} \frac{(-x)^k}{k!} = \frac{n^m x^{\frac{n-m}{2}} \exp(-x/2)}{\sqrt{\Gamma(1+n)\Gamma(1+m)}}\,. \tag{B.33}$$

If n, x are fixed and $n-m$ is not integer we get from (B.8) in the limit $m \to \infty$

$$\Gamma(n-m+k+1) = \frac{(-1)^k \pi}{\sin \pi (n-m)} \left[\Gamma(m-n-k)\right]^{-1}\,.$$

In the same case equation (B.24) can be written in the form

$$I_{n,m}(x) = \frac{\sin \pi(n-m)}{\pi} \sqrt{\Gamma(1+n)\Gamma(1+m)} x^{\frac{n-m}{2}} \exp(-x/2) \sum_{k=0}^{\infty} \frac{\Gamma(m-n-k)}{\Gamma(m-k+1)} \frac{x^k}{k!}\,.$$

Taking into account (B.31), one can write

$$\frac{\Gamma(m-n-k)}{\Gamma(m-k+1)} \approx m^{-n-1}, \quad m \to \infty\,.$$

Thus, in the case under consideration we have

$$I_{n,m}(x) \approx \frac{\sin \pi(n-m)}{\pi m^{n+1}} \sqrt{\Gamma(n+1)\Gamma(m+1)} e^{\frac{x}{2}} x^{\frac{n-m}{2}}\,. \tag{B.34}$$

If $n-m$ is integer then the problem is reduced to (B.33), as one can see from (2.143).

Consider now asymptotic behavior of $I_{\alpha+n,n}(x)$ as $n \to \infty$ and α, x are fixed. It follows from (B.24) that

$$I_{\alpha+n,n}(x) = x^{\frac{\alpha}{2}} e^{-\frac{x}{2}} \sum_{k=0}^{\infty} \sqrt{\frac{\Gamma(n+\alpha+1)\Gamma(n+1)}{\Gamma(n-k+1)\Gamma(n-k+1)}} \frac{(-x)^k}{\Gamma(k+\alpha+1)k!}\,. \tag{B.35}$$

Then, taking into account (B.31), one finds

$$I_{\alpha+n,n}(x) \approx e^{-\frac{x}{2}} (nx)^{\frac{\alpha}{2}} \sum_{k=0}^{\infty} \frac{(-nx)^k}{\Gamma(k+\alpha+1)k!}, \quad n \to \infty\,.$$

The sum in the right hand side of the latter equation can be expressed via the Bessel function ([191], 8.445),

$$I_{\alpha+n,n}(x) \approx e^{-\frac{x}{2}} J_\alpha(2\sqrt{nx})\,. \tag{B.36}$$

If one uses an asymptotic expression for the Bessel function (see equation 8.451 in [191]) then (B.36) can be simplified,

$$I_{\alpha+n,n}(x) \approx \frac{e^{-\frac{x}{2}}}{(\pi^2 nx)^{1/4}} \cos\left(2\sqrt{nx} - \frac{\pi}{2}\alpha - \frac{\pi}{4}\right), \quad n \to \infty\,. \tag{B.37}$$

Finally we prove the asymptotic formula

$$\lim_{p\to\infty} I_{p+\alpha,p+\beta}\left(x^2/4p\right) = J_{\alpha-\beta}(x)\,. \tag{B.38}$$

According to equation (B.25) one can write

$$I_{p+\alpha,p+\beta}\left(x^2/4p\right) = (x/2)^{\alpha-\beta}\exp(x^2/8p)\sum_{k=0}^{\infty}\frac{(-1)^k Q_k}{\Gamma(k+\alpha-\beta+1)k!}\left(\frac{x}{2}\right)^{2k},$$

where

$$Q_k = \frac{\Gamma(p+\alpha+1+k)p^{\frac{\beta-\alpha}{2}-k}}{\sqrt{\Gamma(p+\alpha+1)\Gamma(p+\beta+1)}}\,.$$

It follows from (B.31) that $\lim_{p\to\infty} Q_k = 1$. Then taking into account the series decomposition of the Bessel function, we arrive just at equation (B.38).

(8) Taking into account (B.27) and (B.28) one can see that only the functions $I_{\alpha+n,n}(x)$ with $n \in \mathbb{Z}_+$ and $\alpha > -1$ are square-integrable on the interval $(0,\infty)$. Such functions obey the orthogonality relations

$$\int_0^\infty I_{\alpha+n,n}(x)I_{\alpha+m,m}(x)dx = \delta_{m,n}\,, \tag{B.39}$$

which follow from the corresponding properties of the Laguerre polynomials (see equation 7.414.3 in Ref. [191]). In this case we have a relation, which follows from equation (B.4)

$$I_{\alpha+n,n}(x) = \sqrt{\frac{n!}{\Gamma(n+\alpha+1)}}e^{-\frac{x}{2}}x^{\frac{\alpha}{2}}L_n^\alpha(x)\,. \tag{B.40}$$

(9) The Laguerre polynomials and functions appear in numerous physical calculations. Below we consider examples of important sums with these functions.

One can prove the following relation:

$$F = \sum_{k=0}^{n}\binom{\alpha+\beta+n-k-1}{n-k}L_k^\beta(x) = (-1)^n\sum_{k=0}^{n}(-1)^k\binom{\beta-\alpha}{n-k}L_k^\beta(x) = L_n^\alpha(x)\,. \tag{B.41}$$

To this end we note that the first part of (B.41) is a simple consequence of equation (B.8) for k integer. Then using (B.5) one gets

$$F = \sum_{k=0}^{n}(-1)^{n+k}\binom{\beta-\alpha}{n-k}L_k^\beta(x) = \sum_{k=0}^{n}\sum_{m=0}^{k}(-1)^{m+n+k}\binom{\beta-\alpha}{n-k}\binom{k+\beta}{k-m}\frac{x^m}{m!}\,. \tag{B.42}$$

The last sum can be transformed if one takes into account the following property:

$$\sigma = \sum_{k=0}^{n}\sum_{m=0}^{k}f(k,m) = \sum_{m=0}^{n}\sum_{k=m}^{n}f(k,m) \Rightarrow_{k\to k+m}\sum_{m=0}^{n}\sum_{k=0}^{n-m}f(k+m,m)\,,$$

which is valid for any function $f(k, m)$. Then an identity holds true

$$\sum_{k=0}^{n}\sum_{m=0}^{k} f(k,m) = \sum_{m=0}^{n}\sum_{k=0}^{n-m} f(k+m,m)\,. \tag{B.43}$$

In particular, for $n = \infty$ one gets

$$\sum_{k=0}^{\infty}\sum_{m=0}^{k} f(k,m) = \sum_{m=0}^{\infty}\sum_{k=0}^{\infty} f(k+m,m)\,. \tag{B.44}$$

Applying the identity (B.43) to (B.42), we can write

$$F = \sum_{m=0}^{n}\sum_{k=0}^{n-m} (-1)^{n+k}\binom{\beta-\alpha}{n-m-k}\binom{m+\beta+k}{k}\frac{x^m}{m!}\,. \tag{B.45}$$

Equation (B.8) implies

$$\binom{m+\beta+k}{k} = (-1)^k\binom{-m-\beta-1}{k}\,,$$

which allows one to find

$$F = (-1)^n\sum_{m=0}^{n}\frac{x^m}{m!}\sum_{k=0}^{n-m}\binom{\beta-\alpha}{n-m-k}\binom{-m-\beta-1}{k}\,.$$

The sum over k can be calculated by the help of (B.10)

$$F = (-1)^n\sum_{m=0}^{n}\frac{x^m}{m!}\binom{-m-\alpha-1}{n-m}\,.$$

Using equation (B.8) once again, and making the replacement

$$\binom{-m-\alpha-1}{n-m} = (-1)^{n+m}\binom{n+\alpha}{n-m}\,,$$

we arrive finally at the expression

$$F = \sum_{m=0}^{n}\binom{n+\alpha}{n-m}\frac{(-x)^m}{m!}\,,$$

which proves (B.41) after one takes into account (B.5).

Replacing Laguerre polynomials in (B.41) by the Laguerre functions, and using (B.40), we find, for $n \in \mathbb{Z}_+$,

$$I_{\alpha+n,n}(x) = x^{\frac{\alpha-\beta}{2}}\sum_{k=0}^{n}\binom{\alpha-\beta+n-k-1}{n-k}\sqrt{\frac{\Gamma(1+n)\Gamma(1+\beta+k)}{\Gamma(1+k)\Gamma(1+\alpha+n)}}\,I_{\beta+k,k}(x)\,. \tag{B.46}$$

(10) When studying the Laguerre functions one often encounters the following integrals:

$$\int_0^\infty e^{-ay} y^{m+\frac{\alpha}{2}} I_\alpha\left(2\sqrt{by}\right) dy = \frac{\Gamma(m+\alpha+1)}{\Gamma(\alpha+1)} \frac{e^{\frac{b}{a}} b^{\frac{\alpha}{2}}}{a^{m+\alpha+1}} \Phi\left(-m, 1+\alpha; -\frac{b}{a}\right)$$

$$= \frac{\Gamma(m+\alpha+1)}{\Gamma(\alpha+1)} \frac{b^{\frac{\alpha}{2}}}{a^{m+\alpha+1}} \Phi\left(m+\alpha+1, 1+\alpha; \frac{b}{a}\right), \quad \operatorname{Re} a > 0, \quad \operatorname{Re}(m+\alpha+1) > 0\,, \tag{B.47}$$

and

$$\int_0^\infty e^{-ay} y^{m+\frac{\alpha}{2}} J_\alpha\left(2\sqrt{by}\right) dy = \frac{\Gamma(m+\alpha+1)}{\Gamma(\alpha+1)} \frac{b^{\frac{\alpha}{2}}}{a^{m+\alpha+1}} \Phi\left(m+\alpha+1, 1+\alpha; -\frac{b}{a}\right)$$

$$= \frac{\Gamma(m+\alpha+1)}{\Gamma(\alpha+1)} \frac{e^{-\frac{b}{a}} b^{\frac{\alpha}{2}}}{a^{m+\alpha+1}} \Phi\left(-m, 1+\alpha; \frac{b}{a}\right), \quad \operatorname{Re} a > 0, \quad \operatorname{Re}(m+\alpha+1) > 0\,. \tag{B.48}$$

Here $I_\alpha(x)$ is the Bessel function of an imaginary argument (see equation 966; 8.406 in Ref. [191]) and the Bessel functions $J_\alpha(x)$ are defined in the standard way (see equation 965; 8.402 in Ref. [191]).

To calculate the above integrals, one can use series decomposition of the Bessel functions and the series (B.2). Thus, one can find the first equality in (B.48) and the second equality in (B.47). After that, the first equality in (B.47) and the second equality in (B.48) follow from properties of the confluent hypergeometric function.

In particular, by comparing the second equality in (B.48) with (B.39), one can find the following integral representation:

$$\sqrt{\Gamma(1+n)\Gamma(1+m)} I_{n,m}(x) = e^{\frac{x}{2}} \int_0^\infty e^{-y} y^{\frac{n+m}{2}} J_{n-m}\left(2\sqrt{xy}\right) dy, \quad \operatorname{Re}(n+1) > 0\,. \tag{B.49}$$

The relation (B.23) follows then from (B.49) for $n-m$ integer.

(11) Let us prove the equality

$$\int_0^\infty e^{-y} L_n^\alpha(y) y^{\frac{\nu}{2}} J_\nu\left(2\sqrt{xy}\right) dy = (-1)^n e^{-x} x^{\frac{\nu}{2}} L_n^{\nu-\alpha-n}(x)\,. \tag{B.50}$$

Replacing $L_n^\alpha(y)$ by the second expression in (B.5) and using (B.48) we find

$$\int_0^\infty e^{-y} L_n^\alpha(y) y^{\frac{\nu}{2}} J_\nu\left(2\sqrt{xy}\right) dy = \sum_{k=o}^{n} (-1)^k \binom{n+\alpha}{n-k} \frac{1}{k!} \int_0^\infty e^{-y} y^{k+\frac{\nu}{2}} J_\nu\left(2\sqrt{xy}\right) dy$$
$$= \sum_{k=0}^{n} (-1)^k \binom{n+\alpha}{n-k} \frac{\Gamma(k+\nu+1)}{k!\Gamma(\nu+1)} e^{-x} x^{\frac{\nu}{2}} \Phi(-k, 1+\nu; x)$$
$$= e^{-x} x^{\frac{\nu}{2}} \sum_{k=0}^{n} (-1)^k \binom{n+\alpha}{n-k} L_k^\nu(x) .$$

Expressing the Laguerre polynomials via the Laguerre functions we find

$$\sqrt{\Gamma(n+\alpha+1)} \int_0^\infty e^{-\frac{y}{2}} y^{\frac{\beta+n}{2}} I_{\alpha+n,n}(y) J_{\alpha+\beta+n}\left(\sqrt{4xy}\right) dy$$
$$= (-1)^n \sqrt{\Gamma(n+\alpha+1)} e^{-\frac{x}{2}} x^{\frac{\alpha+n}{2}} I_{\beta+n,n}(x) . \tag{B.51}$$

(12) Substituting (B.4) into (B.23) for $n, m \in \mathbb{N}$, we find

$$(-1)^m \Gamma(1+m) x^n L_m^{n-m}(x) = (-1)^n \Gamma(1+n) x^m L_n^{m-n}(x) . \tag{B.52}$$

(13) The following property of the Laguerre polynomials take place:

$$\frac{d^k}{dx^k} x^\nu L_n^\nu(x) = \frac{\Gamma(n+\nu+1)}{\Gamma(n+\nu+1-k)} x^{\nu-k} L_n^{\nu-k}(x) . \tag{B.53}$$

To prove it one can use the relations

$$\frac{d}{dx} x^\nu L_n^\nu(x) = \nu x^{\nu-1} L_n^\nu(x) + x^\nu \frac{d}{dx} L_n^\nu(x)$$
$$= x^{\nu-1}\left[\nu L_n^\nu(x) + x \frac{d}{dx} L_n^\nu(x)\right] = x^{\nu-1}\left[\nu L_n^\nu(x) + n L_n^\nu(x) - (n+\nu) L_{n-1}^\nu(x)\right]$$
$$= (n+\nu) x^{\nu-1}\left[L_n^\nu(x) - L_{n-1}^\nu(x)\right] = (n+\nu) x^{\nu-1} L_n^{\nu-1}(x) ,$$

which is a consequence of equations (1051) and (8.971.3,5) from [191].

(14) Starting with the well-known properties of the Bessel functions (see equation 982; 8.472.3,4 in [191])

$$\left(\frac{1}{z}\frac{d}{dz}\right)^m z^\nu J_\nu(z) = z^{\nu-m} J_{\nu-m}(z) ,$$
$$\left(\frac{1}{z}\frac{d}{dz}\right)^m z^{-\nu} J_\nu(z) = (-1)^m z^{-\nu-m} J_{\nu+m}(z) , \tag{B.54}$$

and doing a change of the variable $z = 2\sqrt{qx}$, we obtain

$$\left(\frac{d}{dx}\right)^m x^{\frac{\nu}{2}} J_\nu\left(2\sqrt{qx}\right) = q^{\frac{m}{2}} x^{\frac{\nu-m}{2}} J_{\nu-m}\left(2\sqrt{qx}\right) , \tag{B.55}$$

$$\left(\frac{d}{dx}\right)^m x^{-\frac{\nu}{2}} J_\nu\left(2\sqrt{qx}\right) = (-1)^m q^{\frac{m}{2}} x^{-\frac{\nu+m}{2}} J_{\nu+m}\left(2\sqrt{qx}\right) . \tag{B.56}$$

(15) Let us prove the equality

$$\begin{aligned} J &= \int_0^\infty e^{-x} L_m^\alpha(x) L_n^\beta(x) x^{\frac{\alpha+\beta}{2}} J_{\alpha+\beta}\left(2\sqrt{qx}\right) dx \\ &= (-1)^{n+m} e^{-q} q^{\frac{\alpha+\beta}{2}} L_m^{\beta+n-m}(q) L_n^{\alpha+m-n}(q) , \quad \operatorname{Re}(\alpha+\beta+1) > 0 . \end{aligned} \tag{B.57}$$

Using the first equality in (B.5) we transform it to the form

$$J = \frac{1}{m!} \int_0^\infty dx x^\beta L_n^\beta(x) x^{-\frac{\alpha+\beta}{2}} J_{\alpha+\beta}\left(2\sqrt{qx}\right) \frac{d^m}{dx^m} e^{-x} x^{m+\alpha} .$$

Integrating the above integral by parts and using the Leibnitz formula we get

$$\begin{aligned} J &= \frac{(-1)^m}{m!} \int_0^\infty dx\, e^{-x} x^{m+\alpha} \frac{d^m}{dx^m} \left[x^\beta L_n^\beta(x) x^{-\frac{\alpha+\beta}{2}} J_{\alpha+\beta}\left(2\sqrt{qx}\right) \right] \\ &= \frac{(-1)^m}{m!} \sum_{k=0}^m \binom{m}{k} \int_0^\infty e^{-x} x^{m+\alpha} \frac{d^k}{dx^k} \left[x^{-\frac{\alpha+\beta}{2}} J_{\alpha+\beta}\left(2\sqrt{qx}\right) \right] \frac{d^{m-k}}{dx^{m-k}} x^\beta L_n^\beta(x)\, dx . \end{aligned}$$

By means of (B.53) and (B.55) we can reduce J to the form

$$J = (-1)^m \sum_{k=0}^m (-1)^k \binom{n+\beta}{m-k} \frac{q^{\frac{k}{2}}}{k!} \int_0^\infty e^{-x} L_n^{\beta-m+k}(x) x^{\frac{\alpha+\beta+k}{2}} J_{\alpha+\beta+k}\left(2\sqrt{qx}\right)\, dx .$$

Finally, using (B.50), we can write

$$J = (-1)^{n+m} L_n^{\alpha+m-n}(q) e^{-q} q^{\frac{\alpha+\beta}{2}} \sum_{k=0}^m (-1)^k \binom{n+\beta}{m-k} \frac{q^{\frac{k}{2}}}{k!} .$$

This proves the representation (B.57) if one takes into account (B.5). By substituting the Laguerre functions for the Laguerre polynomials in (B.57), we arrive at the relation

$$\int_0^\infty I_{\alpha+m,m}(x) I_{\beta+n,n}(x) J_{\alpha+\beta}\left(2\sqrt{qx}\right) dx = (-1)^{n+m} I_{\beta+n,m}(q) I_{\alpha+m,n}(q) , \tag{B.58}$$

where $\operatorname{Re}(\alpha+\beta+1) > 0$, and $n, m \in \mathbb{Z}_+$.

(16) Let us prove the equality

$$j = \int_0^\infty e^{-x} x^{\frac{\alpha+\beta}{2}} L_m^\alpha(x) L_n^\beta(x) J_{\alpha-\beta}\left(2\sqrt{qx}\right) dx = (-1)^{n+m} \frac{\Gamma(m+\alpha+1)}{\Gamma(\alpha-\beta+m-n+1)n!}$$
$$\times e^{-q} q^{\frac{\alpha-\beta}{2}} L_m^{n-m}(q) \Phi(-n-\beta, \alpha-\beta+m-n+1; q), \quad \mathrm{Re}(\alpha+1) > 0\,. \tag{B.59}$$

As in the previous case we can use the first relation in (B.5) for $L_m^\alpha(x)$. Then after integrating by parts in the right hand side of (B.59), after using the Leibnitz formula, equation (B.56), and equations (1051; 8.971.2) from [191],

$$\left(\frac{d}{dx}\right)^s L_n^\beta(x) = (-1)^s L_{n-s}^{\beta+s}(x)\,, \tag{B.60}$$

we get

$$j = \frac{1}{m!} \sum_{k=0}^m \binom{m}{k} q^{\frac{k}{2}} \int_0^\infty dx e^{-x} x^{m+\alpha-\frac{\alpha-\beta+k}{2}} J_{\alpha-\beta+k}\left(2\sqrt{qx}\right) L_{n-m+k}^{\beta+m-k}(x)\,.$$

Using again the first relation (B.5) for $L_m^\alpha(x)$, and integrating by parts by means of (B.56) we transform the above expression to the form

$$j = (-1)^{n+m} q^{\frac{n-m}{2}} \sum_{k=0}^m \frac{(-q)^k}{(n-m+k)!(m-k)!k!} \int_0^\infty e^{-x} x^{n+\beta-\frac{\alpha-\beta+m-n}{2}} J_{\alpha-\beta+m-n}\left(2\sqrt{qx}\right)\,.$$

Taking into account (B.48) and the second relation in (B.5) we arrive just at (B.59).

Using (B.1) and (B.40) in (B.59), we obtain

$$\int_0^\infty I_{\alpha+m,m}(x) I_{\beta+n,n}(x) J_{\alpha-\beta}\left(2\sqrt{qx}\right) dx = (-1)^{n+m} I_{n,m}(q) I_{\alpha+m,\beta+n}(q)\,, \tag{B.61}$$

where $\mathrm{Re}(\alpha+\beta+1) > 0$, and n, m are nonnegative integers.

(17) Let us prove the relation

$$\sum_{n=0}^\infty \frac{z^n L_n^\alpha(x)}{\Gamma(1+\alpha+n)} = e^z (xz)^{-\frac{\alpha}{2}} J_\alpha(2\sqrt{xz})\,. \tag{B.62}$$

Using (B.5) for the Laguerre polynomials, we find

$$S = \sum_{n=0}^\infty \frac{z^n L_n^\alpha(x)}{\Gamma(1+\alpha+n)} = \sum_{n=0}^\infty \sum_{m=0}^n f(n,m), \quad f(n,m) = \frac{z^n(-x)^m}{\Gamma(1+n-m)\Gamma(1+\alpha+m)\Gamma(1+m)}\,.$$

According to (B.44)

$$S = \sum_{m=0}^\infty \sum_{n=0}^\infty \frac{z^{m+n}(-x)^m}{\Gamma(1+n)\Gamma(1+m)\Gamma(1+\alpha+m)} = \sum_{m=0}^\infty \frac{(-xz)^m}{\Gamma(1+\alpha+m)m!} \sum_{n=0}^\infty \frac{z^n}{n!}\,.$$

If one uses the power series expansion of the Bessel function, then (B.62) follows immediately for any complex x, z, α.

The following relation can be proved if one uses (B.40),

$$\sum_{n=0}^{\infty} \frac{I_{\alpha+n,n}(x) z^n}{\sqrt{\Gamma(1+n)\Gamma(1+\alpha+n)}} = e^{z-\frac{x}{2}} z^{-\frac{\alpha}{2}} J_\alpha\left(2\sqrt{xz}\right) . \tag{B.63}$$

In a similar manner the relations can be proved

$$\sum_{n=0}^{\infty} \frac{(-z)^n L_n^\alpha(x)}{\Gamma(1+\alpha+n)} = e^{-z}(xz)^{-\frac{\alpha}{2}} I_\alpha\left(2\sqrt{xz}\right),$$

$$\sum_{n=0}^{\infty} \frac{I_{\alpha+n,n}(x)(-z)^n}{\sqrt{\Gamma(1+n)\Gamma(1+\alpha+n)}} = e^{-z-\frac{x}{2}} z^{-\frac{\alpha}{2}} I_\alpha(2\sqrt{xz}) . \tag{B.64}$$

(18) Let us prove the following relation:

$$\begin{aligned} Q = \sum_{n=0}^{\infty} \frac{\Gamma(1+\beta+n)}{\Gamma(1+\alpha+n)} L_n^\alpha(x) z^n &= \frac{\Gamma(1+\beta)}{\Gamma(1+\alpha)}(1-z)^{-1-\beta} \Phi\left(1+\beta, 1+\alpha; \frac{xz}{z-1}\right) \\ &= \frac{\Gamma(1+\beta)}{\Gamma(1+\alpha)}(1-z)^{-1-\beta} e^{-\frac{xz}{1-z}} \Phi\left(\alpha-\beta, 1+\alpha; \frac{xz}{1-z}\right), \quad |z|<1, \quad (-\beta) \neq \text{integer} . \end{aligned} \tag{B.65}$$

Using the second equation in (B.5) we find

$$Q = \sum_{n=0}^{\infty} \sum_{m=0}^{n} f(n,m), \quad f(n,m) = \frac{\Gamma(1+\beta+n) z^n (-x)^m}{\Gamma(1+n-m)\Gamma(1+\alpha+m) m!} .$$

Then the equation (B.44) allows one to transform the above expression to the form

$$Q = \sum_{m=0}^{\infty} \frac{(-xz)^m}{\Gamma(1+\alpha+m) m!} \sum_{n=0}^{\infty} \frac{\Gamma(1+\beta+m+n) z^n}{n!} .$$

We use the well-known sum

$$(1-z)^{-p} = \frac{1}{\Gamma(p)} \sum_{n=0}^{\infty} \frac{\Gamma(p+n)}{n!} z^n \tag{B.66}$$

to continue the reduction of Q,

$$Q = (1-z)^{-1-\beta} \sum_{m=0}^{\infty} \frac{\Gamma(1+\beta+m)}{\Gamma(1+\alpha+m) m!} \left(\frac{xz}{z-1}\right)^m .$$

Now one has only to recall equation (B.2) to complete the proof of (B.65).

By means of (B.40) and (B.1) one gets

$$\sum_{n=0}^{\infty} \frac{\Gamma(1+\beta+n)}{\sqrt{\Gamma(1+n)\Gamma(1+\alpha+n)}} I_{\alpha+n,n}(x) z^n$$
$$= \sqrt{\Gamma(1+\beta)\Gamma(1+\beta-\alpha)}\, z^{-\frac{\alpha}{2}} (1-z)^{\frac{\alpha}{2}-\beta-1} e^{\frac{x}{2(z-1)}} I_{\beta,\beta-\alpha}\left(\frac{xz}{1-z}\right), \tag{B.67}$$

where $|z| < 1$, and $(-\beta)$ is not an integer.

If $\alpha = \beta$ then it follows from (B.65) and (B.67) that

$$\sum_{n=0}^{\infty} L_n^{\alpha}(x) z^n = (1-z)^{-1-\alpha} \exp\left(\frac{xz}{z-1}\right), \quad |z| < 1, \tag{B.68}$$

$$\sum_{n=0}^{\infty} \sqrt{\frac{\Gamma(1+\alpha+n)}{\Gamma(1+n)}} I_{\alpha+n,n}(x) z^n = x^{\frac{\alpha}{2}} (1-z)^{-1-\alpha} \exp\left[\frac{(z+1)x}{2(z-1)}\right], \ |z| < 1. \tag{B.69}$$

Using the Cauchy theorem one can find from (B.68), (B.69) that at $n \in \mathbb{Z}_+$ the following relations take place:

$$L_n^{\alpha}(x) = \frac{1}{2\pi i} \oint_{|W|=\rho} \frac{\exp\left(\frac{xW}{W-1}\right) dW}{(1-W)^{1+\alpha} W^{n+1}}, \quad 0 < \rho < 1,$$
$$I_{\alpha+n,n}(x) = \sqrt{\frac{\Gamma(1+n)}{\Gamma(1+\alpha+n)}} \frac{x^{\frac{\alpha}{2}}}{2\pi i} \oint_{|W|=\rho} \frac{\exp\left(\frac{(W+1)x}{(W-1)2}\right) dW}{(1-W)^{1+\alpha} W^{n+1}}, \ 0 < \rho < 1. \tag{B.70}$$

(19) Let us prove the relation

$$P = \sum_{n=0}^{\infty} L_n^{\alpha-n}(x) z^n = (1+z)^{\alpha} \exp(-xz), \quad |z| < 1, \tag{B.71}$$

which implies that

$$L_n^{\alpha-n}(x) = \frac{1}{n!} \frac{\partial^n}{\partial z^n} (1+z)^{\alpha} e^{-xz}\Big|_{z=0}. \tag{B.72}$$

Using the second equation (B.5) and equation (B.44) we get

$$P = \sum_{m=0}^{\infty} \frac{(-xz)^m}{m!} \sum_{n=0}^{\infty} \binom{\alpha}{n} z^n.$$

Taking into account the well-known sum

$$\sum_{n=0}^{\infty} \binom{\alpha}{n} z^n = (1+z)^{\alpha}, \ |z| < 1,$$

we arrive immediately at (B.71). The latter relation is equivalent also to the following one:

$$\sum_{n=0}^{\infty} \frac{I_{\alpha,n}(x) z^n}{\sqrt{n!}} = \frac{(z+\sqrt{x})^{\alpha}}{\sqrt{\Gamma(1+\alpha)}} \exp\left(-\frac{x}{2} - z\sqrt{x}\right), \quad |z|^2 < |x|. \tag{B.73}$$

(20) Let us justify the following expression for the integral:

$$J_a^\alpha(x,y) = \int_0^\infty e^{-ap} J_\alpha(2\sqrt{xp}) J_\alpha(2\sqrt{yp}) dp = a^{-1} e^{-\frac{x+y}{a}} I_\alpha\left(2a^{-1}\sqrt{xy}\right),$$

$$\operatorname{Re}\alpha > -1, \quad \operatorname{Re} a > 0\,. \tag{B.74}$$

To this end we substitute a series decomposition for $J_\alpha(2\sqrt{xp})$ in the integral. Thus, we get

$$J_a^\alpha(x,y) = \sum_{n=0}^\infty \frac{(-1)^n x^{\frac{\alpha}{2}+n}}{\Gamma(1+\alpha+n)n!} \int_0^\infty e^{-ap} p^{n+\frac{\alpha}{2}} J_\alpha(2\sqrt{yp})\, dp\,.$$

Using the second relation (B.48) and (B.1) we can transform $J_a^\alpha(x,y)$ to the form

$$J_a^\alpha(x,y) = a^{-1}\left(\frac{x}{a}\right)^{\frac{\alpha}{2}} e^{-\frac{y}{2a}} \sum_{n=0}^\infty \frac{(-1)^n \left(\frac{x}{a}\right)^n I_{\alpha+n,n}\left(\frac{y}{a}\right)}{\Gamma(1+n)\Gamma(1+\alpha+n)}\,.$$

Taking into account (B.64), we get (B.74).

The following property takes place:

$$\lim_{\alpha\to+0} J_a^\alpha(x,y) = \delta(x-y), \quad x>0, \quad y>0\,. \tag{B.75}$$

To prove it let us recall a definition. If the relation

$$|f(x)| \le Ae^{Bx}, A>0, \quad 0<x<\infty \tag{B.76}$$

holds with A, B being real, then $f(x)$ is called a function of restricted growth. The least possible B is called the index of the growth. Keeping in mind this definition, we consider a function $f(x)$ of restricted growth. Selecting $\alpha < 1/|B|$ in (B.74) we conclude that the integral

$$J_a^\alpha(x) = \int_0^\infty f(y) J_a^\alpha(x,y)\, dy, \operatorname{Re}\alpha > -1 \tag{B.77}$$

exists for any $x > 0$. Performing the change of the integration variable in the right hand side of equation (B.77)

$$y = x + 2\sqrt{ax}t, dy = 2\sqrt{ax}dt, \quad -q < t < \infty, q = \frac{1}{2}\sqrt{\frac{x}{a}},$$

we get

$$J_a^\alpha(x) = \int_{-q}^\infty f\left(x+2\sqrt{ax}t\right) R^\alpha(a;x,t)\, dt, \quad R^\alpha(a;x,t) = 2\sqrt{ax} J_a^\alpha\left(x, x+2\sqrt{ax}t\right)\,. \tag{B.78}$$

There exists an asymptotic formula

$$I_\alpha(x) \approx \frac{e^x}{\sqrt{2\pi x}}, \quad x \to +\infty\,, \tag{B.79}$$

which results in

$$\lim_{a\to+0} R^\alpha\,(a; x, t) = \frac{\exp\,(-t^2)}{\sqrt{\pi}}\,.$$

Thus, we obtain the following relation

$$\lim_{a\to+0} J_a^\alpha(x) = f(x) \int\limits_{-\infty}^{\infty} \frac{\exp\,(-t^2)}{\sqrt{\pi}} dt = f(x)\,,$$

which proves (B.75).

In particular, it follows from (B.74) as $a \to 0$ that

$$\int\limits_0^\infty J_\alpha\,(2\sqrt{xp})\,J_\alpha\,(2\sqrt{yp})\,dp = \delta\,(x-y)\,, \quad x>0, \quad y>0, \quad \operatorname{Re}\alpha > -1\,. \tag{B.80}$$

This relation can be rewritten in the form

$$\int\limits_0^\infty J_\alpha\,(xp)\,J_\alpha\,(yp)\,pdp = \frac{\delta\,(x-y)}{\sqrt{xy}} = \frac{\delta\,(x-y)}{x}, \quad x>0, \quad y>0, \quad \operatorname{Re}\alpha > -1\,. \tag{B.81}$$

Relation (B.74) can be obtained by means of (B.64).

(21) Let us prove the equality

$$\begin{aligned} A_\alpha(z; x, y) &= \sum_{n=0}^{\infty} I_{\alpha+n,n}(x) I_{\alpha+n,n}(y) z^n \\ &= \frac{z^{-\frac{\alpha}{2}}}{1-z} \exp\left[\left(\frac{z+1}{z-1}\right)\left(\frac{x+y}{2}\right)\right] I_\alpha\left(\frac{2\sqrt{xyz}}{1-z}\right), \quad |z|<1, \quad \operatorname{Re}\alpha > -1\,. \end{aligned} \tag{B.82}$$

Using the integral representation (B.49) for $I_{\alpha+n,n}(x)$, and the Equation (B.63), we get

$$\begin{aligned} A_\alpha(z; x, y) &= e^{\frac{x}{2}} \int\limits_0^\infty dp\, e^{-p} p^{\frac{\alpha}{2}} J_\alpha(2\sqrt{xp}) \sum_{n=0}^{\infty} \frac{I_{\alpha+n,n}(y)(pz)^n}{\sqrt{\Gamma(1+n)\Gamma(1+\alpha+n)}} \\ &= e^{\frac{x-y}{2}} z^{-\frac{\alpha}{2}} \int\limits_0^\infty e^{-(1-z)p} J_\alpha(2\sqrt{xp}) J_\alpha(2\sqrt{yzp})\, dp\,. \end{aligned}$$

The relation (B.82) can be now obtained by means of (B.74).

Making the change of the variable $z \to -z$ in (B.82), one derives

$$B_\alpha(z;x,y) = \sum_{n=0}^{\infty} I_{\alpha+n,n}(x) I_{\alpha+n,n}(y)(-z)^n$$
$$= \frac{z^{-\frac{\alpha}{2}}}{1+z} \exp\left[\left(\frac{z-1}{z+1}\right)\left(\frac{x+y}{2}\right)\right] J_\alpha\left(\frac{2\sqrt{xyz}}{1+z}\right), \quad |z|<1, \quad \operatorname{Re}\alpha > -1 . \quad \text{(B.83)}$$

(22) The function $A_\alpha(z;x,y)$ that was introduced in (B.82) has the important property

$$\lim_{z\to 1-0} A_\alpha(z;x,y) = \delta(x-y), \quad x>0, \quad y>0, \quad \operatorname{Re}\alpha > -1 , \quad \text{(B.84)}$$

which follows from (B.75). Thus we have proved that if $z \to 1-0$ the following relation holds:

$$\sum_{n=0}^{\infty} I_{\alpha+n,n}(x) I_{\alpha+n,n}(y) = \delta(x-y), \quad x>0, \quad y>0, \quad \operatorname{Re}\alpha > -1 . \quad \text{(B.85)}$$

It means that the set of functions $I_{\alpha+n,n}(x)$, $\operatorname{Re}\alpha > -1$, is complete in the space of functions $f(x)$ on the interval $0 < x < \infty$, for which

$$C_n = \int_0^{\infty} f(x) I_{\alpha+n,n}(x) dx < \infty .$$

In this case

$$f(x) = \sum_{n=0}^{\infty} C_n I_{\alpha+n,n}(x), \quad x > 0 , \quad \text{(B.86)}$$

if the series is convergent and $f(x)$ is defined and continuous in the point x.

(23) Let us justify the following equality:

$$j = \int_0^{\infty} e^{-ay} I_\alpha\left(2\sqrt{by}\right) I_{\alpha+n,n}(y) dy$$
$$= \frac{2}{2a+1}\left(\frac{2a-1}{2a+1}\right)^{n+\frac{\alpha}{2}} \exp\left(\frac{4ab}{4a^2-1}\right) I_{\alpha+n,n}\left(\frac{4b}{4a^2-1}\right),$$
$$\operatorname{Re}(2a-1) > 0, \quad \operatorname{Re}\alpha > -1 . \quad \text{(B.87)}$$

Using the decomposition (B.25) for the Laguerre functions, we get for j

$$j = [\Gamma(1+n)\Gamma(1+\alpha+n)]^{-\frac{1}{2}} \sum_{k=0}^{\infty} \frac{\Gamma(1+\alpha+n+k)}{\Gamma(1+\alpha+k)k!}(-1)^k \int_0^{\infty} e^{-qy} y^{k+\frac{\alpha}{2}} I_\alpha(2\sqrt{by}) dy ,$$

where $q = a - 1/2$. The integral over y can be calculated according to (B.47). The confluent hypergeometric function is expressed via Laguerre polynomials by means of (B.5) since k are integers. Thus we get

$$j = \frac{b^{\frac{\alpha}{2}} \exp(b/q)}{\sqrt{\Gamma(1+n)\Gamma(1+\alpha+n)}q^{1+\alpha}} \sum_{k=0}^{\infty} \frac{\Gamma(1+\alpha+n+k)}{\Gamma(1+\alpha+k)} (-1/q)^k L_k^{\alpha}(-b/q) \,. \quad \text{(B.88)}$$

Setting $z = -1/q$, $x = -b/q$, $\beta = \alpha + n$ in (B.65) we obtain

$$j = \frac{1}{1+q}\left(\frac{q}{1+q}\right)^{n+\frac{\alpha}{2}} e^{\frac{(1+2q)b}{2q(1+q)}} \sqrt{\frac{\Gamma(1+\alpha+n)}{\Gamma(1+n)}} \frac{e^{-\frac{b}{2q(1+q)}}\left[\frac{b}{q(1+q)}\right]^{\frac{\alpha}{2}}}{\Gamma(1+\alpha)} \Phi\left(-n, 1+\alpha; \frac{b}{b(1+q)}\right) .$$

Substituting the explicit expression for q and using (B.1) we arrive at (B.87). The series (B.88) is formally convergent only for $|q| > 1$, that corresponds to $\operatorname{Re} a > 3/2$. However, the integral in (B.87) exists also at $\operatorname{Re} a > 1/2$ and at any complex n (for which the function $I_{\alpha+n,n}(x)$ is defined). The point $a = 3/2 + i\gamma$, where γ is an arbitrary real number, is not singular, thus the analytic continuation to the above mentioned area is possible.

If $n \in \mathbb{Z}_+$, then the integral exists in the area $\operatorname{Re} a > -1/2$, thus the right hand side can be extended to such an area. In particular, at $a = 1/2$, passing to the limit by means of (B.28) we get

$$\int_0^{\infty} e^{-\frac{y}{2}} I_\alpha(2\sqrt{xy})\, I_{\alpha+n,n}(y) dy = \frac{(-1)^n e^x x^{n+\frac{\alpha}{2}}}{\sqrt{\Gamma(1+\alpha+n)\Gamma(1+n)}}, \quad \operatorname{Re} a > -1, \quad n \in \mathbb{Z}_+ . \tag{B.89}$$

(24) In the similar way we can justify the following representations:

$$\begin{aligned}
&\int_0^{\infty} e^{-ay} J_\alpha\left(2\sqrt{by}\right) I_{\alpha+n,n}(y) dy \\
&= \sqrt{\frac{\Gamma(1+\alpha+n)}{\Gamma(1+n)}} \frac{b^{\frac{\alpha}{2}} \exp\left(\frac{2b}{1-2a}\right)}{\Gamma(1+\alpha)} \left(\frac{2}{2a+1}\right)^{1+\alpha} \left(\frac{2a-1}{2a+1}\right)^n \Phi\left(1+\alpha+n, 1+\alpha; \frac{4b}{4a^2-1}\right) \\
&= \sqrt{\frac{\Gamma(1+\alpha+n)}{\Gamma(1+n)}} \frac{b^{\frac{\alpha}{2}} \exp\left(-\frac{2b}{1+2a}\right)}{\Gamma(1+\alpha)} \left(\frac{2}{2a+1}\right)^{1+\alpha} \left(\frac{2a-1}{2a+1}\right)^n \Phi\left(-n, 1+\alpha; \frac{4b}{1-4a^2}\right) \\
&= e^{i\pi n} \frac{2}{2a+1} \left(\frac{1-2a}{1+2a}\right)^{n+\frac{\alpha}{2}} \exp\left(\frac{4ab}{1-4a^2}\right) I_{\alpha+n,n}\left(\frac{4b}{1-4a^2}\right), \\
&\operatorname{Re}(2a-1) > 0, \quad \operatorname{Re} \alpha > -1 \,.
\end{aligned} \tag{B.90}$$

As before, if $n \in \mathbb{N}$ the latter formula takes place at $\mathrm{Re}(2a+1) > 0$, $\mathrm{Re}\,\alpha > -1$. The case $a = 1/2$ can be obtained as a limit

$$\int_0^\infty e^{-\frac{y}{2}} J_\alpha\left(2\sqrt{xy}\right) I_{\alpha+n,n}(y)dy = \frac{e^{-x}x^{n+\frac{\alpha}{2}}}{\sqrt{\Gamma(1+\alpha+n)\Gamma(1+n)}}, \quad \mathrm{Re}\,\alpha > -1, \quad n \in \mathbb{N}. \tag{B.91}$$

(25) Let us define linear integral operators $\hat{A}$ and $\hat{B}$ by the relations

$$\hat{A}f(x) = \int_0^\infty A_\alpha(z;x,y)f(y)dy, \quad \hat{B}f(x) = \int_0^\infty B_\alpha(z;x,y)f(y)dy, \tag{B.92}$$

where the kernels $A_\alpha(z;x,y)$ and $B_\alpha(z;x,y)$ of the operators are defined by the second relations in (B.82) and (B.83) at $\mathrm{Re}\,\alpha > -1$, and restrictions on the complex parameter z are related to the behavior of the function $f(x)$ as $x \to \infty$. We assume that the function $f(x)$ is selected to provide the existence of the integrals (B.92).

It is easy to see that $A_\alpha(z;x,y) = B_\alpha(-z;x,y)$. Let us set

$$a = \frac{1}{2}\left(\frac{1+z}{1-z}\right), \quad b = \frac{xz}{(1-z)^2}, \tag{B.93}$$

in equation (B.87) and let us multiply the right hand side and left hand side of (B.87) by the factor

$$\frac{z^{-\frac{\alpha}{2}}}{1-z}\exp\left[\left(\frac{z+1}{z-1}\right)\frac{x}{2}\right].$$

Then, taking into account (B.82), we get

$$\int_0^\infty A_\alpha(z;x,y)I_{\alpha+n,n}(y)dy = z^n I_{\alpha+n,n}(x), \quad \mathrm{Re}\,\alpha > -1, \tag{B.94}$$

and the area $\mathrm{Re}\,a > 1/2$ is transformed into the area $|2z-1| < 1$.

Similarly, let us set

$$a = \frac{1}{2}\left(\frac{1-z}{1+z}\right), \quad b = \frac{xz}{(1+z)^2},$$

in the equation (B.90) and let us multiply the right hand sides of (B.90) by the factor

$$\frac{z^{-\frac{\alpha}{2}}}{1+z}\exp\left[\left(\frac{z-1}{z+1}\right)\frac{x}{2}\right].$$

In this case we get

$$\int_0^\infty B_\alpha(z;x,y)I_{\alpha+n,n}(y)dy = (ze^{i\pi})^n I_{\alpha+n,n}(x), \quad \mathrm{Re}\,\alpha > -1, \ |2z+1| < 1. \tag{B.95}$$

Note that the areas $|2z+1| < 1$ and $|2z-1| < 1$ do not intersect.

If $n \in \mathbb{Z}_+$ then the integral (B.94) exists for $\operatorname{Re} z < 1$ and $\operatorname{Re}\alpha > -1$, and the integral (B.95) exists for $\operatorname{Re} z > -1$, $\operatorname{Re}\alpha > -1$. The two latter areas have a common part $-1 < \operatorname{Re} z < 1$, $\operatorname{Re}\alpha > -1$. Thus, the functions $I_{\alpha+n,n}(x)$ for $\operatorname{Re}\alpha > -1$ are eigenvectors for the operators (B.92) and (B.93) with the eigenvalues z^n and $(-z)^n$ respectively.

If one demands $I_{\alpha+n,n}(x)$ to be square-integrable on the interval $(0,\infty)$, then: n has to be nonnegative integer, $n \in \mathbb{Z}_+$; there exists an area common for the operators (B.92) and (B.93), in which $I_{\alpha+n,n}(x)$ is an eigenvector for the both operators; and decompositions (B.82) and (B.83) hold true in accordance with the corresponding theorems for the integral operators.

(26) Let us prove that

$$J = \frac{1}{2}\int_0^\infty \sqrt{\frac{a+1}{x}}\,[I_{a,m}(x)I_{a+1,n}(x) + I_{a+1,m}(x)I_{a,n}(x)]dx = \delta_{m,n}\,, \tag{B.96}$$

if $\operatorname{Re} a > (m+n)/2$ and $n, m \in \mathbb{Z}_+$. Suppose $n \neq m$, then we use the representation (B.46) with $\beta = \alpha - (m+n)/2$ to get

$$\begin{aligned}
&\sqrt{\frac{a+1}{x}}\,[I_{a,m}(x)I_{a+1,n}(x) + I_{a+1,m}(x)I_{a,n}(x)] \\
&= \frac{1}{\Gamma(1+a)}\sum_{k=0}^{n}\sum_{s=0}^{m}\left[\binom{\frac{m+n}{2}-1-k}{n-k}\binom{\frac{m+n}{2}-s}{m-s}\right. \\
&\quad \left. + \binom{\frac{m+n}{2}-1-s}{m-s}\binom{\frac{m+n}{2}-k}{n-k}\right] Q I_{\beta+k,k}(x) I_{\beta+s,s}(x)\,,
\end{aligned}$$

$$Q = \left[\frac{\Gamma(1+n)\Gamma(1+m)\Gamma(1+\beta+k)\Gamma(1+\beta+s)}{\Gamma(1+k)\Gamma(1+s)}\right]^{\frac{1}{2}}.$$

Then using the identity (B.8) we arrive at the relation

$$\begin{aligned}
&\binom{\frac{m+n}{2}-1-k}{n-k}\binom{\frac{m+n}{2}-s}{m-s} + \binom{\frac{m+n}{2}-1-s}{m-s}\binom{\frac{m+n}{2}-k}{n-k} \\
&= \frac{2(k-s)}{n-m}\binom{\frac{m+n}{2}-1-k}{n-k}\binom{\frac{m+n}{2}-1-s}{m-s}\,,
\end{aligned}$$

which allows one to write (at $n \neq m$)

$$J = \sum_{k=0}^{n}\sum_{s=0}^{m}\frac{Q(k-s)}{\Gamma(1+a)(n-m)}\binom{\frac{m+n}{2}-1-k}{n-k}\binom{\frac{m+n}{2}-1-s}{m-s}\int_0^\infty I_{\beta+k,k}(x)I_{\beta+s,s}(x)dx\,.$$

Thus, for $n \neq m$ we have $J = 0$ due to the orthogonality relation (B.39). If $n = m$ then

$$J = \int_0^\infty \sqrt{\frac{a+1}{x}}\,I_{a+1,n}(x)I_{a,n}(x)dx, \quad \operatorname{Re} a > n-1\,.$$

Setting $\alpha = a + 1 - n$, $\beta = a - n$ in (B.46), we get in the case under consideration

$$I_{a+1,n}(x) = \sqrt{x} \sum_{k=0}^{n} \binom{n-k}{n-k} \sqrt{\frac{\Gamma(1+n)\Gamma(1+a-n+k)}{\Gamma(1+k)\Gamma(2+a)}} I_{a-n+k,k}(x) .$$

This leads to the following result:

$$J = \sum_{k=0}^{n} \sqrt{\frac{\Gamma(1+n)\Gamma(1+a-n+k)}{\Gamma(1+k)\Gamma(1+a)}} \int_0^\infty I_{a,n}(x) I_{a-n+k,k}(x) dx .$$

However, it follows from (B.39) that

$$\int_0^\infty I_{a,n}(x) I_{a-n+k,k}(x) dx = \delta_{n,k} ,$$

which means that $J = 1$. Thus, the formula (B.96) is proved.

(27) Consider the sum

$$S(x) = S_{\mu,s}^{m,n}(x) = \sum_{k=0}^{s} I_{k+\mu,m}(x) I_{k+\mu,n}(x) . \tag{B.97}$$

Differentiating it with respect to x (in a supposition that this can be done term by term) we obtain

$$S' = \sum_{k=0}^{s} [I'_{k+\mu,m}(x) I_{k+\mu,n}(x) + I_{k+\mu,m}(x) I'_{k+\mu,n}(x)] . \tag{B.98}$$

Subtracting (B.11) from (B.13), we get

$$2\sqrt{x} I'_{n,m}(x) = \sqrt{n} I_{n-1,m}(x) - \sqrt{n+1} I_{n+1,m}(x) . \tag{B.99}$$

Then we use (B.99) in (B.98),

$$2\sqrt{x} S'(x) = \sum_{k=0}^{s} Q_k - \sum_{k=0}^{s} Q_{k+1},$$

$$Q_k = \sqrt{k+\mu}[I_{k+\mu-1,m}(x) I_{k+\mu,n}(x) + I_{k+\mu,m}(x) I_{k+\mu-1,n}(x)] .$$

Let us perform the shift $k \to k - 1$ in the second sum

$$2\sqrt{x} S'(x) = \sum_{k=0}^{s} Q_k - \sum_{k=1}^{s+1} Q_k = Q_0 + \sum_{k=1}^{s} Q_k - Q_{s+1} - \sum_{k=1}^{s} Q_k = Q_0 - Q_{s+1} .$$

Then we obtain

$$\begin{aligned} \frac{d}{dx} S_{\mu,s}^{m,n}(x) &= \frac{1}{2}\sqrt{\frac{\mu}{x}} [I_{\mu-1,m}(x) I_{\mu,n}(x) + I_{\mu,m}(x) I_{\mu-1,n}(x)] \\ &- \frac{1}{2}\sqrt{\frac{s+\mu+1}{x}} [I_{s+\mu,m}(x) I_{s+\mu+1,n}(x) + I_{s+\mu+1,m}(x) I_{s+\mu,n}(x)] . \end{aligned} \tag{B.100}$$

The sum

$$R_\mu^{m,n}(x) = \sum_{k=0}^{\infty} I_{k+\mu,m}(x) I_{k+\mu,n}(x) \tag{B.101}$$

exists due to (B.33). Considering the limit $s \to \infty$ in (B.100) one finds

$$\frac{d}{dx} R_\mu^{m,n}(x) = \frac{1}{2}\sqrt{\frac{\mu}{x}} \left[I_{\mu-1,m}(x) I_{\mu,n}(x) + I_{\mu,m}(x) I_{\mu-1,n}(x)\right] . \tag{B.102}$$

It is easy to see that

$$S_{\mu,s}^{m,n}(x) = R_\mu^{m,n}(x) - R_{\mu+s+1}^{m,n}(x) . \tag{B.103}$$

If $\mathrm{Re}(2\mu - m - n) > 0$ then $R_\mu^{m,n}(0) = 0$ and

$$R_\mu^{m,n}(x) = \frac{1}{2}\int_0^x \sqrt{\frac{\mu}{y}} \left[I_{\mu-1,m}(y) I_{\mu,n}(y) + I_{\mu,m}(y) I_{\mu-1,n}(y)\right] dy . \tag{B.104}$$

If n, m are integers, one can lift the restriction $\mathrm{Re}(2\mu - m - n) > 0$. Indeed, it follows from (B.103) that

$$R_{\mu+s+1}^{m,n}(x) + S_{\mu,s}^{m,n}(x) = R_\mu^{m,n}(x) . \tag{B.105}$$

However, for any μ, m, n there always exists such an integer s that $\mathrm{Re}(2\mu + 2s + 2 - m - n) > 0$. For such s the relation

$$R_{\mu+s+1}^{m,n}(x) = \frac{1}{2}\int_0^x \sqrt{\frac{\mu+s+1}{y}} [I_{\mu+s,m}(y) I_{\mu+s+1,n}(y) + I_{\mu+s+1,m}(y) I_{\mu+s,n}(y)] dy \tag{B.106}$$

takes place due to (B.104). One can see that $\lim_{x\to\infty} S_{\mu,s}^{m,n}(x) = 0$ for any $n, m \in \mathbb{N}$. In this case it follows from (B.100) that

$$\begin{aligned} S_{\mu,s}^{m,n}(x) = {} & \frac{1}{2}\int_x^\infty \sqrt{\frac{\mu+s+1}{y}} [I_{\mu+s,m}(y) I_{\mu+s+1,n}(y) + I_{\mu+s+1,m}(y) I_{\mu+s,n}(y)] dy \\ & - \frac{1}{2}\int_x^\infty \sqrt{\frac{\mu}{y}} [I_{\mu-1,m}(y) I_{\mu,n}(y) + I_{\mu,m}(y) I_{\mu-1,n}(y)] dy . \end{aligned} \tag{B.107}$$

Substituting (B.106) and (B.107) into (B.105) for any $n, m \in \mathbb{R}_+$, we find by means of (B.96)

$$R_\mu^{m,n}(x) = \delta_{m,n} - \frac{1}{2}\int_x^\infty \sqrt{\frac{\mu}{y}} [I_{\mu-1,m}(y) I_{\mu,n}(y) + I_{\mu,m}(y) I_{\mu-1,n}(y)] dy . \tag{B.108}$$

The relation (B.102) takes place at any μ, n, m, thus, one can write

$$R_\mu^{m,n}(x) = R_\mu^{m,n}(a) - \frac{1}{2}\int_x^a \sqrt{\frac{\mu}{y}} [I_{\mu-1,m}(y) I_{\mu,n}(y) + I_{\mu,m}(y) I_{\mu-1,n}(y)] dy , \tag{B.109}$$

where $a \geq 0$ is arbitrary number. In the general case it is difficult to select any specific points a in which one can find $R_{\mu}^{m,n}(a)$ explicitly. If, for example, $m - \mu, n - \mu$ are integers, or equivalently $m = \mu + p, n = \mu + q$, p, q are integers, then we get, due to (B.23), that $I_{k+\mu,p+\mu}(0) = \delta_{k,p}$ (k, p are integers), and relation (B.104) can be replaced by

$$R_{\mu}^{m,n}(x) = \delta_{m,n} + \frac{1}{2}\int_0^x \sqrt{\frac{\mu}{y}}[I_{\mu-1,m}(y)I_{\mu,n}(y) + I_{\mu,m}(y)I_{\mu-1,n}(y)]dy \tag{B.110}$$

for any μ.

If $\mu = 0$ and $m > 0$ is an integer, then due to (B.19) we get

$$\frac{d}{dx}R_0^{m,n}(x) = (-1)^{m+1}\frac{\sin n\pi}{2\pi}\sqrt{\frac{\Gamma(1+n)}{\Gamma(1+m)}}x^{\frac{m-n}{2}-1},$$

which results in

$$R_0^{m,n}(x) = (-1)^m \frac{\sin n\pi}{\pi(n-m)}\sqrt{\frac{\Gamma(1+n)}{\Gamma(1+m)}}x^{\frac{m-n}{2}}. \tag{B.111}$$

Finally, if $n > 0$ is an integer, then due to the relation

$$(-1)^m \frac{\sin n\pi}{\pi(n-m)} = \delta_{m,n},$$

we obtain for $n, m \in \mathbb{N}$

$$R_0^{m,n}(x) = \sum_{k=0}^{\infty} I_{k,n}(x)I_{k,m}(x) = \delta_{m,n}. \tag{B.112}$$

(28) Let us consider the sum

$$F^{n,m}(z; x, y) = \sum_{k=0}^{\infty} I_{k,n}(x)I_{k,m}(y)z^k \tag{B.113}$$

for any $n, m \in \mathbb{N}$. It follows from (B.71)

$$L_n^{\alpha-n}(x) = \frac{1}{n!}\frac{d^n}{d\tau^n}(1+\tau)^{\alpha}e^{-x\tau}\bigg|_{\tau=0} = \sqrt{\frac{\Gamma(1+\alpha)}{\Gamma(1+n)}}e^{\frac{x}{2}}x^{\frac{n-\alpha}{2}}I_{\alpha,n}(x). \tag{B.114}$$

The latter representation allows one to transform (B.113) to

$$F^{n,m}(z; x, y) = \frac{e^{-\frac{x+y}{2}}x^{-\frac{n}{2}}y^{-\frac{m}{2}}}{\sqrt{\Gamma(1+n)\Gamma(1+m)}}\frac{d^m}{dt^m}\frac{d^n}{d\tau^n}\Phi\bigg|_{\tau=t=0},$$

$$\Phi = \sum_{k=0}^{\infty}\frac{[(1+\tau)(1+t)z\sqrt{xy}]^k}{k!}e^{-x\tau-yt} = \exp[(1+\tau)(1+t)z\sqrt{xy} - x\tau - xt].$$

It is easy to see that

$$\frac{d^n}{d\tau^n}\Phi|_{\tau=0} = [(1+t)z\sqrt{xy} - x]^n e^{-(y-z\sqrt{xy})t+z\sqrt{xy}} ,$$

thus, one can get for (B.113)

$$F^{n,m}(z;x,y) = \frac{y^{-\frac{m}{2}}\exp(z\sqrt{xy} - \frac{x+y}{2})}{\sqrt{\Gamma(1+n)\Gamma(1+m)}}\frac{d^m}{dt^m}(z\sqrt{y} - \sqrt{x} + tz\sqrt{y})^n e^{-(y-z\sqrt{xy})t}\bigg|_{t=0} .$$

Making the replacement

$$t = \frac{z\sqrt{y}-\sqrt{x}}{z\sqrt{y}}\tau, \quad \frac{d}{dt} = \frac{z\sqrt{y}}{z\sqrt{y}-\sqrt{x}}\frac{d}{d\tau}, \quad t=0 \to \tau = 0 ,$$

we transform $F^{n,m}(z;x,y)$ to the form

$$F^{n,m}(z;x,y) = \frac{z^m(z\sqrt{y}-\sqrt{x})^{n-m}\exp(z\sqrt{xy} - \frac{x+y}{2})}{\sqrt{\Gamma(1+n)\Gamma(1+m)}}\frac{d^m}{d\tau^m}(1+\tau)^n e^{-q\tau}\bigg|_{\tau=0} ,$$

$$q = z^{-1}\left(\sqrt{y} - z\sqrt{x}\right)(z\sqrt{y}-\sqrt{x}) = x + y - \sqrt{xy}\left(z+z^{-1}\right) .$$

Using once again (B.114), we get finally for $n, m \in \mathbb{N}$

$$\begin{aligned} F^{n,m}(z;x,y) &= \sum_{k=0}^{\infty} I_{k,n}(x)I_{k,m}(y)z^k \\ &= z^{\frac{m+n}{2}}\left(\frac{z\sqrt{y}-\sqrt{x}}{\sqrt{y}-z\sqrt{x}}\right)^{\frac{n-m}{2}}\exp\left[\frac{\sqrt{xy}}{2}\left(z-z^{-1}\right)\right] \\ &\quad \times I_{n,m}\left[x+y-\sqrt{xy}\left(z+z^{-1}\right)\right] . \end{aligned} \quad \text{(B.115)}$$

In the special case $z = 1$ this relation is transformed to

$$\sum_{k=0}^{\infty} I_{k,n}(x)I_{k,m}(y) = I_{n,m}[(\sqrt{x}-\sqrt{y})^2] . \quad \text{(B.116)}$$

At $x = y$ we return to (B.112).

(29) Consider the sum

$$G^{m,n}_{\mu,s}(x) = \sum_{k=0}^{s} I_{m,k+\mu}(x)I_{n,k+\mu}(x) . \quad \text{(B.117)}$$

In contrast to (B.97), one cannot find here a limit $s \to \infty$ in the general case. However, if $m = \mu + p$, $n = \mu + q$, p, q are integers, then

$$G^{m,n}_{\mu,s}(x) = (-1)^{m-n}S^{n,m}_{\mu,s}(x) \quad \text{(B.118)}$$

due to (B.23), and such a limit exists. It exists also if only one out of the numbers $m = \mu + p$ and p is an integer, while $n - \mu$ is not, but $n - m > 0$.

Combining (B.12) and (B.14) we get

$$2\sqrt{x}I'_{n,m}(x) = \sqrt{m+1}I_{n,m+1}(x) - \sqrt{m}I_{n,m-1}(x)\,. \tag{B.119}$$

In the same manner as in item 28 we can derive a relation

$$\begin{aligned}\frac{d}{dx}G^{m,n}_{\mu,s}(x) &= \frac{1}{2}\sqrt{\frac{\mu+s+1}{x}}\left[I_{m,\mu+s}(x)I_{n,\mu+s+1}(x) + I_{m,\mu+s+1}(x)I_{n,\mu+s}(x)\right] \\ &- \frac{1}{2}\sqrt{\frac{\mu}{x}}\left[I_{m,\mu-1}(x)I_{n,\mu}(x) + I_{m,\mu}(x)I_{n,\mu-1}(x)\right]\,,\end{aligned} \tag{B.120}$$

which allows one to restore $G^{m,n}_{\mu,s}(x)$ by an integration. In particular, at $\mu = 0$, $n > m$, m integer we obtain

$$\sum_{k=0}^{\infty} I_{m,k}(x)I_{n,k}(x) = 0\,; \tag{B.121}$$

with account of (B.20).

(30) Here we are going to consider a class of functions, which are closely related to Laguerre functions, and which appear often in various problems of mathematical physics.

As it follows from the Equation (B.17), the Laguerre functions are eigenvectors for the following operator:

$$\begin{aligned}R_\alpha &= \frac{\alpha^2}{4x} + \frac{x}{4} - \frac{d}{dx} - x\frac{d^2}{dx^2}, \quad \alpha = \text{conts}, \\ R_\alpha\psi &= \lambda\psi, \quad 0 < x < \infty\,.\end{aligned} \tag{B.122}$$

The general solution of this equation has the form

$$\psi(x) = aI_{n,m}(x) + bI_{m,n}(x), \quad \alpha = n - m, \quad 2\lambda = n + m + 1\,, \tag{B.123}$$

where a, b are arbitrary constants. In the general case the function $\psi(x)$ vanishes at $x \to \infty$ only if one of the numbers n or m is positive and integer. However, one can provide such a behavior at any n, m for some special values of a, b. Consider the function

$$\psi_{\lambda,\alpha}(x) = x^{-1/2}W_{\lambda,\frac{\alpha}{2}}(x), \quad \psi_{\lambda,\alpha}(x) = \psi_{\lambda,-\alpha}(x)\,, \tag{B.124}$$

where $W_{\lambda,\mu}(x)$ is a Whittaker function (see [191], 9.220.4). The function $\psi_{\lambda,\alpha}(x)$ can be expressed via the confluent hypergeometric function

$$\begin{aligned}\psi_{\lambda,\alpha}(x) = e^{-x/2}\Bigg[&\frac{\Gamma(-\alpha)\,x^{\alpha/2}}{\Gamma\left(\frac{1-\alpha}{2} - \lambda\right)}\Phi\left(\frac{1+\alpha}{2} - \lambda, 1+\alpha; x\right) \\ &+ \frac{\Gamma(\alpha)\,x^{-\alpha/2}}{\Gamma\left(\frac{1+\alpha}{2} - \lambda\right)}\Phi\left(\frac{1-\alpha}{2} - \lambda, 1-\alpha; x\right)\Bigg]\,,\end{aligned} \tag{B.125}$$

and, using (B.1), via the Laguerre functions

$$\psi_{\lambda,\alpha}(x) = \frac{\sqrt{\Gamma(1+n)\Gamma(1+m)}}{\sin(n-m)\pi}\left[\sin n\pi I_{n,m}(x) - \sin m\pi I_{m,n}(x)\right],$$
$$\alpha = n-m, \quad 2\lambda = 1+n+m, \quad n = \lambda - \frac{1-\alpha}{2}, \quad m = \lambda - \frac{1+\alpha}{2}. \tag{B.126}$$

Using (B.11)–(B.16) the following properties of the function $\psi_{\lambda,\alpha}(x)$ can be established,

$$\psi_{\lambda,\alpha}(x) = \sqrt{x}\psi_{\lambda-1/2,\alpha-1}(x) + \frac{1+\alpha-2\lambda}{2}\psi_{\lambda-1,\alpha}(x),$$
$$\psi_{\lambda,\alpha}(x) = \sqrt{x}\psi_{\lambda-1/2,\alpha+1}(x) + \frac{1-\alpha-2\lambda}{2}\psi_{\lambda-1,\alpha}(x),$$
$$2x\psi'_{\lambda,\alpha}(x) = (2\lambda-1-x)\psi_{\lambda,\alpha}(x) + \frac{1}{2}(2\lambda-1-\alpha)(2\lambda-1+\alpha)\psi_{\lambda-1,\alpha}(x),$$
$$2x\psi'_{\lambda,\alpha}(x) = (\alpha-x)\psi_{\lambda,\alpha}(x) + (2\lambda-1-\alpha)\sqrt{x}\psi_{\lambda-\frac{1}{2},\alpha+1}(x). \tag{B.127}$$

As a consequence of these properties we get

$$A_\alpha\psi_{\lambda,\alpha}(x) = \frac{2\lambda-1+\alpha}{2}\psi_{\lambda-\frac{1}{2},\alpha-1}(x), \quad \overset{+}{A}_\alpha\psi_{\lambda-\frac{1}{2},\alpha-1}(x) = \psi_{\lambda,\alpha}(x),$$
$$A_\alpha = \frac{x+\alpha}{2\sqrt{x}} + \sqrt{x}\frac{d}{dx}, \quad \overset{+}{A}_\alpha = \frac{x+\alpha-1}{2\sqrt{x}} - \sqrt{x}\frac{d}{dx}. \tag{B.128}$$

The operator R_α can be expressed via the operators A_α and $\overset{+}{A}_\alpha$,

$$R_\alpha = \overset{+}{A}_\alpha A_\alpha + \frac{1-\alpha}{2}, \quad R_{\alpha-1} = A_\alpha\overset{+}{A}_\alpha - \frac{\alpha}{2}. \tag{B.129}$$

Since (B.126) is a particular case of (B.123), then $\psi_{\lambda,\alpha}(x)$ is also an eigenfunction for the operator R_α.

Using well-known asymptotics of the Whittaker function (see [191], 9.227), we get

$$\psi_{\lambda,\alpha}(x) \sim x^{\lambda-\frac{1}{2}}e^{-\frac{x}{2}}, \quad x \to \infty,$$
$$\psi_{\lambda,\alpha}(x) \sim \frac{\Gamma(|\alpha|)}{\Gamma\left(\frac{1+|\alpha|}{2}-\lambda\right)}x^{-\frac{|\alpha|}{2}}, \quad x \sim 0, \quad \alpha \neq 0. \tag{B.130}$$

The function $\psi_{\lambda,0}(x)$ has a logarithmic singularity at $x \sim 0$. It is important to stress that the functions $\psi_{\lambda,\alpha}(x)$ are well defined and infinitely differentiable at $0 < x < \infty$ and at any complex λ, α. In this respect one can mention that the Laguerre functions are not defined at negative integer n, m. In particular cases, when one of the numbers n, m is a nonnegative integer, the function $\psi_{\lambda,\alpha}(x)$ coincides (up to a constant factor) with a Laguerre function. Thus, $\psi_{\lambda,\alpha}(x)$ is an eigenvector of the operator R_α which vanishes at $x \to \infty$.

According to (B.130) the function $\psi_{\lambda,\alpha}(x)$ is square-integrable on the interval $0 < x < \infty$ only, if $|\alpha| < 1$, and is not if $|\alpha| \geq 1$. The corresponding integrals are known (see [191], 7.611),

$$\int_0^\infty \psi_{\lambda,\alpha}(x)\psi_{\lambda',\alpha}(x)dx = \frac{\pi}{(\lambda' - \lambda)\sin\alpha\pi}\left\{\left[\Gamma\left((1+\alpha)/2 - \lambda'\right)\Gamma\left((1-\alpha)/2-\lambda\right)\right]^{-1}\right.$$
$$\left. - \left[\Gamma\left((1-\alpha)/2 - \lambda'\right)\Gamma\left((1+\alpha)/2-\lambda\right)\right]^{-1}\right\}, \quad |\alpha| < 1\,, \tag{B.131}$$

$$\int_0^\infty |\psi_{\lambda,\alpha}(x)|^2 dx = \frac{\pi}{\sin\alpha\pi}\frac{\psi\left((1+\alpha)/2-\lambda\right) - \psi\left((1-\alpha)/2-\lambda\right)}{\Gamma\left((1+\alpha)/2-\lambda\right)\Gamma\left((1-\alpha)/2-\lambda\right)}, \quad |\alpha| < 1\,, \tag{B.132}$$

$$\int_0^\infty |\psi_{\lambda,0}(x)|^2 dx = \frac{\psi'\left(1/2-\lambda\right)}{\Gamma^2\left(1/2-\lambda\right)}, \quad \int_0^\infty |\psi_{n+1/2,0}(x)|^2 dx = \Gamma^2\left(1+n\right).$$

Here $\psi(x)$ stands for the logarithmic derivative of the Γ-function (see [191], 8.360).

At $|\alpha| \geq 1$ the situation is the following: the only eigenvectors of the operator R_α which are quadratically integrable are Laguerre functions, they form a complete set. As to the functions $\psi_{\lambda,\alpha}(x)$, they are orthogonal in the case when arguments of the Γ-function in (B.131) are negative and integers. This corresponds to the case when n, m are nonnegative integers, thus, we face again Laguerre functions according to (B.126). If $|\alpha| < 1$ then the functions $\psi_{\lambda,\alpha}(x)$ and $\psi_{\lambda',\alpha}(x)$, $\lambda' \neq \lambda$, are not orthogonal, in the general case, as it follows from (B.131). This is a reflection of the fact that the operator R_α is no longer self-adjoint at these values of α, thus, their eigenvectors, which belong to different eigenvalues, must not be orthogonal. However, in such a case, for any given real λ there exists an enumerable set of real numbers λ'_s, $s = 1, 2, \ldots$, $(\lambda'_1 < \lambda'_2 < \lambda'_3 < \cdots)$, so that the set of functions $\psi_{\lambda,\alpha}$, $\psi_{\lambda'_s,\alpha}$ is orthogonal. One can see that $\lambda'_s > 0$, $s \geq 2$. Indeed, let one of the numbers n, m be not an integer (if they both are integer we deal with Laguerre functions). Consider the condition (B.131) for two orthogonal functions $\psi_{\lambda,\alpha}$, and $\psi_{\lambda',\alpha}$. Let us multiply the right side of this condition (which is zero) by a finite number $\Gamma((1+\alpha)/2 - k)\Gamma((1+\alpha)/2 - \lambda)$. Then the right hand side is equal to zero if

$$\frac{\Gamma\left((1+\alpha)/2-\lambda\right)}{\Gamma\left((1-\alpha)/2-\lambda\right)} = a = \frac{\Gamma\left((1+\alpha)/2-\lambda'\right)}{\Gamma\left((1-\alpha)/2-\lambda'\right)}\,, \tag{B.133}$$

where a is a constant. In fact (B.133) can be considered as a transcendental equation for λ'. Studying this equation one can see that there exists an enumerable set of real roots of this equation, and λ is one of them. Only one root can be negative since the right hand side of (B.133) is a monotonous function at $\lambda' < 0$ (at $\lambda' > 0$ it is not monotonous). Asymptotically at large s we get $k_{s+1} \approx k_s + 1$. Thus, at $|\alpha| < 1$ any function $\psi_{\lambda,\alpha}$ belongs to an enumerable set of orthogonal functions $\psi_{\lambda'_s,\alpha}$.

B.2 Hermite polynomials and Hermite functions

(1) We define *Hermitian polynomials* in the standard manner (see [191], 8.950) by the Rodriguez formula

$$H_n(x) = (-1)^n e^{x^2} \frac{d^n}{dx^n} e^{-x^2} . \tag{B.134}$$

(2) From this definition one can derive the simple properties

$$H_n(x) = (-1)^n H_n(-x) , \tag{B.135}$$

$$H_0(x) = 1, \quad H_1(x) = 2x, \quad H_2(x) = 4x^2 - 2, \quad H_3(x) = 8x^3 - 12x . \tag{B.136}$$

By fulfilling the following calculations:

$$\begin{aligned}
H_{n+1}(x) &= (-1)^{n+1} e^{x^2} \frac{d^{n+1}}{dx^{n+1}} e^{-x^2} = (-1)^{n+1} e^{x^2} \frac{d^n}{dx^n} \frac{d}{dx} e^{-x^2} \\
&= (-1)^n e^{x^2} \frac{d^n}{dx^n} 2x e^{-x^2} = 2(-1)^n e^{x^2} \left[x \frac{d^n}{dx^n} e^{-x^2} + n \frac{d^{n-1}}{dx^{n-1}} e^{-x^2} \right] \\
&= 2x(-1)^n e^{x^2} \frac{d^n}{dx^n} e^{-x^2} - 2n(-1)^{n-1} e^{x^2} \frac{d^{n-1}}{dx^{n-1}} e^{-x^2} \\
&= 2xH_n(x) - 2nH_{n-1}(x) ,
\end{aligned}$$

we arrive at a recurrent relation

$$H_{n+1}(x) - 2xH_n(x) + 2nH_{n-1}(x) = 0 . \tag{B.137}$$

Differentiating with respect to x one finds

$$\begin{aligned}
\frac{dH_n(x)}{dx} &= 2x(-1)^n e^{x^2} \frac{d^n}{dx^n} e^{-x^2} - (-1)^{n+1} e^{x^2} \frac{d^{n+1}}{dx^{n+1}} e^{-x^2} \\
&= 2xH_n(x) - H_{n+1}(x) = 2nH_{n-1}(x) .
\end{aligned}$$

The latter equation is just (B.137). Thus, we get

$$\frac{dH_n(x)}{dx} = 2xH_n(x) - H_{n+1}(x) = 2nH_{n-1}(x) . \tag{B.138}$$

(3) The following integral representation takes place:

$$H_n(x) = \frac{2^n}{\sqrt{\pi}} \int_{-\infty}^{\infty} (x + it)^n e^{-t^2} dt . \tag{B.139}$$

To prove it, one can, for example, check first that (B.139) obeys the recurrent relation (B.137)

$$H_{n+1}(x) - 2xH_n(x) + 2nH_{n-1}(x)$$
$$= \frac{1}{\sqrt{\pi}} \int_{-\infty}^{\infty} dte^{-t^2} \left[2^{n+1}(x+it)^{n+1} - 2x2^n(x+it)^n + 2n2^{n-1}(x+it)^{n-1}\right]$$
$$= \frac{2^n}{\sqrt{\pi}} \int_{-\infty}^{\infty} dte^{-t^2} (x+it)^{n-1}\left(n + 2ixt - 2t^2\right) = i\frac{2^n}{\sqrt{\pi}} \int_{-\infty}^{\infty} dt\frac{d}{dt}\left[(x+it)^n e^{-t^2}\right] = 0\,.$$

Then one can verify that it reproduces the particular cases (B.136). Thus, (B.139) really defines the Hermitian polynomials.

(4) A differential equation for Hermitian polynomials can be found by a differentiation of the first part of (B.138) with respect to x,

$$\frac{d^2H_n(x)}{dx^2} = 2x\frac{dH_n(x)}{dx} + 2H_n(x) - \frac{dH_{n+1}(x)}{dx}\,.$$

The derivative H'_{n+1} can be found by the use of the second part of (B.138). Finally we get an equation for the Hermitian polynomials

$$H''_n(x) - 2xH'_n(x) + 2nH_n(x) = 0\,. \tag{B.140}$$

(5) Some particular values: It follows from (B.139) that

$$H_{2n}(0) = (-1)^n\frac{2^{2n}\Gamma\left(n+\frac{1}{2}\right)}{\sqrt{\pi}}, \quad H_{2n+1}(0) = 0\,. \tag{B.141}$$

Then one can find from (B.138)

$$H'_{2n+1}(0) = (-1)^n\frac{2^{2(n+1)}\Gamma\left(n+\frac{3}{2}\right)}{\sqrt{\pi}}, \quad H'_{2n}(0) = 0\,. \tag{B.142}$$

Taking into account the relation for the Γ-function

$$\Gamma(2x) = \frac{2^{2x-1}}{\sqrt{\pi}}\Gamma(x)\Gamma\left(x+\frac{1}{2}\right), \tag{B.143}$$

one can rewrite (B.141) and (B.142) in the form

$$H_{2n}(0) = (-1)^n\frac{\Gamma(2n+1)}{\Gamma(n+1)}, \quad H'_{2n+1}(0) = (-1)^n\frac{\Gamma(2n+3)}{\Gamma(n+2)},$$
$$H_{2n+1}(0) = H'_{2n}(0) = 0\,. \tag{B.144}$$

(6) One can find relations between the Hermitian polynomials, the confluent hypergeometric function, and the Laguerre polynomials:

$$H_{2n}(x) = (-1)^n \frac{\Gamma(2n+1)}{\Gamma(n+1)} \Phi\left(-n; \frac{1}{2}; x^2\right),$$
$$H_{2n+1}(x) = (-1)^n \frac{\Gamma(2n+3)}{\Gamma(n+2)} x \Phi\left(-n; \frac{3}{2}; x^2\right). \tag{B.145}$$

Here $\Phi(\alpha, \beta; x)$ is the confluent hypergeometric function (see [191], 9.210). One can substitute (B.145) into the equation (B.140) and see that the latter appears to be the confluent hypergeometric equation. Taking into account the the initial conditions (B.144) one can confirm the validity of (B.145).

The following relations can be established by means of the equation (B.5):

$$H_{2n}(x) = (-1)^n 2^{2n} \Gamma(n+1) L_n^{-\frac{1}{2}}\left(x^2\right),$$
$$H_{2n+1}(x) = (-1)^n 2^{2n+1} \Gamma(n+1) x L_n^{\frac{1}{2}}\left(x^2\right). \tag{B.146}$$

(7) The function

$$\varphi(x, y) = \sum_{k=0}^{\infty} \frac{y^k H_k(x)}{k!}$$

is a generating function for the Hermitian polynomials. To get a closed expression for for it we use equation (B.139)

$$\varphi(x, y) = \frac{1}{\sqrt{\pi}} \int_{-\infty}^{\infty} dt e^{-t^2} \sum_{k=0}^{\infty} \frac{[2y(x+it)]^k}{k!} = \frac{1}{\sqrt{\pi}} \int_{-\infty}^{\infty} dt\, e^{-(t-iy)^2 - y^2 + 2xy} = e^{-y^2+2xy}.$$

Thus, we get

$$e^{-y^2+2xy} = \sum_{k=0}^{\infty} \frac{y^k H_k(x)}{k!}, \tag{B.147}$$

and

$$H_k(x) = \left. \frac{\partial^k}{\partial y^k} e^{-y^2+2xy} \right|_{y=0}. \tag{B.148}$$

(8) The following integral with Hermitian polynomial can be calculated:

$$\int_{-\infty}^{\infty} \exp\left(2xy - \alpha^2 x^2\right) H_n(x) dx$$
$$= \frac{\sqrt{\pi}\left(\alpha^2 - 1\right)^{\frac{n}{2}}}{\alpha^{n+1}} \exp\left(\frac{y^2}{\alpha^2}\right) H_n\left(\frac{y}{\alpha\sqrt{\alpha^2 - 1}}\right), \quad \operatorname{Re}\alpha^2 > 0. \tag{B.149}$$

Indeed, one can write by means of (B.148)

$$J = \int_{-\infty}^{\infty} \exp\left(2xy - \alpha^2 x^2\right) H_n(x) dx = \frac{\partial^n}{\partial t^n} \int_{-\infty}^{\infty} \exp\left(2xy + 2xt - \alpha^2 x^2 - t^2\right) dx \bigg|_{t=0} .$$

Using the Gaussian integral

$$\int_{-\infty}^{\infty} \exp\left(2\beta x - \alpha^2 x^2\right) dx = \frac{\sqrt{\pi}}{2} \exp \frac{\beta^2}{\alpha^2}, \quad \operatorname{Re} \alpha^2 > 0 , \tag{B.150}$$

we get

$$J = \frac{\partial^n}{\partial t^n} \frac{\sqrt{\pi}}{2} \exp\left[\frac{y^2 + 2yt - (\alpha^2 - 1)t^2}{\alpha^2}\right]\bigg|_{t=0} .$$

Doing a change of integration variable, $t \to q$, $\alpha q = t\sqrt{\alpha^2 - 1}$, and $t = 0 \Rightarrow q = 0$, we get

$$J = \frac{\sqrt{\pi}\,(\alpha^2 - 1)^{\frac{n}{2}}}{\alpha^{n+1}} \frac{\partial^n}{\partial q^n} \exp\left(\frac{2yq}{\alpha\sqrt{\alpha^2 - 1}} - q^2\right)\bigg|_{q=0} .$$

Recalling (B.148), we arrive just at (B.149).

(9) The Hermitian polynomials can be presented in the form

$$H_n(x) = \hat{b}^n \cdot 1, \quad \hat{b} = 2x - \frac{d}{dx} . \tag{B.151}$$

This representation follows from (B.134). Indeed,

$$\begin{aligned} H_n(x) &= \underbrace{\left(-e^{x^2} \frac{d}{dx} e^{-x^2}\right)\left(-e^{x^2} \frac{d}{dx} e^{-x^2}\right) \ldots \left(-e^{x^2} \frac{d}{dx} e^{-x^2}\right)}_{n} \cdot 1 \\ &= \hat{b}^n \cdot 1, \quad \hat{b} = -e^{x^2} \frac{d}{dx} e^{-x^2} = 2x - \frac{d}{dx} . \end{aligned} \tag{B.152}$$

(10) Often the *Hermitian functions* $U_n(x)$ are used

$$\begin{aligned} U_n(x) &= \left(2^n n! \sqrt{\pi}\right)^{-\frac{1}{2}} \exp\left(-x^2/2\right) H_n(x) , \\ H_n(x) &= \left(2^n n! \sqrt{\pi}\right)^{\frac{1}{2}} \exp\left(x^2/2\right) U_n(x) . \end{aligned} \tag{B.153}$$

(11) An analog of the recurrent relation (B.137) takes place

$$\begin{aligned} &\sqrt{\frac{n+1}{2}} U_{n+1}(x) - x U_n(x) + \sqrt{\frac{n}{2}} U_{n-1}(x) = 0, \\ &U_0(x) = \pi^{-\frac{1}{4}} \exp\left(-\frac{x^2}{2}\right), \quad U_1(x) = \left(\frac{4}{\pi}\right)^{\frac{1}{4}} x \exp\left(-\frac{x^2}{2}\right) . \end{aligned} \tag{B.154}$$

(12) Using the corresponding formulas for the Hermitian polynomials we get

$$\begin{aligned} U_n'(x) &= \sqrt{2n}U_{n-1}(x) - xU_n(x) = xU_n(x) - \sqrt{2(n+1)}U_{n+1}(x) \\ &= \sqrt{\frac{n}{2}}U_{n-1}(x) - \sqrt{\frac{n+1}{2}}U_{n+1}(x)\,. \end{aligned} \tag{B.155}$$

Important relations follow from (B.155)

$$\begin{aligned} \hat{a}U_n(x) &= \sqrt{n}U_{n-1}(x), \quad \hat{a}^+U_n(x) = \sqrt{n+1}U_{n+1}(x)\,, \\ \hat{a}^+\hat{a}U_n(x) &= nU_n(x), \quad \hat{a}\hat{a}^+U_n(x) = (n+1)U_n(x)\,, \end{aligned} \tag{B.156}$$

where annihilation and creation operators are introduced,

$$\hat{a} = \frac{1}{\sqrt{2}}\left(x + \frac{d}{dx}\right), \quad \hat{a}^+ = \frac{1}{\sqrt{2}}\left(x - \frac{d}{dx}\right), \quad [\hat{a}, \hat{a}^+] = 1\,. \tag{B.157}$$

(13) The Hermitian function obeys the differential equation

$$U_n''(x) + \left(2n + 1 - x^2\right)U_n(x) = 0\,. \tag{B.158}$$

(14) The Fourier transform of the Hermitian function is again a Hermitian function,

$$\frac{1}{\sqrt{2\pi}}\int_{-\infty}^{\infty} e^{ixy}U_n(x)dx = i^nU_n(y)\,. \tag{B.159}$$

This formula can be checked if one substitutes (B.152) into the left hand side of (B.159) and takes into account (B.149).

(15) The following integral of Hermitian functions takes place:

$$\int_{-\infty}^{\infty} U_n(x+a)\,U_m(x+b)\,e^{\alpha x}dx = \left(\frac{b-a+\alpha}{b-a-\alpha}\right)^{\frac{m-n}{2}} e^{-\frac{\alpha}{2}(a+b)} I_{m,n}\left[\frac{(b-a)^2-\alpha^2}{2}\right], \tag{B.160}$$

where $I_{m,n}$ are Laguerre functions (B.4). To justify this result one can use expressions (B.152) and (B.148) for Hermitian functions and polynomials, respectively,

$$J = \int_{-\infty}^{\infty} e^{\alpha x}U_n(x+a)\,U_m(x+b)\;dx = (2^{m+n}n!m!\pi)^{-\frac{1}{2}}\frac{\partial^n}{\partial t^n}\frac{\partial^m}{\partial \tau^m}\int_{-\infty}^{\infty} e^{-Q}dx\Bigg|_{t=\tau=0}\,.$$

$$\begin{aligned} Q &= \frac{1}{2}(x+a)^2 + \frac{1}{2}(x+b)^2 + t^2 + \tau^2 - 2t(x+a) - 2\tau(x+b) - \alpha x \\ &= \left(x + \frac{a+b-\alpha-2t-2\tau}{2}\right)^2 - \bar{Q} + P\,, \end{aligned}$$

$$\bar{Q} = \tau(b-a+2\tau+\alpha) + t(a-b+\alpha)\,, \quad 4P = (a-b)^2 - \alpha^2 + 2\alpha(a+b)\,.$$

The integral over x can be easily done, thus,

$$J = \left(2^{m+n} n! m!\right)^{-\frac{1}{2}} e^{-P} \frac{\partial^n}{\partial t^n} \frac{\partial^m}{\partial \tau^m} e^{\tilde{\theta}} \Big|_{\tau=\tau=0} .$$

Differentiating with respect to τ we get

$$J = \left(2^{m+n} n! m!\right)^{-\frac{1}{2}} e^{-P} \frac{\partial^n}{\partial t^n} (b - a + \alpha + 2t)^m e^{t(a-b+\alpha)} \Big|_{t=0} .$$

Now we replace t by q

$$2t = q(b - a + \alpha), \quad \partial_t = \frac{2}{b - a + \alpha} \partial_q, \quad t = 0 \Rightarrow q = 0 .$$

Then

$$J = 2^{\frac{n-m}{2}} (n! m!)^{-\frac{1}{2}} (b - a + \alpha)^{m-n} e^{-P} \frac{\partial^n}{\partial q^n} (1 + q)^m e^{-\frac{(b-a)^2 - \alpha^2}{2} q} \Big|_{q=0} .$$

At this step we use the formula (B.72) to get

$$J = \sqrt{\frac{\Gamma(n+1)}{\Gamma(m+1)}} \left(\frac{b - a + \alpha}{\sqrt{2}}\right)^{m-n} e^{-\frac{(b-a)^2 - \alpha^2 + 2\alpha(a+b)}{4}} L_n^{m-n} \left[\frac{(b-a)^2 - \alpha^2}{2}\right] . \tag{B.161}$$

Expressing Laguerre polynomials via the corresponding functions, we arrive just at the result (B.160).

Setting $\alpha = 0$ we derive from (B.160)

$$\int_{-\infty}^{\infty} U_n(x + a) U_m(x + b)\, dx = I_{m,n} \left[\frac{(b-a)^2}{2}\right] . \tag{B.162}$$

At $a = b$ we get orthogonality relations for the Hermitian functions

$$\int_{-\infty}^{\infty} U_n(x + a) U_m(x + a)\, dx = \int_{-\infty}^{\infty} U_n(x) U_m(x) dx = \delta_{m,n} . \tag{B.163}$$

(16) Consider the sum

$$\begin{aligned} \sigma(z; x, y) &= \sum_{n=0}^{\infty} z^n U_n(x) U_n(y) \\ &= \frac{1}{\sqrt{\pi(1 - z^2)}} \exp\left[\frac{4xyz - (x^2 + y^2)(1 + z^2)}{2(1 - z^2)}\right], \quad |z| < 1 . \end{aligned} \tag{B.164}$$

To justify this result, let us use (B.152)

$$\sigma(z; x, y) = \frac{1}{\sqrt{\pi}} \exp\left(-\frac{x^2 + y^2}{2}\right) \sum_{n=0}^{\infty} \frac{H_n(x) H_n(y) z^n}{2^n n!} .$$

Using the representation (B.139) and equation (B.147), we get

$$\sigma(z;x,y) = \frac{1}{\pi}\exp\left(-\frac{x^2+y^2}{2}\right)\int_{-\infty}^{\infty} dte^{-t^2}\sum_{n=0}^{\infty}\frac{[(x+it)z]^n H_n(y)}{n!}$$

$$= \frac{1}{\pi}\int_{-\infty}^{\infty} dt\exp\left[-\frac{x^2+y^2}{2}+2yz(x+it)-z^2(x+it)^2-t^2\right].$$

The integral over t can be done by means of Equation (B.150). Thus, we arrive at (B.164).

(17) The following limit takes place:

$$\lim_{z\to 1-0}\sigma(z;x,y) = \delta(x-y), \quad -\infty < x < \infty, \quad -\infty < y < \infty. \tag{B.165}$$

To prove it we consider the integral

$$J = \int_{-\infty}^{\infty}\sigma(z;x,y)f(x)dx,$$

where $f(x)$ is an arbitrary function. Taking into account the explicit representation for $\sigma(z;x,y)$ let us perform the replacement

$$x = \sqrt{1-z^2}t + \frac{2yz}{1+z^2}, \quad dx = \sqrt{1-z^2}dt.$$

Thus, we get

$$J = \frac{1}{\sqrt{\pi}}\exp\left[-\frac{(1-z^2)y^2}{2(1+z^2)}\right]\int_{-\infty}^{\infty}\exp\left(-\frac{1+z^2}{2}t^2\right)f\left(\sqrt{1-z^2}t+\frac{2yz}{1+z^2}\right)dt. \tag{B.166}$$

If we let z tend to zero from the left, $\lim_{z\to 1-0} J = f(y)$, we justify relation (B.165). This relation can be treated as a completeness relation for the Hermitian functions,

$$\sum_{n=0}^{\infty}U_n(x)U_n(y) = \delta(x-y). \tag{B.167}$$

It follows from (B.166) at $f(x) = 1$ that

$$\int_{-\infty}^{\infty}\sigma(z;x,y)dy = \sqrt{\frac{2}{1+z^2}}\exp\left[-\frac{(1-z^2)x^2}{2(1+z^2)}\right], \quad |z| < 1. \tag{B.168}$$

(18) The Hermitian functions obey the following integral equation:

$$\int_{-\infty}^{\infty}\sigma(z;x,y)U_n(y)dy = z^n U_n(x), \quad |z| < 1. \tag{B.169}$$

This fact can be checked by substituting expressions (B.152) and (B.164) in the integral (B.169) and using equation (B.149). At $z = i$ we get equation (B.159).

(19) The following sum takes place:

$$F_n(a, b; z) = \sum_{s=0}^{\infty} z^s U_s(a) I_{s,n}(b) = z^n U_n\left[a - \sqrt{\frac{b}{2}}\left(z + \frac{1}{z}\right)\right] e^Q,$$
$$Q = \left[a - \frac{1}{2}\sqrt{\frac{b}{2}}\left(z + \frac{1}{z}\right)\right]\sqrt{\frac{b}{2}}\left(z - \frac{1}{z}\right), \tag{B.170}$$

where $I_{s,n}$ are Laguerre functions (B.4). To justify it one may first use first equations (B.152) and (B.139)

$$F_n(a, b; z) = \frac{1}{\sqrt{\pi\sqrt{\pi}}} \exp\left(-\frac{a^2}{2}\right) \int_{-\infty}^{\infty} dt e^{-t^2} \sum_{s=0}^{\infty} \frac{\left[\sqrt{2}z(a + it)\right]^s}{\sqrt{s!}} I_{s,n}(b).$$

Then, by means of the sum (B.73) and by taking (B.23) into account we reduce the above equation to the form

$$F_n(a, b; z) = \int_{-\infty}^{\infty} \frac{\left[\sqrt{2}z(a + it) - \sqrt{b}\right]^n}{\sqrt{n!\pi\sqrt{\pi}}}$$
$$\times \exp\left[\sqrt{2b}az - \left(t - i\sqrt{\frac{b}{2}}z\right)^2 - \frac{a^2}{2} - \frac{b}{2}\left(1 + z^2\right)\right] dt$$
$$= \frac{z^n \exp\left[\sqrt{2b}az - \frac{a^2}{2} - \frac{b}{2}(1 + z^2)\right]}{\sqrt{2^n n!\sqrt{\pi}}} \frac{2^n}{\sqrt{\pi}}$$
$$\times \int_{-\infty}^{\infty} dt e^{-t^2} \left[a - \sqrt{\frac{b}{2}}\left(z + \frac{1}{z}\right) + it\right]^n.$$

Using again (B.139) we get

$$F_n(a, b; z) = \frac{z^n \exp\left[\sqrt{2b}az - \frac{a^2}{2} - \frac{b}{2}(1 + z^2)\right]}{\sqrt{2^n n!\sqrt{\pi}}} H_n\left[a - \sqrt{\frac{b}{2}}\left(z + \frac{1}{z}\right)\right]. \tag{B.171}$$

Finally, recalling (B.153) we arrive at (B.170).

Bibliography

[1] D. I. Abakarov and V. P. Oleinik, *Electron in the field of a quantized electromagnetic wave and in a homogeneous magnetic field*, Theor. Math. Phys. **12** (1) (1972) 77–87.

[2] A. Akhmeteli, *On Real Functions Instead of the Dirac Spinor Function*, Journ. Math. Phys. **52** (2011) 082303 (5 pages).

[3] A. D. Alhaidari, *The Three-Dimensional Dirac Oscillator in the Presence of Aharonov – Bohm and Magnetic Monopole Potentials*, Found. Phys. Lett. **18** (7) (2005) 651–664.

[4] S. T. Ali, J-P. Antoine and J.-P. Gazeau, *Coherent States, Wavelets and Their Generalizations* (Springer-Verlag, New York 2000).

[5] S. T. Ali, B. Heller and J. P. Gazeau, *Coherent states and Bayesian duality*, Journ. Phys. A: Math. Theor. **41**, 365302 (2008) 1–22.

[6] T. M. Aliev, V. Ya. Fainberg and N. K. Pak, *Path integral for spin: A new approach*, Nucl. Phys. B **429** (2) (1994) 321–343.

[7] Y. Aharonov and D. Bohm, *Significance of electromagnetic potentials in quantum theory*, Phys. Rev. **115** (1959) 485–491.

[8] A. I. Akhiezer and V. B. Berestetskiĭ, *Quantum Electrodynamics* (Interscience Publishers, New York 1965).

[9] J. Ambjørn, B. Durhuus and T. Jonsson, *A random walk representation of the Dirac propagator*, Nucl. Phys. **330** (1990) 509–522.

[10] V. G. Bagrov and V. A. Bordovitsyn, *Motion of an electron in electromagnetic fields of special forms*, ZhVMMF **8** (3) (1968) 691–695.

[11] V. G. Bagrov and V. A. Bordovitsyn, *Motion of a polarized electron, possessing the vacuum magnetic moment in crossed electromagnetic fields*, Izv. Tomsk. Polytekn. Inst. **156** (1969) 215–223.

[12] V. G. Bagrov, I. M. Ternov and B. V. Kholomay, *On certain exact solutions of the Dirac equation*, Vestn. MGU, ser. Fiz. Astron. **3** (1970) 275–280.

[13] V. G. Bagrov and V. A. Bordovitsyn, *Spin precession of an electron moving in crossed fields*, Izv. Tomsk. Polytekhn. Inst. **184** (Tomsk 1970) 20–23.

[14] V. G. Bagrov, P. V. Bozrikov and D. M. Gitman, *Charge in a quantized plane-wave field*, Radiophysics and Quantum Electronics **16** (1) (1973) 97–105.

[15] V. G. Bagrov, P. V. Bozrikov and D. M. Gitman, *Electron in a quantized electromagnetic plane-wave field*, Theor. Math. Phys. **14** (2) (1973) 202–210.

[16] V. G. Bagrov, P. V. Bozrikov, D. M. Gitman and P. M. Lavrov, *An Electron in a Field of a Plane Quantized Monochromatic Electromagnetic Wave*, Soviet Physics Journ. **16** (1973) 1082–1085.

[17] V. G. Bagrov, D. M. Gitman and V. A. Kuchin, *Electron in the Parallel-Aligned Fields of Classical and Quantized Plane Waves*, Soviet Physics Journ. **17** (1974) 952–956.

[18] V. G. Bagrov, P. V. Bozrikov and D. M. Gitman, *Fermion with Anomalous Moment in a Field of Quantized Plane Wave*. Soviet Physics Journ. **17** (1974) 864–866.

[19] V. G. Bagrov, P. V. Bozrikov, D. M. Gitman, Yu. I. Klimenko and A. I. Khudomyasov, *Scattering of Electromagnetic Waves on the Neutral Fermion with Electric and Magnetic Moments*, Soviet Physics Journ. **17** (1974) 1027–1028.

[20] V. G. Bagrov, P. V. Bozrikov and V. N. Zadorozhny, *Electron in a classical and quantum field of a special form*, Deposited with VINITI, 1984, No. 3021-84 Dep.

[21] V. G. Bagrov, I. L. Buchbinder, D. M. Gitman and P. M. Lavrov, *Coherent states of an electron in a quantized electromagnetic wave*, Theor. Math. Phys. **33** (3) (1977) 419–426.

[22] V. G. Bagrov, D. M. Gitman, V. A. Kuchin and P. M. Lavrov, *Basics of Electrodynamics of Electrons in a Quantized Field of Plane Wave. I., II.* Soviet Physics Journ. **17** (1974) 1709–1712; **18** (1975) 909–912.

[23] V. G. Bagrov, D. M. Gitman and Sh. M. Shvartsman, *Electron in Quantized Plane Wave Field and Classical Field of Longitudinal Electric Wave*, Soviet Physics Journ. **18** (1975) 347–350.

[24] V. G. Bagrov, D. M. Gitman and V. A. Kuchin, *Eigenfunctions of Linear Combinations of Creation and Annihilation Operators*, Soviet Physics Journ. **18** (1975) 1218–1223.

[25] V. G. Bagrov, D. M. Gitman and A. V. Yushin, *Solutions for the motion of an electron in electromagnetic fields*, Phys. Rev. D **12** (10) (1975) 3200–3202.

[26] V. G. Bagrov, V. N. Shapovalov and G. G. Eckle, *Complete sets and variable separation in the Dirac equation*, Deposited with *VINITI*, 1975, No. 405-75. Dep. p. 17.

[27] V. G. Bagrov, A. G. Meshkov, V. N. Shapovalov and A. V. Shapovalov, *Separation of Variables in the K–G Equation. I, II, III, IV*, Soviet Physics **16** (1973) 1533–1538; 1659–1665; **17** (1974) 812–815; **18** (1975) 431–432.

[28] V. G. Bagrov, Yu. I. Klimenko, V. N. Shapovalov, V. M. Shakhmatov and G. G. Eckle, *New exact solutions of the Dirac equation. I.* Soviet Physics Journ. **16** (1973) 969–974.

[29] V. G. Bagrov, D. M. Gitman, V. N. Zadorozhny, P. M. Lavrov and V. N. Shapovalov, *New exact solutions of the Dirac equation. II.* Soviet Physics Journ. **18** (1975) 461–464.

[30] V. G. Bagrov, D. M. Gitman, V. N. Zadorozhny, P. M. Lavrov and V. N. Shapovalov, *New exact solutions of the Dirac equation. III.* Soviet Physics Journ. **18** (1975) 905–908.

[31] V. G. Bagrov, D. M. Gitman, A. G. Meshkov, V. I. Simanchuk, N. I. Fedosov, A. V. Shapovalov and V. N. Shapovalov, *New exact solutions of the Dirac equation. IV*, Soviet Physics Journ. **18** (1975) 1123–1127.

[32] V. G. Bagrov, N. N. Byzov, D. M. Gitman, Yu. I. Klimenko, A. G. Meshkov, V. N. Shapovalov and V. M. Shakhmatov, *New exact solutions of the Dirac equation. V.* Soviet Physics Journ. **18** (1975) 1304–1309.

[33] V. G. Bagrov, D. M. Gitman, A. G. Meshkov and V. N. Shapovalov, *Addition to the papers "New exact solutions of the Dirac equation. II, III."* Soviet Physics Journ. 20 (1977) 103–104.

[34] V. G. Bagrov, I. L. Buchbinder and D. M. Gitman, *Coherent states of relativistic particles*, Soviet Physics Journ. **18** (1975) 1180–1181.

[35] V. G. Bagrov, I. L. Buchbinder and D. M. Gitman, *Coherent states of a relativistic particle in an external electromagnetic field*, Journ. Phys. A **9** (11) (1976) 1955–1965.

[36] V. G. Bagrov, D. M. Gitman and V. A. Kuchin, *External field in quantum electrodynamics and coherent states* In *Actual problems of theoretical physics* (Moscow State University, Moscow, 1976) pp. 334–342.

[37] V. G. Bagrov and D. M. Gitman, *Exact solutions of the relativistic wave equations in external electromagnetic fields*. In *Quantum Electrodynamics with External Field* (Tomsk University Press, Tomsk 1977) 7–10.

[38] V. G. Bagrov, D. M. Gitman, V. N. Zadorozhny and P. M. Lavrov, *Some Properties of Exact Solutions in the Problem of Electron in a Quantized Plane-Wave Field*, Soviet Physics Journ. **20** (1977) 283–289.

[39] V. G. Bagrov, D. M. Gitman, A. V. Shapovalov and V. N. Shapovalov, *New exact solutions of the Dirac equation. VI.* Soviet Physics Journ. **20** (1977) 783–790.

[40] V. G. Bagrov, D. M. Gitman, N. B. Sukhomlin, A. V. Shapovalov and V. N. Shapovalov, *New exact solutions of the Dirac equation. VII.* Soviet Physics **20** (1977) 871–876.

[41] V. G. Bagrov, D. M. Gitman and A. V. Shapovalov, *Integrals of Motion in the Problem of Electron in a Quantized Plane Electromagnetic Wave*, Soviet Physics Journ. **20** (1977) 233–237.

[42] V. G. Bagrov, D. M. Gitman, V. N. Zadorozhny, N. B. Sukhomlin and V. N. Shapovalov, *New exact solutions of the Dirac equation. VIII.* Soviet Physics Journ. **21** (1978) 142–146.

[43] V. G. Bagrov, D. M. Gitman, V. N. Zadorozhny, A. V. Shapovalov and V. N. Shapovalov, *New exact solutions of the Dirac equation. IX.* Soviet Physics Journ. **21** (1978) 304–307.

[44] V. G. Bagrov, P. M. Lavrov and G. I. Flesher, *Electron with anomalous moments in an electromagnetic field.* Soviet Physics Journ. **22** (1979) 419–425.

[45] V. G. Bagrov, P. M. Lavrov and G. I. Flesher, *Motion of Charged Particles with Anomalous Moments in an Electromagnetic Field.* Soviet Physics Journ. **23** (1980) 91–92.

[46] V. G. Bagrov, D. M. Gitman and V. N. Zadorozhny, *New exact solutions of the Dirac equation. X.* Soviet Physics Journ. **23** (1980) 276–281.

[47] V. G. Bagrov, D. M. Gitman and V. N. Shapovalov, *Electron motion in longitudinal electromagnetic fields.* Journ. Math. Phys. **23** (12) (1982) 2558–2561.

[48] V. G. Bagrov and M. D. Noskov, *New exact solutions of the Dirac equation. XI.* Soviet Physics Journ. **27** (1984) 1030–1034.

[49] V. G. Bagrov, J. C. Barata, D. M. Gitman and W. F. Wreszinski, *Aspects of two-level systems under external time-dependent fields*, J. Phys. A**34**, 10 869 (2001).

[50] V. G. Bagrov, M. C. Baldiotti, D. M. Gitman and V. V. Shamshutdinova, *Darboux transformation for two-level system*, Annalen der Physik **14** (2005) 390–397.

[51] V. G. Bagrov, M. C. Baldiotti, D. M. Gitman and A. D. Levin, *Spin equation and its solutions*, Ann. der Physik **14**(11–12) (2005) 764–789.

[52] P. V. Bozrikov, V. N. Zadorozhny *Partially-coherent states of the system composed of an electron plus a quantized plane-wave electromagnetic field*, Deposited with *VINITI*, 1985, No. 643-85 Dep.

[53] V. G. Bagrov, M. D. Noskov and V. N. Shapovalov, *New exact solutions of the Dirac equation. XII.* Soviet Physics Journ. **28** (1985) 70–74.

[54] V. G. Bagrov, N. I. Fedosov, G. F. Kopytov, S. S. Oxsyzyan and V. B. Tlyachev, *Radiational Self – polarization of Electrons Moving in the Electromagnetic Plane-Wave Field*, Nuovo Cimento **103** (5) (1989) 549–560.

[55] V. G. Bagrov and D. M. Gitman, *Exact Solutions of Relativistic Wave Equations*, Math. its Appl., Sov. ser. **39** (Kluwer, Dordrecht 1990).

[56] V. G. Bagrov, D. M. Gitman and V. B. Tlyachev, *Solutions of relativistic wave equations in superpositions of Aharonov-Bohm, magnetic, and electric fields*, J. Math. Phys. **42** (2001) 1933–1959.

[57] V. G. Bagrov, D. M. Gitman, A. Levin and V. B. Tlyachev, *Impact of Aharonov-Bohm Solenoid on Particle Radiation in Magnetic Field*, Mod. Phys. Lett. A **16** (2001) 1171–1179.

[58] V. G. Bagrov, D. M. Gitman, A. Levin and V. B. Tlyachev, *Aharonov-Bohm effect in cyclotron and synchrotron radiations*, Nucl. Phys. B **605** (2001) 425–454.

[59] V. G. Bagrov, M. C. Baldiotti, D. M. Gitman and I. V. Shirokov, *New Solutions of Relativistic Wave Equations in Magnetic Field and Longitudinal Fields*, Journ. Math. Phys. **43** (5) (2002) 2284–2305.

[60] V. G. Bagrov, D. M. Gitman and V. B. Tlyachev, *l-dependence of particle radiation in magnetic-solenoid field and Aharonov-Bohm effect*, Int. Journ. Mod. Phys. A**17** (2002) 1045–1048.

[61] V. G. Bagrov and D. M. Gitman, *Non–Volkov Solutions for a Charge in a Plane Wave*, Ann. der Phys. **14** (8) (2005) 467–478.

[62] V. G. Bagrov, M. C. Baldiotti and D. M. Gitman, *Charged particles in crossed and logitudinal electromagnetic fields and beam guides*, Journ. Math. Phys. **48**, 8 (2007) 082305 (15 pages).

[63] V. G. Bagrov, S. P. Gavrilov, D. M. Gitman and D. P. Meira Filho, *Coherent states of non-relativistic electron in magnetic–solenoid field*, Journ. Physics. A**43** (2010) 3540169; *Coherent and semiclassical states in magnetic field in the presence of the Aharonov-Bohm solenoid*, J. Phys. A: Math. Theor. **44** (2011) 055301.

[64] V. G. Bagrov, S. P. Gavrilov, D. M. Gitman and K. Gorska, *Completeness for coherent states in magnetic-solenoid field*, Journ. Phys. A **45** (2012) 244008 (11pp).

[65] V. G. Bagrov, D. M. Gitman and A. D. Levin, *Structure of the Electromagnetic Field Allowing Exact Solution of the Schrödinger Equation in Superposition with an Aharonov–Bohm Field*, Russ. Phys. Journ. **54** (8) (2011) 855–866.

[66] V. G. Bagrov, D. M. Gitman and A. Levin, *Quantum motion in superposition of Aharonov-Bohm with some additional electromagnetic fields*, Journ. Math. Phys. **55**/5, 052304 (2012) (16 pages).

[67] V. G. Bagrov, D. M. Gitman, E. S. Macedo and A. S. Pereira, *Coherent states of inverse oscillator and related problems*, Journ. Phys. A: Math. Theor. **46** (2013) 325305 (13pp).

[68] V. G. Bagrov, D. M. Gitman and A. Pereira, *Coherent and semiclassical states of a free particle*, submitted to UFN 28/01/14 (2014).

[69] V. N. Baier, V. M. Katkov and V. S. Fadin, *Radiation of relativistic electrons* (Atomizdat, Moscow 1973).

[70] V. N. Baier and A. I. Mil'shtein, *Operatorial Green's function of electron in the Coulomb field*, DAN SSSR **235** (1977) 67–69.

[71] V. N. Baier, V. M. Katkov and V. M. Strakhovenko, *Operator approach to quantum electrodynamics in an external field. Electron loops*, Zh. Eksp. Teor. Fiz. **68** (1975) 405–420; Sov. Phys.–JETP **41** (1975) 198–204.

[72] V. N. Baier and V. M. Katkov, *Pair creation by a photon in a strong magnetic field*, Phys. Rev. D **75**: 073009 (2007).

[73] I. G. Baranov, *Wave Functions of a Nucleon in a Plane Electromagnetic Wave*, Soviet Physics **10** (1967) 23–24.

[74] J. C. Barata and D. A. Cortez, *Time evolution of two-level systems driven by periodic fields*, Phys. Lett. A **301** (2002) 350–360.

[75] J. C. Barata, *On Formal Quasi-Periodic Solutions of the Schrödinger Equation for a Two-Level System with a Hamiltonian Depending Quasi-Periodically in Time*, Rev. Math. Phys. **12** (2000) 25–64.

[76] B. M. Barbashov, *Functional integrals in quantum electrodynamics and infrared asymptotic behaviour of Green's functions*, Zh. Eksp. Teor. Fiz. **48** (1965) 607–621.

[77] K. Bardakci and M. B. Halpern, *Theories of infinite momentum*, Phys. Rev. **176** (5) (1968) 1686–1699.

[78] V. Bargmann and E. P. Wigner, *Group theoretical discussion of relativistic wave equations*, Proc. Nat. Acad. Sci., USA **34** (5) (1948) 211–223.

[79] V. Bargmann, I. L. Michel and V. L. Telegdi, *Presession of the polarization of particles moving in a homogeneous electromagnetic field*. Phys. Rev. Lett. **2** (10) (1959) 435–436.

[80] A. O. Barut and J. Kraus, *Solution of the Dirac equation with Coulomb and magnetic moment interactions*, Journ. Math. Phys. **17** (4) (1976) 506–508.

[81] I. A. Batalin and E. S. Fradkin, *Quantum electrodynamics in external fields*, Theor. Math. Phys. **5** (1970) 190–218.

[82] H. A. Bethe and E. E. Salpeter, *Quantum mechanics of atoms with one and two electrons*, Handbuch der Physik (Bd. XXXV/1, Berlin 1957); *Quantum Mechanics of One- and Two-Electron Systems*, Encyclopedia of Physics, Vol. XXXV/1 (Julius Springer, Berlin 1957).

[83] W. Becker and H. Mitter, *Relativistic theory of nucleons in laser fields*, Journ. Phys. A **7** (11) (1974) 1266–1273.

[84] V. V. Belov, *Exact Green's function of equations of quantum mechanics in an electromagnetic field. I*, Soviet Physics Journ. **18** (1975) 1536–1539.

[85] V. V. Belov, *Exact Green's function of equations of quantum mechanics in an electromagnetic field. II*, Soviet Physics Journ. **18** (1975) 1540–1545.

[86] V. B. Berestetsky, E. M. Lifshits and L. P. Pitayevsky, *Relativistic Quantum Theory. I.* (Nauka, Moscow 1968).

[87] F. A. Berezin, *General concept of quantization*, Comm. Math. Phys. **40** (1975) 153–174.

[88] F. A. Berezin and M. S. Marinov, Pisma Zh. Eksp. Theor. Fiz. **21** (1975) 678 [JETP Lett. **21** (1975) 320]; Ann. Phys. (N.Y.) **104** (1977) 336–362.

[89] F. A. Berezin, *The Method of Second Quantization* (Academic Press, New York 1966).

[90] F. A. Berezin and M. A. Shubin, *Schrödinger Equation* (Kluwer Publ., New York 1991).

[91] P. G. Bergmann, *Introduction to the Theory of Relativity* (Prentice-Hall, Inc., New York 1942).

[92] I. Ya. Berson, *Electron in a quantized field of a monochromatic electromagnetic field*, Zh. Eksp. Teor. Fiz. **56** (5) (1969) 1627–1633.

[93] I. Ya. Berson, *Motion of an electron in an electromagnetic wave and a magnetic field parallel to it*, Izv. AN Latv. SSR. Fiz. **5** (1969) 3–8.

[94] I. Ya. Berson, *Electron in a quantized field of a plane electromagnetic wave*, Izv. AN Latv. SSR. Fiz. **3** (1970) 3–8.

[95] I. Berson and J. Vaidmanis, *Electron in the field of two monochromatic electromagnetic waves*, Journ. Math. Phys. **14** (10) (1973) 1481–1484.

[96] I. Bialynicki-Birula, *On internal degrees of freedom of particles in quantum field theory*, Bull. Acad. Polon. Sci., cl. 3,5, No.9 (1957) 905–909; I. Bialnynicki-Birula and Z. Bialynicki-Birula, *On the polarization and magnetic moment of spinor particles in quantum field theory*, Bull. Acad. Polon. Sci., cl. 3,5, No. 12 (1957) 1119–1121; I. Bialynicki-Birula, *On operators characterising the internal degrees of freedom of the interacting fields*. Bull. Acad. Polon. Sci., cl. 3,5, No.12 (1957) 1123–1126.

[97] I. Bialynicki-Birula, *Particle Beams Guided by Electromagnetic Vortices: New Solutions of the Lorentz, Schrödinger, Klein–Gordon, and Dirac Equations*, Phys. Rev. Lett. **93** (2) (2004) 020402 (4 pages).

[98] F. Bloch and A. Siegert, *Magnetic Resonance for Nonrotating Fields*, Phys. Rev. **57** (1940) 522–527.

[99] I. D. Bjorken, I. B. Kogut and D. E. Soper, *Quantum electrodynamics at infinite momentum: Scattering from an external field*, Phys. Rev. **3** (6) (1971) 1382–1399.

[100] N. N. Bogoliubov and D. V. Shirkov, *Introduction to the Theory of Quantized Fields*, 3rd edn. (John Wiley & Sons, New York 1980).

[101] A. A. Borgard and D. Ya. Karpenko, *Electron in a uniform electromagnetic field and in the field of an arbitrarily polarized plane wave*, Ukr. Fiz. Zhurn. **19** (1974) 227–236.

[102] N. V. Borisov and P. P. Kulish, *Path integral in superspace for a relativistic spinor particle in an external gauge field*, Teor. Math. Fiz. **51** (1982) 335–540.

[103] A. A. Borgard and D. Ya. Karpenko, *Hyperaxial symmetry of the relativistic wave equations*. Ukr. Fiz. Zhurn. **20** (4) (1975) 566–573.

[104] R. Brauer and H. Weyl, *Spinors in n dimensions*, Amer. J. Math. **57** (1935) 425–449.

[105] M. L. Brillouin, *Les Moments de Rotation et le Magnétisme Dans la Mécanique Ondulatorie*, Le Journal de Physique et le Radium **VIII**, No 2 (1927) 74–84.

[106] F. Calogero, *Covariant spin operators and associated conservation laws for a spinor field*, Nuovo Cimento **20** (2) (1961) 280–296.

[107] V. Canuto and C. Chiuderi, *Solution of the Dirac equation in orthogonal electric and magnetic fields*, Lett. Nuovo Cimento **2** (6) (1969) 223–227.

[108] A. Chakrabarti, *Exact solution of the Dirac–Pauli equation for a class of fields: Precession of polarization*. Nuovo Cimento, A **56** (3) (1968) 604–624.

[109] C. Cohen-Tannoudji, B. Diu and F. Laloë, *Quantum Mechanics* (Wiley, New York 1977).

[110] V. V. Rabin (ed.), *Coherent States in Quantum Theory* (collection of translations into Russian) (Mir, Moscow 1972).

[111] F. A. B. Coutinho and J. F. Perez, *Boundary-conditions in the Aharonov-Bohm scattering of Dirac particles and the effect of Coulomb interaction*, Phys. Rev. D **48** (1993) 932–939.

[112] Ramesh Chand and G. Szamosi, *Solution of the Dirac equation with anomalous magnetic moment*, Lett. Nuovo Cim. **22** (16) (1978) 660–664.

[113] A. S. Davydov, *Quantum Mechanics* 2nd edn. (Pergamon Press, Oxford, New York 1976).

[114] Yu. A. Danishevsky and Sh. M. Shvartsman, *Green's Function of Charged Scalar Particles Moving in the Field of a Traveling Electric Wave.* Soviet Physics Journ. **21** (1978) 1021–1023.

[115] G. Darboux, *Sur une proposition relative aux équations linéaires*, C. R. Acad. Sci. **94** (1882) 1456–1459.

[116] C. G. Darvin, *The wave equations of the electron*, Proc. Roy. Soc. London A **118** (780) (1928) 654–680.

[117] M. Demeur, *Solutions singulieres des equations de Klein–Gordon et de Dirac, Tenant compte d'un champ electrique exterieur*, Physica **17** (1951) 933–937.

[118] P. A. M. Dirac, *Quantized singularities in the electromagnetic field*, Proc. Roy. Soc. London A **133** (1931) 60–72.

[119] P. A. M. Dirac, *The Theory of Magnetic Poles*, Phys. Rev. **74** (7) (1948) 817–830.

[120] P. A. M. Dirac, *The Quantum theory of the Electron. I, II.* Proc. Roy. Soc. A **117** (778) (1928) 610–624; Proc. Roy. Soc. A **118** (779) (1928) 351–361.

[121] P. A. M. Dirac, *The principles of quantum mechanics* (Clarendon Press, Oxford 1958).

[122] V. V. Dodonov, I. A. Malkin and V. I. Man'ko, *Coherent states and Green's functions of relativistic quadratic systems*, Physica **82**A (1976) 113–133.

[123] V. V. Dodonov, I. A. Malkin and V. I. Man'ko, *Green's functions for relativistic particles in non-uniform external fields*, Journ. Phys. A **9** (1976) 1791–1796.

[124] V. V. Dodonov, I. A. Malkin and V. I. Man'ko, *The Green's function of the stationary Schrödinger equation for a particle in a uniform magnetic field*, Phys. Lett. A **51** (1975) 133–134.

[125] V. V. Dodonov and V. I. Man'ko, *Invariants and correlated states of nonstationary quantum systems.* In *Invariants and the Evolution of Nonstationary Quantum Systems*, Proceedings of Lebedev Physics Institute **183**, M. A. Markov (ed.) (Nauka, Moscow 1987) 71–181, [translated by Nova Science, Commack, New York, 1989, pp. 103–261].

[126] Th. de Donder and Fr. H. van den Dungen, *La quantification deduite de la Gravifique Einsteinienne.* Compt. rend. sci. **183** (183) (1926) 22–26.

[127] G. G. Eckle, *Algebraic properties of the Dirac equation*, Soviet Physics Journ. **15** (1972) 220–224.

[128] W. Ehrenberg and R. E. Siday, *The Refractive Index in Electron Optics and the Principles of Dynamics*, Proc. Phys. Soc. London B **62** (1949) 8–21.

[129] V. Ya. Fainberg and A. V. Marshakov, *Representation of the propagator of a Dirac particle in an external gravitational field as a sum over paths*, JETP Lett. **47** (1988) 565–568; *A propagator for the fermionic string*, Phys. Lett. B **211** (1988) 81–85; *Local supersymmetry and Dirac particle propagator as a path integral*, Nucl. Phys. B **306** (1988) 659–676.

[130] M. V. Fedorov and A. E. Kazakov, *An electron in a quantized plane wave and in a constant magnetic field*, Zeit. Phys. B **261** (3) (1973) 191–202.

[131] M. V. Fedorov, *Charge Symmetry of Wave Functions for an Electron in a Quantized Electromagnetic Wave Field*, Soviet Physics Journ. **18** (1975) 773–776.

[132] M. V. Fedorov, *Green's Function for an Electron in the Field of a Quantized Wave*, Soviet Physics Journ. **18** (1975) 836–839.

[133] F. I. Fedorov, A. F. Radyuk and V. G. Zhulego, *Exact solutions for a vector meson in quantized field of a monochromatic electromagnetic plane wave*, Acta Physica Polonica B **9** (9) (1978) 787–791.

[134] R. P. Feynman and A. R. Hibbs, *Quantum Mechanics and Path Integrals* (McGraw-Hill, New York 1965).

[135] P. Filipowicz, *Exact solution of the Dirac equation for the electron interacting with a quantized plane electromagnetic wave*, Bull. Acad. Pol. Sci. Phys. et Astron. **28** (2) (1980) 79–82.

[136] R. P. Feynman, *Space-Time Approach to Non-Relativistic Quantum Mechanics*, Rev. Mod. Phys. **20** (2) (1948) 367–387.

[137] R. P. Feynman, *Mathematical formulation of the quantum theory of electromagnetic interaction*, Phys. Rev. **80** (2) (1950) 440–457.

[138] R. P. Feynman, *An Operator Calculus Having Applications in Quantum Electrodynamics*, Phys. Rev. **84** (1951) 108–128.

[139] P. Filipowicz, *Relativistic electron in a quantized plane wave*, Journ. Phys. A **18** (10) (1985) 1675–1684.

[140] V. A. Fock, *Zur Schrödingerschen Wellenmechanik*. Zeit. Phys. **38** (3) (1926) 242–250; *Über die invariante Form der Wellen – und der Bewegungsgleichungen für einen geladenen Massenpunkt*, Zeit. Phys. **39** (2–3) (1926) 226–232.

[141] R. P. Feynman, *Simulating Physics with Computers*, Int. J. Theor. Phys. **21** 467–488 (1982); *Quantum Mechanical Computers*, Found. Phys. **16**, 507–531 (1986); D. Deutsch, *Quantum Theory, the Church-Turing Principle and the Universal Quantum Computer*, Proc. Roy. Soc. London A **400** (1985) 97–117.

[142] R. P. Feynman and F. L. Vernon, *Geometrical Representation of the Schrödinger Equation for Solving Maser Problems*, Journ. Applied Phys. **28** (1957) 49–52.

[143] V. Fock, *Die Eigenzeit in der Klassischen und in der Quantenmechanik*, Sow. Phys. **12** (1937) 404–425.

[144] E. S. Fradkin, *Application of functional methods in quantum field theory and quantum statistics (II)*. Nucl. Phys. **76** (1966) 588–624.

[145] E. S. Fradkin, D. M. Gitman and Sh. M. Shvartsman, *Quantum Electrodynamics with Unstable Vacuum* (Springer-Verlag, Berlin 1991).

[146] E. S. Fradkin and D. M. Gitman, *Furry picture for quantum electrodynamics with pair-creating external field*, Fortschr. Phys. **29** (1981) 381–411.

[147] E. S. Fradkin and D. M. Gitman, *Path integral representation for the relativistic particle propagators and BFV quantization*, Phys. Rev., D **44** (1991) 3230–3236.

[148] D. M. Fradkin and R. H. Good, *Electron polarization operators*, Rev. Mod. Phys. **33** (2) (1961) 343–352; *Tensor operator for electron polarization*, Nuovo Cimento **22** (3) (1961) 643–649.

[149] W. H. Furry, *On Bound States and Scattering in Positron Theory*, Phys. Rev. **81** (1951) 115–124.

[150] S. P. Gavrilov, D. M. Gitman and Sh. M. Shvartsman, *The Green's functions in an external electric field and in its combination with a magnetic field and a plane-wave field*, Kratk. Soob. Fiz., No. 2 (1979) 22–25.

[151] S. P. Gavrilov, D. M. Gitman and Sh. M. Shvartsman, *Green's functions in an external electric field*, Yad. Fiz. **29** (1979) 1097–1110 (Soviet Journal of Nuclear Physics **29** (1979) 567–572).

[152] S. P. Gavrilov, D. M. Gitman and Sh. M. Shvartsman, *Green's functions in an external electric field and in its combination with a magnetic and a plane-wave field*, Yad. Fiz. **29** (1979) 1392–1405.

[153] S. P. Gavrilov and D. M. Gitman, *Processes of pair creation and particle scattering in a constant and plane-wave field*. Soviet Physic Journ. **24** (1981) 108–111.

[154] S. P. Gavrilov and D. M. Gitman, *Proper time and path integral representations for the commutation function*, Journ. Math. Phys. **37** (7) (1996) 3118–3130.

[155] S. P. Gavrilov, D. M. Gitman and A. A. Smirnov, *Dirac equation in the magnetic-solenoid field*, Europ. Phys. Journ. C **30** (2003) 009; Europ. Phys. Journ. C direct **30** (2003) 009; **32** Suppl.

(2003) 119–142; *Self-adjoint extensions of Dirac Hamiltonian in magnetic-solenoid field and related exact solutions*, Phys. Rev. A **67** (2003) 024103(4).
[156] S. P. Gavrilov, D. M. Gitman and A. A. Smirnov, *Green functions of the Dirac equation with magnetic-solenoid field*, J. Math. Phys. **45** (2004) 1873–1886.
[157] S. P. Gavrilov, D. M. Gitman, A. A. Smirnov and B. L. Voronov, *Dirac fermions in a magnetic-solenoid field*, "Focus on Mathematical Physics Research", Charles V. Benton (ed.) (Nova Science Publishers, New York, 2004) pp. 131–168.
[158] S. P. Gavrilov and D. M. Gitman, *Vacuum instability in external fields*, Phys. Rev. D **53** (1996) 7162–7175.
[159] S. P. Gavrilov, D. M. Gitman and J. L. Tomazelli, *Density matrix of a quantum field in a particle-creating background*, Nucl. Phys. B **795** [FS] (2008) 645–677.
[160] S. P. Gavrilov and D. M. Gitman, *One-loop energy-momentum tensor in QED with electric-like background*, Phys. Rev. D**78**, 045017(35) (2008); *Energy-momentum tensor in thermal strong-field QED with unstable vacuum*, Journ. Phys. A: Math. Theor. 41 (2008) 164046.
[161] S. P. Gavrilov and D. M. Gitman, *Consistency Restrictions on Maximal Electric-Field Strength in Quantum Field Theory*, Phys. Rev. Lett. **101**, 130403(4) (2008).
[162] S. P. Gavrilov and D. M. Gitman, *Quantization of the Relativistic Particle*, Class. Quant. Grav. **17** issue 19 (2000) L133-L139; *Quantization of Point-Like Particles and Consisitent Relativistic Quantum Mechanics*, Int. J. Mod. Phys. A**15** (2000) 4499–4538.
[163] A. D. Gazazian, *Light scattering from a quantum harmonic oscillator*, Theor. Math. Phys. **10** (3) (1972) 388–398.
[164] J. P. Gazeau, *Coherent States in Quantum Physics* (Wiley-VCH, Berlin, 2009); S. T. Ali, J-P. Antoine and J-P. Gazeau, *Coherent States, Wavelets and Their Generalizations* (Springer-Verlag, New York, Berlin, Heidelberg, 2000).
[165] J. P. Gazeau and J. Klauder, *Coherent States for Systems with Discrete and Continuous Spectrum*, J. Phys. A: Math. Gen. **32** (1999) 123–132.
[166] J. Geheniau, *Les fonctions singulieres de l'equation de Klein–Gordon, Tenant compte d'un champ magnetique exterieur*, Physica **16** (1950) 822–830.
[167] J. Geheniau and M. Demeur, *Solutions singulieres des equations de Dirac, Tenant compte d'un champ magnetique exterieur*, Physica **17** (1951) 71–75.
[168] J. Ben Geloun and J. R. Klauder, *Ladder operators and coherent states for continuous spectra, J. Phys. A: Math. Theor.* **42**, 375209 (2009); J. Ben Geloun, J. Hnybida and J. R. Klauder, *Coherent states for continuous spectrum operators with non-normalizable fiducial states, J. Phys. A: Math. Theor.* **45** (2012) 085301 (14pp).
[169] R. Gilmore, *Geometry of symmetrized states*, Ann. Phys. *(NY)* **74** (1972) 391–463.
[170] D. M. Gitman, V. M. Shakhmatov and Sh. M. Shvartsman, *Completeness and orthogonality on the null plane of one class of solutions of the relativistic wave equations*, Soviet Physics Journ. **18** (8) (1975) 1097–1101.
[171] D. M. Gitman, *Processes of arbitrary order in quantum electrodynamics with a pair-creating external field*, Journ. Phys. A **10** (1977) 2007–2020.
[172] D. M. Gitman, *Path integrals and pseudoclassical description for spinning particles in arbitrary dimensions*, Nucl. Phys. B **488** (1997) 490–512.
[173] D. M. Gitman and P. M. Lavrov, *Electron motion in a homogeneous magnetic field and in quantized electromagnetic plane-wave field*, Soviet Physics Journ. **18** (1975) 29–34.
[174] D. M. Gitman and I. V. Tyutin, *Canonical quantization of fields with constraints* (Nauka, Moscow 1986); *Quantization of Fields with Constraints* (Springer-Verlag, Berlin 1990).
[175] D. M. Gitman, M. D. Noskov and Sh. M. Shvartsman, *Quantum effects in a combination of a constant uniform field and a plane wave field*, Intern. Journ. Mod. Phys. A **6** (1991) 4437–4489.

[176] D. M. Gitman, I. V. Tyutin and B. L. Voronov, *Self-adjoint Schrödinger and Dirac operators with Aharonov-Bohm and magnetic-solenoid fields*, Phys. Scr. **85** (2012) 045003.
[177] D. M. Gitman, I. V. Tyutin and B. L. Voronov, *Self-adjoint Extensions in Quantum Mechanics. General theory and applications to Schrödinger and Dirac equations with singular potentials* (Birkhäuser, Boston 2012).
[178] D. M. Gitman and A. L. Shelepin, *Coherent States of the $SU(N)$ groups*, Journ. Phys. A **26** (1993) 313–327.
[179] D. M. Gitman and A. L. Shelepin, *Coherent States of SU(l,1) Groups*, Journ. Phys. A **26** (1993) 7003–7018.
[180] D. M. Gitman, D. A. Petrusevich and A. L. Shelepin, *Semiclassical description of quantum rotator in terms of SU(2) coherent states*, Phys. Scr. **88** (2013) 045005.
[181] D. M. Gitman and Sh. M. Shvartsman, *Spinor and isospinor structure of relativistic particle propagator*, Phys. Lett. B **318** (1993) 122–126; Errata, Phys. Lett. B **331** (1994) 449,450.
[182] D. M. Gitman and S. I. Zlatev, *Spin factor in path integral representation for Dirac propagator in external field*, Phys. Rev. D**55** (1997) 7701–7714.
[183] D. M. Gitman, S. I. Zlatev and W. da Cruz, *Spin Factor and Spinor Structure of Dirac Propagator in Constant Field*, Bras. Journ. Phys. **26** (1996) 419–425.
[184] R. J. Glauber, *The quantum theory of optical coherence, Phys. Rev.* **130** (6) (1963) 2529–2539; *Coherent and incoherent states of the radiation field*, Phys. Rev. **131** (6) (1963) 2766–2788.
[185] D. M. Gitman, S. I. Zlatev and P. B. Barros, *Application of Path Integration to Operator Calculus*, Journ. Phys. A: Math. Gen. **31** (1998) 7791–7799.
[186] R. H. Good Jr. and M. E. Rose, *Relation between Foldy–Wouthuysen and Lorentz transformations.* Nuovo Cimento **24** (5) (1962) 864–869.
[187] W. Gordon, *Der Comptoneffekt nach der Schrödingerschen Theorie*, Zeit. Phys. **40** (1–2) (1926) 117–133.
[188] W. Gordon, *Die Energieniveaus des Wasserstoffatoms nach der Dirackschen Quanten Theorie des Electrons*, Zs. Phys. **48** (102) (1928) 11–14.
[189] E. U. Gordon and G. H. Shortley, *The Theory of Atomic Spectra* (Cambridge University Press, Cambridge 1935).
[190] E. P. Gordov and S. D. Tvorogov, *Eigenfunctions of the Vector-Potential Operator of the Electromagnetic Field and Semiclassical Electrodynamics*, Phys. Rev. D **8** (1973) 3286–3293.
[191] I. S. Gradshtein and I. W. Ryzhik, *Table of Integrals, Series, and Products* (Academic Press, New York 1994).
[192] W. Greiner, *Relativistic Quantum Mechanics. Wave Equations* (Springer-Verlag, Berlin 1997).
[193] J. Guerrero, F. F. Lûpez-Ruiz, V. Aldaya and F. Cossio, *Harmonic states for the free particle, J. Phys. A* **44** (2011) 1–26, 445307.
[194] N. D. Sen Gupta, *On the solution of the equation of electron in the field of electromagnetic radiation*, Bull. Calcutta Math. Soc. **44** (3) (1952) 132–136.
[195] J. Hamilton, *Aharonov-Bohm and Other Cyclic Phenomena*, Springer Tracts in Modern Physics (Springer-Verlag, New York 1997).
[196] Harish-Candra, *Motion of electron in the field of a magnetic pole*, Phys. Rev. **74** (8) (1948) 883–887.
[197] A. Hautot, *New magnetic field for Dirac's equations*, Phys. Lett. A **35** (2) (1971) 129–130.
[198] A. Hautot, *The exact motion of a charged particle in the magnetic field* $\mathbf{B} = (x^2 + y^2)^{-1/2} (-\frac{\gamma y}{x^2+y^2}, \frac{\gamma x}{x^2+y^2}, 0)$, Physica **58** (1) (1972) 37–44.
[199] A. Hautot, *About the solution of Dirac's equations in the presence of new magnetic fields*, Journ. Math. Phys. **13** (5) (1972) 710–714.
[200] A. Hautot, *Motion in radial magnetic fields*, Journ. Math. Phys. **14** (12) (1973) 2011–2017.
[201] W. Heitler, *The Quantum Theory of Radiation* (Oxford Univ. Press, London, 1936).

[202] M. Henneaux and C. Teitelboim, *Relativistic quantum mechanics of supersymmetric particles*, Ann. Phys. **143** (1982) 127–159.

[203] Le Van Hoang, Lu Xuan Hai, L. T. Komarov and T. S. Romanova, *Relativistic analogy of the Aharonov-Bohm effect in the presence of Coulomb field and magnetic charge*, J. Phys. A **25** (23) (1992) 6461–6469.

[204] J. W. van Holten, *Propagators and path integrals*, Nucl. Phys. B **457** (1995) 375–407.

[205] J. Hilgevoord and S. A. Wouthusen, *On the spin angular momentum of the Dirac particle*, Nucl. Phys. **40** (1) (1963) 1–12.

[206] M. Hongoh, *Coherent states associated with the continuous spectrum of noncompact groups*, *J. Math. Phys.* **18**, 2081–2085 (1977).

[207] E. D'Hoker and D. G. Gagné, *Worldline path integrals for fermions with scalar, pseudoscalar and vector couplings*, Nucl. Phys. B **467** (1996) 272–296.

[208] L. D. Huff, *The motion of a Dirac electron in a magnetic field*, Phys. Rev. **38** (3) (1931) 501–512.

[209] D. J. Hylton, *The reduced Dirac Green's function for the Coulomb potential.* J. Math. Phys. **25** (1984) 1125–1132.

[210] N. J. Ionescu-Pallas, *Generalizing the Dirac equation*, Rev. Roumaine Phys. **15** (3) (1970) 397–401.

[211] A. Jannussis, *Die relativistische Bewegung eines Electrons in außeren homogenen Magnetfeld*, Zeit. Phys. **190** (2) (1966) 129–134.

[212] A. E. Kazakov and M. V. Fedorov, *On the theory of electron motion in a constant magnetic field and in the field of a quantized electromagnetic wave*, Kratk. Soob. Fiz. No. 11 (1972) 42–46.

[213] A. E. Kazakov and M. V. Fedorov, *A particle in the field of a quantized electromagnetic wave, propagating along a constant magnetic field*, Kratk. Soob. Fiz. No. 6 (1973) 3–8.

[214] V. R. Khalilov, *Relativistic Aharonov-Bohm Effect in the Presence of Planar Coulomb Potentials*, Phys. Rev. **71** (1) (2005) 012105 (6 pages).

[215] I. A. Khokhlov, *Electron Green's function in the field of a quantized plane wave*, Soviet Physics Journ. **8** (1983) 71–74.

[216] J. R. Klauder and E. C. G. Sudarshan, *Fundamentals of Quantum Optics* (W. A. Benjamin, New York 1968); I. R. Klauder and B. S. Skagerstam, *Coherent States. Applications in Physics and Mathematical Physics* (World Scientific, Singapore, 1985).

[217] J. R. Klauder, *Quantization Without Quantization*, Ann. Phys. (NY) **237** (1995) 147–160.

[218] J. R. Klauder, *Beyond Conventional Quantization* (Cambridge University Press, Cambridge 2000).

[219] O. Klein, *Die Reflexion von Electronen an einem Potentialsprung nach der relativistischen Dynamik von Dirac*, Zeit. Phys. **53** (3–4) (1929) (157–165).

[220] O. Klein, *Quantentheorie und fünfdimensionale Relativitäts-theorie*, Zeit. Phys. **37** (12) (1926) 895–906.

[221] Yu. I. Klimenko, *Motion of neutral Fermi-particles with anomalous magnetic and electric moments in external fields.* Soviet Physics Journ. **17** (1974) 1481–1485.

[222] Yu. I. Klimenko, *Motion of charged Fermi-particles with anomalous moments in external fields*, Deposited with VINITI, 1975, No. 477-75 Dep.

[223] I. V. Komarov, L. I. Ponomariov and S. Yu. Slavianov, *Spheroidal and Coulomb Spheroidal Functions* (Nauka, Moscow 1976).

[224] G. M. Korchemsky, Phys. Lett. B **322** (1989) 334; Int. J. Mod. Phys. A **7** (1992) 339; J. Grundberg, T. H. Hansson and A. Karlhede, Phys. Rev. D **41** (1990) 2642; Nucl. Phys. B **347** (1990) 420.

[225] S. I. Kruglov, *Exact solutions of the equation for composite scalar particles in quantized electromagnetic waves*, Can. Journ. Phys. **81** (2003) 953–969.

[226] I. Kudar, *Zur vierdimensionalen Formulierung der ondulatorischen Mechanik*, Ann. der Phys. **81** (21) (1926) 632–636.

[227] A. B. Kukanov and N. D. Naumov, *Motion of a neutral particle in a nonuniform magnetic field*, Vestn. MGU, ser. Fiz. Astron. **16** (1) (1975) 116–119.

[228] N. Kubota, M. Yamaguchi and G. Iwata, *Schrödinger equations soluble in terms of confluent hypergeometric functions*, Natur. Sci. Rept. Ochanomizu Univ. **20** (2) (1969) 39–44.

[229] J. B. Kogut and D. E. Soper, *Quantum electrodynamics in the infinite momentum frame.* Phys. Rev. D **1** (10) (1970) 2901–2913.

[230] L. D. Landau and E. M. Lifshitz, *Mechanics* (Pergamon Press, Oxford 1960).

[231] L. D. Landau and E. M. Lifshitz, *Classical Theory of Fields* (Butterworth-Heinemann, 1975).

[232] L. D. Landau and E. M. Lifshitz, *Quantum Mechanics: Non-Relativistic Theory* (Pergamon Press, Oxford 1977).

[233] Lui Lam, *New exact solutions of the Dirac equations*, Can. Journ. Phys. **48** (16) (1970) 1935–1937; *Dirac electron in orthogonal electric and magnetic fields*, Lett. Nuovo Cimento **3** (9) (1970) 292–294.

[234] Lui Lam, *Motion in electric and magnetic fields*, Journ. Math. Phys. **12** (2) (1971) 299–303.

[235] R. R. Lewis, *Aharonov-Bohm effect for trapped ions.* Phys. A **28** (1983) 1228–1236.

[236] R. G. Littlejohn, *The semiclassical evolution of wave packets*, Phys. Rep. **138** (4,5), 193–291.

[237] Yu. M. Loskutov and V. P. Leventuyev, *Effect of the Electric Dipole Moment of the polarization of a Spinor Particle in a Magnetic Field*, Soviet Physics Journ. **12** (1969) 823–824.

[238] Yu. M. Loskutov and V. P. Leventuyev, *Effect of the anomalous magnetic and (hypothetical) electric moments on the Fermi-particle behaviour in a magnetic field*, Vestn. MGU, ser. Fiz. Astron. **11** (5) (1970) 554–559.

[239] I. A. Malkin and V. I. Man'ko, *Coherent states of a charged particle in a magnetic field*, Zh. Eksp. Teor. Fiz. **55** (3) (1968) 1014–1025.

[240] I. A. Malkin, V. I. Man'ko and D. A. Trifonov, *Coherent states and transition probabilities in a time-dependent electromagnetic field*, Phys. Rev. **2** (8) (1970) 1371–1385.

[241] I. A. Malkin, V. I. Man'ko and D. A. Trifonov, *Coherent states and Green's function for nonstationary quadratic Hamiltonians*, Kratk. Soob. Fiz. No. 5 (1971) 20–26.

[242] I. A. Malkin and V. I. Man'ko, *Dynamical Symmetries and Coherent States of Quantum Systems* (Nauka, Moscow 1979).

[243] V. P. Maslov, *Operational Methods* (Nauka, Moscow 1973).

[244] P. T. Matthews, *The relativistic quantum theory of elementary particle interactions* (Rochester, N.Y. 1957).

[245] V. B. Matveev and M. A. Salle, *Darboux transformations and solutions* (Springer-Verlag, Berlin 1991).

[246] A. G. Meshkov, *On a method for solving the Dirac equation*, Soviet Physics Journ. **23** (1980) 1019–1021.

[247] N. F. Mott, *The scattering of fast electrons by atomic nuclei*, Proc. Roy. Soc. London A **124** (793) (1929) 425–442.

[248] N. B. Narozhny and A. I. Nikishov, *Solutions of the Klein–Gordon and Dirac equations for a particle in a constant electric field and in a plane electromagnetic wave propagating along it*, Theor. Math. Phys. **26** (1976) 16–34.

[249] N. B. Narozhny and A. I. Nikishov, *A particle in a constant electric field and a plane electromagnetic field propagating along it.* In *Quantum Electrodynamics of Intense Fields*, Proceedings of Lebedev Physics Institute **168**, V. L. Ginzburg (ed.) (Nauka, Moscow 1986) pp. 175–199 (English translation: Nova Science Publishers, 1987).

[250] G. A. Nagorsky, *Particle with an anomalous magnetic moment in the field of a plane electromagnetic wave*, Izv. AN Arm. SSR, Fizika **1** (5) (1966) 340–342.

[251] I. M. Nieto, A. A. Pecheritsin and B. F. Samsonov, *Intertwining technique for the one-dimensional stationary Dirac equation*, Ann. Phys. **305** (2) (2003) 151–189.

[252] M. Nielsen and I. Chuang, *Quantum Computation and Quantum Information* (Cambridge University Press, Cambridge, England, 2000).

[253] A. I. Nikishov, *Pair formation by a constant external field*, Zh. Eksp. Teor. Fiz. **57** (1969) 1210–1216.

[254] A. I. Nikishov, *Problems of intense external field in quantum electrodynamics.* In *Quantum Electrodynamics of Phenomena in Intense Field*, Proceedings of Lebedev Physics Institute **111**, N. G. Basov (ed.) (Nauka, Moscow, 1979) 152–171.

[255] Von K. Nikolsky, *Das Oszillatorproblem nach der Diracschen Theorie*, Zeit. Phys. **62** (9–10) (1930) 677–681.

[256] H. M. Nussenzveig, *Introduction to Quantum Optics* (Gordon and Breach, New York 1973).

[257] A. T. Ogielski and J. Sobczuk, *Path integral quantization of the supersymmetric model of the Dirac particle*, Journ. Math. Phys. **22** (1981) 2060–2065.

[258] S. Olariu and I. I. Popescu, *The quantum effects af electromagnetic fluxes*, Rev. Mod. Phys. **57** (1985) 339–436.

[259] V. P. Oleinik, *The Green's function and quasi-energy spectrum of an electron in a plane-wave electromagnetic field and a homogeneous magnetic field*, Ukr. Fiz. Zhurn. **13** (1968) 1205–1214.

[260] L. Page, *Deflection of electrons by a magnetic field in the wave mechanics*, Phys. Rev. **36** (4) (1930) 444–456.

[261] W. Pauli, *Zur Quantenmechanik des magnetischen Elektrons*, Zeit. Phys. **43** (1927) 601–623.

[262] W. Pauli, *Relativistic field theories of elementary particles*, Rev. Mod. Phys. **13** (1941) 203–232.

[263] A. M. Perelomov, *Coherent states for arbitrary Lie groups*, Commun. Math. Phys. **26** (1972) 222–236; *Generalized Coherent States and Their Applications* (Springer-Verlag, Berlin 1986).

[264] M. Peskin and A. Tonomura, *The Aharonov-Bohm Effect* (Springer-Verlag, Berlin 1989).

[265] M. Planck, *The Principle of Relativity and the Fundamental Equations of Mechanics*, Verhandlungen Deutsche Physikalische Gesellschaft. **8** (1906) 136–141.

[266] M. S. Plesset, *Relativistic wave mechanics of electrons deflected by a magnetic field*, Phys. Rev. **36** (12) (1930) 1728–1731.

[267] A. M. Polyakov, in *Fields, Strings and Critical Phenomena*, Proceedings of the Les Houches Summer School, Les Houches, France, 1988, E. Brézin and J. Zinn-Justin (eds.), Les Houches Summer School Proceedings, Vol. 49 (North-Holland, Amsterdam, 1990); *Fermi-Bose transmutations induced by Gauge-fields*, Mod. Phys. Lett. A **3** (3) (1988) 325–328; *Gauge Fields and Strings* (Harwood, Chur, Switzerland, 1987).

[268] I. I. Rabi, *Das freie Electron in Homogenen Magnetfeld nach der Diracschen Theorie*, Zeit. Phys. **49** (7–8) (1928) 507–511.

[269] I. I. Rabi, *Space Quantization in a Gyrating Magnetic Field*, Phys. Rev. **51** (1937) 652–654.

[270] I. I. Rabi, N. F. Ramsey and J. Schwinger, *Use of Rotating Coordinates in Magnetic Resonance Problems*, Rev. Mod. Phys. **26** (1945) 167–171.

[271] A. F. Radyuk and F. I. Fedorov, *Fermion with an anomalous magnetic moment in the field of a plane electromagnetic wave.* DAN SSR **211** (5) (1973) 1091–1093.

[272] P. K. Rashevsky, *On mathematical grounds of quantum electrodynamics.* Usp. Mat. Nauk **13** (3) (1958) 3–110.

[273] P. J. Redmond, *Solution of the Klein–Gordon and Dirac equations for a particle in a plane electromagnetic wave and a parallel magnetic field*, Journ. Math. Phys. **6** (7) (1965) 1163–1169.

[274] C. S. Roberts and S. J. Buchsbaum, *Motion of a Charged Particle in a Constant Magnetic Field and a Transverse Electromagnetic Wave Propagating along the Field*, Phys. Rev. A **135** (2) (1964) 381–389.

[275] F. Rohrlich, *Theory of photons and electrons on null planes*, Acta Phys. Austr. **32** (1) (1970) 87–106.

[276] F. Rohrlich and R. A. Neville, *Quantum electrodynamics on null planes and applications to lasers*. Phys. Rev. D **3** (8) (1971) 1692–1707.

[277] M. E. Rose, *Relativistic Electron Theory* (Wiley, New York 1961).

[278] C. S. Roberts and S. Buchsbaum, *Motion of a charged particle in a constant magnetic field and a transverse electromagnetic wave propagating along the field*, Phys. Rev. **135** (2A) (1964) 381–389.

[279] H. B. Reiss and J. H. Eberly, *Green's function in intense-field electrodynamics*, Phys. Rev. **151** (1966) 1058–1066.

[280] V. I. Ritus, *The mass operator and exact electron Green's function in intense field*, Pis'ma Zh. Eksp. Teor. Fiz. **12** (1970) 416–418.

[281] V. I. Ritus, *Radiative corrections to the motion of an electron and photons in an intense field, and their analytic properties*. In *Problems of Theoretical Physics* (Nauka, Moscow, 1972) 298.

[282] V. I. Ritus, *Quantum effects of elementary particles interaction with intense electromagnetic field*. In *Quantum Electrodynamics of Phenomena in Intense Field*, Proceedings of Lebedev Physics Institute **111**, N. G. Basov (ed.) (Nauka, Moscow 1979) 5–151.

[283] V. I. Ritus, *On diagonality of the electron mass operator in a constant field*, Pis'ma Zh. Eksp. Teor. Fiz. **20** (1974) 135–138.

[284] V. I. Ritus, *The eigenfunction method and the mass operator in quantum electrodynamics of a constant field*, Zh. Eksp. Teor. Fiz. **75** (1978) 1560–1583.

[285] V. I. Ritus, *Electron mass shift in an intense field*. In *Quantum Electrodynamics of Intense Field*, Proceedings of Lebedev Physics Institute. **168**, V. L. Ginzburg (ed.) (Nauka, Moscow, 1986) pp. 52–119 (English translation: Nova Science Publishers, 1987).

[286] V. I. Ritus, *Radiative effects and their enhancement in an intense electromagnetic field*, In *Quantum Electrodynamics of Intense Field*, Proceedings of Lebedev Physics Institute. **168**, V. L. Ginzburg (ed.) (Nauka, Moscow, 1986) pp. 141–155 (English translation: Nova Science Publishers, 1987).

[287] J. Saavedra, *Proper-time formulation of the Dirac electron theory*, Nucl. Phys. **77** (3) (1966) 673–679.

[288] F. Sauter, *Über das Verhalten eines Electrons im homogenen electrischen Feld nach der relativistischen Theorie Diracs*, Zeit. Phys. **69** (11–12) (1931) 742–764; *Zum "Kleinschen Paradoxen"*, Zeit. Phys. **73** (7–8) (1931) 547–552.

[289] E. Schrödinger, *Der stetige Übergang von der Mikro- zur Makro-mechanik*, Naturwissenschaften, **14** (28) (1926) 664–666.

[290] E. Schrödinger, *Sitzungsberichte Preus. Acad. Wiss., Phys.-Math. Klasse* **19** 296 (Berlin 1930); H. P. Robertson, *Phys. Rev.* **35** (1930) 667.

[291] E. Schrödinger, *Quantisierung als Eigenwertprobleme. IV*, Ann. der Phys., **81** (18) (1926) 109–139.

[292] S. Schweber, *An Introduction to Relativistic Quantum Field Theory* (Harper & Row, New York 1961).

[293] J. Schwinger, *The theory of quantized fields. III*, Phys. Rev. **82** (5) (1951) 664–675.

[294] J. Schwinger, *On gauge invariance and vacuum polarization*, Phys. Rev. **82** (1951) 664–679.

[295] J. Schwinger, *Magnetic charge and quantum field theory*, Phys. Rev. **144** (1966) 1087–1093.

[296] A. E. Shabad, *On a class of solutions of the Dirac equation*, Yad. Fiz. **7** (3) (1969) 685–689.

[297] V. N. Shapovalov, *Calculation of the symmetry algebra for the Dirac equation*, Soviet Physics Journ. **11** (1968) 98–99.

[298] V. N. Shapovalov and A. V. Shapovalov, *Separation of variables in the Klein–Gordon equation for a superposition of a quantized and classical field*, Soviet Physics Journ. **19** (1976) 23–27.

[299] V. N. Shapovalov, V. G. Bagrov and G. G. Eckle, *Symmetry properties and variable separation in the stationary Dirac equation*, In Proceedings of the 8th Scientific Conference in Physics, Khabarovsk (1971) 20–25.

[300] V. N. Shapovalov, V. G. Bagrov and A. G. Meshkov, *Separation of Variables in the Schrödinger Stationary Equation*, Soviet Physics Journ. **15** (1972) 1115–1119.

[301] V. N. Shapovalov and G. G. Eckle, *Separation of Variables in the Dirac Equation*, Soviet Physics Journ. **16** (1973) 818–823.

[302] V. N. Shapovalov and N. B. Sukhomlin, *Separation of variables in a nonstationary Schrödinger equation*, Soviet Physics Journ. **17** (1974) 1718–1722.

[303] V. N. Shapovalov, *Symmetry and separation of variables in a linear second-order differential equation. I, II*, Soviet Physics Journ. **21** (1978) 645–650; 693–695; *Symmetry and separation of variables in the Hamilton–Jacobi equation. I, II*, Soviet Physics Journ. **21** (1978) 1124–1129; 1130–1132.

[304] L. I. Schiff, *Quantum Mechanics*, 2nd edn. (McGraw–Hill, New York 1955).

[305] Yu. M. Shirokov, *On the problem of intersection of the new type spin-1/2 particles with an external field*, DAN SSSR **99** (5) (1954) 737–740.

[306] Yu. M. Shirokov, *Relativistic theory of spin*, Zh. Eksp. Teor. Fiz. **21** (6) (1951) 748–760.

[307] A. A. Sokolov, *On the polarization of electron waves*, Journ. Phys. **9** (5) (1945) 363–372.

[308] A. A. Sokolov and M. M. Kolesnikova, *On scattering of transversely polarized fermions*, Zh. Eksp. Teor. Fiz. **38** (6) (1960) 1778–1785.

[309] A. A. Sokolov, *Über das Verhalten des Spins von Dirac – Teilchen bei Streuungprozessen*, Ann. der Phys. **8** (5–6) (1961) 237–259.

[310] A. A. Sokolov and I. M. Ternov, *Synchrotron Radiation* (Akad. Verlag, Berlin 1968); *Radiation from Relativistic Electrons* (American Institute of Physics, New York 1986); I. M. Ternov, V. V. Mihailin and V. R. Khalilov, *Synchrotron Radiation and its Application* (Harwood Academic Publishers, New York 1985).

[311] A. A. Sokolov and I. M. Ternov, *Relativistic Electron* (Nauka, Moscow 1974).

[312] V. I. Stragev and L. M. Tomilchik, *Electrodynamics with the magnetic charge* (Nauka i Technika, Minsk 1975).

[313] F. Strocchi, *Relativistic particles in a constant magnetic field*, Nuovo Cimento **37** (3) (1965) 1079–1085.

[314] P. V. Strokov and G. I. Flesher, *Green's function of the generalized Dirac equation for the plane-wave and longitudinal electric wave fields.* Soviet Physics Journ. **28** (1985) 36–39.

[315] B. Stech, *Zur Streuung von Teilchen mit dem Spin 1/2. Teil I: Relativistische Behandlung der Streuung*, Zeit. Phys. **144** (1–3) (1956) 214–218.

[316] G. N. Stanciu, *Solvable vector and scalar potentials for the Dirac equation*, Phys. Lett. **23** (3) (1966) 232–233.

[317] G. N. Stanciu, *Further exact solution of the Dirac equation*, Journ. Math. Phys. **8** (10) (1967) 2043–2047.

[318] N. B. Sukhomlin and V. N. Shapovalov, *Integral Sets and Variable Separation in the Klein–Cordon equation.* Soviet Physics Journ. **16** (1973) 1584–1587.

[319] A. A. Sokolov (ed.), *Synchrotron Radiation* (Nauka, Moscow 1966).

[320] I. M. Ternov, V. G. Bagrov and V. A. Bordovitsyn, *Covariant description of spin states of Fermi-particles moving in external fields*, Soviet Physics Journ. **10** (1967) 20–22.

[321] I. E. Tamm, *Die verallgemeinerten Kugelfunktionen und die Wellenfunktionen eines Elektrons im Felde eines Magnetpoles*, Zeit. Phys. **71** (3–4) (1931) 141–150.

[322] J. H. Ten and F. Rohrlich, *Equivalence of null-plane and conventional quantum electrodynamics*, Phys. Rev. D **9** (8) (1974) 2237–2251.

[323] I. M. Ternov, V. G. Bagrov, V. A. Bordovitsyn and Yu. A. Markin, *Scattering of electromagnetic waves from neutral Fermi-particles*, Zh. Eksp. Thor. Fiz. **52** (6) (1967) 1584–1591.

[324] I. M. Ternov, V. G. Bagrov and Yu. I. Klimenko, *Motion and Radiation of an Electron, Exhibiting a Vacuum Magnetic Moment, in the Field of a Plane Electromagnetic Wave*, Soviet Physics Journ. **11** (1968) 29–33.

[325] I. M. Ternov, V. G. Bagrov and P. V. Bozrikov, *Motion of Dirac particles with anomalous magnetic moment in a centrally-symmetric electric field*, Soviet Physics Journ. **14** (1971) 1489–1494.

[326] I. M. Ternov, V. G. Bagrov, R. A. Rzayev and Yu. I. Klimenko, *Motion of a Polarized Electron with Vacuum Magnetic Moment*, Izw. VUZov USSR, Fisika **6** (1964) 111–121.

[327] I. M. Ternov, V. G. Bagrov and A. M. Khapaev, *Radiation of a relativistic charge in a plane-wave electromagnetic field*, Annalen der Physik **22** (7) (1968) 25–32.

[328] I. M. Ternov and V. S. Tumanov, *On the effect of the vacuum fluctuations on the polarization of electrons moving in a magnetic field*, Zh. Eksp. Teor. Fiz. **37** (4) (1959) 1137–1139.

[329] I. M. Ternov and V. S. Tumanov, *On the motion of polarized electrons in a magnetic field*, Izw. VUZov USSR, Fizika, No. 1 (1960) 155–163.

[330] B. Thaller, *The Dirac Equation* (Springer-Verlag, Berlin 1992).

[331] Wu-yang Tsai and T. Erber, *Photon pair creaition in intense magnetic fields*, Phys. Rev. D **10** (1974) 492–499.

[332] I. V. Tyutin, *Electron scattering on a solenoid*, Preprint FIAN (P. N. Lebedev Physical Institute, Moscow) No. 27 (1974); arXiv:0801.2167 (quant-ph).

[333] A. N. Vasiliev, *Functional Methods in Quantum Field Theory and Statistical Physics* (Leningrad State University Press, Leningrad 1976).

[334] V. M. Villalba, *Exact solutions of the Dirac equation for a Coulomb and scalar potential in the presence of an Aharonov-Bohm and magnetic monopole fields*, J. Math. Phys. **36** (7) (1995) 3332–3344.

[335] P. A. Vlasov, *Charged particle with a spin in a uniform magnetic field.* Ukr. Fiz. Zhurn. **30** (11) (1985) 1605–1609.

[336] D. M. Volkov, *Solution exacte de l'equation de Dirac pour une onde plane de frequence determinee*, Comptes Rendus de L'Académie des Sciences de L'URSS. **1** (9) (1935) 608–610.

[337] D. M. Volkov, *Über eine Klasse von Lösungen der Diracschen Gleichung*, Zeit. Phys. **94** (3–4) (1935) 250–260.

[338] D. M. Volkov, *Electron in the Field of Plane Nonpolarized Electromagnetic Waves as Viewed Upon in Light of the Dirac Equation*, Zh. Eksp. Teor. Fiz. (JETP) **7** (11) (1937) 1286–1289.

[339] B. L. Voronov, D. M. Gitman and I. V. Tyutin, *The Dirac Hamiltonian with a superstrong Coulomb field*, Theoretical and Mathematical Physics **150** (1) (2007) 34–72; D. M. Gitman, A. D. Levin, I. V. Tyutin and B. L. Voronov, *Electronic Structure of Superheave Atoms. Revisited*, Physica Scripta (2013).

[340] Yu. Yu. Vol'fengaut, S. P. Gavrilov, D. M. Gitman and Sh. M. Shvartsman, *Radiative processes in a pair-creating external electromagnetic field*, Yad. Fiz. **33** (1981) 743–757.

[341] B. S. DeWitt, *Dynamical Theory in Curved Spaces*, Rev. Mod. Phys. **29** (1957) 377–397.

[342] H. Yamasaci, *An extension of Fennman–Bunge–Corben's relation regarding the position operator of Dirac electron to arbitrary operators I, II.* Progr. Theor. Phys. **31** (2) (1964) 322–323; 324–325; *The Frenkel–Kramers model and the Zitterbewegung in the Heisenberg picture of the Dirac electron*, Progr. Theor. Phys. **39** (2) (1968) 372–392.

[343] V. N. Zadorozhny and Sh. M. Shvartsman, *The Green's function of the system composed of a charged particle plus a quantized plane-wave field*, Soviet Physics Journ. **21** (1978) 1023–1027.

[344] S. I. Zlatev, *Alternative pseudoclassical action and spin factor*, Phys. Rev. D **62** (2000) 105020.

[345] B. A. Zon, N. L. Manakov and L. N. Rapoport, *Coulomb Green's functions in the x-representation, and relativistic polarizability of the hydrogen-like atoms*, Yad. Fiz. **15** (1972) 508–517.

Index

De Gruyter Studies in Mathematical Physics

Volume 23
Boris A. Arbuzov
Non-perturbative Effective Interactions in the Standard Model, 2014
ISBN 978-3-11-030292-9, e-ISBN 978-3-11-030521-0, Set-ISBN 978-3-11-030522-7

Volume 22
Sergei M. Kopeikin (Ed.)
Frontiers in Relativistic Celestial Mechanics, Volume 2: Applications and Experiments, 2014
ISBN 978-3-11-034545-2, e-ISBN 978-3-11-034566-7, Set-ISBN 978-3-11-034567-4

Volume 21
Sergei M. Kopeikin (Ed.)
Frontiers in Relativistic Celestial Mechanics, Volume 1: Theory, 2014
ISBN 978-3-11-033747-1, e-ISBN 978-3-11-033749-5, Set-ISBN 978-3-11-035932-9

Volume 20
Valery A. Slaev, Anna G. Chunovkina, Leonid A. Mironovsky
Metrology and Theory of Measurement, 2013
ISBN 978-3-11-028473-7, e-ISBN 978-3-11-028482-9, Set-ISBN 978-3-11-028483-6

Volume 19
Edward A. Bormashenko
Wetting of Real Surfaces, 2013
ISBN 978-3-11-025853-0, e-ISBN 978-3-11-025879-0, Set-ISBN 978-3-11-220375-0

Volume 18
Michael V. Sadovskii
Statistical Physics, 2012
ISBN 978-3-11-027031-0, e-ISBN 978-3-11-027037-2, Set-ISBN 978-3-11-027038-9

Bei Fragen zur Produktsicherheit wenden Sie sich bitte an:
If you have any questions regarding product safety,
please contact:

Walter de Gruyter GmbH
Genthiner Straße 13
10785 Berlin
productsafety@degruyterbrill.com